Periodic Table of the Elements with the Gmelin System Numbers

1 H 2																1 H 2	2 He 1
3 Li 20	4 Be 26											5 B 13	6 C 14	7 N 4	8 O 3	9 F 5	10 Ne 1
11 Na 21	12 Mg 27											13 Al 35	14 Si 15	15 P 16	16 S 9	17 Cl 6	18 Ar 1
19 * K 22	20 Ca 28	21 Sc 39	22 Ti 41	23 V 48	24 Cr 52	25 Mn 56	26 Fe 59	27 Co 58	28 Ni 57	29 Cu 60	30 Zn 32	31 Ga 36	32 Ge 45	33 As 17	34 Se 10	35 Br 7	36 Kr 1
37 Rb 24	38 Sr 29	39 Y 39	40 Zr 42	41 Nb 49	42 Mo 53	43 Tc 69	44 Ru 63	45 Rh 64	46 Pd 65	47 Ag 61	48 Cd 33	49 In 37	50 Sn 46	51 Sb 18	52 Te 11	53 I 8	54 Xe 1
55 Cs 25	56 Ba 30	57** La 39	72 Hf 43	73 Ta 50	74 W 54	75 Re 70	76 Os 66	77 Ir 67	78 Pt 68	79 Au 62	80 Hg 34	81 Tl 38	82 Pb 47	83 Bi 19	84 Po 12	85 At 8a	86 Rn 1
87 Fr 25a	88 Ra 31	89*** Ac 40	104 71	105 71													

* NH_4 23

**Lanthanides 39	58 Ce	59 Pr	60 Nd	61 Pm	62 Sm	63 Eu	64 Gd	65 Tb	66 Dy	67 Ho	68 Er	69 Tm	70 Yb	71 Lu
***Actinides	90 Th 44	91 Pa 51	92 U 55	93 Np 71	94 Pu 71	95 Am 71	96 Cm 71	97 Bk 71	98 Cf 71	99 Es 71	100 Fm 71	101 Md 71	102 No 71	103 Lr 71

A Key to the Gmelin System is given on the Inside Back Cover

Gmelin Handbook of Inorganic Chemistry

8th Edition

Volumes published on "Molybdenum" (Syst. No. 53)

* **Molybdenum**
Main Volume – 1935

** **Molybdenum Suppl. Vol. A 1**
Metal. Technology – 1977

Molybdenum Suppl. Vol. A 2a
Element. Physical Properties, Pt. 1 – 1985

Molybdenum Suppl. Vol. A 3
Metal. Chemical Reactions – 1983

** **Molybdenum Suppl. Vol. B 1**
Compounds with Noble Gases, Hydrogen, and Oxygen. Anhydrous Antimony, Bismuth, and Alkali Molybdates – 1975

** **Molybdenum Suppl. Vol. B 2**
Anhydrous Compounds of Molybdenum Oxides with Oxides of Other Metals – 1976

Molybdenum Suppl. Vol. B 4
Hydrous Molybdates of Groups VA to VIB Metals (System Nos. 18 to 52) – 1985
(present volume)

* in German
** in German, with English reviews and marginalia

Gmelin Handbook of Inorganic Chemistry

8th Edition

Gmelin Handbuch der Anorganischen Chemie

Achte, völlig neu bearbeitete Auflage

Prepared and issued by
Gmelin-Institut für Anorganische Chemie
der Max-Planck-Gesellschaft
zur Förderung der Wissenschaften
Director: Ekkehard Fluck

Founded by Leopold Gmelin

8th Edition: 8th Edition begun under the auspices of the Deutsche Chemische Gesellschaft by R. J. Meyer

Continued by E. H. E. Pietsch and A. Kotowski, and by Margot Becke-Goehring

Springer-Verlag Berlin Heidelberg GmbH 1985

Gmelin-Institut für Anorganische Chemie
der Max-Planck-Gesellschaft zur Förderung der Wissenschaften

Gmelin Handbook of Inorganic Chemistry

8th Edition

Mo
Molybdenum

Supplement Volume B 4

With 85 illustrations

Hydrous Molybdates of Groups VA to VIB Metals
(System Nos. 18 to 52)

AUTHORS Karl-Heinz Tytko, Universität Göttingen
Wolf-Dietrich Fleischmann, Dieter Gras, Eberhard Warkentin, Gmelin-Institut, Frankfurt am Main

EDITORS Hartmut Katscher, Friedrich Schröder

CHIEF EDITOR Hartmut Katscher

System Number 53

Springer-Verlag Berlin Heidelberg GmbH 1985

The polymolybdate hydrates of antimony, the alkali and alkaline earth metals on pp. 1/213 are written by Karl-Heinz Tytko.

LITERATURE CLOSING DATE: 1982
IN MANY CASES MORE RECENT DATA HAVE BEEN CONSIDERED

Library of Congress Catalog Card Number: Agr 25-1383

ISBN 978-3-662-07838-9 ISBN 978-3-662-07836-5 (eBook)
DOI 10.1007/978-3-662-07836-5

Originally published by Springer-Verlag Berlin Heidelberg New York Tokyo in 1985
Softcover reprint of the hardcover 8th edition 1985

Preface

This volume contains the hydrous oxo compounds of the metals Sb to Cr (System Nos. 18 to 52) with molybdenum. (The corresponding anhydrous compounds have been described in the volumes "Molybdän" Erg.-Bd. B 1 and B 2.) With these metals, molybdenum forms monomolybdates, isopolymolybdates, peroxomolybdates, and molybdometalates.

Antimony forms only the compound $H_3SbMo_{12}O_{40} \cdot nH_2O$ and with bismuth no hydrous oxo compounds are known.

More than half the volume describes the monomolybdates, isopolymolybdates, and peroxomolybdates of the alkali and alkaline earth metals, including ammonium and organic cations. First, a detailed review on the structures, spectra, and other properties of the various isopolymolybdate types is given. It also contains a comprehensive description on photochromism, a property characteristic of many of the organic ammonium polymolybdates. Then follows the description of the individual alkali compounds. The large group of molybdates with organic cations is placed after the caesium molybdates. The photochromism and the photogalvanic effect are investigated at length for $(i\text{-}C_3H_7NH_3)_6[Mo_7O_{24}] \cdot 3H_2O$. The alkaline earth molybdates are treated briefly compared with the alkali molybdates, e.g., no new data are available on beryllium molybdates.

In the following chapters concerning the compounds of Mo with Zn to Cr a great variety of types can be found. Some of the metals form mainly molybdate hydrates, e.g., the rare earth metals, or peroxomolybdates, e.g., Pb. There are metals which are able to form molybdate hydrates as well as molybdometalates, e.g., Zn and In. However, for most of them the molybdometalates are the only type of compounds known and they have been investigated in detail, see e.g., Ti, Ge, V, and Cr.

Frankfurt am Main, August 1985 Hartmut Katscher

Table of Contents

Page

Molybdenum and Oxygen (continued)

The Mo-O system and the molybdenum oxides can be found in the volume "Molybdän" Erg.-Bd. B 1, 1975, pp. 21/144. The anhydrous molybdates with Sb to Cs (System Nos. 18 to 25) are described in "Molybdän" Erg.-Bd. B 1, 1975, pp. 145/241, those with Be to Cr (System Nos. 26 to 52) in "Molybdän" Erg.-Bd. B 2, 1976. The molybdenum oxide hydrates and the molybdate ions can be found in "Molybdenum" Suppl. Vol. B 3 (to be published).

Hydrous Molybdates of Groups VA to VIB Metals (System Nos. 18 to 52)

1 Molybdate Hydrates with Main Group 5 Metals

1.1 Antimony Polymolybdate Hydrates

The compounds of antimony(V) and molybdate formed on acidification of the solution are molybdoantimonates, i.e., heteropolymolybdates. An additional metal (which could act as a cation) being excluded, the compounds under consideration are heteropolymolybdic acids and their hydrates.

1.1.1 $H_3SbMo_{12}O_{40}\cdot nH_2O$ ($=Sb_2O_5\cdot 24MoO_3\cdot(2n+3)H_2O$), n = 48, 13, 10 12-Molybdoantimonic Acid Hydrates

To prepare the 48-hydrate 50 mL of 0.05 M molybdic acid solution were refluxed with 80 mL of 0.0133 M potassium diantimonate solution for 45 to 60 min and then evaporated overnight to yield fine, needle-shaped, cream-colored crystals [1, 2]. The compound was recrystallized from hot water [2].

The compound is characterized by elemental analysis [1, 2]. The Sb : Mo mole ratio was also determined by potentiometric and thermometric titrations [2].

From potentiometric and conductometric measurements the compound was found to be tribasic; one of the three hydrogen atoms is bonded differently [1, 2]. The pH of a 0.05 M solution of the acid is 4.62 [2].

Single crystal X-ray analysis showed the crystals to be orthorhombic [1, 2], lattice parameters a = 10.78, b = 34.85, c = 10.56 Å; Z = 4 (in contrast to these data a tetragonal space group is given in the paper) [2]. The measured density is 4.64 g/cm^3. From these data the molecular weight was determined to be 2772 which agrees closely with the theoretical molecular weight of 2781. Molecular weight determination by cryoscopic method in pure water as the solvent gave 2572 [1, 2]. On the basis of these results the formula is written $H_3[SbMo_{12}O_{40}]\cdot 48H_2O$ [1, 2] which is the formula of Keggin [3].

Isothermal dehydration at 35°C showed a 13-hydrate and a 10-hydrate to exist. Isothermal dehydration at 45°C led only to the 10-hydrate [2].

Previous reports cited in "Molybdän", 1935, p. 373, are inadequate [1, 2].

References:

[1] G. C. Bhattacharya, S. K. Roy (J. Indian Chem. Soc. **50** [1973] 359). – [2] S. K. Roy, G. C. Bhattacharya (J. Indian Chem. Soc. **53** [1976] 1074/6). – [3] J. F. Keggin (Proc. Roy. Soc. [London] A **144** [1934] 75/100).

2 Molybdate Hydrates with Alkali Metals and Ammonium

2.1 Review and Comparative Data for the Alkali and Alkaline Earth Polymolybdate Hydrates

2.1.1 Compounds with Mo of Oxidation State <6

There are three mixed-valence compounds containing a large proportion of the molybdenum in the oxidation state 5: $[(C_2H_5)_4N]_2[H_2Mo^{V}_2Mo^{VI}_4O_{19}]$ [5], $Rb_2Mo^{V}_2Mo^{VI}_4O_{18} \cdot 10H_2O$, and $Rb_2HMo^{V}_3Mo^{VI}_3O_{18} \cdot 12H_2O$ [1]. The reduction of the molybdenum(VI) has been carried out chemically in these cases. The polymolybdate ions of the rubidium salts most probably also have the Mo_6O_{19} structure and therefore the compounds should be formulated $Rb_2[H_2Mo^{V}_2Mo^{VI}_4O_{19}] \cdot 9H_2O$ and $Rb_2[H_3Mo^{V}_3Mo^{VI}_3O_{19}] \cdot 11H_2O$ [2]. The presence of direct metal-metal bonds in these types of molybdenum blue has been postulated from the reciprocal relation between the absorption band maximum (of the dissolved species) and the number of reduced molybdenum atoms in the species using the free electron model [4] (for details see the paper).

Another molybdenum blue compound, $[(i\text{-}C_3H_7)NH_3^+]_4[Mo_{13}O_{40}^{4-}]_{0.33}[H_4Mo^{VI}_{12}O_{40}^{4-}]_{0.67}$, contains much less molybdenum in the oxidation state 5. It is obtained by photoreduction of the isopropylammonium heptamolybdate complex. The two anions have the so-called Keggin type structure, $[Mo_{13}O_{40}]^{4-}$ containing the reduced molybdenum atoms and an MoO_4 tetrahedron in the center of the $Mo_{12}O_{40}$ group [6]. Note added in proof: According to [7, 8] there is no MoO_4 tetrahedron in the center of the $Mo_{12}O_{40}$ group but an SiO_4 tetrahedron.

Another kind of mixed-valence polymolybdate contains only small amounts of molybdenum in the oxidation state 5 and is produced by γ irradiation of ammonium paramolybdate: $(NH_4)_{6-n}[Mo^{V}_{n}Mo^{VI}_{7-n}O_{24-n}] \cdot 4H_2O$ or $(NH_4)_6[Mo^{V}_{n}Mo^{VI}_{7-n}O_{24-n/2}] \cdot 4H_2O$ [3].

Finally, it should be mentioned that the colored, photochromic, organic ammonium polymolybdates contain some molybdenum in the oxidation state 5 (see p. 127ff).

References:

[1] S. Ostrowetsky (Bull. Soc. Chim. France **1964** 1003/11). – [2] M. T. Pope (Inorg. Chem. **11** [1972] 1973/4). – [3] I. Pascaru, O. Constantinescu, M. Constantinescu, D. Arizan (J. Chim. Phys. **62** [1965] 1283/8). – [4] E. E. Kriss, V. K. Rudenko, K. B. Yatsimirskii (Zh. Neorgan. Khim. **16** [1971] 2147/53; Russ. J. Inorg. Chem. **16** [1971] 1146/50). – [5] C. Heitner-Wirguin, D. Hall (J. Inorg. Nucl. Chem. **36** [1974] 3870/1).

[6] T. Yamase, T. Ikawa, Y. Ohashi, Y. Sasada (J. Chem. Soc. Chem. Commun. **1979** 697/8). – [7] H. T. Evans, Jr., M. T. Pope (Inorg. Chem. **23** [1984] 501/4). – [8] V. I. Spitsyn, L. P. Kazanskii, E. A. Torchenkova (Soviet Sci. Rev. B **3** [1981] 111/96, 183).

2.1.2 Formulations and Nomenclature of Polymolybdates(VI)

There is a need for different formulations to express the knowledge of a polymolybdate adequately by a formula.

If there are discrete polymolybdate ions the formula should express the sort and number of atoms building up the polyanion and also certain connections between the atoms (e.g., OH groups, coordinated H_2O molecules). In case of chain-like structures the length of the identity period should be stated. Two- or three-dimensional structures should be treated in the same way. Thus there are formulae such as $K_6[Mo_7O_{24}] \cdot 4H_2O$ or $K_8[Mo_{36}O_{112}(H_2O)_{16}] \cdot 36$ to $40H_2O$ for polymolybdates with discrete ions, and $[Rb_4Mo_6O_{20} \cdot 2H_2O]_\infty$ or $[KMo_5O_{15}(OH)(H_2O) \cdot H_2O]_\infty$ for polymolybdates with chain-like, two- or three-dimensional network anions.

If the degree of aggregation or the length of the identity period are unknown, the formulation according to the "base:acid" ratio in the salt is advisable and indeed often used, e.g., $Na_2O \cdot 4MoO_3 \cdot 6H_2O$ (compare [1, 2]). However, frequently the elemental constituents are arranged simply side by side (see, e.g., [2 to 4]), as for $Na_2Mo_4O_{13} \cdot 6H_2O$ or $K_2Mo_2O_7 \cdot H_2O$, thus suggesting polymolybdates with discrete anions to be present although the structure is unknown (as in case of the sodium salt) or of the chain-like type having a tetrameric identity period (as in case of the potassium salt). To avoid misunderstandings, in this article discrete polymolybdate ions are always formulated in brackets as shown in the above examples unless this is irrelevant for the matter under discussion.

For the nomenclature of the polymolybdates there is also a need for different designations to express the knowledge about a polymolybdate adequately by a name.

Polymolybdates containing discrete anions are named using the degree of aggregation as a prefix, e.g., potassium heptamolybdate tetrahydrate in the case of the salt $K_6[Mo_7O_{24}] \cdot 4H_2O$. Correspondingly, polymolybdates with a chain-like polyanion can be named using the addendum "poly", e.g., ammonium polyoctamolybdate tetrahydrate in the case of the salt $[(NH_4)_6Mo_8O_{27} \cdot 4H_2O]_\infty$.

If only the analytical base:acid ratio is known, polymolybdates are named according to this ratio, e.g., sodium (1:4)-molybdate hexahydrate in case of the salt $Na_2O \cdot 4MoO_3 \cdot 6H_2O$. This nomenclature is applicable also to the above examples, in that case, however, implying a loss of information.

It has become common practice to name polymolybdates according to their acid:base ratio (note the inversion of the sequence!) using the proportional number of the acid as a prefix and omitting the proportional number "1" of the base (cf. [1, 2, 4, 5]). Hence, in the preceding example the compound is named sodium "tetramolybdate" hexahydrate. In conjunction with certain practices of formulation this nomenclature is a source of frequent misunderstandings [5]. For example, four different kinds of "tetramolybdates" have been reported in the literature: (1) the tetramolybdate ion $[Mo_4O_{12}(OH)_4]^{4-}$ (discrete ion, thus correctly named); (2) a "tetramolybdate" $(NH_4)_2O \cdot 4MoO_3 \cdot 2.5H_2O$, in reality an octamolybdate containing the discrete octamolybdate ion $[Mo_8O_{26}]^{4-}$; (3) a "tetramolybdate" $K_2Mo_4O_{13}$ that does not contain a discrete polyanion and, according to its structure, can be regarded as polymeric octamolybdate containing the ion $[Mo_8O_{26}^{4-}]_\infty$; (4) "tetramolybdates" of different cations and hitherto unknown structures but obviously chain-like or two-dimensional network polymolybdates; additionally, there is (5) a polytetramolybdate $[K_4Mo_4O_{14} \cdot 2H_2O]_\infty$ (correctly so-called) named potassium "dimolybdate" hydrate.

In this article the numerical prefix is used to indicate the degrees of aggregation (the number of molybdenum atoms) of the polymolybdate ions. In any other sense it is used between quotation marks.

The (3:7)-molybdates (heptamolybdates) are frequently also named "paramolybdates" (compare [1]) and the (1:4)-molybdates sometimes "metamolybdates" (see, e.g., [6, 7]).

A procedure frequently applied to describe polymolybdate ions in solution is also applicable to solid polymolybdates; it is the use of the stoichiometric coefficients p and q of H^+ and MoO_4^{2-}, respectively, in the overall equation of formation for the polymolybdate ions (see, e.g., [5]). Thus, for example, the compound $K_6[Mo_7O_{24}] \cdot 4H_2O$ is named potassium (8,7)-molybdate tetrahydrate. The quotient p/q allows a direct arrangement of the polymolybdate ions in a sequence of increasing acidity.

References:

[1] Gmelin Handbuch "Molybdän", 1935, pp. 113/9, 212/98. – [2] I. Lindqvist (Nova Acta Regiae Soc. Sci. Upsaliensis [4] **15** No. 1 [1950] 1/22). – [3] Gmelin Handbuch "Molybdän" Erg.-Bd. B 1, 1975, pp. 181/241. – [4] Y. Sasaki, L. G. Sillén (Arkiv Kemi **29** [1969] 253/77, 255, 275). – [5] K. H. Tytko, O. Glemser (Advan. Inorg. Chem. Radiochem. **19** [1976] 239/315, 243/5).

[6] K. F. Jahr, J. Fuchs (Angew. Chem. **78** [1966] 725/35; Angew. Chem. Intern. Ed. Engl. **5** [1966] 689/99). – [7] I. Lindqvist (Acta Chem. Scand. **4** [1950] 551/2).

2.1.3 Methods of Preparation

The classic method of preparing isopolymolybdates is crystallization of the alkali (including ammonium) and sometimes also alkaline earth salts from acidified aqueous molybdate solutions (compare [1]). Hydrochloric and nitric acid are commonly used to acidify the nearly neutral monomolybdate solution but also perchloric, sulphuric, and other (including organic) acids have been applied. Usage of hydrochloric acid leads to complications at higher degrees of acidification ($>2.5\,H^+/MoO_4^{2-}$ [2]) due to the formation of chloro complexes [3 to 6]. A variant is the acidification with "molybdic acid" ($MoO_3\cdot 2H_2O$, $MoO_3\cdot H_2O$, or MoO_3) allowing crystallization from solutions of higher concentrations since in this case there is no interference from the salt of the acid that might be used otherwise. Adjustment of the conditions for the formation of polymolybdates with a base starting from the stage of "molybdic acid" or with an acid or a base starting from another polymolybdate are further variants [1]. One of the decisive factors for the formation of a certain polymolybdate type is the acidity (see 2.1.5, p. 6) of the molybdate solution, and the operations described so far serve mainly to establish the corresponding conditions. A second decisive factor is the kind of the cation(s) present; for example, Na^+, K^+, and NH_4^+ deposit quite different polymolybdate types from solutions acidified to the "metamolybdate" stage. (Other important factors are the temperature, time, etc.) The cation may form a solid compound with the polymolybdate ion which is present in the solution as the main component or may induce the formation of other, discrete or infinite (chain-like, two- or three-dimensional network) polyanions which are extensively formed due to their insolubility [8]. (The reasons for the insolubility are discussed in 2.1.12, p. 45.) Most polymolybdates with inorganic and many with organic cations have been prepared in one of these ways.

The reaction of "molybdic acid" with an (organic) base is also carried out in organic solvents (or in aqueous mixtures of such) to prepare polymolybdates with organic cations. The solvent may be the (liquid) base itself, see, e.g., [7, 9, 10]. Decisive of the occurring polymolybdate types is apparently the number of hydrogen bonds that the organic cation can form and not the basicity of the amine [13]. Occasionally direct treatment of an (inorganic) monomolybdate in an organic solvent with acid and subsequent addition of an organic cation to precipitate the polymolybdate has also been applied [11]. By controlled hydrolysis of molybdic acid tetraesters in organic solvents in the presence of organic bases or organic salts, less hydrated or anhydrous polymolybdates with organic cations are obtained [7]. Many of the polymolybdates with organic cations have been prepared using these methods operating with organic solvents. Mixtures of polymolybdates with organic cations containing discrete polyanions may be separated or purified using various organic solvents [7]. Double exchange in aqueous solutions [1] or suspensions, and also in organic solvents [7], is another method to obtain polymolybdates with a definite cation.

Some polymolybdates are accessible by thermal dehydration, decomposition, or conversion (sometimes occurring at room temperature [8]) of other polymolybdates (compare p. 41).

Some few polymolybdates were prepared using special methods (see, e.g., 2.3.2.12, p. 67, 2.8.2.2, pp. 133 and 134, and 2.8.2.25, p. 169), and occasionally crystals of polymolybdates separate from unlikely reaction mixtures [12].

All these methods may yield not only polymolybdates containing water of crystallization or solvent molecules but also anhydrous salts. The preparation of polymolybdates from melts by reaction of molybdenum trioxide with oxides, hydroxides, or carbonates of the alkali or alkaline earth metals yields only anhydrous products and is mentioned here only to complete the list of preparation methods. The anhydrous polymolybdates are described in "Molybdän" Erg.-Bd. B1, 1975 (alkali polymolybdates) and Erg.-Bd. B 2, 1976 (alkaline earth polymolybdates).

References:

[1] Gmelin Handbuch "Molybdän", 1935, pp. 117/9, 212/98. – [2] I. Böschen (Diss. Kiel, West Germany, 1974, p. 69). – [3] J. Aveston, E. W. Anacker, J. S. Johnson (Inorg. Chem. **3** [1964] 735/46, 739, 746). – [4] H. M. Neumann, N. C. Cook (J. Am. Chem. Soc. **79** [1957] 3026/30). – [5] W. P. Griffith, T. D. Wickins (J. Chem. Soc. A **1967** 675/9).

[6] W. P. Griffith, P. J. B. Lesniak (J. Chem. Soc. A **1969** 1066/71). – [7] J. Fuchs (Z. Naturforsch. **28b** [1973] 389/404). – [8] K. H. Tytko, B. Schönfeld (Z. Naturforsch. **30b** [1975] 471/84). – [9] M. J. Weill (Bull. Soc. Chim. France **1960** 1136/8). – [10] F. Arnaud-Neu, M. J. Schwing-Weill (Bull. Soc. Chim. France **1973** 3225/32).

[11] M. Che, M. Fournier, J. P. Launay (J. Chem. Phys. **71** [1979] 1954/60). – [12] C. D. Garner, N. C. Howlader, F. E. Mabbs, A. T. McPhail, et al. (J. Chem. Soc. Dalton Trans. **1978** 1582/9). – [13] J. Fuchs, A. Thiele (Z. Naturforsch. **34b** [1979] 155/9).

2.1.4 Identification and Uniformity of the Compounds

The existence of many polymolybdates described in the older literature is dubious. Due to inadequate methods of characterization mixtures of different compounds have often been taken for new compounds and identical products have often been taken for different compounds due to erroneous analytical results [1, 2]. Not until about 1950 did the investigations become more reliable, at least with respect to the polymolybdates occurring in the range $\leqq 1.5\,H^+/MoO_4^{2-}$ of the aqueous solutions [1]. The relations of the polymolybdates existing in the range $>1.5\,H^+/MoO_4^{2-}$ were not explained until about 1975 [2, 3]. There are, however, even now open questions about many compounds concerning their uniformity and identity, in particular about the polymolybdates of calcium and strontium (see pp. 192 and 202) and about the "dimolybdates" (= "bimolybdates") (see p. 24).

Remarkably, there are some polymolybdates which have been prepared only a few times (e.g., the compound $Na_2O \cdot 4MoO_3 \cdot 6H_2O$ [2, 4, 5]) or on one occasion only (e.g., the compounds $[K_4Mo_4O_{14} \cdot 2H_2O]_\infty$ [6] and $3MgO \cdot 8MoO_3 \cdot 24H_2O$ [7]) and which are difficult to reproduce but nevertheless are well characterized, in one case [6] even by a complete crystal structure determination. Presumably, the reason for this is that a parameter important for the preparation has not yet been detected and described. Such a parameter controlling the reaction seems to have been found [7] in the case of the magnesium (3:8)-molybdate (see pp. 183 and 184).

A polymolybdate is ideally characterized when a complete crystal structure determination can be performed or at least the structure of the Mo-O skeleton can be determined. It can be regarded as sufficiently characterized if there is a "finger-print", e.g., Raman or infrared spectra or X-ray powder diagram [2]. Since most polymolybdates occur as "types" with different cations and variable proportion of water of crystallization, other polymolybdates of the type can be regarded as well characterized even if there is a complete crystal structure determination of only one example of the type.

The statement of a reliable stoichiometric formula (i.e., the ratio M^I_2O or $M^{II}O:MoO_3$; the degree of aggregation, a common factor of M^I_2O or $M^{II}O$ and MoO_3, and the proportion of H_2O are not considered here) for a polymolybdate is often very difficult if there is no complete determination of its structure. For example, there was a controversy for many decades whether paramolybdates should be formulated as (3:7)- or (5:12)-molybdates since the analytical values (5:12=0.417 and 3:7=0.429) are nearly the same (compare "Molybdän", 1935, pp. 113/9, 210/98). The differences were finally resolved in 1937 [8] in favor of the (3:7) formulation by determination of the unit cell and density of the polymolybdate $(NH_4)_6[Mo_7O_{24}]\cdot 4H_2O$ (see p. 96). This is the only way (if no complete determination of the crystal structure is forthcoming) to establish correctly the ratio M^I_2O or $M^{II}O:MoO_3$. If the salt is composed of discrete polymolybdate ions, an (approximative) determination of its molecular weight allows calculation of the common factor of M^I_2O or $M^{II}O$ and MoO_3 and thus the complete formula of the polymolybdate ion (with the exception of its proportion of constitutional water) [2, 9].

References:

[1] I. Lindqvist (Nova Acta Regiae Soc. Sci. Upsaliensis [4] **15** No. 1 [1950] 1/22). – [2] K. H. Tytko, B. Schönfeld (Z. Naturforsch. **30b** [1975] 471/84). – [3] B. Krebs, I. Paulat-Böschen (Acta Cryst. B **32** [1976] 1697/704). – [4] J. Byé (Bull. Soc. Chim. France [5] **10** [1943] 239/44). – [5] Y. Sasaki, L. G. Sillén (Arkiv Kemi **29** [1969] 253/77, 275).

[6] B. M. Gatehouse (J. Less-Common Metals **54** [1977] 283/8). – [7] G. Teller (Bull. Soc. Chim. France **1959** 1535/6). – [8] J. H. Sturdivant (J. Am. Chem. Soc. **59** [1937] 630/1). – [9] K. H. Tytko, B. Schönfeld, B. Buss, O. Glemser (Angew. Chem. **85** [1973] 305/7; Angew. Chem. Intern. Ed. Engl. **12** [1973] 330/2).

2.1.5 Polymolybdate Types

Since methods for identifying the uniformity of compounds are more generally available, there are only a few [1 to 4] systematic investigations concerning the solids which crystallize from aqueous molybdate solutions in relation to the degree of acidification. Nevertheless, today we have a rather clear picture of the polymolybdate types which crystallize from such solutions.

There is the rule that the ratio $Z^+ = p/q$ of the polymolybdates crystallizing from aqueous solution corresponds to the acidity Z (Z = molar ratio of *reacted* H^+ to *initially* present MoO_4^{2-} ions) of the solution (see, e.g., [3]). Since up to Z = 1.5 in the concentration range suitable for preparative work this value is equal to the degree of acidification P of the solution [5] (P = molar ratio of H^+ ions *introduced* to MoO_4^{2-} ions *initially* present), adjustment of the conditions for the preparation of the different polymolybdate types is in general relatively easy. Certain polymolybdates have a broad range of formation (cf. $K_2O\cdot 3MoO_3\cdot 3H_2O$, p. 80). The conditions of formation for the general polymolybdate types [3] (i.e., polymolybdates structurally typified) are listed in Table 1.

The α-octamolybdate type has been obtained only with organic cations [11 to 14] and in the form of pure products only from organic solvents [11 to 13].

The hexamolybdate type has also been prepared only from aqueous solution with organic cations [6]. Alkali salts have been obtained using a special technique of preparation [7].

Table 1

General Types of Polymolybdates (structurally typified polymolybdates) Obtainable from Aqueous Solution (and occasionally from organic solvents) and Dependence on the Conditions in the Solution.

polymolybdate type[1]	$p/q = Z^+$	a:b[2]	optimal conditions for formation			name most commonly used
			P	Z	pH	
$M_6[Mo_7O_{24}] \cdot nH_2O$	8/7 = 1.143	3:7	1.0 to 1.1	1.0 to 1.1	6 to 5	heptamolybdate
$[M_6Mo_8O_{27} \cdot nH_2O]_\infty$	10n/8n = 1.250	3:8	~1.2	~1.2	~5	(3:8)-molybdate
$[M_4Mo_6O_{20} \cdot nH_2O]_\infty$	8n/6n = 1.333	1:3	1.2 to 1.4	1.2 to 1.4	5 to 3	"trimolybdate"
$M_4[Mo_8O_{26}] \cdot nH_2O$	12/8 = 1.500	1:4	~1.5	~1.5	~2.5	β-octamolybdate
$M_4[Mo_8O_{26}] \cdot nH_2O$[3]	12/8 = 1.500	1:4				α-octamolybdate
$M_2[Mo_6O_{19}] \cdot nH_2O$[3]	10/6 = 1.667	1:6	~1.8	~1.7	~2	hexamolybdate
$M_8[Mo_{36}O_{112}(H_2O)_{16}] \cdot nH_2O$	64/36 = 1.778	1:9	1.8 to 1.9	1.78	~1	36-molybdate
$[MMo_5O_{15}(OH)(H_2O) \cdot H_2O]_\infty$	9n/5n = 1.800	1:10	2 to 5	1.8	<0, 20 to 40°C	"decamolybdate"[4]
0.04 to 0.10 $M_2O \cdot MoO_3 \cdot$ 0.3 to 0.7 H_2O	1.8 to 1.9	1:10 to 25	2 to 5		<0, 100°C	"phase C" polymolybdate[4]

[1] M = M^I or $M^{II}/2$. – [2] Ratio $M_2O:MoO_3$; $Z^+ = 2 - 2a/b$ [5]. – [3] This type mostly crystallizes without water. – [4] Note added in proof: The "decamolybdate" and "phase C" polymolybdate types have recently [20] been shown to be identical. All statements on these compounds are correct; however, the cation H_3O^+ has to be considered in addition to alkali and alkaline earth metal cations. Hence, $Z^+ = 1.800$ has to be regarded as the Z^+ value of the polyanion and the Z^+ values calculated from the a:b ratios are only hypothetic.

There are further polymolybdate types which are merely characterized by their analytical $M_2O:MoO_3$ ratio but not by a definite structure, see Table 2. Both these types often crystallize without water. Some individual polymolybdates are given in Table 3.

Table 2
Polymolybdates Typified Only by $M_2O:MoO_3$ Ratio Obtainable from Aqueous Solution and Dependence on the Conditions in the Solution.

polymolybdate type	Z^+	a:b	optimal conditions for formation P	Z	pH	name most commonly used
$M_2O \cdot 2MoO_3 \cdot nH_2O$	1	1:2	1)			"dimolybdate"
$M_2O \cdot 4MoO_3 \cdot nH_2O$	1.5	1:4	~1.5	~1.5	~2.5	"tetramolybdate"

1) Special conditions ($(NH_4)_2O \cdot 2MoO_3$ [3, 15, 16]) or conditions not reported ($K_2O \cdot 2MoO_3 \cdot H_2O$ [17]).

Table 3
Individual Polymolybdate Types Obtainable from Aqueous or Organic Solutions.

polymolybdate	$p/q = Z^+$	a:b	name most commonly used
$[(C_4H_9)_4N]_2[Mo_2O_7]$ 1)	2/2 =1.00	1:2	dimolybdate
$(NH_4)_8[Mo_{10}O_{34}]$ 2)	12/10 =1.20	2:5	decamolybdate
$(i\text{-}C_3H_7NH_3)_6[Mo_8O_{26}(OH)_2] \cdot 2H_2O$ 2)	10/8 =1.25	3:8	dihydrogenoctamolybdate

1) This type is obtainable only from organic solvents. – 2) Special conditions for formation ($(NH_4)_8[Mo_{10}O_{34}]$ [18], $(i\text{-}C_3H_7NH_3)_6[Mo_8O_{26}(OH)_2] \cdot 2H_2O$ [19]).

As a comparison with the older literature shows, the existence of "octa-", "16-", and other polymolybdates (compare [8]) could not be confirmed as polymolybdate *types*. These have been found to be, like the "hexamolybdates" [9] and many other suggestions made in the meantime, either of the 36-molybdate or "decamolybdate" type. But many of the "decamolybdates" described in the older literature are also not "decamolybdates" but 36-molybdates (see pp. 67 and 69).

While aqueous solutions deposit both polymolybdates with discrete and with chain-like, two- or three-dimensional network polyanions, organic solvents deposit only polymolybdates with discrete polyanions. Most of the polymolybdate types prepared using organic solvents can also be obtained from aqueous solutions. However, there is one type which can be prepared only from organic solvents, $[(C_4H_9)_4N]_2[Mo_2O_7]$ [10].

Polymolybdate types not prepared from solutions will not be considered here.

References:

[1] I. Lindqvist (Nova Acta Regiae Soc. Sci. Upsaliensis [4] **15** No. 1 [1950] 1/22). – [2] Y. Sasaki, L. G. Sillén (Arkiv Kemi **29** [1969] 253/77). – [3] K. H. Tytko, B. Schönfeld (Z. Naturforsch. **30b** [1975] 471/84). – [4] I. Böschen (Diss. Kiel, West Germany, 1974). – [5] K. H. Tytko, O. Glemser (Advan. Inorg. Chem. Radiochem. **19** [1976] 239/315, 241, 244).

[6] J. Fuchs, K. F. Jahr (Z. Naturforsch. **23b** [1968] 1380). – [7] A. M. Golubev, E. A. Torchenkova, V. I. Spitsyn (Dokl. Akad. Nauk SSSR **217** [1974] 345/7; Dokl. Chem. Proc. Acad. Sci. USSR **214/219** [1974] 495/7). – [8] Gmelin Handbuch "Molybdän", 1935, pp. 113/9, 210/98. – [9] J. Byé

(Ann. Chim. [Paris] [11] **20** [1945] 463/550, 518/28). – [10] V. W. Day, M. F. Fredrich, W. G. Klemperer, W. Shum (J. Am. Chem. Soc. **99** [1977] 6146/8).

[11] J. Fuchs, H. Hartl (Angew. Chem. **88** [1976] 385/6; Angew. Chem. Intern. Ed. Engl. **15** [1976] 375). – [12] W. G. Klemperer, W. Shum (J. Am. Chem. Soc. **98** [1976] 8291/3). – [13] J. Fuchs, I. Brüdgam (Z. Naturforsch. **32b** [1977] 853/7). – [14] V. W. Day, M. F. Fredrich, W. G. Klemperer, W. Shum (J. Am. Chem. Soc. **99** [1977] 952/3). – [15] W. D. Hunnius (Z. Naturforsch. **29b** [1974] 599/602).

[16] C. J. Hallada (J. Less-Common Metals **36** [1974] 103/10). – [17] B. M. Gatehouse (J. Less-Common Metals **54** [1977] 283/8). – [18] W. D. Hunnius (Z. Naturforsch. **30b** [1975] 63/5). – [19] T. Yamase, T. Ikawa (Bull. Chem. Soc. Japan **50** [1977] 746/9). – [20] G. Baethe (Diss. Göttingen, West Germany, 1985, pp. 96/101, 118/20, 156).

2.1.6 Structures and "Finger-Prints" of the General Types of Polymolybdates
(Structurally Typified Polymolybdates)

The characteristic feature of the polymolybdate structures (as for other polymetalates) are MoO_6 octahedra sharing edges. Sometimes such groups of MoO_6 octahedra sharing edges are connected via common corners. Coordination polyhedra other than MoO_6 octahedra and other modes of connection are seldom found. For further structural features see pp. 18 and 20 and for more detailed considerations, p. 27.

Data which may serve as "finger-prints" to identify a particular species in different circumstances are mainly Raman and infrared spectra and X-ray powder diffraction patterns. Since the Raman and infrared spectra do not appreciably differ whether the salt of a given polymolybdate ion is a hydrate or anhydrous, and since solutions may deposit hydrated (or otherwise solvated) or anhydrous salts, some references will also be given in this section to anhydrous compounds to give access to the "finger-prints" of as many reference substances as possible. Thus, when writing this article it was possible to identify a number of polymolybdates by comparison of their "finger-prints" (and of some other data) given in the literature with those of reference substances.

2.1.6.1 The Heptamolybdate Type, $M_6[Mo_7O_{24}]\cdot nH_2O$ ($=3M_2O\cdot7MoO_3\cdot nH_2O$)

Compounds of this type are termed in the literature (approximately in the order of the frequency) heptamolybdates, paramolybdates, (3:7)-molybdates, (8,7)-molybdates. In the literature up to about 1937 [1] this type is incorrectly formulated also as (5:12)-molybdate.

This type is known for nearly all cations under discussion. All compounds hitherto characterized have the same polyanion whose structure is shown in **Fig. 1**: $Na_6[Mo_7O_{24}]\cdot14H_2O$ [2], $K_6[Mo_7O_{24}]\cdot4H_2O$ [3, 4], $(NH_4)_6[Mo_7O_{24}]\cdot4H_2O$ [4 to 8, 24], $(C_3H_7NH_3)_6[Mo_7O_{24}]\cdot3H_2O$, $(i\text{-}C_3H_7NH_3)_6[Mo_7O_{24}]\cdot3H_2O$ [36], $(CH_6N_3)_6[Mo_7O_{24}]\cdot H_2O$ (CH_5N_3 = guanidine) [9].

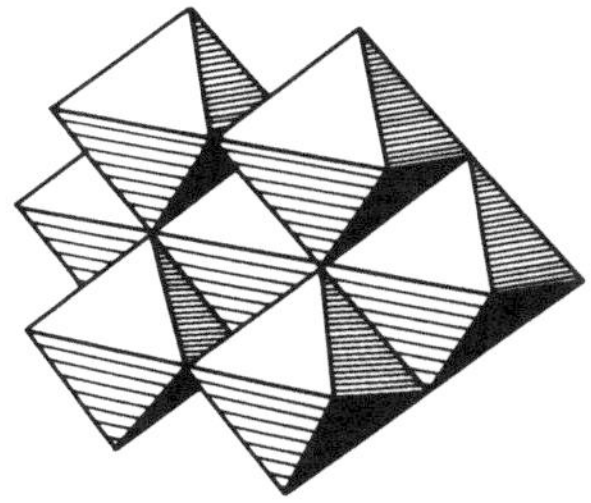

Fig. 1. Structure of the heptamolybdate ion $[Mo_7O_{24}]^{6-}$ in the sodium [2], potassium [3, 4], ammonium [4 to 8, 24], monopropyl- and monoisopropylammonium [36], and guanidinium [9] salts (from [33]).

The Raman spectra (**Fig. 2**) and infrared spectra (**Fig. 3**) of the heptamolybdates closely resemble each other. Raman spectra are reported for $Na_6[Mo_7O_{24}]\cdot 23H_2O$ [10], $Na_6[Mo_7O_{24}]\cdot 14H_2O$ [10, 11, 31], $K_6[Mo_7O_{24}]\cdot 4H_2O$ [10], and $(NH_4)_6[Mo_7O_{24}]\cdot 4H_2O$ [10, 12, 25, 34]. For the spectra of a single crystal of $Na_6[Mo_7O_{24}]\cdot 14H_2O$ in different geometrical orientations with respect to the exciting laser beam see [37]. Infrared spectra are reported for $Na_6[Mo_7O_{24}]\cdot 14H_2O$ [13, p. 44], [32] (the spectrum attributed to the 22-hydrate [14] is most probably that of the 14-hydrate generated by dehydration of the 22-hydrate), $K_6[Mo_7O_{24}]\cdot 4H_2O$ [13, p. 44], [15], $(NH_4)_6[Mo_7O_{24}]\cdot 4H_2O$ [13, p. 44], [14, 16, 26 to 30, 34, 35, 38, 39] and $1H_2O$ [38], $[(C_2H_5)NH_3]_6[Mo_7O_{24}]\cdot 3H_2O$ and H_2O, $[(CH_3)_3NH]_6[Mo_7O_{24}]\cdot 7H_2O$ [17], $(C_5H_{11}NH)_6[Mo_7O_{24}]\cdot 6H_2O$ and $3H_2O$ ($C_5H_{11}N$ = piperidine), $(C_6H_{13}NH)_6[Mo_7O_{24}]\cdot 2H_2O$ ($C_6H_{13}N$ = methylpiperidine) [14], and $(C_4H_9NH)_6[Mo_7O_{24}]\cdot H_2O$ (C_4H_9N = pyrrolidine) [17].

Fig. 2

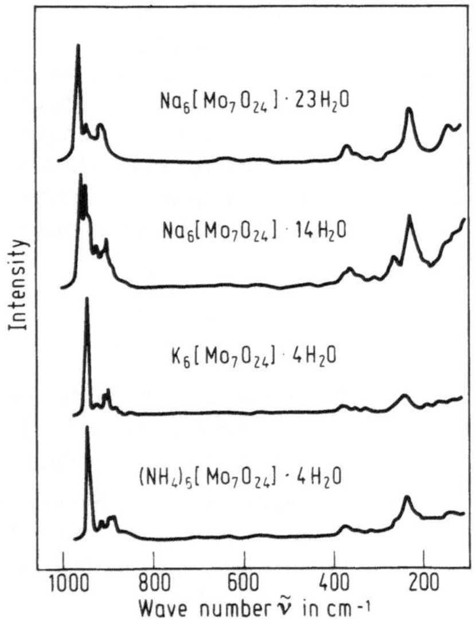

Fig. 3

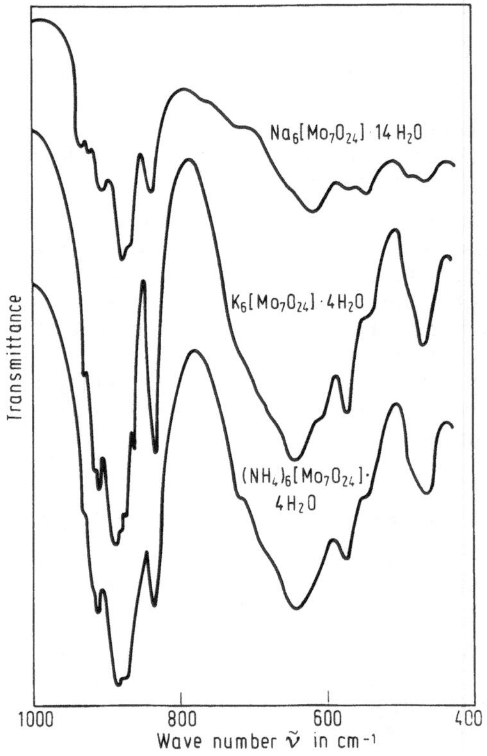

Raman spectra of $Na_6[Mo_7O_{24}]\cdot 23H_2O$, $Na_6[Mo_7O_{24}]\cdot 14H_2O$, $K_6[Mo_7O_{24}]\cdot 4H_2O$, and $(NH_4)_6[Mo_7O_{24}]\cdot 4H_2O$ [10] (left) and infrared spectra of $Na_6[Mo_7O_{24}]\cdot 14H_2O$, $K_6[Mo_7O_{24}]\cdot 4H_2O$, and $(NH_4)_6[Mo_7O_{24}]\cdot 4H_2O$ [13, p. 44] (right).

X-ray powder diffraction diagrams are cited for $Na_6[Mo_7O_{24}]\cdot 14H_2O$, $K_6[Mo_7O_{24}]\cdot 4H_2O$ [13, p. 84], $(NH_4)_6[Mo_7O_{24}]\cdot 4H_2O$ [13, p. 84], [18, 19, 38], $2H_2O$ [20], and $1H_2O$ [38], $(C_5H_{11}NH)_6[Mo_7O_{24}]\cdot 6H_2O$ and $3H_2O$ ($C_5H_{11}N$ = piperidine) [21], $(C_6H_{13}NH)_6[Mo_7O_{24}]\cdot 2H_2O$ ($C_6H_{13}N$ = 4-methylpiperidine) [22], and $Ba_3[Mo_7O_{24}]\cdot xH_2O$ [23].

References:

[1] J. H. Sturdivant (J. Am. Chem. Soc. **59** [1937] 630/1). – [2] K. Sjöbom, B. Hedman (Acta Chem. Scand. **27** [1973] 3673/91). – [3] B. M. Gatehouse, P. Leverett (Chem. Commun. **1968** 901/2). – [4] H. T. Evans, Jr., B. M. Gatehouse, P. Leverett (J. Chem. Soc. Dalton Trans. **1975** 505/14). – [5] I. Lindqvist (Arkiv Kemi **2** [1951] 325/41).

[6] E. Shimao (Nature **214** [1967] 170/1). – [7] E. Shimao (Bull. Chem. Soc. Japan **40** [1967] 1609/13). – [8] H. T. Evans (J. Am. Chem. Soc. **90** [1968] 3275/6). – [9] A. Don, T. J. R. Weakley (Acta Cryst. B **37** [1981] 451/3). – [10] K. H. Tytko, B. Schönfeld (Z. Naturforsch. **30b** [1975] 471/84).

[11] G. Johansson, L. Petterson, N. Ingri (Acta Chem. Scand. A **33** [1979] 305/12). – [12] J. Aveston, E. W. Anacker, J. S. Johnson (Inorg. Chem. **3** [1964] 735/46, 744). – [13] B. Schönfeld (Diss. Göttingen, West Germany, 1973). – [14] M. J. Schwing-Weill, F. Arnaud-Neu (Bull. Soc. Chim. France **1970** 853/60). – [15] A. La Ginestra, G. Rubino (Atti Accad. Nazl. Lincei Classe Sci. Fis. Mat. Nat. Rend. [8] **41** [1966] 510/20).

[16] T. Yamase, T. Ikawa (Bull. Chem. Soc. Japan **50** [1977] 746/9). – [17] F. Arnaud-Neu, M. J. Schwing-Weill (Bull. Soc. Chim. France **1973** 3225/32). – [18] E. Ma (Bull. Chem. Soc. Japan **37** [1964] 171/5, 648/53). – [19] M. J. Schwing-Weill (Bull. Soc. Chim. France **1967** 3795/8). – [20] A. Louisy, J. M. Dunoyer (J. Chim. Phys. **67** [1970] 1390/4).

[21] M. J. Schwing-Weill (Bull. Soc. Chim. France **1967** 3801/5). – [22] F. Arnaud-Neu, M. J. Schwing-Weill (Bull. Soc. Chim. France **1971** 60/8). – [23] M. Haeringer, G. Goldstein, P. Lagrange, J. P. Schwing (Bull. Soc. Chim. France **1967** 723/8). – [24] I. Lindqvist (Acta Cryst. **3** [1950] 159/60). – [25] J. Gupta (Indian J. Phys. **12** [1938] 223/32).

[26] F. A. Miller, C. H. Wilkins (Anal. Chem. **24** [1952] 1253/94, 1254, 1258, 1288). – [27] M. Hacskaylo (Anal. Chem. **26** [1954] 1410/2). – [28] F. A. Miller, G. L. Carlson, F. F. Bentley, W. H. Jones (Spectrochim. Acta **16** [1960] 135/235, 141, 156, 226). – [29] A. Kiss, S. Holly, E. Hild (Magy. Kem. Folyoirat **77** [1971] 418/26; C.A. **75** [1971] No. 135415). – [30] A. W. Armour, M. G. B. Drew, P. C. H. Mitchell (J. Chem. Soc. Dalton Trans. **1975** 1493/6).

[31] L. Lyhamn, L. Pettersson (Chem. Scr. **12** [1977] 142/52). – [32] L. Lyhamn (Chem. Scr. **12** [1977] 153/61). – [33] K. H. Tytko, O. Glemser (Advan. Inorg. Chem. Radiochem. **19** [1976] 239/315, 266). – [34] W. P. Griffith, P. J. B. Lesniak (J. Chem. Soc. A **1969** 1066/71). – [35] A. B. Kiss, P. Gadó, I. Asztalos, A. J. Hegedüs (Acta Chim. [Budapest] **66** [1970] 235/49).

[36] Y. Ohashi, K. Yanagi, Y. Sasada, T. Yamase (Bull. Chem. Soc. Japan **55** [1982] 1254/60). – [37] L. Lyhamn (Acta Chem. Scand. A **36** [1982] 595/603). – [38] Z. M. Hanafi, M. A. Khilla, M. H. Askar (Thermochim. Acta **45** [1981] 221/32). – [39] J. Fuchs, A. Thiele (Z. Naturforsch. **34b** [1979] 155/9).

2.1.6.2 The (3:8)-Molybdate Type, $[M_6Mo_8O_{27} \cdot nH_2O]_\infty$ ($=3M_2O \cdot 8MoO_3 \cdot nH_2O$)

The compounds of this type are named (3:8)-molybdates in the literature, and the ammonium salt is called polyoctamolybdate.

This type is known for a small number of cations only. However, it seems that in some cases compounds of this type have been erroneously assigned to the "trimolybdate" type or possibly to the paramolybdate type. The interrelations between para-, (3:8)-, and "tri"-molybdates with respect to the conditions of their formation are reflected in [1] (the case of the ammonium salts) and [2] (the case of the magnesium salts).

Only one compound of this type has been structurally characterized: $[(NH_4)_6Mo_8O_{27}\cdot 4H_2O]_\infty$ [3, 4], see **Fig. 4**. However, infrared spectra and the conditions of preparation suggest the same polymolybdate ion to be present in all compounds of this type.

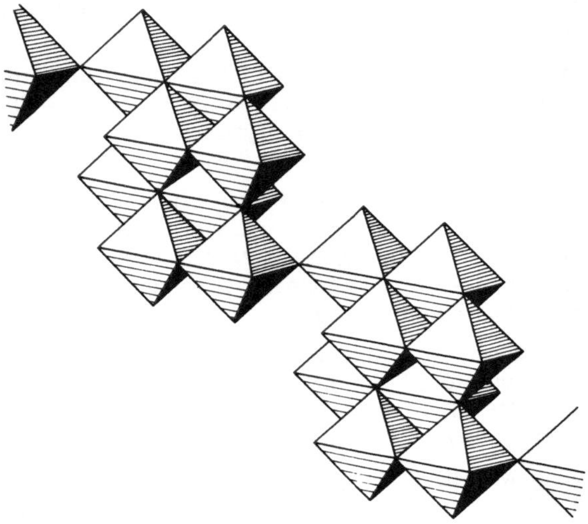

Fig. 4. Structure of the polyoctamolybdate ion $[Mo_8O_{27}^{6-}]_\infty$ in the ammonium salt $[(NH_4)_6Mo_8O_{27}\cdot 4H_2O]_\infty$ [3, 4] (from [12]).

The Raman spectrum is only reported for the ammonium salt [1, 5 to 7] (in [5] incorrectly stated to be a 3-hydrate) (**Fig. 5**). Infrared spectra are reported for $[(NH_4)_6Mo_8O_{27}\cdot 4H_2O]_\infty$ [5 to 8] and $Ba_3Mo_8O_{27}\cdot 9H_2O$ (described under the headings "$\sim BaMo_3O_{10}\cdot 3H_2O$" and "barium trimolybdates" but assumed to be (3:8)-molybdates) [9] (**Fig. 6**).

Fig. 5

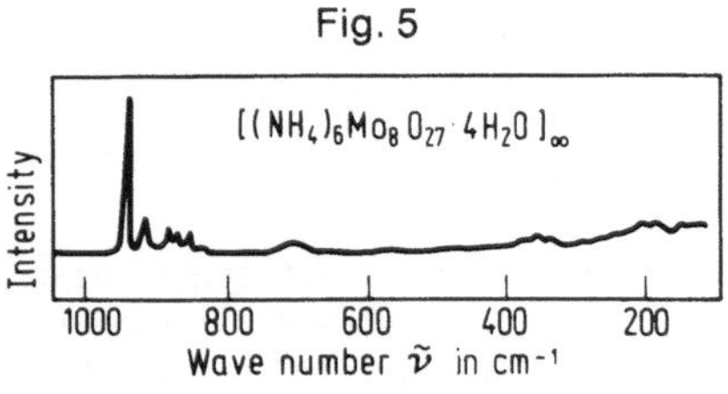

Fig. 6

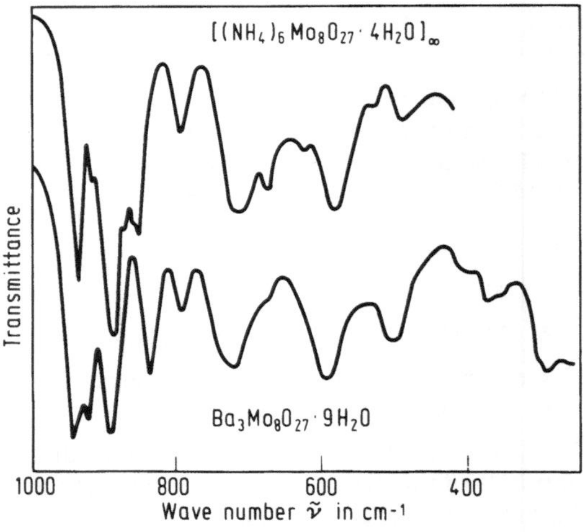

Raman spectrum of $[(NH_4)_6Mo_8O_{27}\cdot 4H_2O]_\infty$ [1] (left) and infrared spectra of $[(NH_4)_6Mo_8O_{27}\cdot 4H_2O]_\infty$ [8] and $Ba_3Mo_8O_{27}\cdot 9H_2O$ [9] (right).

X-ray powder diffraction diagrams (or interplanar distances) are quoted for $3(NH_4)_2O \cdot 8MoO_3 \cdot 3H_2O$ [10] (as in the case of [5], this compound is most probably a 4-hydrate, see above), $3MgO \cdot 8MoO_3 \cdot 24H_2O$ [2], and $Ba_3Mo_8O_{27} \cdot 9H_2O$ [9] (previously formulated as $BaO \cdot 3MoO_3 \cdot nH_2O$ [11]). The compounds $Ba_3Mo_8O_{27} \cdot 9H_2O$ and $BaO \cdot 3MoO_3 \cdot nH_2O$ are described to be identical [9].

References:

[1] K. H. Tytko, B. Schönfeld (Z. Naturforsch. **30b** [1975] 471/84). – [2] G. Teller (Bull. Soc. Chim. France **1959** 1535/6). – [3] I. Böschen, B. Buss, B. Krebs (Acta Cryst. B **30** [1974] 48/56). – [4] I. Böschen, B. Buss, B. Krebs, O. Glemser (Angew. Chem. **85** [1973] 409; Angew. Chem. Intern. Ed. Engl. **12** [1973] 409). – [5] O. Glemser, G. Wagner, B. Krebs (Angew. Chem. **82** [1970] 639; Angew. Chem. Intern. Ed. Engl. **9** [1970] 639).

[6] W. D. Hunnius (Z. Naturforsch. **30b** [1975] 63/5). – [7] I. Böschen (Diss. Kiel, West Germany, 1974, p. 167). – [8] B. Schönfeld (Diss. Göttingen, West Germany, 1973, p. 49). – [9] J. Meullemeestre (Bull. Soc. Chim. France **1978** I 236/42). – [10] M. J. Schwing-Weill (Bull. Soc. Chim. France **1967** 3795/8).

[11] M. Haeringer, G. Goldstein, P. Lagrange, J. P. Schwing (Bull. Soc. Chim. France **1967** 723/8). – [12] K. H. Tytko, O. Glemser (Advan. Inorg. Chem. Radiochem. **19** [1976] 239/315, 266).

2.1.6.3 The "Trimolybdate" Type, $[M_4Mo_6O_{20} \cdot 2nH_2O]_\infty$ ($= M_2O \cdot 3MoO_3 \cdot nH_2O$)

The compounds are named in the literature "trimolybdates" or (1:3)-molybdates. It has been proposed (see p. 2) to call these compounds polyhexamolybdates.

This type is known for nearly all cations in question. Only one compound of this type has been structurally characterized, $[Rb_4Mo_6O_{20} \cdot 2H_2O]_\infty$ [1], see **Fig. 7**. However, Raman and infrared spectra, X-ray powder diagrams, and the characteristic shape of the crystals ("very fine needles", "fibrous crystals", "cotton-like product") indicate the presence of the same chain-like polymolybdate ion in all compounds of this type. Note added in proof: Recently [19] a second compound of this type, $[((CH_3)_2NH_2)_4Mo_6O_{20} \cdot 2H_2O]_\infty$, has been structurally characterized. The

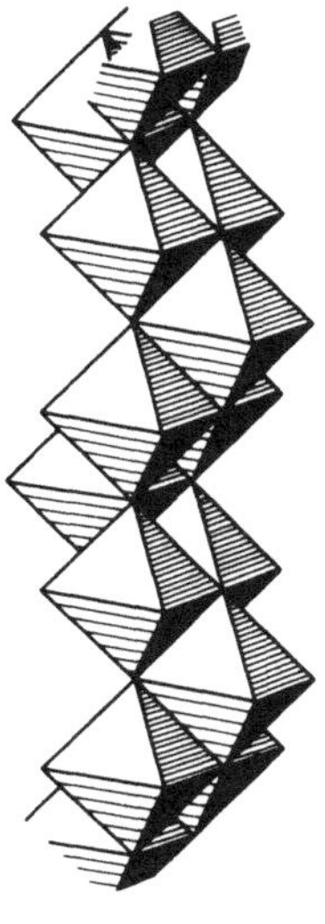

Fig. 7. Structure of the polyhexamolybdate anion $[Mo_6O_{20}^{4-}]_\infty$ in the rubidium salt $[Rb_4Mo_6O_{20} \cdot 2H_2O]_\infty$ [1].

structure of the polyanion is somewhat different from that represented in Fig. 7 (see Fig. 55, p. 134). The infrared spectrum [10] of the compound is a typical "trimolybdate" spectrum without any irregularities and the length of the "b axis" [19] is also that of the "trimolybdates". This new "trimolybdate" structure containing oxygen of coordination number two is in disagreement, however, with the interpretation of the vibrational spectra of the "trimolybdates" by [1] according to which the absence of Mo-O vibrational bands in the range 875 to 670 cm^{-1} is due to the absence of oxygen of coordination number two.

The Raman spectra (**Fig. 8**) and infrared spectra (**Fig. 9**) of the "trimolybdates" closely resemble each other. Raman spectra are reported for $Na_2O \cdot 3MoO_3 \cdot 4H_2O$, $K_2O \cdot 3MoO_3 \cdot 3H_2O$, $(NH_4)_2O \cdot 3MoO_3 \cdot 9H_2O$ [2], and $Rb_2O \cdot 3MoO_3 \cdot H_2O$ [1]. Infrared spectra are reported for $Li_2O \cdot 3MoO_3 \cdot 5.7H_2O$ [3], $Na_2O \cdot 3MoO_3 \cdot 4H_2O$ [4 to 6] and $3H_2O$ [7], $K_2O \cdot 3MoO_3 \cdot 3H_2O$ [4 to 6, 8, 9], $(NH_4)_2O \cdot 3MoO_3 \cdot 9H_2O$ [5] and $1H_2O$ [18], $[(CH_3)_2NH_2]_2O \cdot 3MoO_3 \cdot H_2O$ [10], $[(C_2H_5)_2NH_2]_2O \cdot 3MoO_3 \cdot H_2O$ [6, 10], $(C_5H_{11}NH)_2O \cdot 3MoO_3 \cdot 2H_2O$ ($C_5H_{11}N$ = piperidine) [4], $(C_4H_{10}N_2H_2)O \cdot 3MoO_3 \cdot H_2O$ ($C_4H_{10}N_2$ = piperazine), $(C_4H_9NH)_2O \cdot 3MoO_3 \cdot H_2O$ (C_4H_9N = pyrroli-

Fig. 8

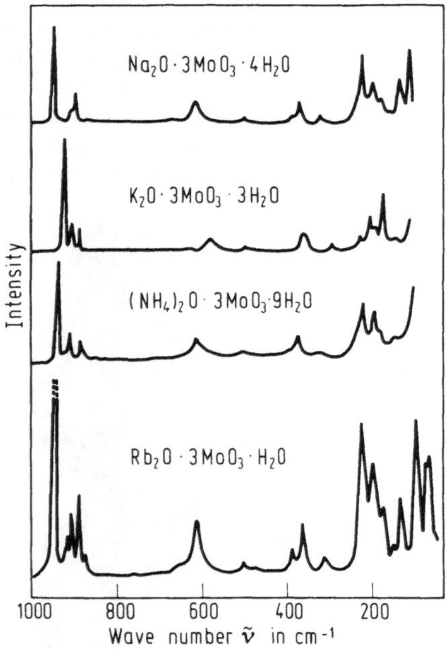

Fig. 9

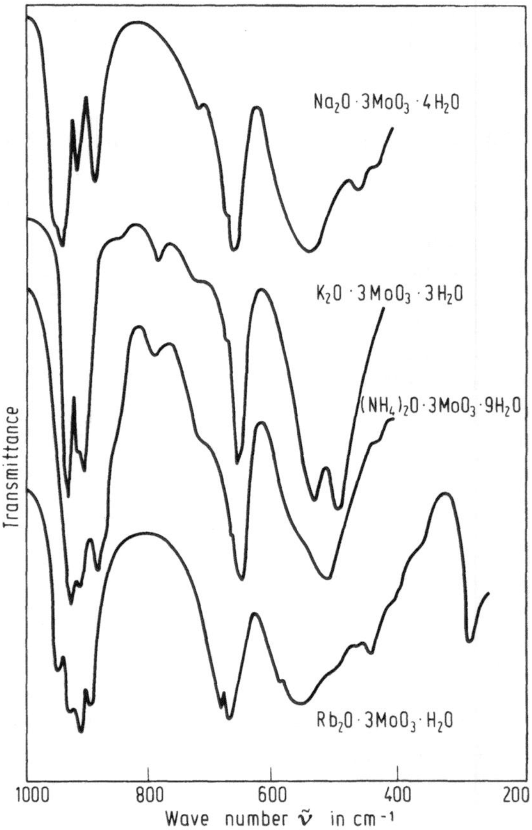

Raman spectra of $Na_2O \cdot 3MoO_3 \cdot 4H_2O$, $K_2O \cdot 3MoO_3 \cdot 3H_2O$, $(NH_4)_2O \cdot 3MoO_3 \cdot 9H_2O$ [2], and $Rb_2O \cdot 3MoO_3 \cdot H_2O$ [1] (left) and infrared spectra of $Na_2O \cdot 3MoO_3 \cdot 4H_2O$, $K_2O \cdot 3MoO_3 \cdot 3H_2O$, $(NH_4)_2O \cdot 3MoO_3 \cdot 9H_2O$ [5], and $Rb_2O \cdot 3MoO_3 \cdot H_2O$ [1] (right).

dine) [6], $Rb_2O \cdot 3MoO_3 \cdot H_2O$ [1, 20], $MgO \cdot 3MoO_3 \cdot 10H_2O$, $7H_2O$, and $5H_2O$ [11], $CaO \cdot 3MoO_3 \cdot 6H_2O$, $3H_2O$, and $1H_2O$, $SrO \cdot 3MoO_3 \cdot 4H_2O$, $3H_2O$, and $1H_2O$ [12]. The spectra of the compounds $(NH_4)_2O \cdot 3MoO_3 \cdot 2H_2O$, $[(CH_3)_3NH]_2O \cdot 3MoO_3 \cdot H_2O$ [6], $(C_6H_{13}NH)_2O \cdot 3MoO_3 \cdot H_2O$ ($C_6H_{13}N$ = 4-methylpiperidine), $(C_7H_{15}NH)_2O \cdot 3MoO_3 \cdot 3H_2O$ ($C_7H_{15}N$ = 2,6-dimethylpiperidine), and $(C_6H_{13}NH)_2O \cdot 3MoO_3 \cdot H_2O$ ($C_6H_{13}N$ = N-methylpiperidine) [4], however, show strong deviations from the "trimolybdate" spectra.

X-ray powder diffraction diagrams (or interplanar distances) are cited for $Li_2O \cdot 3MoO_3 \cdot 5.7H_2O$ [13], $Na_2O \cdot 3MoO_3 \cdot 3H_2O$ [7], $K_2O \cdot 3MoO_3 \cdot 3H_2O$ [14], $(C_5H_{11}NH)_2O \cdot 3MoO_3 \cdot 2H_2O$ ($C_5H_{11}N$ = piperidine) [15], $Rb_2O \cdot 3MoO_3 \cdot H_2O$ [20], $Cs_2O \cdot 3MoO_3 \cdot H_2O$ [21], $MgO \cdot 3MoO_3 \cdot 10H_2O$, $7H_2O$ [11, 17], and $5H_2O$ [11], $CaO \cdot 3MoO_3 \cdot 6H_2O$, $3H_2O$, and $1H_2O$, $SrO \cdot 3MoO_3 \cdot 4H_2O$, $3H_2O$, and $1.75H_2O$ [12]. The data reported for $(C_6H_{13}NH)_2O \cdot 3MoO_3 \cdot H_2O$ ($C_6H_{13}N$ = 4-methylpiperidine) and $(C_6H_{13}NH)_2O \cdot 3MoO_3 \cdot H_2O$ ($C_6H_{13}N$ = N-methylpiperidine) [16] do not belong to "trimolybdates" since these compounds do not show the "trimolybdate" spectra (see above). The characteristic length of the (shortest) "b axis" which extends parallel to the chains is 7.61 Å [1].

References:

[1] H. U. Kreusler, A. Förster, J. Fuchs (Z. Naturforsch. **35b** [1980] 242/4). – [2] K. H. Tytko, B. Schönfeld (Z. Naturforsch. **30b** [1975] 471/84). – [3] S. Hodorowicz, S. Sagnowski (Acta Phys. Polon. A **50** [1976] 817/21). – [4] M. J. Schwing-Weill, F. Arnaud-Neu (Bull. Soc. Chim. France **1970** 853/60). – [5] B. Schönfeld (Diss. Göttingen, West Germany, 1973, p. 47).

[6] F. Arnaud-Neu, M. J. Schwing-Weill (Bull. Soc. Chim. France **1973** 3225/32). – [7] J. Chojnacki, S. Hodorowicz (Roczniki Chem. **47** [1973] 2213/9). – [8] A. La Ginestra, G. Rubino (Atti Accad. Nazl. Lincei Classe Sci. Fis. Mat. Nat. Rend. [8] **41** [1966] 510/20). – [9] S. Hodorowicz (Roczniki Chem. **50** [1976] 1031/3). – [10] T. Yamase, T. Ikawa (Bull. Chem. Soc. Japan **50** [1977] 746/9).

[11] J. Meullemeestre (Bull. Soc. Chim. France **1978** I 231/5). – [12] J. Meullemeestre (Bull. Soc. Chim. France **1978** I 236/42). – [13] S. Hodorowicz, W. Surga (Roczniki Chem. **51** [1977] 411/5). – [14] J. Chojnacki, S. Hodorowicz (Roczniki Chem. **49** [1975] 679/82). – [15] M. J. Schwing-Weill (Bull. Soc. Chim. France **1967** 3801/5).

[16] F. Arnaud-Neu, M. J. Schwing-Weill (Bull. Soc. Chim. France **1971** 60/8). – [17] G. Teller (Bull. Soc. Chim. France **1959** 1533/5). – [18] W. D. Hunnius (Diss. Berlin F.U. 1970, pp. 50/1). – [19] H. Toraya, F. Marumo, T. Yamase (Acta Cryst. B **40** [1984] 145/50). – [20] W. Surga, S. Sagnowski, S. Hodorowicz (J. Inorg. Nucl. Chem. **43** [1981] 1821/5).

[21] S. Hodorowicz, E. Hodorowicz, S. Sagnowski, W. Surga (Pol. J. Chem. **54** [1980] 1859/64).

2.1.6.4 The β-Octamolybdate Type, $M_4[Mo_8O_{26}] \cdot nH_2O$ (= $M_2O \cdot 4MoO_3 \cdot 0.5nH_2O$)

The compounds of this type are termed in the literature octamolybdates, "tetramolybdates", (1:4)-molybdates, metamolybdates, (12,8)-molybdates, β-octamolybdates (the addendum "β" is used to make a distinction between a second octamolybdate, named α-octamolybdate, having the same formula but a different structure).

This type is formed only with certain cations, usually ammonium and cations of organic bases. The structure of the polymolybdate ion as found in the compounds $(NH_4)_4[Mo_8O_{26}] \cdot 5H_2O$ [1, 2], $(NH_4)_4[Mo_8O_{26}] \cdot 4H_2O$ [3 to 5], $[(CH_3)_2NH_2]_4[Mo_8O_{26}] \cdot 2C_3H_7NO$ (C_3H_7NO = N,N-dimethylformamide) [21], $[(CH_3)_4N]_2Na_2[Mo_8O_{26}] \cdot 2H_2O$ [6], $(C_6H_7NH)_4[Mo_8O_{26}]$ (C_6H_7N = 3-methylpyridine [22], $(C_7H_9NH)_4[Mo_8O_{26}]$ (C_7H_9N = 3-ethylpyridine [7], 4-ethylpyridine [8], 2-ethylpyridine

[9]) is shown in **Fig. 10**. However, the four alkylpyridinium salts do not contain water of crystallization. There are other (1:4)-molybdates not containing the β-octamolybdate ion (e.g., the compounds $Na_2O \cdot 4\,MoO_3 \cdot 6\,H_2O$ (yellowish) and $(NH_4)_2O \cdot 4\,MoO_3$, compare Figs. 11 and 12).

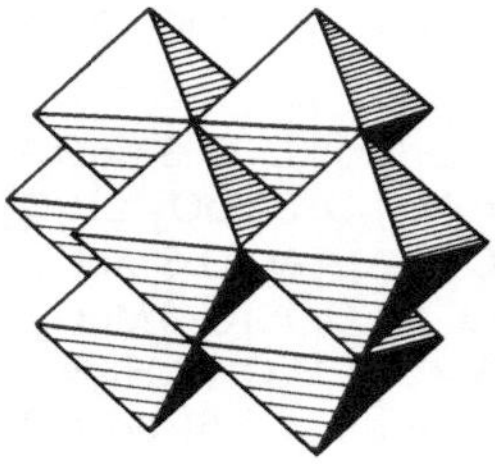

Fig. 10. Structure of the β-octamolybdate ion $[Mo_8O_{26}]^{4-}$ in the ammonium [1 to 5], dimethylammonium [21], bis(tetramethylammonium)-disodium [6], 3-methylpyridinium [22], 2- [9], 3- [7], and 4-ethylpyridinium [8] salts (from [19]).

Raman spectra [10] (left) and infrared spectra [13] (right) of β-$(NH_4)_4[Mo_8O_{26}] \cdot 4(5)\,H_2O$ and two (1:4)-molybdates of a different type: $Na_2O \cdot 4\,MoO_3 \cdot 6\,H_2O$ (yellowish) and $(NH_4)_2O \cdot 4\,MoO_3$.

Fig. 11

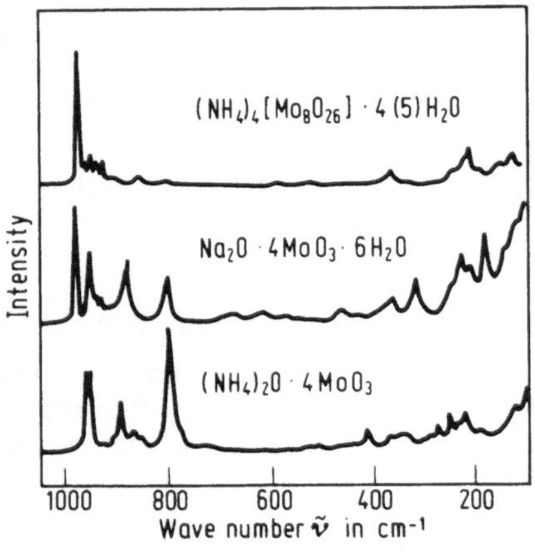

Fig. 12

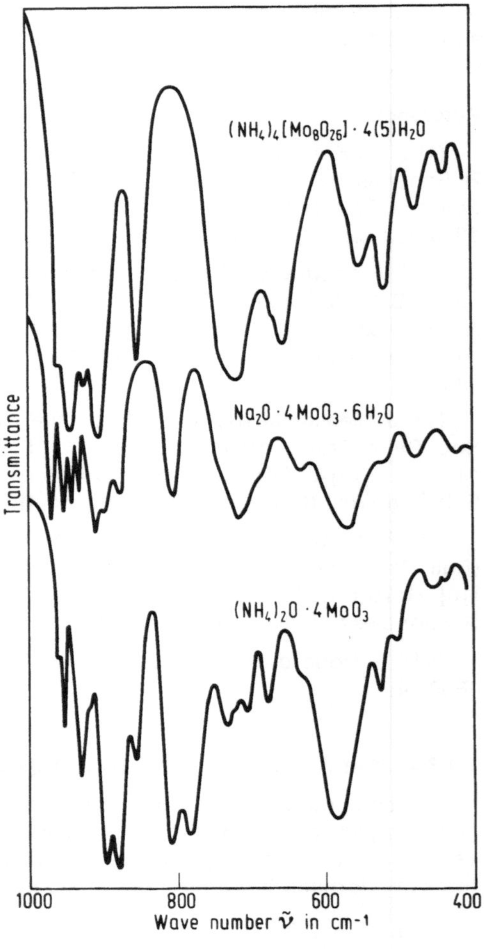

The Raman spectrum is reported for the compounds $(NH_4)_4[Mo_8O_{26}] \cdot 4(5)\,H_2O$ [10, 11, 20] (see **Fig. 11**), $[(C_3H_7)_4N]_2O \cdot 4\,MoO_3 \cdot n\,H_2O$, and $[(CH_3)_4N]_2Na_2[Mo_8O_{26}] \cdot 2\,H_2O$ [14] (erroneously described as $[(CH_3)_4N]_2O \cdot 8\,MoO_3 \cdot 5\,H_2O$ [6]). Infrared spectra are reported for $K_2O \cdot$

$4MoO_3 \cdot 3H_2O$ [12], $(NH_4)_4[Mo_8O_{26}] \cdot 4(5)H_2O$ [13, 20] (see **Fig. 12**), $[(C_3H_7)_4N]_2O \cdot 4MoO_3 \cdot nH_2O$ [14], $(C_5H_{11}NH)_2Mo_4O_{13} \cdot 2H_2O$ ($C_5H_{11}N$ = piperidine), $(C_6H_{13}NH)_2Mo_4O_{13} \cdot H_2O$ ($C_6H_{13}N$ = N-methylpiperidine), $(C_9H_{19}NH)_2Mo_4O_{13} \cdot H_2O$ ($C_9H_{19}N$ = 2,2,6,6-tetramethylpiperidine) [15], $[(CH_3)_4N]_2Na_2[Mo_8O_{26}] \cdot 2H_2O$ [14] (this compound is erroneously described as $[(CH_3)_4N]_2O \cdot 8MoO_3 \cdot 5H_2O$ [6]), and $[(n\text{-}C_4H_9)_4N]_3K[Mo_8O_{26}] \cdot 2H_2O$ [16]. From the similarity of the spectra it is assumed that the same polymolybdate ion is present in all compounds of this list. See also [23].

X-ray powder diffraction patterns are reported for the compounds $(NH_4)_4[Mo_8O_{26}] \cdot 4H_2O$ [18] and $(C_5H_{11}NH)_2O \cdot 4MoO_3 \cdot 2H_2O$ ($C_5H_{11}N$ = piperidine) [17].

References:

[1] I. Lindqvist (Arkiv Kemi **2** [1951] 349/55). – [2] T. J. R. Weakley (Polyhedron **1** [1982] 17/9). – [3] L. O. Atovmyan, O. N. Krasochka (Zh. Strukt. Khim. **13** [1972] 342/3; J. Struct. Chem. [USSR] **13** [1972] 319/20). – [4] B. M. Gatehouse (J. Less-Common Metals **54** [1977] 283/8). – [5] H. Vivier, J. Bernard, H. Djomaa (Rev. Chim. Minerale **14** [1977] 584/604).

[6] J. Fuchs, I. Knöpnadel (Z. Krist. **158** [1982] 165/79). – [7] P. Román, J. Jaud, J. Galy (Z. Krist. **154** [1981] 59/68). – [8] P. Román, M. Martínez-Ripoll, J. Jaud (Z. Krist. **158** [1982] 141/7). – [9] P. Román, A. Vegas, M. Martínez-Ripoll, S. García-Blanco (Z. Krist. **159** [1982] 291/5). – [10] K. H. Tytko, B. Schönfeld (Z. Naturforsch. **30b** [1975] 471/84).

[11] J. Aveston, E. W. Anacker, J. S. Johnson (Inorg. Chem. **3** [1964] 735/46, 744). – [12] A. La Ginestra, G. Rubino (Atti Accad. Nazl. Lincei Classe Sci. Fis. Mat. Nat. Rend. [8] **41** [1966] 510/20). – [13] B. Schönfeld (Diss. Göttingen, West Germany, 1973, p. 51). – [14] J. Fuchs (Z. Naturforsch. **28b** [1973] 389/404). – [15] M. J. Schwing-Weill, F. Arnaud-Neu (Bull. Soc. Chim. France **1970** 853/60).

[16] W. G. Klemperer, W. Shum (J. Am. Chem. Soc. **98** [1976] 8291/3). – [17] M. J. Schwing-Weill (Bull. Soc. Chim. France **1967** 3801/5). – [18] E. Ma (Bull. Chem. Soc. Japan **37** [1964] 171/5). – [19] K. H. Tytko, O. Glemser (Advan. Inorg. Chem. Radiochem. **19** [1976] 239/315, 266). – [20] W. P. Griffith, P. J. B. Lesniak (J. Chem. Soc. A **1969** 1066/71).

[21] A. J. Wilson, V. McKee, B. R. Penfold, C. J. Wilkins (Acta Cryst. C **40** [1984] 2027/30). – [22] P. Román, M. E. Gonzales-Aguado, C. Esteban-Calderón, M. Martínez-Ripoll, S. García-Blanco (Z. Krist. **165** [1983] 271/6). – [23] J. Fuchs, A. Thiele (Z. Naturforsch. **34b** [1979] 155/9).

2.1.6.5 The α-Octamolybdate Type, $M_4[Mo_8O_{26}] \cdot nH_2O$ (= $M_2O \cdot 4MoO_3 \cdot 0.5nH_2O$)

The compounds of this type are termed in the literature α-octamolybdates. This type is formed only with certain cations of organic bases. It is usually prepared from organic solvents and, hence, crystallizes mostly without water. The structure of the polymolybdate ion as found in the compounds $[(C_4H_9)_4N]_4[Mo_8O_{26}]$ [1, 2] and $[(C_3H_7)(C_6H_5)_3P]_4[Mo_8O_{26}] \cdot H_2O \cdot CH_3CN$ [3] is shown in **Fig. 13**.

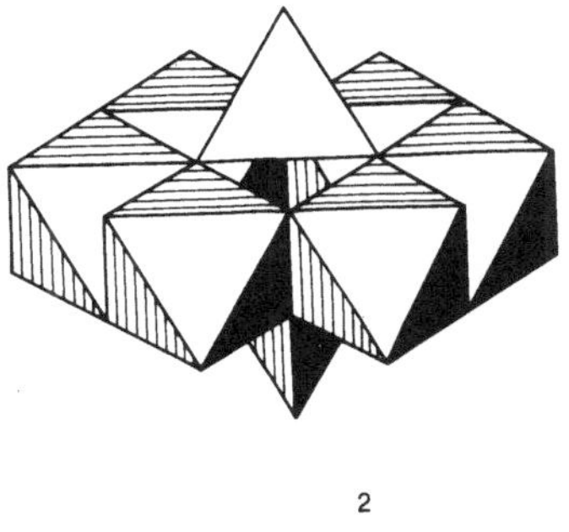

Fig. 13. Structure of the α-octamolybdate ion $[Mo_8O_{26}]^{4-}$ in the compounds $[(C_4H_9)_4N]_4[Mo_8O_{26}]$ [1, 2] and $[(C_3H_7)(C_6H_5)_3P]_4[Mo_8O_{26}] \cdot H_2O \cdot CH_3CN$ [3].

The Raman spectrum is only reported for the tetrabutylammonium salt $[(C_4H_9)_4N]_4[Mo_8O_{26}]$ [5], see **Fig. 14**. Infrared spectra are also only reported for the tetrabutylammonium salt [4, 5] (**Fig. 15**).

Fig. 14

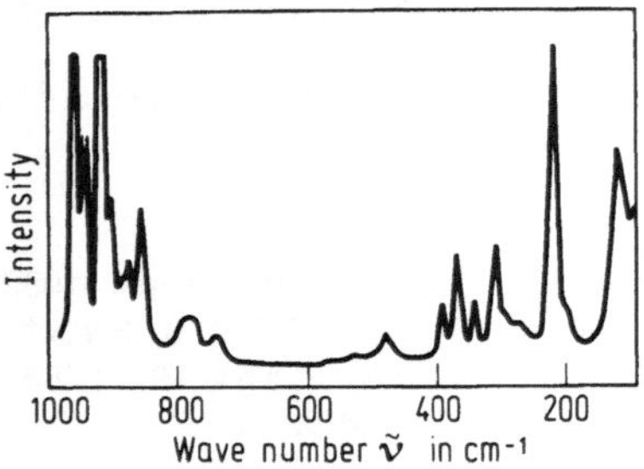

Fig. 15

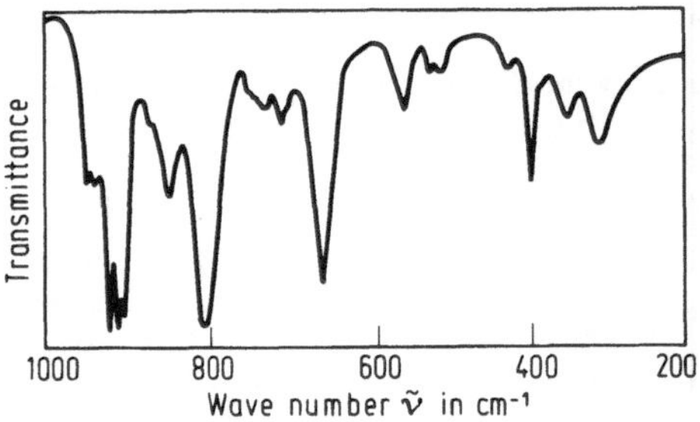

Raman (left) and infrared spectrum (right) of α-$[(C_4H_9)_4N]_4[Mo_8O_{26}]$ [5].

References:

[1] J. Fuchs, H. Hartl (Angew. Chem. **88** [1976] 385/6; Angew. Chem. Intern. Ed. Engl. **15** [1976] 375). – [2] M. F. Fredrich, V. W. Day, W. Shum, W. G. Klemperer (Am. Cryst. Assoc. Summer Meeting, 1976, Paper M 5 from [4]). – [3] V. W. Day, M. F. Fredrich, W. G. Klemperer, W. Shum (J. Am. Chem. Soc. **99** [1977] 952/3). – [4] W. G. Klemperer, W. Shum (J. Am. Chem. Soc. **98** [1976] 8291/3). – [5] J. Fuchs, I. Brüdgam (Z. Naturforsch. **32b** [1977] 853/7).

2.1.6.6 The Hexamolybdate Type, $M_2[Mo_6O_{19}]\cdot nH_2O$ ($= M_2O\cdot 6MoO_3\cdot nH_2O$)

Compounds of this type are termed in the literature hexamolybdates (meaning a discrete polyanion), (1:6)-molybdates, "hexamolybdates" (meaning an (1:6)-molybdate), or (10,6)-molybdates.

This type is formed only with certain cations, often cations of organic bases, and crystallizes mostly without water of crystallization. The structure of the polymolybdate ion as found in the compounds $(HN_3P_3[N(CH_3)_2]_6)_2[Mo_6O_{19}]$ ($N_3P_3[N(CH_3)_2]_6$ = hexakis(dimethylamino)cyclotriphosphazene) [1, 2], $[Mo(S_2CN(C_2H_5)_2)_4]_2[Mo_6O_{19}]$ ($Mo(S_2CN(C_2H_5)_2)_4$ = tetrakis(diethyldithiocarbamato)molybdenum(V)) [3], and $(C_{12}H_{24}O_6)_2\cdot K_2[Mo_6O_{19}]\cdot H_2O$ ($C_{12}H_{24}O_6$ = 18-crown-6) [4] by X-ray structural analysis is shown in **Fig. 16**.

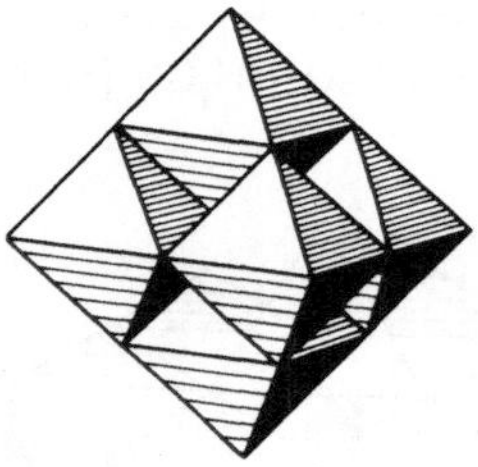

Fig. 16. Structure of the hexamolybdate ion $[Mo_6O_{19}]^{2-}$ in the compounds $(HN_3P_3[N(CH_3)_2]_6)_2[Mo_6O_{19}]$ [1, 2], $[Mo(S_2CN(C_2H_5)_2)_4]_2$-$[Mo_6O_{19}]$ [3], and $(C_{12}H_{24}O_6)_2K_2[Mo_6O_{19}]\cdot H_2O$ [4] (from [11]).

The Raman spectrum is only reported for the tetrabutylammonium salt $[(n\text{-}C_4H_9)_4N]_2$-$[Mo_6O_{19}]$ [5 to 8], see **Fig. 17**, containing no water of crystallization. Infrared spectra (**Fig. 18**) are reported for $Na_2[Mo_6O_{19}]\cdot 5H_2O$, $(NH_4)_2[Mo_6O_{19}]\cdot H_2O$ [9], $[(n\text{-}C_4H_9)_4N]_2[Mo_6O_{19}]$ [5, 6, 8, 10, 12], and $Cs_2[Mo_6O_{19}]$ [9].

Fig. 18

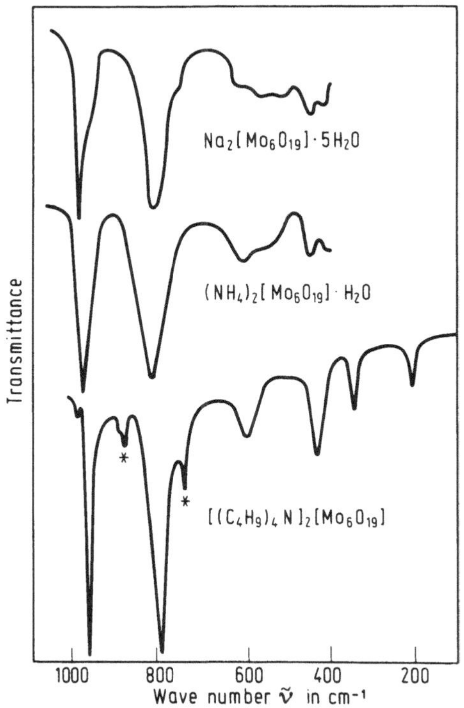

Fig. 17

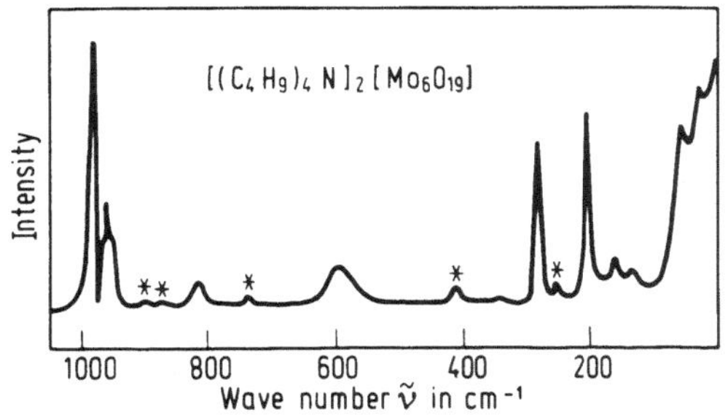

Raman spectrum of $[(n\text{-}C_4H_9)_4N]_2[Mo_6O_{19}]$ [8] (left) and infrared spectra of $Na_2[Mo_6O_{19}]\cdot 5H_2O$, $(NH_4)_2[Mo_6O_{19}]\cdot H_2O$ [9], and $[(C_4H_9)_4N]_2[Mo_6O_{19}]$ [8] (right).
* Vibration modes of the cation.

References:

[1] H. R. Allcock, E. C. Bissell, E. T. Shawl (J. Am. Chem. Soc. **94** [1972] 8603/4). – [2] H. R. Allcock, E. C. Bissell, E. T. Shawl (Inorg. Chem. **12** [1973] 2963/8). – [3] C. D. Garner, N. C. Howlader, F. E. Mabbs, A. T. McPhail, R. W. Miller, K. D. Onan (J. Chem. Soc. Dalton Trans. **1978** 1582/9). – [4] O. Nagano, Y. Sasaki (Acta Cryst. B **35** [1979] 2387/9). – [5] R. Mattes, H. Bierbüsse, J. Fuchs (Z. Anorg. Allgem. Chem. **385** [1971] 230/42).

[6] J. Fuchs (Z. Naturforsch. **28b** [1973] 389/404). – [7] K. H. Tytko, B. Schönfeld (Z. Naturforsch. **30b** [1975] 471/84). – [8] C. Rocchiccioli-Deltcheff, R. Thouvenot, M. Fouassier (Inorg. Chem. **21** [1982] 30/5). – [9] A. M. Golubev, E. A. Torchenkova, V. I. Spitsyn (Dokl. Akad. Nauk SSSR **217** [1974] 345/7; Dokl. Chem. Proc. Acad. Sci. USSR **214/219** [1974] 495/7). – [10] J. Fuchs, K. F. Jahr (Z. Naturforsch. **23b** [1968] 1380).

[11] K. H. Tytko, O. Glemser (Advan. Inorg. Chem. Radiochem. **19** [1976] 239/315, 266). – [12] M. Che, M. Fournier, J. P. Launay (J. Chem. Phys. **71** [1979] 1954/60).

2.1.6.7 The 36-Molybdate Type, $M_8[Mo_{36}O_{112}(H_2O)_{16}] \cdot nH_2O$ ($= M_2O \cdot 9MoO_3 \cdot (4+0.25n)H_2O$)

The compounds of this type are named in the literature 36-molybdates or (64, 36)-molybdates. The terms (1:9)-molybdates or "nonamolybdates" are not in use since the correct description of the analytical ratio $M_2O:MoO_3$ and the degree of aggregation were found at the same time. However, this type is described in the literature prior to 1973 erroneously as "hexamolybdate", "octamolybdate", "decamolybdate", 19-molybdate, and others [1, 2] (compare p. 67).

Taking into account the erroneously described polymolybdates [3] obviously being 36-molybdates, this type is known for nearly all cations under discussion. Only one compound of this type has been structurally characterized: $K_8[Mo_{36}O_{112}(H_2O)_{16}] \cdot 36$ to $40H_2O$ [4, 5], see **Fig. 19**. However, Raman and infrared spectra suggest the same polymolybdate ion to be present in all the compounds of this type [1, 2]. The structure consists of, in addition to MoO_6 octahedra, pentagonal MoO_7 bipyramids. Other features uncommon for isopolymolybdates are groups of corner-sharing MoO_6 octahedra forming three- and four-membered "rings", and H_2O molecules coordinated to Mo atoms, some of them coordinated to two Mo atoms simultaneously.

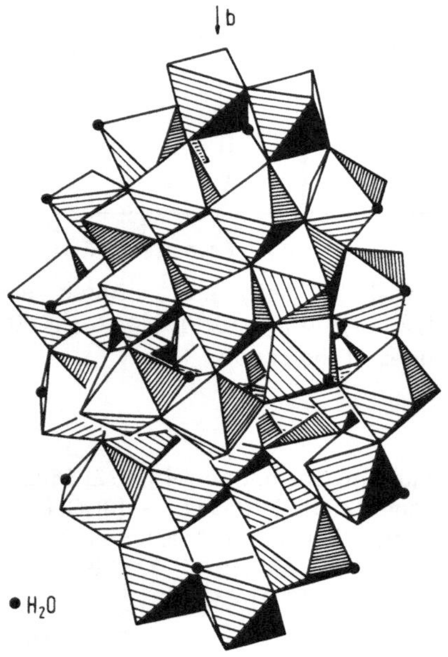

Fig. 19. Structure of the 36-molybdate ion $[Mo_{36}O_{112}(H_2O)_{16}]^{8-}$ in the potassium salt $K_8[Mo_{36}O_{112}(H_2O)_{16}] \cdot 36$ to $40H_2O$ [4, 5].

Raman spectra are reported for $Na_8[Mo_{36}O_{112}(H_2O)_{16}] \cdot \sim 64H_2O$, $K_8[Mo_{36}O_{112}(H_2O)_{16}] \cdot \sim 64H_2O$ (probably identical with the 36 to 40-hydrate) [1, 2], $K_8[Mo_{36}O_{112}(H_2O)_{16}] \cdot 36$ to $40H_2O$ [6], $(NH_4)_8[Mo_{36}O_{112}(H_2O)_{16}] \cdot \sim 64H_2O$ [1, 2], and $Ba_4[Mo_{36}O_{112}(H_2O)_{16}] \cdot \sim 64H_2O$ [2], see **Fig. 20**. Infrared spectra are reported for $Na_8[Mo_{36}O_{112}(H_2O)_{16}] \cdot \sim 64H_2O$ [7] (**Fig. 21**) and $K_8[Mo_{36}O_{112}(H_2O)_{16}] \cdot 36$ to $40H_2O$ [6]. The infrared spectra of the following compounds are similar to these spectra thus showing the presence of the 36-molybdate ions: $K_2O \cdot 8MoO_3 \cdot 12H_2O$ [8], $(NH_4)_2O \cdot 8MoO_3 \cdot nH_2O$, $(C_7H_{15}NH)_2O \cdot 8MoO_3 \cdot nH_2O$ ($C_7H_{15}N$ = 2,6-dimethylpiperidine) [9], "acid salts" of monomethylamine, monoethylamine, and diethylamine [10].

Fig. 20

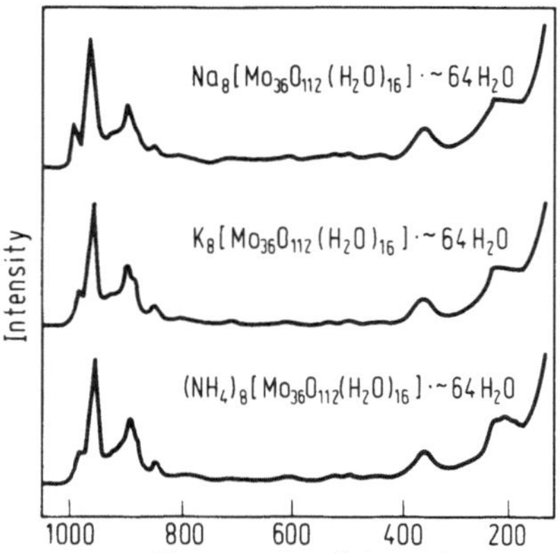

Fig. 21

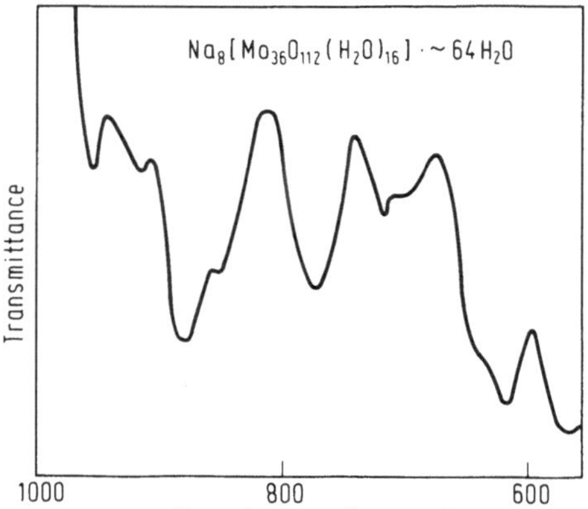

Raman spectra of $M_8[Mo_{36}O_{112}(H_2O)_{16}]\cdot\sim 64H_2O$, M = Na, K, NH_4 [1] (left) and infrared spectrum of $Na_8[Mo_{36}O_{112}(H_2O)_{16}]\cdot\sim 64H_2O$ [7] (right).

References:

[1] K. H. Tytko, B. Schönfeld (Z. Naturforsch. **30b** [1975] 471/84). – [2] K. H. Tytko, B. Schönfeld, B. Buss, O. Glemser (Angew. Chem. **85** [1973] 305/7; Angew. Chem. Intern. Ed. Engl. **12** [1973] 330/2). – [3] Gmelin Handbuch "Molybdän", 1935, pp. 210/98. – [4] I. Paulat-Böschen (J. Chem. Soc. Chem. Commun. **1979** 780/2). – [5] B. Krebs, I. Paulat-Böschen (Acta Cryst. B **38** [1982] 1710/8).

[6] I. Böschen (Diss. Kiel, West Germany, 1974, p. 169). – [7] B. Schönfeld (Diss. Göttingen, West Germany, 1973, p. 55). – [8] A. La Ginestra, G. Rubino (Atti Accad. Nazl. Lincei Classe Sci. Fis. Mat. Nat. Rend. [8] **41** [1966] 510/20). – [9] M. J. Schwing-Weill, F. Arnaud-Neu (Bull. Soc. Chim. France **1970** 853/60). – [10] F. Arnaud-Neu, M. J. Schwing-Weill (Bull. Soc. Chim. France **1973** 3225/32).

2.1.6.8 The "Decamolybdate" Type, $[MMo_5O_{15}(OH)(H_2O)\cdot H_2O]_\infty$ (= $M_2O\cdot 10MoO_3\cdot 5H_2O$)

The compounds of this type are termed in the literature "decamolybdates" or (1:10)-molybdates. It has been proposed (see p. 2) to call these compounds polypentamolybdates. Compounds of this type are described in the literature prior to 1976 erroneously as "hexamolybdates", "octamolybdates", "12-molybdates", "16-molybdates", "phase C", "hexagonal hydrate of molybdenum trioxide", or $MoO_3\cdot H_2O$ [1, 2].

Taking into account the erroneously described polymolybdates [3] obviously being polypentamolybdates, this type is known for nearly all cations under discussion. The structure of the polymolybdate ion as found in the compounds $[NaMo_5O_{15}(OH)(H_2O)\cdot H_2O]_\infty$ [4] and $[KMo_5O_{15}(OH)(H_2O)\cdot H_2O]_\infty$ [2] is shown in **Fig. 22**, p. 22. It is only distantly related to the structures of the other typical polymolybdates. The structure consists of double chains made of edge-sharing MoO_6 octahedra, and the double chains are linked by common corners of octahedra to form a three-dimensional network with tunnels having a minimal diameter of ~2.9 Å, in which the cations are located. The double chains are not perfect due to statistical nonoccupation of one of the six equivalent Mo positions in the unit cell. In the neighborhood of a vacant Mo position the O atoms are partially replaced by coordinated H_2O molecules, isolated H_2O molecules (water of crystallization), and OH groups. This structure shows close relations to the structures of α-$MoO_3 \cdot H_2O$ and of MoO_3 [2].

Fig. 22a

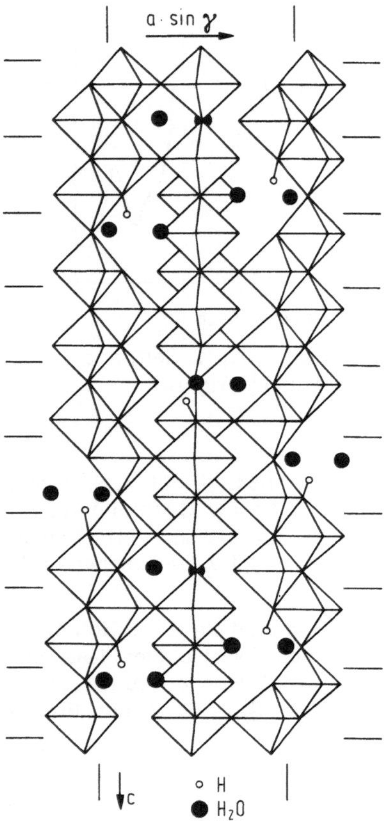

Fig. 22b

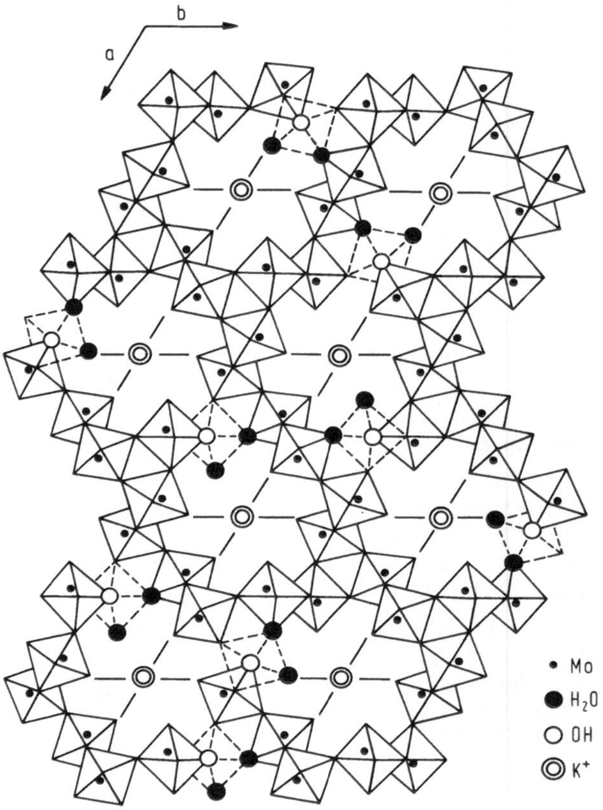

Structure of the polypentamolybdate ion $[Mo_5O_{15}(OH)(H_2O) \cdot H_2O^-]_\infty$ in the potassium salt $[KMo_5O_{15}(OH)(H_2O) \cdot H_2O]_\infty$ [2]. (a) Projection of three double chains of MoO_6 octahedra parallel to [010]. The vacant octahedra and accordingly the coordinated H_2O molecules and OH groups are statistically distributed. Some of the coordinated H_2O molecules belong to neighboring coordination octahedra not drawn. (b) Several unit cells of the structure in the projection parallel to [001]. Vacant, statistically distributed octahedra are drawn with dotted lines.

The Raman spectra (**Fig. 23**) and infrared spectra (**Fig. 24**) of the "decamolybdates" closely resemble each other. Raman spectra are reported for $[NaMo_5O_{15}(OH)(H_2O) \cdot H_2O]_\infty$, $[KMo_5O_{15}(OH)(H_2O) \cdot H_2O]_\infty$ [1, 6], and $[NH_4Mo_5O_{15}(OH)(H_2O) \cdot H_2O]_\infty$ [1]. The Raman spectrum of $[(CH_3)_4N]_2O \cdot 10\,MoO_3 \cdot 7\,H_2O$ [7] shows some similarities. Infrared spectra are reported for $[NaMo_5O_{15}(OH)(H_2O) \cdot H_2O]_\infty$ and $[KMo_5O_{15}(OH)(H_2O) \cdot H_2O]_\infty$ [6]. The spectra of a compound reported to be $(NH_4)_2O \cdot 10.6$ to $12\,MoO_3 \cdot n\,H_2O$ [8] and of another compound reported to be a "hexagonal hydrate of molybdenum trioxide" [9] are nearly identical, and the spectrum of $[(CH_3)_4N]_2O \cdot 10\,MoO_3 \cdot 7\,H_2O$ [7] shows some similarities to the reference spectra.

Fig. 23

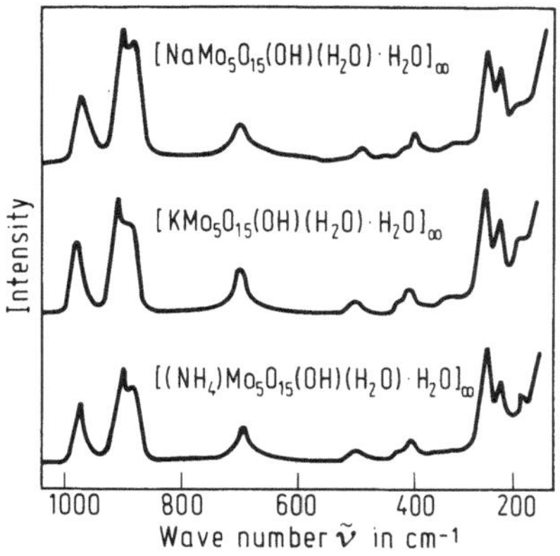

Fig. 24

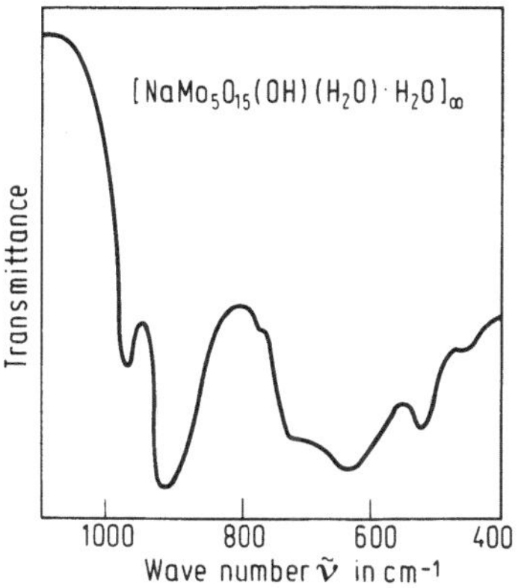

Raman spectra of [$MMo_5O_{15}(OH)(H_2O)\cdot H_2O]_\infty$, M = Na, K, NH_4 [1] (left) and infrared spectrum of [$NaMo_5O_{15}(OH)(H_2O)\cdot H_2O]_\infty$ [6] (right).

X-ray powder diffraction patterns are cited for [$NaMo_5O_{15}(OH)(H_2O)\cdot H_2O]_\infty$ and [$KMo_5O_{15}(OH)(H_2O)\cdot H_2O]_\infty$ [6]. The diffraction pattern reported for a "hexagonal hydrate of molybdenum trioxide" [9] is identical with the reference diffraction patterns [2].

Note added in proof: It has recently been shown that the "decamolybdate" type as defined by Raman and infrared spectra and X-ray powder diffraction patterns (or interplanar distances) and, hence, by the structure of the polyanion, is present in polymolybdates possessing the analytical ratio $M_2O:MoO_3 = 1:10$ to ~ 30. Thus, the polymolybdates which have been stated above to be erroneously described in the literature are obviously analytically correctly described in the case of the "b-molybdates" with $b \geqq 10$ (a = 1). However, in contrast to the normal case the term "b-molybdate" does not express (as assumed in the literature) the presence of a definite polymolybdate type but only the analytical composition of the polymolybdate. This difference has been explained by the incorporation of varying quantities of H_3O^+ in addition to the alkali and alkaline earth metal cations into the crystal [10].

References:

[1] K. H. Tytko, B. Schönfeld (Z. Naturforsch. **30b** [1975] 471/84). – [2] B. Krebs, I. Paulat-Böschen (Acta Cryst. B **32** [1976] 1697/704). – [3] Gmelin Handbuch "Molybdän", 1935, pp. 210/98. – [4] B. Hedman, R. Strandberg (from [5]). – [5] B. Krebs, I. Paulat-Böschen (Acta Cryst. B **38** [1982] 1710/8).

[6] I. Böschen (Diss. Kiel, West Germany, 1974, pp. 170/83). – [7] J. Fuchs (Z. Naturforsch. **28b** [1973] 389/404). – [8] M. J. Schwing-Weill, F. Arnaud-Neu (Bull. Soc. Chim. France **1970** 853/60). – [9] N. Sotani (Bull. Chem. Soc. Japan **48** [1975] 1820/5). – [10] G. Baethe (Diss. Göttingen, West Germany, 1985, pp. 96/101, 118/20, 156).

2.1.6.9 The "Phase C" Type Polymolybdates, 0.03 to 0.10 $M_2O\cdot MoO_3\cdot$0.3 to 0.7 H_2O

Phase C type polymolybdates are polymolybdates found in attempts to prepare the well-known hydrates of molybdenum trioxide $MoO_3\cdot 2H_2O$, $MoO_3\cdot H_2O$ (white), and $MoO_3\cdot H_2O$

(yellow) by acidification of solutions of normal molybdates under certain conditions [1, 2]. They are the predominant precipitates from hot solutions of high concentrations of the alkali salts at acid concentrations below 2.5 N [3]. By reason of their small amounts of alkali (or alkaline earth) metals they are readily taken to be the trioxide hydrates from which they can be distinguished with certainty only by X-ray investigation [1, 2]. They are, however, also confused with the "decamolybdates" since the formulae of the latter (e.g., $M_2O \cdot 10MoO_3 \cdot 5H_2O$ corresponds to $0.1M_2O \cdot MoO_3 \cdot 0.5H_2O$) are covered by those of the phase C polymolybdates. The name [3] is used in analogy with the so-called tungsten C phases.

Phase C polymolybdates form in the presence of easily polarizable cations of large size and low ionic charge (Na^+, K^+, NH_4^+, Ba^{2+}) [2, 3]. The ammonium compounds form most rapidly and are the most stable. The less polarizable Li^+, Mg^{2+}, and Ca^{2+} form very unstable phase C compounds. Ethylene diammonium molybdate also produces a relatively stable phase C. The barium phase C compound, however, was obtained by a somewhat modified way [3].

The structures of the phase C compounds are not known. However, the cations must be an integral part of the phase C structure since it does not exist in their absence [3]. The compounds crystallize in a body-centered cubic lattice, the lattice parameters being 13.03(2) [1], 13.01(1) Å [2] for the sodium, 12.98(2) Å for the potassium and the ammonium [1] phase C. See also the discussion on the crystal symmetry of ammonium (1:14)- and (1:22)-molybdate in [4] (compare p. 100). The X-ray diffraction patterns are almost unchanged by substitution of cations or by large variations in composition. The molybdenum and tungsten phase C compounds are not isomorphous [2, 3].

X-ray powder diffraction diagrams (or interplanar distances) are reported for the phase C compounds $Na_2O \cdot 10$ to $30MoO_3 \cdot 4$ to $12H_2O$ [2], $Na_2O \cdot 10MoO_3 \cdot 4H_2O$ [3], $0.074Na_2O \cdot MoO_3 \cdot 0.54H_2O$ [1], $K_2O \cdot 15MoO_3 \cdot 5H_2O$ [3], 0.058 to $0.060K_2O \cdot MoO_3 \cdot 0.54$ to $0.65H_2O$ [1], $(NH_4)_2O \cdot 12MoO_3 \cdot 5H_2O$ [3], 0.075 to $0.085(NH_4)_2O \cdot MoO_3 \cdot 0.32$ to $0.42H_2O$ [1], and $BaO \cdot 22MoO_3 \cdot 14H_2O$ [3].

Note added in proof: As has been stated in the note added in proof of the preceding section polymolybdates possessing the analytical ratio $M_2O:MoO_3 = 1:10$ to ~ 30 are of the "decamolybdate" type. Hence the "phase C" polymolybdates are of the "decamolybdate" type.

References:

[1] H. Peters, L. Till, K. H. Radeke (Z. Anorg. Allgem. Chem. **365** [1969] 14/21). – [2] M. L. Freedman, S. Leber (J. Less-Common Metals **7** [1964] 427/32). – [3] M. L. Freedman (J. Chem. Eng. Data **8** [1963] 113/6). – [4] A. B. Kiss, P. Gadó, I. Asztalos, A. J. Hegedüs (Acta Chim. [Budapest] **66** [1970] 235/49).

2.1.7 Structures and "Finger-Prints" of the Polymolybdates Typified only by Their $M_2O:MoO_3$ Ratio

2.1.7.1 The (1:2)-Molybdate Type, $M_2O \cdot 2MoO_3 \cdot nH_2O$

$[K_4Mo_4O_{14} \cdot 2H_2O]_\infty$ ($= K_2O \cdot 2MoO_3 \cdot H_2O$). The structure of the polyanion in this compound obtained from aqueous solution is a chain of MoO_6 octahedra and MoO_5 square pyramids [1], see **Fig. 25**. "Finger-prints" are not reported.

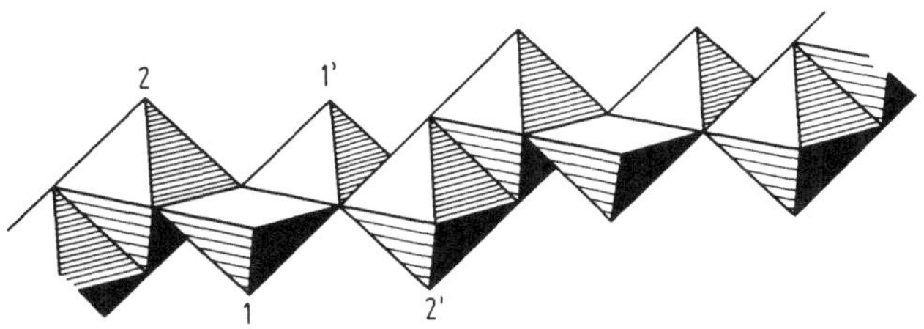

Fig. 25. Idealized representation of the chain of MoO_6 octahedra (2 and 2') and square MoO_5 pyramids (1 and 1') of the polyanion in $[K_4Mo_4O_{14} \cdot 2H_2O]_\infty$ [1].

$[(NH_4)_4Mo_4O_{14}]_\infty$ ($=(NH_4)_2O \cdot 2MoO_3$). This compound is also obtained from aqueous solutions [3 to 5], but does not contain water of crystallization. The structure of the polyanion is different from that of the compound $[K_4Mo_4O_{14} \cdot 2H_2O]_\infty$ and has a chain of MoO_6 octahedra and MoO_4 tetrahedra [6, 7]. It is characterized by Raman [4, 5] and infrared spectra [4]. For X-ray diffraction data compare [8, 9]. The compound is described in "Molybdän" Erg.-Bd. B 1, 1975, pp. 227/8.

Other Compounds. The compounds reported by [10] to be "dimolybdates" ($R_2Mo_2O_7 \cdot 2H_2O$, R = alkylammonium) are, according to their infrared spectra, heptamolybdates, and the compounds reported by [2] to be "dimolybdates" ($R_2Mo_2O_7 \cdot nH_2O$, R = organic ammonium cation) should, according to their infrared spectra, be regarded as "trimolybdates".

References:

[1] B. M. Gatehouse (J. Less-Common Metals **54** [1977] 283/8). – [2] F. Arnaud-Neu, M. J. Schwing-Weill (Bull. Soc. Chim. France **1973** 3225/32). – [3] C. J. Hallada (J. Less-Common Metals **36** [1974] 103/10). – [4] W. D. Hunnius (Z. Naturforsch. **29b** [1974] 599/602). – [5] K. H. Tytko, B. Schönfeld (Z. Naturforsch. **30b** [1975] 471/84).

[6] I. Knöpnadel, H. Hartl, W. D. Hunnius, J. Fuchs (Angew. Chem. **86** [1974] 894/5; Angew. Chem. Intern. Ed. Engl. **13** [1974] 823). – [7] A. W. Armour, M. G. B. Drew, P. C. H. Mitchell (J. Chem. Soc. Dalton Trans. **1975** 1493/6). – [8] I. Lindqvist (Nova Acta Regiae Soc. Sci. Upsaliensis [4] **15** No. 1 [1950] 1/22, 5). – [9] M. L. Freedman, S. Leber (J. Less-Common Metals **7** [1964] 427/32). – [10] T. Yamase, T. Ikawa (Bull. Chem. Soc. Japan **50** [1977] 746/9).

2.1.7.2 The (1:4)-Molybdate Type, $M_2O \cdot 4MoO_3 \cdot nH_2O$

$Na_2O \cdot 4MoO_3 \cdot 6H_2O$ (yellowish). The structure of this compound is still unknown. The salt is characterized by its Raman [1] and infrared spectrum [2], see Figs. 11 and 12, p. 16.

$(NH_4)_2O \cdot 4MoO_3$. A second compound of this type with unknown structure also obtained from aqueous solution is $(NH_4)_2O \cdot 4MoO_3$ [3]; however, it contains no water of crystallization. The salt is characterized by its Raman [1, 4] and infrared [2, 4, 5] spectrum (see Figs. 11 and 12, p. 16) and X-ray powder diagram [6]. In [4] this compound is unjustifiably reported to be $(NH_4)_4Mo_8O_{26}$. The compound is described in "Molybdän" Erg.-Bd. B1, 1975, pp. 228/9.

Other Compounds. There are a number of (1:4)-molybdate hydrates of the alkali and alkaline earth metal cations and some hydrous and anhydrous (1:4)-molybdates of organic ammonium cations prepared from aqueous solution which are very poorly characterized. The β-octamolybdates described on p. 15 are also (1:4)-molybdates, according to their stoichiometry. They form, however, a separate group because they are structurally typified.

References:

[1] K. H. Tytko, B. Schönfeld (Z. Naturforsch. **30b** [1975] 471/84). – [2] B. Schönfeld (Diss. Göttingen, West Germany, 1973, p. 51). – [3] Gmelin Handbuch "Molybdän", 1935, p. 258. – [4] W. D. Hunnius (Z. Naturforsch. **30b** [1975] 63/5). – [5] A. Kiss, P. Gadó, I. Asztalos, A. J. Hegedüs (Acta Chim. [Budapest] **66** [1970] 235/49).

[6] M. J. Schwing-Weill (Bull. Soc. Chim. France **1967** 3795/8).

2.1.8 Structures and "Finger-Prints" of Individual Polymolybdates

$[(C_4H_9)_4N]_2[Mo_2O_7]$ (= $[(C_4H_9)_4N]_2O \cdot 2MoO_3$). The structure of this genuine anhydrous dimolybdate, obtainable only from organic solvents, is shown in **Fig. 26**; for the infrared spectrum see the paper [1].

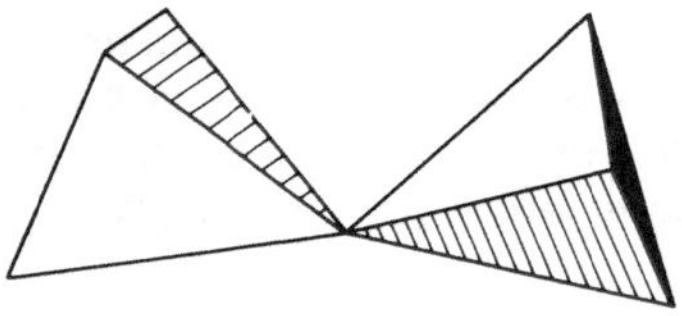

Fig. 26. Structure of the $[Mo_2O_7]^{2-}$ ion in $[(C_4H_9)_4N]_2[Mo_2O_7]$ [7].

$(NH_4)_8[Mo_{10}O_{34}]$ (= $2(NH_4)_2O \cdot 5MoO_3$). This anhydrous polymolybdate, a genuine decamolybdate, is obtained by thermal decomposition of $(NH_4)_6[Mo_7O_{24}] \cdot 4H_2O$ (see pp. 41 and 99) and can also be prepared from aqueous solution [2]. Its structure [3] is shown in **Fig. 27** and Raman [2] and infrared vibrations [2, 4] as well as X-ray diffraction data [4, 5] are reported. The compound is described in "Molybdän" Erg.-Bd. B1, 1975, p. 228.

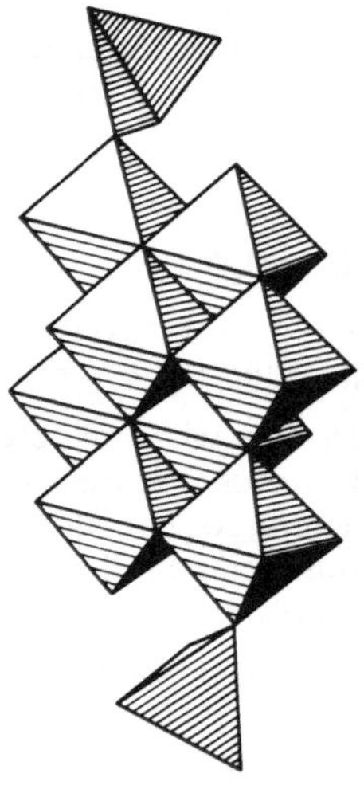

Fig. 27. Structure of the $[Mo_{10}O_{34}]^{8-}$ ion in $(NH_4)_8[Mo_{10}O_{34}]$ [4].

$(i\text{-}C_3H_7NH_3)_6[Mo_8O_{26}(OH)_2]\cdot 2H_2O$ $(=3(i\text{-}C_3H_7NH_3)_2O\cdot 8MoO_3\cdot 3H_2O)$. This dihydrogenoctamolybdate, obtained from aqueous solution, is a (3:8)-molybdate like the ammonium polyoctamolybdate. Its structure is shown in **Fig. 28** [6], its infrared spectrum in **Fig. 29** [7].

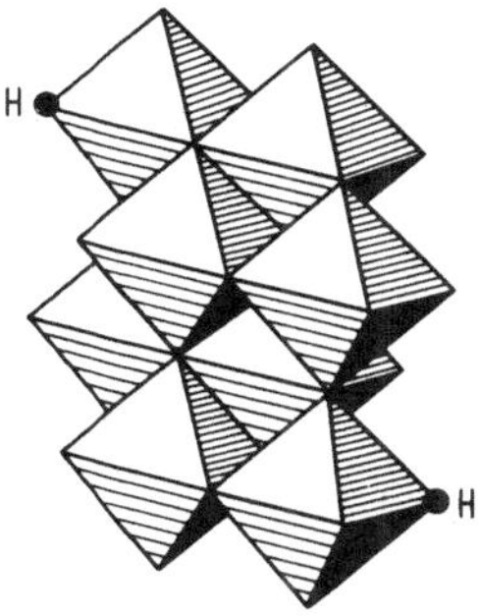

Fig. 28. Structure of the $[Mo_8O_{26}(OH)_2]^{6-}$ ion in $(i\text{-}C_3H_7NH_3)_6[Mo_8O_{26}(OH)_2]\cdot 2H_2O$ [1].

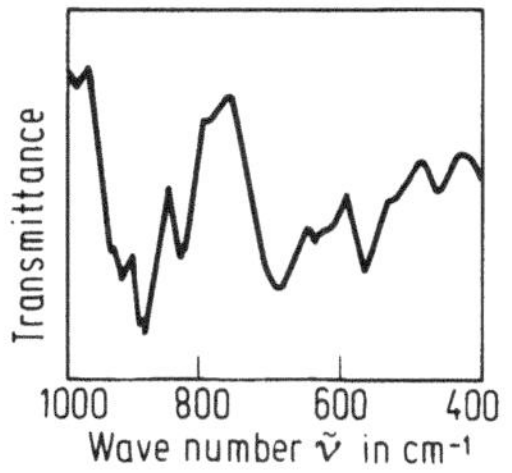

Fig. 29. Infrared spectrum of $(i\text{-}C_3H_7NH_3)_6[Mo_8O_{26}(OH)_2]\cdot 2H_2O$ [2].

References:

[1] V. W. Day, M. F. Fredrich, W. G. Klemperer, W. Shum (J. Am. Chem. Soc. **99** [1977] 6146/8). – [2] W. D. Hunnius (Z. Naturforsch. **30b** [1975] 63/5). – [3] J. Fuchs, H. Hartl, W. D. Hunnius, S. Mahjour (Angew. Chem. **87** [1975] 634/5; Angew. Chem. Intern. Ed. Engl. **14** [1975] 644). – [4] A. B. Kiss, P. Gadó, I. Asztalos, A. J. Hegedüs (Acta Chim. [Budapest] **66** [1970] 235/49, 239/40). – [5] M. J. Schwing-Weill (Bull. Soc. Chim. France **1967** 3795/8).

[6] M. Isobe, F. Marumo, T. Yamase, T. Ikawa (Acta Cryst. B **34** [1978] 2728/31). – [7] T. Yamase, T. Ikawa (Bull. Chem. Soc. Japan **50** [1977] 746/9).

2.1.9 Bonding and Structure of the Polymolybdates

2.1.9.1 Coordination Numbers of the Molybdenum and Oxygen Atoms

The coordination number of molybdenum in the polymolybdates varies from four in the tetrahedral MoO_4 base units, five in the tetragonal-pyramidal MoO_5 base units, six in the octahedral MoO_6 base units, up to seven in a pentagonal bipyramidal MoO_7 unit.

The coordination number of oxygen in the Mo-O skeletons varies from one (terminal oxygen atoms O_t) to six (bridging oxygen atoms O_b and central oxygen atoms O_c). The higher coordination numbers four, five, and six occur only in connection with MoO_6 octahedra (cf. [1 to 4]); the coordination number four is found additionally in connection with MoO_7 polyhedra [5]. The terminal oxygen atoms occur either isolated forming a free corner of a polyhedron or two neighboring (cis-)terminal oxygen atoms form a free edge of a polyhedron. Three neighboring (cis-)terminal oxygen atoms have been observed so far only in the case of the compounds

$(NH_4)_8[Mo_{10}O_{34}]$ and $[(C_4H_9)_4N]_2[Mo_2O_7]$ forming part of an MoO_4 tetrahedron. In the case of the compounds $(i\text{-}C_3H_7NH_3)_6[Mo_8O_{26}(OH)_2]\cdot 2H_2O$ and $M_8[Mo_{36}O_{112}(H_2O)_{16}]\cdot nH_2O$ there are octahedra having three free corners (corners not shared with other octahedra) but not three terminal oxygen atoms (O atoms of coordination number one); compare pp. 9/26 and [1 to 4].

References:

[1] F. A. Schröder (Acta Cryst. B **31** [1975] 2294/309). – [2] J. Fuchs, I. Knöpnadel (Z. Krist. **158** [1982] 165/79). – [3] M. A. Porai-Koshits, L. O. Atovmyan (Koord. Khim. **1** [1975] 1271/81; Soviet J. Coord. Chem. **1** [1975] 1065/74). – [4] E. M. Shustorovich, M. A. Porai-Koshits, Yu. A. Buslaev (Coord. Chem. Rev. **17** [1975] 1/98, 70/81). – [5] B. Krebs, I. Paulat-Böschen (Acta Cryst. B **38** [19C2] 1710/8).

2.1.9.2 Bond Lengths

The molybdenum-oxygen bond lengths vary in a broad range from 1.66 [1] (1.61 (?) [2]) to 2.56 Å [3, 4]. They are related to the coordination number of the oxygen atoms (compare [1, 3, 5 to 9, 12]) and to the type of the Mo-O-Mo bridges (compare [5 to 7]). Short Mo-O bonds (1.69 to 1.75 Å) are always in cis-position to each other (angle O-Mo-O close to 104°, cf. [4, 6 to 8, 11 to 14, 25]) and belong to oxygen atoms of the lowest bridge-multiplicity: either to terminal or (in the absence of terminal) linear or (in the absence of linear) nonlinear bridges with a coordination number two. MoO_6 octahedra containing only bridging oxygen atoms are, however, very rare. The oxygen atoms of maximum bridge multiplicity are always found in a trans-position to terminal oxygen atoms or (in the absence of the latter) to oxygen atoms of minimum bridge multiplicity [6, 7] (see **Fig. 30**) and form the longest Mo-O bonds (2.13 to 2.56 Å). The remaining Mo-O bonds are of medium length (1.89 to 2.03 Å, O-Mo-O angle 142° to 153°) (compare [1, 3, 4, 6, 7, 11, 13]) (type II polymetalates according to Pope [26]). In the case of the Mo_6O_{19} hexamolybdates the MoO_6 octahedra each contain one short terminal Mo-O bond (1.68 Å), four equal medium Mo-O bonds (1.93 Å), and one long Mo-O bond (2.32 Å) (type I polymetalate [26]).

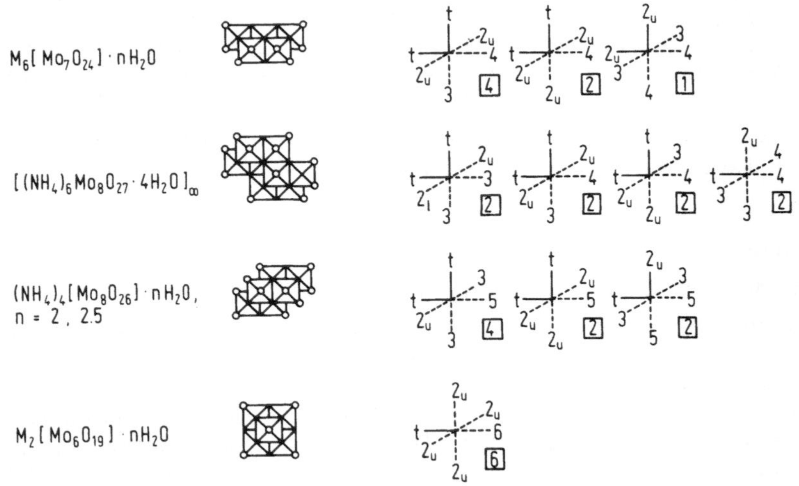

Fig. 30. Linkage of octahedral vertices in some isopolymolybdates. t terminal oxygen atoms; 2,3,... oxygen atoms of coordination number 2,3,...; l, u linear, nonlinear oxygen bridge; solid lines: bonds of high multiplicity; [4] frequency of the type [6, 7].

Table 4
Mean Mo-O Bond Lengths (in Å) of Differently Coordinated Oxygen Atoms in Some Polymolybdates (from [8], completed).

$M_2O:MoO_3$	1:1	1:2	1:2	1:2	3:7	2:5	3:8	3:8	1:3	1:3	1:4	1:4	1:6	1:9
Θ/Mo atom	2.00	1.00	1.00	1.00	0.86	0.80	0.75	0.75	0.67	0.67	0.50	0.50	0.33	0.22
compound	I	II	III	IV	V	VI	VII	VIII	IX	X	XI	XII	XIII	XIV
Mo=O	1.77[a)]	1.73[c, f)]	1.73[c, f)]	1.73*)[b)]	1.73[f)]	1.72[b, f, g)]	1.72[f, g)]	1.72[f, g)]	1.71[f)]	1.71[e, f)]	1.69*)[d, f)]	1.70[f, g)]	1.68[g)]	1.69
Mo–O–Mo		1.99	1.99	1.87*)	1.99	1.99	1.98	1.98		1.94	1.90*)	1.96	1.93	1.95
Mo—O(—Mo)—Mo					2.13	2.10	2.11	2.11	2.09	2.07	2.22*)	2.09		2.06
Mo—O(—Mo)(—Mo)—Mo					2.19	2.19	2.19	2.21						2.17
Mo—O(—Mo)(—Mo)(—Mo)—Mo												2.33		
Mo—O(—Mo)(—Mo)(—Mo)(—Mo)—Mo													2.32	

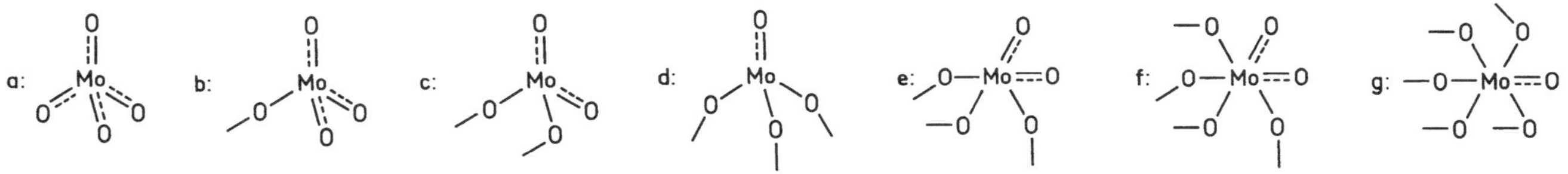

Θ/Mo atom: relative charge = mean negative charge per Mo atom.

*) Less stable polymolybdates being decomposed in aqueous solution.

I $Na_2[MoO_4]\cdot 2H_2O$ [15]
II $[(NH_4)_4Mo_4O_{14}]_\infty$ [16]
III $Na_2Mo_2O_7$ [17]
IV $[(C_4H_9)_4N]_2[Mo_2O_7]$ [18]
V $Na_6[Mo_7O_{24}]\cdot 14H_2O$ [3]
VI $(NH_4)_8[Mo_{10}O_{34}]$ [19]
VII $[(NH_4)_6Mo_8O_{27}\cdot 4H_2O]_\infty$ [9]
VIII $[C_3H_7NH_3]_6[Mo_8O_{26}(OH)_2]\cdot 2H_2O$ [20]
IX $[Rb_4Mo_6O_{20}\cdot 2H_2O]_\infty$ [21]
X $Rb_2Mo_3O_{10}$ [22]
XI α-$[(C_4H_9)_4N]_4[Mo_8O_{26}]$ [23]
XII β-$[(CH_3)_4N]_2Na_2[Mo_8O_{26}]\cdot 2H_2O$ [8]
β-$(NH_4)_4[Mo_8O_{26}]\cdot 4H_2O$ [13]
XIII $(HN_3P_3[N(CH_3)_2]_6)_2[Mo_6O_{19}]$ [24]
XIV $K_8[Mo_{36}O_{112}(H_2O)_{16}]\cdot 36$ to $40H_2O$ [1]

Averaged molybdenum-oxygen distances in relation to the coordination number are given in Table 4, p. 29 [8].

The Mo-O-Mo angle has only weak influence on the bond lengths [10]. The coordination number of the molybdenum atom has almost no influence on the Mo-O distances: tetrahedral, tetragonal pyramidal, and octahedral units show no significant differences in the distances [8] (cf. also [5]). The Mo-O_t distances are independent of the number of O_t in an MoO_6 unit. There are, however, small differences depending on the relative charge z/x of the polymolybdate ion $Mo_xO_y^{z-}$ [8]. Bond lengths in Mo-O-Mo bridges can be very different due to an asymmetry of the bridge (cf. [5 to 7]).

The mean Mo-O distances for Mo-OH_2 (terminal) bonds are 2.37 Å [1, 11] and for Mo-OH_2 (bridging) bonds 2.42 Å [1] (case of the "decamolybdates" and 36-molybdates).

References:

[1] B. Krebs, I. Paulat-Böschen (Acta Cryst. B **38** [1982] 1710/8). – [2] L. O. Atovmyan, O. N. Krasochka (Zh. Strukt. Khim. **13** [1972] 342/3; J. Struct. Chem. [USSR] **13** [1972] 319/20). – [3] K. Sjöbom, B. Hedman (Acta Chem. Scand. **27** [1973] 3673/91). – [4] A. Don, T. J. R. Weakley (Acta Cryst. B **37** [1981] 451/3). – [5] F. A. Schröder (Acta Cryst. B **31** [1975] 2294/309).

[6] M. A. Porai-Koshits, L. O. Atovmyan (Koord. Khim. **1** [1975] 1271/81; Soviet J. Coord. Chem. **1** [1975] 1065/74). – [7] E. M. Shustorovich, M. A. Porai-Koshits, Yu. A. Buslaev (Coord. Chem. Rev. **17** [1975] 1/98, 70/81). – [8] J. Fuchs, I. Knöpnadel (Z. Krist. **158** [1982] 165/79). – [9] I. Böschen, B. Buss, B. Krebs (Acta Cryst. B **30** [1974] 48/56). – [10] J. Fuchs (Z. Naturforsch. **28b** [1973] 389/404).

[11] B. Krebs, I. Paulat-Böschen (Acta Cryst. B **32** [1976] 1697/704). – [12] H. T. Evans, Jr., B. M. Gatehouse, P. Leverett (J. Chem. Soc. Dalton Trans. **1975** 505/14). – [13] H. Vivier, J. Bernard, H. Djomaa (Rev. Chim. Minerale **14** [1977] 584/604). – [14] T. J. R. Weakley (Polyhedron **1** [1982] 17/9). – [15] K. Matsumoto, A. Kobayashi, Y. Sasaki (Bull. Chem. Soc. Japan **48** [1975] 1009/13).

[16] I. Knöpnadel, H. Hartl, W. D. Hunnius, J. Fuchs (Angew. Chem. **86** [1974] 894/5; Angew. Chem. Intern. Ed. Engl. **13** [1974] 823). – [17] J. Fuchs, H. Richter (from [8]). – [18] V. W. Day, M. F. Fredrich, W. G. Klemperer, W. Shum (J. Am. Chem. Soc. **99** [1977] 6146/8). – [19] J. Fuchs, H. Hartl, W. D. Hunnius, S. Mahjour (Angew. Chem. **87** [1975] 634/5; Angew. Chem. Intern. Ed. Engl. **14** [1975] 644). – [20] M. Isobe, F. Marumo, T. Yamase, T. Ikawa (Acta Cryst. B **34** [1978] 2728/31).

[21] H. U. Kreusler, A. Förster, J. Fuchs (Z. Naturforsch. **35b** [1980] 242/4). – [22] J. Fuchs, A. Förster (from [8]). – [23] J. Fuchs, H. Hartl (Angew. Chem. **88** [1976] 385/6; Angew. Chem. Intern. Ed. Engl. **15** [1976] 375). – [24] H. R. Allcock, E. C. Bissel, E. T. Shawl (Inorg. Chem. **12** [1973] 2963/8). – [25] H. T. Evans Jr. (Perspect. Struct. Chem. **4** [1971] 1/59, 55).

[26] M. T. Pope (Inorg. Chem. **11** [1972] 1973/4).

2.1.9.3 Mo-O Bond Length/Bond Order Relationship

Böschen, Buss, and Krebs [1] and Brown and Wu [10] made use of a relationship stated by Donnay and Allmann [2] and calculated bond orders (BO) from bond lengths (d_{Mo-O}) according to $BO = (1.898/d_{Mo-O})^5$ [1], $BO = (1.882/d_{Mo-O})^{6.0}$ [10]. Allmann [7] later proposed the relationship $d_{Mo-O} = 1.90 - 0.76 \log BO$. Schröder [3] produced a bond distance/bond order correlation diagram from a large number of molybdenum compounds (including polymolybdates), see **Fig. 31**. The best results for the polymolybdates are obtained using the equation $d_{Mo-O} =$

1.874 − 0.600 log BO. This relation is based on the following points of definition: (1) The characteristic Mo-O_t distances of three neighboring O_t, 1.738 Å, have a bond order of 1.67 each. (2) The characteristic Mo-O_t distance of the MoO_4^{2-} tetrahedron, 1.77 Å, has a bond order of 1.5. (3) A bond order of 0.33 is assigned to the value of the Mo-O_b distance of 2.17 Å. [(4) A bond order of 0.1 is assigned to the value of Mo-O_b distance of 2.32 Å; however, it is not used for the above equation.] For details of the points of definition (3) and (4) see the paper [3]. Yet another relationship between the bond order for differently coordinated oxygen atoms and Mo-O distances, d_{Mo-O} = 1.915 − 0.80 log BO, (see **Fig. 32**) was deduced by Fuchs and Knöpnadel [4] assuming that the molybdenum atoms occupy the centers of charge of the coordination polyhedra. This means a uniform distribution of the ionic charge over the oxygen atoms. However, from investigations by [1, 8, 9] (and also by [5] for the polytungstate case) it has been shown that the ionic charge (to be compensated by cations) is mainly distributed over the terminal oxygen atoms [11] (see p. 37).

Fig.31 Fig.32

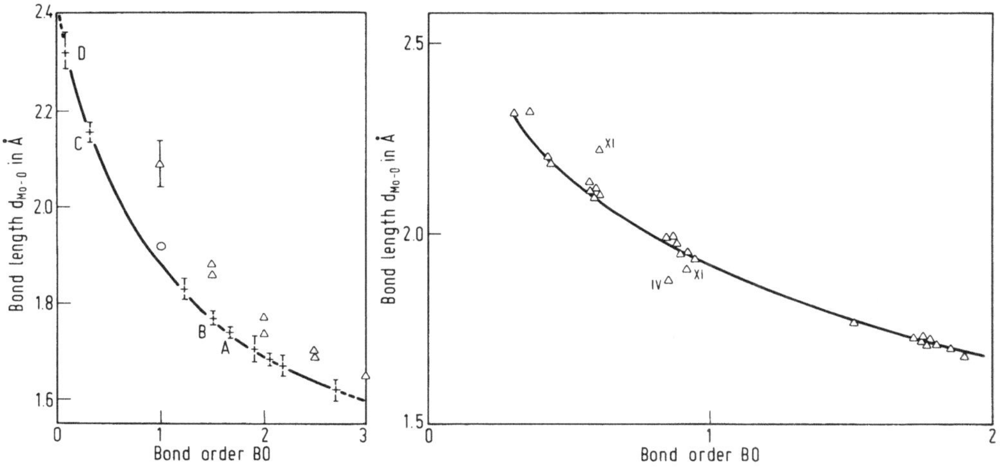

Mo-O bond lengths vs. bond order according to [3] (left) and [4] (right). In Fig. 31 the vertical lines show estimated errors. A, B, C, D are points of definition. The points △ are from F. A. Cotton, R. M. Wing (Inorg. Chem. **4** [1965] 867/73), ○ from J. M. Le Carpentier, A. Mitschler, R. Weiss (Acta Cryst. B **28** [1972] 1288/98). The marked points in Fig. 32 originate from the two less stable polyanions $[Mo_2O_7]^{2-}$ (IV) and α-$[Mo_8O_{26}]^{4-}$ (XI) [4].

According to [4], the graph of the BO/d_{Mo-O} diagram represents a stability condition for polymolybdate ions: a polymolybdate structure is only capable of existence if the bond distances given by the graph can be realized, at least approximatively.

The characteristic bond length of an Mo-O double bond is ~1.67 Å and of a single bond ~1.89 Å (mean values derived from the above equations). Oxygen atoms are assumed to be coordinated to a given molybdenum atom if the distance Mo-O is ≦2.7 Å [7].

Each oxygen atom forms, summed up over the different bonds, two covalent bonds unless its charge is compensated ionically by cations. Each molybdenum atom forms, summed up over the different bonds, six covalent bonds, cf. [1, 4, 6, 8].

References:

[1] I. Böschen, B. Buss, B. Krebs (Acta Cryst. B **30** [1974] 48/56). – [2] G. Donnay, R. Allmann (Am. Mineralogist **55** [1970] 1003/15). – [3] F. A. Schröder (Acta Cryst. B **31** [1975] 2294/309). – [4] J. Fuchs, I. Knöpnadel (Z. Krist. **158** [1982] 165/79). – [5] H. d'Amour, R. Allmann (Z. Krist. **136** [1972] 23/47).

[6] B. Krebs, I. Paulat-Böschen (Acta Cryst. B **38** [1982] 1710/8). – [7] R. Allmann (Monatsh. Chem. **106** [1975] 779/93). – [8] H. Vivier, J. Bernard, H. Djomaa (Rev. Chim. Minerale **14** [1977] 584/604). – [9] H. T. Evans, Jr., B. M. Gatehouse, P. Leverett (J. Chem. Soc. Dalton Trans. **1975** 505/14). – [10] I. D. Brown, K. K. Wu (Acta Cryst. B **32** [1976] 1957/9).

[11] K. H. Tytko (Habilitationsschrift, Göttingen, West Germany, 1977, p. 105).

2.1.9.4 Force Constants

Due to the size and complexity of the polymolybdate structures assignments of infrared and Raman spectra are difficult and have been possible only for the highly symmetrical $[Mo_6O_{19}]^{2-}$ structure by means of a normal-coordinate analysis [1, 2] and for the $[Mo_7O_{24}]^{6-}$ anion [3, 4].

For the $[Mo_6O_{19}]^{2-}$ anion total isotopic substitution (^{18}O, ^{92}Mo, ^{100}Mo) [2] allowed erroneous assignments in [1] to be corrected. For the assignments of the observed vibration frequencies see the papers. The stretching force constants $k_t(Mo\text{-}O_t)$, $k_b(Mo\text{-}O_b)$, and $k_c(Mo\text{-}O_c)$ were found to be [in mdyn/Å ($=10^{-2}$ N/m)]:

k_t	k_b	k_c	Ref.
7.30	3.50	0.83	[1]
7.46	2.35	0.47	[2]

For the complete force fields see the papers [1, 2].

The problem of the force field analysis of the $[Mo_7O_{24}]^{6-}$ anion was solved by treating fragments of the molecule as separate units. For details of the method, assignments of the observed vibration frequencies, and for the force constants see the paper [3]. For a later detailed analysis of Raman spectra obtained from an $Na_6[Mo_7O_{24}]\cdot 14H_2O$ single crystal in different geometrical orientations with respect to the exciting laser beam see [4].

References:

[1] R. Mattes, H. Bierbüsse, J. Fuchs (Z. Anorg. Allgem. Chem. **385** [1971] 230/42). – [2] C. Rocchiccioli-Deltcheff, R. Thouvenot, M. Fouassier (Inorg. Chem. **21** [1982] 30/5). – [3] L. E. Lyhamn, S. J. Cyvin (Z. Naturforsch. **34a** [1979] 867/75). – [4] L. Lyhamn (Acta Chem. Scand. A **36** [1982] 595/603).

2.1.9.5 Character of the Different Mo-O Bonds

The short $Mo\text{-}O_t$ bonds are covalent bonds having some π character, compare [1 to 3, 6, 12 to 14].

Some of the bridging Mo-O_b bonds can be considered to be derived from Mo-O_t bonds, thus also showing some π character (short Mo-O_b bonds) (type I) [1 to 3], or from dative O_t bonds, thus forming medium long and long Mo-O_b bonds (type II) [1]:

$$Mo{=}O_t \; Mo \rightarrow Mo \overset{\cdot\cdot}{=} O_b \cdots Mo$$
$$\text{I} \quad \text{II}$$

The value of the force constant k_t of the $[Mo_6O_{19}]^{2-}$ ion agrees with a rather short bond with some multiple character (Mo-O_t double bond [4, 5]) and fits well with the correlation curve (k_{Mo-O} vs. Mo-O bond length) established by [6]. The smaller value as compared to that of the $[W_6O_{19}]^{2-}$ ion ($k_t = 8.1$ mdyn/Å) is a consequence of a weaker π bond character [4, 7].

The k_b value of the $[Mo_6O_{19}]^{2-}$ ion agrees with that calculated for a single covalent bond according to Siebert's [8] formula [5]. Thus the Mo-O_b bonds can be considered as covalent single bonds [4, 5, 7]. (The bond length Mo-O_b is, according to the bond length/bond order diagram [1], also in accordance with a single bond [5].)

The very low k_c value of the $[Mo_6O_{19}]^{2-}$ ion allows very loose Mo-O_c bonds to be assumed, which is consistent with the rather long Mo-O_c distance. Because of the unusually coordinated O_c atom it is difficult to describe the nature of the Mo-O_c bonds. They could be assumed to be purely ionic bonds or, in an alternative description, delocalized multicenter bondings involving the six molybdenum atoms. In both descriptions the O_c atom is quite loosely bonded [5]. A largely ionic character of the Mo-O_c bonds is also assumed by reason of the small value of k_c by [4, 7] (cf., however, p. 37).

For the hexamolybdates, metal-metal bonds, as they are known from the structurally related halogen-niobium and -tantalum compounds of the type $M_6X_{12}^{z+}$ [10, 11], are not expected because of the large Mo-Mo distances [7] and other conditions (d^0 configuration, small intensity of the low-frequency Raman bands) which do not allow Mo-Mo interactions [4], and since a normal-coordinate analysis of the corresponding niobate and tantalate isopolyanions did not show any metal-metal interactions [9].

References:

[1] F. A. Schröder (Acta Cryst. B **31** [1975] 2294/309). – [2] M. A. Porai-Koshits, L. O. Atovmyan (Koord. Khim. **1** [1975] 1271/81; Soviet J. Coord. Chem. **1** [1975] 1065/74). – [3] E. M. Shustorovich, M. A. Porai-Koshits, Yu. A. Buslaev (Coord. Chem. Rev. **17** [1975] 1/98, 70/81). – [4] R. Mattes, H. Bierbüsse, J. Fuchs (Z. Anorg. Allgem. Chem. **385** [1971] 230/42). – [5] C. Rocchiccioli-Deltcheff, R. Thouvenot, M. Fouassier (Inorg. Chem. **21** [1982] 30/5).

[6] F. A. Cotton, R. M. Wing (Inorg. Chem. **4** [1965] 867/73). – [7] J. Fuchs (Z. Naturforsch. **28b** [1973] 389/404). – [8] H. Siebert (Z. Anorg. Allgem. Chem. **273** [1953] 170/82). – [9] F. J. Farrell, V. A. Maroni, T. G. Spiro (Inorg. Chem. **8** [1969] 2638/42). – [10] H. Schäfer, H. G. Schnering (Angew. Chem. **76** [1964] 833/49).

[11] R. Mattes (Z. Anorg. Allgem. Chem. **364** [1969] 279/89). – [12] H. T. Evans, Jr., B. M. Gatehouse, P. Leverett (J. Chem. Soc. Dalton Trans. **1975** 505/14). – [13] I. Böschen, B. Buss, B. Krebs (Acta Cryst. B **30** [1974] 48/56). – [14] M. T. Pope (Heteropoly and Isopoly Oxometalates, Springer, Berlin 1983, pp. 128/30).

2.1.9.6 Characteristic Vibrational Frequencies of the Mo-O Skeletal Modes

Regions of group frequencies arising from different building groups according to specifications made for isopolymolybdates [1 to 7, 15] are given in the following table. Additionally, some results from other compounds [8 to 11] are included. See also [16, 17] for some assignments.

wave number in cm^{-1}	building group	assignment	Ref.
1000 to 950	Mo=O	ν_s, ν_{as}	[1 to 5, 7]
950 to 875	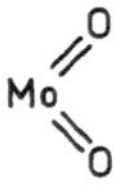	ν_s, ν_{as}	[3, 5, 6, 12, 13, 15, 17]
870 to 750	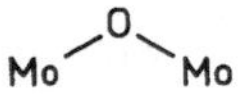	ν_{as}	[1, 2, 4, 6, 7, 15]
780 to 650	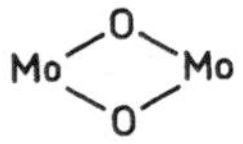		[3, 6]
650 to 400	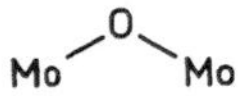	ν_s	[1, 4, 15]
450 to 350	Mo–O_c	ν_{as}	[1, 4, 7]
400 to 350	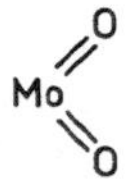	δ	[8 to 10, 15]
near 200	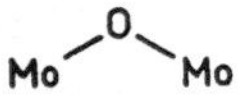	δ	[15]
560 to 530	Mo–OH_2	ν	[11]

The influence of the different cations on the positions of the Mo-O bands in the vibrational spectra is small, as is the influence of different crystal symmetries [3]. However, one cannot compare assignments for small structures with those of large and very condensed ones [14].

It has been pointed out that it is not possible to isolate parts such as

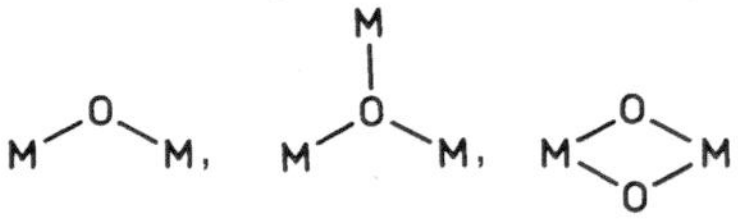

and assign them to unique vibrational modes. The vibrational conditions and peak positions depend upon the way in which such units are bonded to the rest of the complex. It is also not possible to talk in terms of group frequencies for condensed complexes such as $[Mo_7O_{24}]^{6-}$. It is mostly the outermost atoms, or atoms with few bonds, that do have quite assignable vibrational modes [14].

For normal-coordinate treatments of the $[Mo_6O_{19}]^{2-}$ anion see [1, 7] and for a complete vibrational analysis for the $[Mo_7O_{24}]^{6-}$anion see [13]. Tentative assignments of the vibrational modes for Raman spectra obtained from an $Na_6[Mo_7O_{24}]\cdot 14H_2O$ single crystal in different geometrical orientations with respect to the exciting laser beam have been performed by comparing structure and orientation of the Mo_7O_{24} groups within the unit cell. They are given in the following table [14]:

modes	wave number in cm^{-1}
ν_s $Mo(O_t)_2$	952
ν_{as} $Mo(O_t)_2$	939
δ $Mo(O_t)_2$	224
ν_s Mo_2O_2, ν_{as} Mo_2O_2	850 to 900
δ Mo_2O_2	120 to 300
"Mo_3O_3 modes"[1)]	370
"breathing modes" –Mo–O–Mo–	550 to 750

[1)] See the paper [14].

References:

[1] R. Mattes, H. Bierbüsse, J. Fuchs (Z. Anorg. Allgem. Chem. **385** [1971] 230/42). – [2] A. Kiss, S. Holly (Magy. Kem. Folyoirat **77** [1971] 418/26). – [3] J. Fuchs (Z. Naturforsch. **28b** [1973] 389/404). – [4] A. M. Golubev, E. A. Torchenkova, V. I. Spitsyn (Dokl. Akad. Nauk SSSR **217** [1974] 345/7; Dokl. Chem. Proc. Acad. Sci. USSR **214/219** [1974] 495/7). – [5] J. Fuchs, I. Knöpnadel, I. Brüdgam (Z. Naturforsch. **29b** [1974] 473/5).

[6] W. D. Hunnius (Z. Naturforsch. **29b** [1974] 599/602). – [7] C. Rocchiccioli-Deltcheff, R. Thouvenot, M. Fouassier (Inorg. Chem. **21** [1982] 30/5). – [8] W. P. Griffith, T. D. Wickins (J. Chem. Soc. A **1968** 400/4). – [9] W. P. Griffith (J. Chem. Soc. A **1969** 211/8). – [10] R. Mattes, G. Müller, H. J. Becher (Z. Anorg. Allgem. Chem. **389** [1972] 177/87).

[11] F. A. Schröder, B. Krebs, R. Mattes (Z. Naturforsch. **27b** [1972] 22/5). – [12] L. Lyhamn (Chem. Scr. **12** [1977] 153/61). – [13] L. E. Lyhamn, S. J. Cyvin (Z. Naturforsch. **34a** [1979] 867/75). – [14] L. Lyhamn (Acta Chem. Scand. A **36** [1982] 595/603) – [15] W. P. Griffith, P. J. B. Lesniak (J. Chem. Soc. A **1969** 1066/71).

[16] Z. M. Hanafi, M. A. Khilla, M. H. Askar (Thermochim. Acta **45** [1981] 221/32). – [17] A. J. Wilson, V. McKee, B. R. Penfold, C. J. Wilkins (Acta Cryst. C **40** [1984] 2027/30).

2.1.9.7 Distortions and Types of MoO_6 Octahedra

The molybdenum atoms are generally displaced from the centers of the octahedra (polyhedra) in the direction of the free corners or edges (or faces). This has been explained in different ways. Baker et al. [1, 2] have argued that polarization of oxygen atoms by the small, highly charged molybdenum atoms results in strong ion-induced dipole forces that are responsible for atom displacements. Exterior terminal oxygen atoms are polarized in one direction only by a layer of molybdenum atoms just beneath the surface. These molybdenum atoms have moved away from the center of the anion and leave interior oxygen atoms, with positive atoms on all sides, less polarized in anyone direction. According to Kepert [3], the short distances between the molybdenum atoms in neighboring octahedra sharing edges introduce unfavorable Coulombic repulsions between the molybdenum atoms, which can be partly accommodated by allowing the molybdenum atoms to move away from the centers of their octahedral cages of oxygen atoms in the direction of the free corners of the octahedra. Evans [4] and Böschen et al. [5] (and others) agree with this view; however, they assume additional factors: the need to balance charges on the inner, multiply linked oxygen atoms [4], and the different Mo-O bond strengths [5]. According to Fuchs and Knöpnadel [6], the molybdenum atoms tend to occupy the centers of charge of their coordination polyhedra thus reaching an approximate equidistribution of the negative charge over all oxygen atoms (compare, however, pp. 31 and 37). According to Porai-Koshits et al. [7, 8] the distortions are due to the double bond character of the cis-oriented $Mo\text{-}O_t$ bonds and to additional weakening of the bonds in trans-position to the terminal, doubly bonded oxygen atoms (trans-effect of multiple bonds), which is enhanced by the fact that precisely those oxygen atoms being the weakest (hardest) donors (i.e., those of the highest coordination number) are always situated in the trans-position relative to the multiply bonded terminal oxygen atoms. Linnett [19] argued that an electronic formula in which each metal atom participates in one three-electron, four two-electron, and one one-electron bonds (for the case of type I octahedra see below) does account for the observed short $M\text{-}O_t$, the (medium) $M\text{-}O_b$, and the long $M\text{-}O_c$ bond lengths. According to Lipscomb [18], in octahedral coordination with O^{2-} the metal cations are unsymmetrically displaced thus localizing the lines of force to reduce the number of $M^{n+}\text{-}O^{2-}$ bonds.

Oxygen atoms forming common edges between MoO_6 octahedra show generally shorter distances than those forming free edges, compare [9 to 13].

Another type of distortion causes asymmetric Mo-O-Mo bridges and occurs in $[\text{-Mo-O-}]_n$ rings forming part of the polymolybdate structures. Thus, in the structure of $[Mo_6O_{19}]^{2-}$ the Mo atoms are displaced away from one another giving rise to alternately longer and shorter distances between molybdenum and bridging oxygen atoms [14 to 16].

According to the kind of displacement of the molybdenum atom in its MoO_6 octahedron, two types of MoO_6 octahedra are observed in polymolybdates. In type I the molybdenum atom is displaced towards one terminal oxygen atom, in type II towards two cis, usually terminal oxygen atoms (an exception is, e.g., the central MoO_6 octahedron of the heptamolybdate structure), see also p. 27. Type I octahedra can accommodate molybdenum atoms with d^0, d^1, and d^2 electronic configurations, whereas type II octahedra are restricted to d^0 molybdenum atoms. Hence, only type I polymolybdates are reducible to give molybdenum blues [17].

References:

[1] L. C. W. Baker (Proc. 6th Intern. Conf. Coord. Chem., Detroit, Mich., 1961, pp. 604/12). – [2] L. C. W. Baker, L. Lebioda, J. Grochowski, H. G. Mukherjee (J. Am. Chem. Soc. **102** [1980] 3274/6). – [3] D. L. Kepert (Inorg. Chem. **8** [1969] 1556/8). – [4] H. T. Evans Jr. (Perspect. Struct. Chem. **4** [1971] 1/59, 54). – [5] I. Böschen, B. Buss, B. Krebs, O. Glemser (Angew. Chem. **85** [1973] 409; Angew. Chem. Intern. Ed. Engl. **12** [1973] 409).

[6] J. Fuchs, I. Knöpnadel (Z. Krist. **158** [1982] 165/79). – [7] M. A. Porai-Koshits, L. O. Atovmyan (Koord. Khim. **1** [1975] 1271/81; Soviet J. Coord. Chem. **1** [1975] 1065/74). – [8] E. M. Shustorovich, M. A. Porai-Koshits, Yu. A. Buslaev (Coord. Chem. Rev. **17** [1975] 1/98, 70/81). – [9] I. Böschen, B. Buss, B. Krebs (Acta Cryst. B **30** [1974] 48/56). – [10] B. Krebs, I. Paulat-Böschen (Acta Cryst. B **32** [1976] 1697/704).

[11] B. Krebs, I. Paulat-Böschen (Acta Cryst. B **38** [1982] 1710/8). – [12] B. M. Gatehouse, P. Leverett (Chem. Commun. **1968** 901/2). – [13] H. Vivier, J. Bernard, H. Djomaa (Rev. Chim. Minerale **14** [1977] 584/604). – [14] J. Fuchs (Z. Naturforsch. **28b** [1973] 389/404). – [15] R. Mattes, H. Bierbüsse, J. Fuchs (Z. Anorg. Allgem. Chem. **385** [1971] 230/42).

[16] J. Fuchs, W. Freiwald, H. Hartl (Acta Cryst. B **34** [1978] 1764/70). – [17] M. T. Pope (Inorg. Chem. **11** [1972] 1973/4). – [18] W. N. Lipscomb (Inorg. Chem. **4** [1965] 132/4). – [19] J. W. Linnett (J. Chem. Soc. **1961** 3796/803).

2.1.9.8 Other Structural Features Arising from the Diversity of the Mo-O Bonds

A group of Russian authors assumes that the cis-orientation of Mo-O multiple bonds and the trans-effect of the multiple bonds cause the formation of the block polymetalate complexes with terminal oxygen atoms on the outside and highly coordinated oxygen atoms within. They concede, however, that it might be a trivial consequence of the block type structures that terminal oxygen atoms are found outside the block, whereas oxygen atoms with maximum bridge multiplicity are found within the block. However, the block structure itself is thought to be a consequence of the cis-arrangement of the π bonds and the trans-effect of multiple bonds [1 to 3]. The formulated rules do not make it possible to predict the structures of individual concrete polyanions [1].

According to the same authors isopolymetalate ions cannot be centrosymmetric with the inversion center on a molybdenum atom as a consequence of the very tendency for cis-location of the π interaction and trans-effect in each of the MoO_6 octahedra. In the octahedron of the centrosymmetric molybdenum atom all of the trans-oriented bonds would be equivalent in pairs [1, 3]. Isopolymetalate ions may be, however, centrosymmetric if oxygen occupies the inversion center. This sort of oxygen atom acts as a low-charge, weakly bonded anion in the structure [1].

References:

[1] M. A. Porai-Koshits, L. O. Atovmyan (Koord. Khim. **1** [1975] 1271/81; Soviet J. Coord. Chem. **1** [1975] 1065/74). – [2] E. M. Shustorovich, M. A. Porai-Koshits, Yu. A. Buslaev (Coord. Chem. Rev. **17** [1975] 1/98, 70/81). – [3] M. A. Porai-Koshits, L. O. Atovmyan (Zh. Neorgan. Khim. **26** [1981] 3171/80; Russ. J. Inorg. Chem. **26** [1981] 1697/703).

2.1.9.9 Distribution of the Negative Charge over the Oxygen Atoms

The negative charge of a polymetalate ion is distributed over the oxygen atoms. Since according to X-ray photoelectron spectroscopic (XPS) investigations [1] the energy of the O1s level of a bridge oxygen atom is higher than that of a terminal oxygen atom, oxygen atoms of high coordination number bear a lower effective negative charge. That means that the decrease in the transfer of electron density from the oxygen atom to each of its bonds with the metal is compensated by an increase in the number of its bonds, so that the overall negative charge on the oxygen atom decreases [2]. From the bond orders of the oxygen atoms in several polymolybdate structures [5 to 7] (and also in a polytungstate structure [8]) it has also been concluded that the ionic charge (to be compensated by cations) is mainly distributed over the terminal oxygen atoms [9]. Location of the charge electrons on the terminal oxygen atoms of polymetalate ions has furthermore been considered in a theoretical study to explain some features of heteropoly compounds (terminal oxygen atoms behave as if they are bigger than the other oxygen atoms and thus have a decisive influence on the heteropolymetalate structure that is formed) [10]. The distribution of the ionic charge mainly over the terminal oxygen atoms is, however, in disagreement with XPS results on heteropoly compounds [4] and with assumptions by [3] (cf. pp. 31 and 35). (For polymetalate ions in aqueous solution there is also a rather controversial discussion on this point; cf. "Molybdenum" Suppl. Vol. B 3, to be published.)

References:

[1] D. G. Tisley, R. A. Walton (J. Mol. Struct. **17** [1973] 401/9). – [2] M. A. Porai-Koshits, L. O. Atovmyan (Koord. Khim. **1** [1975] 1271/81; Soviet J. Coord. Chem. **1** [1975] 1065/74). – [3] J. Fuchs, I. Knöpnadel (Z. Krist. **158** [1982] 165/79). – [4] L. P. Kazanskii, A. S. Saprykin, A. M. Golubev, V. I. Spitsyn (Dokl. Akad. Nauk SSSR **233** [1977] 405/8; Dokl. Phys. Chem. Proc. Acad. Sci. USSR **232/237** [1977] 282/4). – [5] I. Böschen, B. Buss, B. Krebs (Acta Cryst. B **30** [1974] 48/56).

[6] H. Vivier, J. Bernard, H. Djomaa (Rev. Chim. Minerale **14** [1977] 584/604). – [7] H. T. Evans, Jr., B. M. Gatehouse, P. Leverett (J. Chem. Soc. Dalton Trans. **1975** 505/14). – [8] H. d'Amour, R. Allmann (Z. Krist. **136** [1972] 23/47). – [9] K. H. Tytko (Habilitationsschrift, Göttingen, West Germany, 1977, p. 105). – [10] J. W. Linnett (J. Chem. Soc. **1961** 3796/803).

2.1.9.10 Packing of the Oxygen Atoms in Polymolybdates

Polymetalate ions are sections of distorted (mostly cubic) closest spherical packings in which the molybdenum atoms occupy some of the octahedral (sometimes tetrahedral) gaps, compare [1 to 5, 9]. The oxygen-oxygen distances amount to ~2.60 Å. The peripherally bound oxygen atoms show somewhat larger distances [3]. Cations such as Na^+ and NH_4^+, and molecules of water of crystallization take part in the packing [1, 2]. A packing density of ~20.6 Å^3/oxygen atom has been found in the compound $Na_8[Mo_{36}O_{112}(H_2O)_{16}]\cdot\sim 64\,H_2O$ [2]. Polymetalate ions may also be considered as fragments of the sodium chloride arrangement of transition metal atoms and oxygen atoms [7, 8].

The fact that molybdenum (addenda) atoms are decidedly small relative to the octahedral pockets which enclose them, so that they can move off-center (cf. p. 35), is assumed to be of paramount importance in explaining the very existence of discrete polyanions with definite structures [6].

References:

[1] I. Böschen, B. Buss, B. Krebs (Acta Cryst. B **30** [1974] 48/56). – [2] K. H. Tytko, B. Schönfeld, B. Buss, O. Glemser (Angew. Chem. **85** [1973] 305/7; Angew. Chem. Intern. Ed. Engl. **12** [1973] 330/2). – [3] J. Fuchs, I. Knöpnadel (Z. Krist. **158** [1982] 165/79). – [4] H. Vivier, J. Bernard, H. Djomaa (Rev. Chim. Minerale **14** [1977] 584/604). – [5] L. C. W. Baker (Proc. 6th Intern. Conf. Coord. Chem., Detroit, Mich., 1961, pp. 604/12).

[6] L. C. W. Baker, L. Lebioda, J. Grochowski, H. G. Mukherjee (J. Am. Chem. Soc. **102** [1980] 3274/6). – [7] H. T. Evans, Jr. (Inorg. Chem. **5** [1966] 967/77). – [8] H. T. Evans, Jr. (Perspect. Struct. Chem. **4** [1971] 1/59, 53/6). – [9] J. W. Linnett (J. Chem. Soc. **1961** 3796/803).

2.1.9.11 Interrelations Between Cations and Polymolybdate Ions, the Role of Water of Crystallization, and the Formation of Solids

Cations and polymolybdate anions are held together by electrostatic interactions between the positively and negatively charged particles, and by extended hydrogen bridge bond systems in three dimensions between mainly terminal oxygen atoms of the polymolybdate ions, molecules of water of crystallization, and, if present, ammonium or organic ammonium ions, cf. [1, 2, 7, 8]. Na^+ and NH_4^+ ions are assumed to be coordinated to a given oxygen atom if the distance Na-O is $\leqq 3.2$ Å and NH_4-O$\leqq 3.4$ Å, respectively [6]. H_2O molecules are assumed to be involved in hydrogen bridging systems if H_2O-X (X = O, NH_4, or others) is $\leqq 3.4$ Å [1, 8 to 10]. Hydrogen bridges occur often as bifurcated [1, 8] and even as trifurcated [12] bridges.

The packing of the large polymetalate ions in the crystal along with atomic cations having considerably smaller dimensions and high charge concentration is, however, fraught with certain difficulties and realized only under certain conditions. Therefore, crystal hydrates with a large amount of water are almost always formed during crystallization from aqueous solutions [3, 11]. Some of the water molecules coordinate with the cations, thus increasing the size of the cationic "island" and reducing the charge concentration on them. Other water molecules act as a buffer between the polyanions and the hydrated cations. However, when the partner of the polyanion is a large organic cation, the necessity of the solvent molecules in the crystal vanishes. In particular, this pertains to $[(C_4H_9)_4N]_2[Mo_6O_{19}]$ [3].

The crystallization of the chain-like polymolybdate types is a time-consuming process since the oligomeric species forming the chains do not exist in the solution equilibria in significant concentrations. In particular, the formation of the crystal nucleus is very difficult since chains of the oligomeric species of a certain length first have to form [4, 5] and then assemble in parallel orientation together with the cations and, if necessary to avoid vacancies, with water molecules [5]. Once formed, the growth of the crystals can occur much more rapidly by coupling of oligomeric (or possibly monomeric) units in the solution with those in the crystal. If there are oligomeric units having several sites for connection, crystallization is still more complicated [4, 5]. Several criteria have been proposed to estimate the probability of different oligomeric units forming chain-like polymolybdates. Those oligomeric units have the greatest chance to polymerize, (1) which can combine with each other in a single manner only (thus formation of a regular chain being guaranteed), (2) which give a short period of identity (thus correct packing of the parallel chains being facilitated), (3) which avoid direct contacts between the cations, and between the terminal oxygen atoms of neighboring chains, i.e., those on which the negative charge is concentrated, (4) which allow a close packing of the oxygen atoms, water molecules, and cations, and (5) which give a chain of some flexibility [5]. Similarly (discrete) polymetalate ions of symmetrical structure can crystallize readily [14].

In the case of the chain-like "trimolybdates" the growth of the crystals has been shown to be one-dimensional [13].

References:

[1] I. Böschen, B. Buss, B. Krebs (Acta Cryst. B **30** [1974] 48/56). – [2] A. Don, T. J. R. Weakley (Acta Cryst. B **37** [1981] 451/3). – [3] M. A. Porai-Koshits, L. O. Atovmyan (Koord. Khim. **1** [1975] 1271/81; Soviet J. Coord. Chem. **1** [1975] 1065/74). – [4] K. H. Tytko (Z. Naturforsch. **28b** [1973] 272/5). – [5] K. H. Tytko (Z. Naturforsch. **31b** [1976] 737/48, 744/7).

[6] R. Allmann (Monatsh. Chem. **106** [1975] 779/93). – [7] H. T. Evans, Jr., B. M. Gatehouse, P. Leverett (J. Chem. Soc. Dalton Trans. **1975** 505/14). – [8] H. Vivier, J. Bernard, H. Djomaa (Rev. Chim. Minerale **14** [1977] 584/604). – [9] B. Krebs, I. Paulat-Böschen (Acta Cryst. B **32** [1976] 1697/704). – [10] B. Krebs, I. Paulat-Böschen (Acta Cryst. B **38** [1982] 1710/8).

[11] I. Lindqvist (Nova Acta Regiae Soc. Sci. Upsaliensis [4] **15** No. 1 [1950] 1/22, 9). – [12] Y. Ohashi, K. Yanagi, Y. Sasada, T. Yamase (Bull. Chem. Soc. Japan **55** [1982] 1254/60). – [13] S. Hodorowicz (Krist. Tech. **12** [1977] 431/4). – [14] M. L. Freedman (J. Am. Chem. Soc. **80** [1958] 2072/7).

2.1.10 Photochromism of Organic Ammonium Polymolybdates. Photogalvanic Effect

Many organic ammonium polymolybdates (hydrates and anhydrous compounds) show photochromism. When the white polycrystalline material [1, 2] or a transparent colorless crystal [3, 11, 13] are irradiated with ultraviolet light, the crystals become colored; a return to the original white color is observed in the dark in the presence of oxygen. This color change can be repeated many times. In general, coloration requires some minutes, decoloration several hours [1, 2]. Observed colors range from intense pink, pink-brown, reddish brown, reddish violet to blue.

Secondary amines [1, 2], cyclic or aliphatic [1], and primary amines [2], but also quaternary amines (see the list of the photochromic polymolybdate hydrates on p. 130) yield photochromic molybdates. Polymolybdate types established to be photochromic are mainly "tri-" and heptamolybdates, but a (3:8)-octamolybdate and a "decamolybdate" have also been found to be photochromic (cf. p. 130). The "dimolybdates" cited to be photochromic [1, 2] have to be regarded as "trimolybdates" (those in [1]) or heptamolybdates (those in [2]), according to their infrared spectra as published in [2, 8]. The most sensitive and rapid (in both directions) photochromic systems are the "trimolybdates", that is hydrated dimethylammonium, diethylammonium [1, 2], piperidinium, piperazinium, and pyrrolidinium "trimolybdates" [1].

Infrared spectra and X-ray powder patterns of the irradiated compounds are unchanged [1, 2]. The colored species are produced only in very small quantities at the surface of the irradiated samples [1]. They have been characterized by their diffuse reflectance spectra in the visible range [1, 2] (λ_{max} = 470 to 510 nm [2]) and the resolution of these spectra into Gaussian curves [1], and by EPR spectra obtained from powders [1, 2, 12] and single crystals [3, 11, 13] at ambient temperature.

All reflectance spectra can be resolved into three, four, or five Gaussian components centered at (1) 750 to 1000, (2) approximately 560, (3) 470, (4) 400, and (5) 360 nm. The two most intense Gaussian components (2) and (3) are always present [1, 9]. A detailed kinetic analysis of decoloration by means of the Gaussian components (mostly several simultaneous first-order reactions), as well as a matrix rank treatment of the optical densities at different wavelengths and different times after exposure, showed the presence of more than one colored species in the case of the compounds $[(CH_3)_2NH_2]_2O \cdot 3MoO_3 \cdot H_2O$ and $[(C_2H_5)_2NH_2]_2O \cdot 3MoO_3 \cdot H_2O$. For a free electron model enabling assignments of the different Gaussian curves, which constitute the absorption spectra of the irradiated salts, see the papers [1, 9, 10].

All EPR spectra of the polycrystalline material have the same shape and correspond to the same value of the central g factor ($g_2 \approx 1.925$ [1, 9, 10], 1.930 [12, 14]), showing that the same paramagnetic species are produced in the salts [1, 9]. (Only in the case of the compound $[(C_2H_5)_2NH_2]_2O \cdot 3MoO_3 \cdot H_2O$ is there an additional signal with $g_2 = 1.911$ [9].) Since the same signal is observed at 77 K and for any period of irradiation and during the bleaching, it is assumed that it is due to a single type of paramagnetic center [1], which was identified as Mo^V on the basis of g values reported in the literature [4 to 6] for Mo^V [1, 2]. From single crystal EPR studies the site of the Mo^V formation in the polyanions has been deduced [3, 11, 13, 15].

Bleaching is accelerated under pure oxygen or at more elevated temperatures (30 to 40°C [1], 50°C [2]). No bleaching is observed under nitrogen. Hence bleaching results from the oxidation of Mo^V to Mo^{VI} [1, 2].

The primary reaction step in the formation of the colored species upon irradiation is assumed to be a classical photolysis reaction, $R_2NH_2^+ \xrightarrow{h\nu} H\cdot + R_2\dot{N}H^+$, following either the direct absorption of photons of wavelengths shorter than 250 nm by the cation or absorption of photons of wavelengths shorter than 350 nm by the molybdate anion. The anion would then act as a photosensitizer towards the cation. The free radicals thus formed, or the hydrazinium ion formed by recombination, then reduce the polymolybdate anion [1, 9]. According to [3, 11, 13, 15] the photoreduction of Mo^{VI} to Mo^V proceeds via UV-induced charge transfer in a terminal Mo=O bond of an Mo^{VI} site with an accompanying transfer of a hydrogen-bonding proton from an alkylammonium nitrogen to a bridging oxygen atom followed by an interaction of the non-bonding electrons of the amino nitrogen with a terminal oxo group leading to a charge transfer complex (cf. Eq. (1) on p. 147). The concept that only hexameric molybdate anions (which were assumed to be present in "trimolybdates" and "dimolybdates") can be reduced [1, 9], is untenable in view of the experimental facts (cf. the list of the polymolybdate types on p. 130).

The differences in photosensitivity, reversibility, etc. among the various molybdates in the solid state are assumed to be connected with the crystal structures of the compounds [2].

Photochromism may be restricted to the solid state (e.g., with the compound $(C_5H_{11}NH)_6[Mo_7O_{24}] \cdot 6H_2O$). Solution studies [2, 11 to 13] show at low concentrations a yellow and at high concentrations a blue coloration [13] and EPR signal(s) due to Mo^V and hence presence of various reduced species. The blue form shows thermochromism (blue $\rightleftharpoons$ yellowish green) in the range 0 to 70°C with a heat of reaction $\Delta H \approx 15$ kcal/mol [2, 13], and bleaching on exposure to oxygen (oxidation of Mo^V to Mo^{VI}) [2]. Intermediate species have been observed by flash photolysis, short-lived free radicals ($^{\bullet}OH$, $HO_2^{\bullet}$) by spin-trapping experiments [13].

Primary aliphatic amines often yield photosensitive salts with para- and "tri"-molybdates but the color produced on irradiation is stable (in all cases according to [1], sometimes according to [2]) in the dark and hence the behavior of these salts does not correspond to the definition of photochromism [1, 2]. The same is true for the "acidic" polymolybdates [1, 8] (presumably 36-molybdates) of primary and secondary amines.

Polymolybdates of unsubstituted and tertiary ammonium cations are not photosensitive [8]. "Tetramolybdates" have never been found to be photochromic [1, 9].

The photogalvanic effect has been studied with aqueous solutions of the compound $(i\text{-}C_3H_7NH_3)_6[Mo_7O_{24}]\cdot 3H_2O$, see pp. 148/50.

References:

[1] F. Arnaud-Neu, M. J. Schwing-Weill (J. Less-Common Metals **36** [1974] 71/8; Chem. Uses Molybdenum, Proc. 1st Conf., Reading, Engl., 1973 [1974], pp. 35/8). – [2] T. Yamase, T. Ikawa (Bull. Chem. Soc. Japan **50** [1977] 746/9). – [3] T. Yamase (J. Chem. Soc. Dalton Trans. **1978** 283/5). – [4] N. S. Garif'yanov, N. S. Kucheryavenko, V. N. Fedotov (Dokl. Akad. Nauk SSSR **150** [1963] 802/4; Dokl. Chem. Proc. Acad. Sci. USSR **148/153** [1963] 452/4). – [5] P. Rabette, C. Ropars, J. P. Grivet (Compt. Rend. C **265** [1967] 153/5).

[6] I. Pascaru, O. Constantinescu, M. Constantinescu, D. Arizan (J. Chim. Phys. **62** [1965] 1283/8). – [7] H. U. Kreusler, A. Förster, J. Fuchs (Z. Naturforsch. **35b** [1980] 242/4). – [8] F. Arnaud-Neu, M. J. Schwing-Weill (Bull. Soc. Chim. France **1973** 3225/32). – [9] F. Arnaud-Neu, M. J. Schwing-Weill (Bull. Soc. Chim. France **1973** 3239/46). – [10] F. Arnaud-Neu, M. J. Schwing-Weill (Bull. Soc. Chim. France **1973** 3233/9).

[11] T. Yamase (J. Chem. Soc. Dalton Trans. **1982** 1987/91). – [12] T. Yamase, H. Hayashi, T. Ikawa (Chem. Letters **1974** 1055/6). – [13] T. Yamase, R. Sasaki, T. Ikawa (J. Chem. Soc. Dalton Trans **1981** 628/34). – [14] T. Yamase, T. Ikawa, H. Kokado (Chem. Letters **1973** 615/6). – [15] Y. Ohashi, K. Yanagi, Y. Sasada, T. Yamase (Bull. Chem. Soc. Japan **55** [1982] 1254/60).

2.1.11 Behavior on Heating

Two types of reaction may occur when a polymolybdate hydrate is heated:

1. Partial or total separation of the water of crystallization (dehydration).
2. Dismutation into an acid-rich and base-rich compound.
   a) Decomposition to a polymolybdate richer in the acid and a second one richer in the base.
   b) Decomposition to a polymolybdate richer in the acid and the pure base.
   c) Decomposition to a polymolybdate richer in the base and the pure acid (MoO_3).

The reactions according to (1) and (2) may occur successively or simultaneously. In general, the dehydration reactions take place first.

The polymolybdate hydrates often lose their water of crystallization step by step on heating. If delivery of the water occurs at low temperatures, the structure of the polyanion is mostly preserved for a long period as in the case of the salt $Na_6[Mo_7O_{24}]\cdot 22H_2O$ [1] which readily effloresces [1, 2] to give the 14-hydrate (case 1).

The 36-molybdates, which also contain large amounts of water of crystallization, lose a great deal of their water also at room temperature, with simultaneous conversion to "decamolybdates" [1] (and small amounts of the base MOH) (case 2b).

Alkaline earth polymolybdates react – after dehydration – mostly to form the normal molybdate and molybdenum trioxide [3 to 5] (case 2c).

Reactions according to case (2a) seem to be rather rare.

The most important thermal decomposition reactions are those of the polymolybdates of ammonium and of the organic cations. Due to the volatility of the base that forms the cation, decomposition proceeds with increasing temperature to give more and more acid polymolybdates and finally MoO_3 (case 2b). Intermediate ammonium polymolybdates (see **Fig. 33**, p. 42)

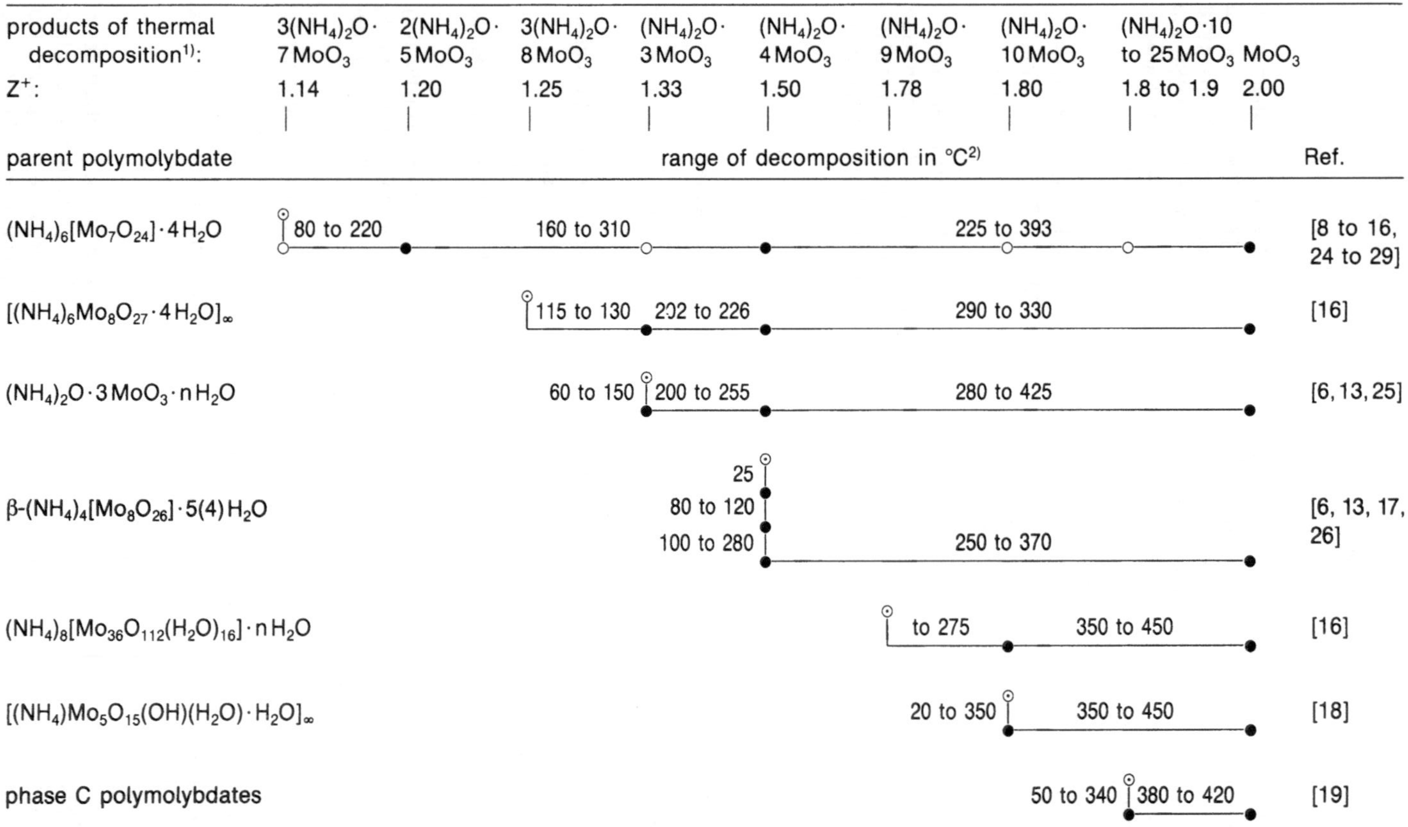

Fig. 33. Course of the thermal decomposition of the different ammonium polymolybdate hydrates in air at normal pressure. ⊙ Hydrated parent polymolybdate, ● intermediate stage and final product, ○ intermediate stage postulated only occasionally.

[1] The decomposition products $(NH_4)_2O \cdot 10MoO_3$ and $(NH_4)_2O \cdot 10$ to $25MoO_3$ are obviously of the "decamolybdate" type and therefore should also contain some water. – [2] Lowest temperature stated in the literature for the beginning and highest temperature stated for the termination of the conversion reaction.

parent polymolybdate		start	1.14 to 1.33	1.33 to 1.50	1.50 to 1.78	1.78 to 2.00	Ref.
products of thermal decomposition:		$3M_2O \cdot 7MoO_3$	$M_2O \cdot 3MoO_3$	$M_2O \cdot 4MoO_3$	$M_2O \cdot 9MoO_3$	MoO_3	
Z^+:		1.14	1.33	1.50	1.78	2.00	
				range of decomposition in °C			
$[(CH_3)NH_3]_2O \cdot 5.87$ to $9.36 MoO_3 \cdot nH_2O$		80				132 to 348	[20]
$[(CH_3)_2NH_2]_6[Mo_7O_{24}] \cdot H_2O$			110 to 124	138 to 150	250 to 380		[20]
$[(CH_3)_2NH_2]_2O \cdot 3MoO_3 \cdot H_2O$				168	255 to 380		[20]
$[(CH_3)_3NH]_6[Mo_7O_{24}] \cdot 7H_2O$			20 to158		234 to 434		[20]
β-$[(CH_3)_4N]_2Na_2[Mo_8O_{26}] \cdot 2H_2O$		309			to 437		[21]
$[(C_2H_5)NH_3]_6[Mo_7O_{24}] \cdot 3H_2O$		38 to 98	to 126		212 to 360		[20]
$[(C_2H_5)_2NH_2]_2O \cdot 3MoO_3 \cdot H_2O$		50 to 82		137 to 154	212 to 382		[20]
$(C_5H_{11}NH)_6[Mo_7O_{24}] \cdot 6H_2O$	[a]		55 to 145	155 to 185	225 to 333		[22]
$(C_5H_{11}NH)_6[Mo_7O_{24}] \cdot 3H_2O$	[a]	60 to 80	135 to 154	168 to 190	240 to 365		[22]

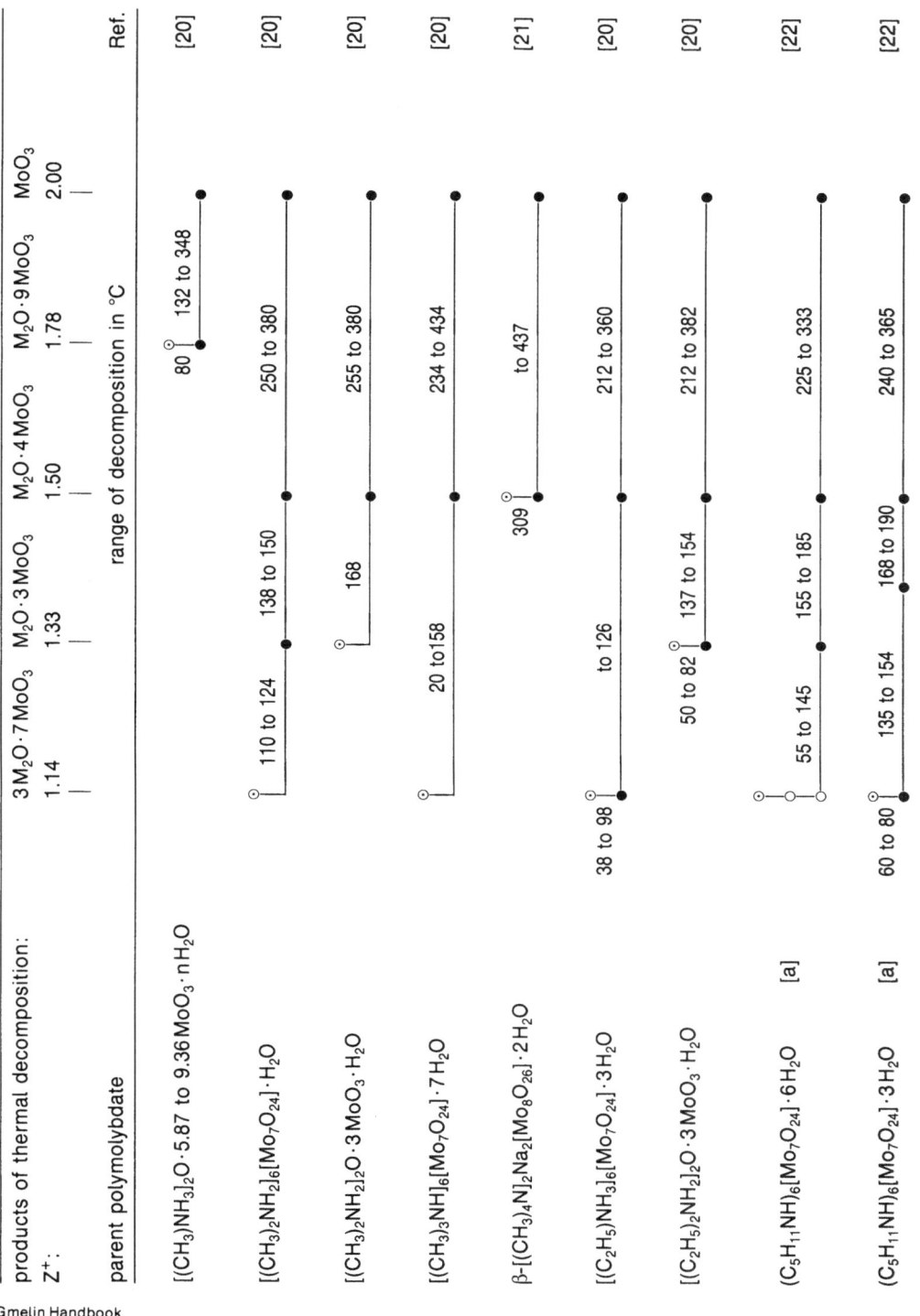

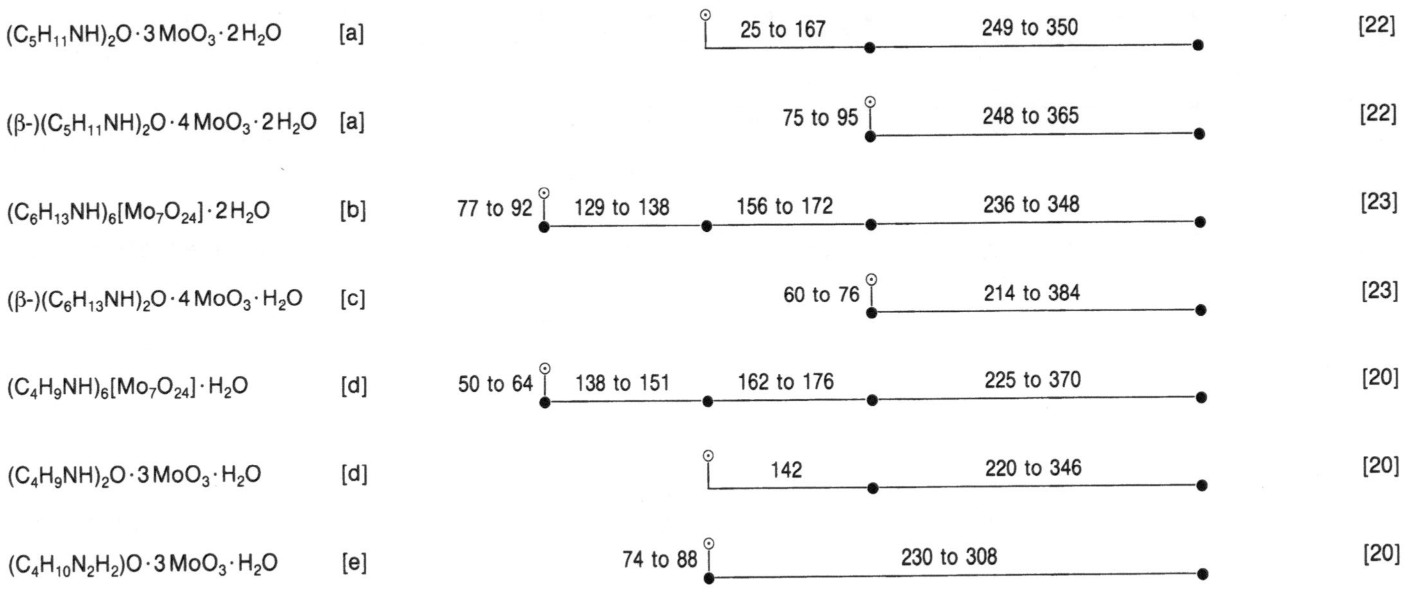

[a] $C_5H_{11}N$ = piperidine. – [b] $C_6H_{13}N$ = 4-methylpiperidine. – [c] $C_6H_{13}N$ = N-methylpiperidine. – [d] C_4H_9N = pyrrolidine. – [e] $C_4H_{10}N_2$ = piperazine.

Fig. 34. Course of the thermal decomposition of some polymolybdates with organic cations.
⊙ Hydrated parent polymolybdate, ● intermediate stage and final product, ○ indication of an intermediate stage.

are – apart from the partially and/or completely dehydrated parent polymolybdates – the anhydrous compounds $2(NH_4)_2O \cdot 5MoO_3$ ($=(NH_4)_8[Mo_{10}O_{34}]$), $(NH_4)_2O \cdot 3MoO_3$, $(NH_4)_2O \cdot 4MoO_3$, $(NH_4)_2O \cdot 10MoO_3$, and $(NH_4)_2O \cdot 10$ to $25MoO_3$ (for Z^+ see p. 6). The dehydrated parent compound does not occur in all cases. According to [6 to 8] MoO_3 is sometimes reduced to a lower oxide, while NH_3 is oxidized to N_2.

Intermediate polymolybdates of organic cations (see **Fig. 34**, pp. 43/4) are – apart from the partially and/or completely dehydrated parent polymolybdates – the anhydrous compounds $M_2O \cdot 3MoO_3$ and $M_2O \cdot 4MoO_3$; in one case a mixed (1:3)- and (1:4)-molybdate was observed. The stage of the anhydrous parent polymolybdate is frequently not observed. The step of the (1:3)-intermediate is also frequently not observed, while the (1:4)-intermediate is observed in nearly all cases. Fig. 34 demonstrates only those polymolybdates whose identity has been proved unambiguously.

References:

[1] K. H. Tytko, B. Schönfeld (Z. Naturforsch. **30b** [1975] 471/84, 476/7). – [2] Gmelin Handbuch "Molybdän", 1935, p. 221. – [3] J. Meullemeestre (Bull. Soc. Chim. France **1978** I 231/5). – [4] M. V. Mokhosoev, Z. E. Batura, V. I. Krivobok (Izv. Akad. Nauk SSSR Neorgan. Materialy **3** [1967] 133/7; Inorg. Materials **3** [1967] 108/11). – [5] J. Meullemeestre (Bull. Soc. Chim. France **1978** I 236/42).

[6] E. Ma (Bull. Chem. Soc. Japan **37** [1964] 171/5). – [7] I. H. Park (Bull. Chem. Soc. Japan **45** [1972] 2745/8). – [8] I. K. Bhatnagar, D. K. Chakrabarty, A. B. Biswas (Indian J. Chem. **10** [1972] 1025/8). – [9] I. Sălăgeanu, D. Trestianu, E. Segal (Rev. Roumaine Chim. **18** [1973] 1537/46). – [10] C. Duval (Anal. Chim. Acta **15** [1956] 223/5).

[11] A. B. Kiss, P. Gadó, I. Asztalos, A. J. Hegedüs (Acta Chim. [Budapest] **66** [1970] 235/49). – [12] W. Romanowski (Chem. Stosowana **7** [1963] 105/23, 122; C.A. **59** [1963] 14623). – [13] E. Ma (Bull. Chem. Soc. Japan **37** [1964] 648/53). – [14] A. Louisy, J. M. Dunoyer (J. Chim. Phys. **67** [1970] 1390/4). – [15] C. J. Hallada (J. Less-Common Metals **36** [1974] 103/10, 110).

[16] M. J. Schwing-Weill (Bull. Soc. Chim. France **1967** 3795/8). – [17] H. Vivier, J. Bernard, H. Djomaa (Rev. Chim. Minerale **14** [1977] 584/604). – [18] N. Sotani (Bull. Chem. Soc. Japan **48** [1975] 1820/5). – [19] H. Peters, L. Till, K. H. Radeke (Z. Anorg. Allgem. Chem. **365** [1969] 14/21). – [20] F. Arnaud-Neu, M. J. Schwing-Weill (Bull. Soc. Chim. France **1973** 3225/32).

[21] J. Fuchs, I. Knöpnadel, I. Brüdgam (Z. Naturforsch. **29b** [1974] 473/5). – [22] M. J. Schwing-Weill (Bull. Soc. Chim. France **1967** 3801/5). – [23] F. Arnaud-Neu, M. J. Schwing-Weill (Bull. Soc. Chim. France **1971** 60/8). – [24] W. D. Hunnius (Z. Naturforsch. **30b** [1975] 63/5). – [25] K. Funaki, T. Segawa (J. Electrochem. Soc. Japan **18** [1950] 152/4).

[26] E. Ya. Rode, V. N. Tverdokhlebov (Zh. Neorgan. Khim. **3** [1958] 2343/6; Russ. J. Inorg. Chem. **3** No. 10 [1958] 157/62). – [27] I. H. Park (Bull. Chem. Soc. Japan **45** [1972] 2739/44). – [28] Z. M. Hanafi, M. A. Khilla, M. H. Askar (Thermochim. Acta **45** [1981] 221/32). – [29] C. Duval (Inorganic Thermogravimetric Analysis, 2nd Ed., Elsevier, Amsterdam 1963, pp. 96/7, 462).

2.1.12 Behavior of the Polymolybdates in (Aqueous) Solution

The polymolybdates prepared from aqueous solution can be divided into three groups according to their behavior in water [1].

Polymolybdates with a chain-like, two- or three-dimensional network polyanion are water-insoluble for structural reasons. Some of them dissolve (with destruction) when heated or a large quantity of water is applied (ammonium "dimolybdate", ammonium polyoctamolybdate, "trimolybdates"). The solutions contain precisely those (discrete) polyanions which correspond to the conditions in the solution (Z value, pH, (total) molybdate concentration) [1] (see "Molybdenum" Suppl. Vol. B 3, to be published). The "decamolybdates" [1] and phase C polymolybdates [2] are slightly soluble also in hot water.

The water-insoluble polymolybdates with discrete polyanions (α- and β-octamolybdates, hexamolybdates, ammonium decamolybdate) have two characteristics: (1) they form only with certain cations (ammonium, alkylammonium, and other organic cations); (2) the polymolybdate ion under consideration does not occur (in significant concentration) as a component in the solution equilibria. It is assumed that the small solubility of these types is due to a high lattice energy, low hydration energy of the ions, and/or to the three-dimensional network formed by hydrogen bridges, which allow formation of the solid from the small, undetectable amounts of these polymolybdate ions in the solution by continuous removal from equilibrium. The solutions of the compounds contain also in these cases precisely those polyanions which correspond to the conditions in the solutions (see above) [1]. (For a rather controversial discussion concerning the question whether the existence of the β-octamolybdate ion [Lindqvist's octamolybdate ion] in aqueous solution can be seen as unequivocally proven, see "Molybdenum" Suppl. Vol. B 3, to be published.)

The heptamolybdates and the 36-molybdates are the only soluble polymolybdates with a discrete polyanion. The formation of these polymolybdates is not restricted to certain cations. Only in these two cases is there the same polymolybdate ion in solution and in the solid [1].

Polymolybdates with inorganic cations are slightly soluble in organic solvents whereas those with the large organic cations are slightly soluble in water [3]. However, while from aqueous solution both polymolybdates with a discrete polyanion and those with an infinite polyanion can be prepared, organic solvents allow only preparation of polymolybdates with discrete polyanions.

The many slightly soluble compounds formed by polyanions which do not exist in solution have no counterpart in the polytungstate system in which such compounds do not occur. Explanations for this different behavior of the systems are given in [4].

References:

[1] K. H. Tytko, B. Schönfeld (Z. Naturforsch. **30b** [1975] 471/84, 479/81). – [2] M. L. Freedman (J. Chem. Eng. Data **8** [1963] 113/6). – [3] J. Fuchs (Z. Naturforsch. **28b** [1973] 389/404). – [4] V. Cordis, K. H. Tytko, O. Glemser (Z. Naturforsch. **30b** [1975] 834/41, 838).

2.1.13 Ion Exchange Properties of the Phase C Polymolybdates[1)]

All compounds of this polymolybdate type have the same X-ray diffraction patterns and lattice parameters despite the wide range of composition and the differences in cations [1 to 3] (see 2.1.6.9, p. 23). This behavior was interpreted to mean that the compounds have a structure suited for cation exchange, and this was experimentally confirmed by heating the phase C compounds in solutions 0.67 N in HNO_3 and 0.33 M in M^I or M^{II} nitrate to reflux for 1 h. The total quantity of cations per g of exchanger remained approximately constant (~1 mval/g) which is seen as a proof that these are true cation-exchange processes and not adsorption effects. Reversibility was also proven. The bonding power of the cations to the polymolybdate anions increases in the sequence $Li^+ < Ca^{2+} < Na^+ < Sr^{2+} < K^+ < Ba^{2+} < NH_4^+$ which is the order

[1)] See the notes added in proof on pp. 7 and 24.

of increasing cation radii and polarizability (and the order of stability found in the preparative work, see 2.1.6.9). In concentrated HNO_3 the interchangeability of the cations and the total quantity of cations per g of exchanger are distinctly reduced, due to the conversion of the polymolybdate into MoO_3 [4].

References:

[1] M. L. Freedman (J. Chem. Eng. Data **8** [1963] 113/6). – [2] M. L. Freedman, S. Leber (J. Less-Common Metals **7** [1964] 427/32). – [3] H. Peters, L. Till, K. H. Radeke (Z. Anorg. Allgem. Chem. **365** [1969] 14/21). – [4] H. Peters, K. H. Radeke (Monatsber. Deut. Akad. Wiss. Berlin **11** [1969] 351/4).

2.2 Lithium Molybdate Hydrates

Older data are given in "Molybdän", 1935, pp. 210/4.

2.2.1 Lithium Molybdate Hydrates with Mo of Oxidation State <6

Compounds of this type have not been reported.

2.2.2 Lithium Molybdate(VI) Hydrates

2.2.2.1 The Li_2MoO_4-LiOH-H_2O System

This system at 25°C is of the simple eutonic form, the constituents being the only compounds (see figure in the paper [1]), but also the compound $Li_4MoO_5 \cdot 4H_2O$ (see below) has been reported [2, 3]. The eutonic solution contains 4.02 wt% LiOH and 40.10 wt% Li_2MoO_4. Equilibrium compositions of the solutions and observed solid phases are reported as follows (selected values); solution composition: LiOH and Li_2MoO_4 in wt%; solid phases: I = Li_2MoO_4, II = $LiOH \cdot H_2O$ [1]:

[LiOH]	0.0	0.56	2.32	4.02	4.01	4.04	4.01	4.23	6.30	8.51	9.91	11.19
[Li_2MoO_4] ..	44.42	43.73	41.54	39.95	40.20	39.94	40.23	38.03	23.83	12.25	5.72	0.0
solids	I	I	I	I+II	I+II	I+II	I+II	II	II	II	II	II

Physical properties of the saturated solutions in the system are shown in a table and a graph in the paper [1]. The limiting values for 100% Li_2MoO_4 (0% LiOH) and 0% Li_2MoO_4 (100% LiOH) are the following: the density ranges from 1.4935 to 1.1135 g/cm^3, the surface tension from 88.1×10^{-3} to 74.1×10^{-3} N/m, the viscosity from 11.668×10^{-3} to 3.4044×10^{-3} $N \cdot s \cdot m^{-2}$ (with a maximum of about 25.4×10^{-3} near the eutonic composition), specific electrical conductivity from 5.275 to 46.534 $\Omega^{-1} \cdot cm^{-1}$ (with a value of about 7.4 at the eutonic), and the index of refraction from 1.4389 to 1.3730 (maximum at the eutonic 1.4466). Also given are the ionic strengths (11.45 to 5.20) and pH values (7.21 to 12.92) [1].

References:

[1] Z. G. Karov, N. D. Tkhashokov, M. I. Shavaev (Khim. Tekhnol. Molibdena Vol'frama No. 3 [1976] 126/34; Ref. Zh. Khim. **1977** No. 14B848). – [2] G. Guiter (Compt. Rend. **218** [1944] 406/8). – [3] G. Guiter (Bull. Soc. Chim. France [5] **11** [1944] 537/8).

2.2.2.2 $Li_4MoO_5 \cdot 4H_2O$ (= $2Li_2O \cdot MoO_3 \cdot 4H_2O$)

This compound crystallizes at pH>12.5, at optimum pH 13, from solutions containing LiOH and molybdate. $Li_4MoO_5 \cdot 4H_2O$ forms white, triclinic tablets.

G. Guiter (Compt. Rend. **218** [1944] 406/8; Bull. Soc. Chim. France [5] **11** [1944] 537/8).

2.2.2.3 $Li_2MoO_4 \cdot 0.75H_2O$ (= $Li_2O \cdot MoO_3 \cdot 0.75H_2O$)?

For a compound of this composition experimental (and calculated) solubility values in H_2O are reported: 4.74 (4.74) at 0°C, 4.63 (4.68) at 20°C, and 4.24 (4.20) mol/kg H_2O at 80°C.

S. S. Chin (Zh. Fiz. Khim. **26** [1952] 960/9, 965).

2.2.2.4 Aqueous Solutions of Li_2MoO_4

The enthalpy changes of the following processes were determined in a solution calorimeter: $MoO_3(s) + 2LiOH(aq, 0.2\ mol/L) \rightarrow Li_2MoO_4(aq) + H_2O(l)$ with $\Delta H = -(18.54 \pm 0.04)$ kcal/mol and $Li_2MoO_4(s) \rightarrow Li_2MoO_4(aq)$ with $\Delta H = -(7.61 \pm 0.01)$ kcal/mol [1].

Physical properties of the solutions of Li_2MoO_4 in H_2O were investigated at temperatures from 10 to 98.2°C, the concentrations ranging from 0.01 to 7 mol/L (0.087 to 41.811 wt% Li_2MoO_4). The solubility was found to vary linearly between 45.28 and 41.00 wt% at temperatures from −2.5 to 100°C. The densities of the solutions with concentrations between 0.01 and 7 mol/L are between 0.9971 and 1.4569 g/cm³ at 10°C, between 0.9971 and 1.4533 g/cm³ at 25°C, and between 0.9654 and 1.4295 g/cm³ at 90°C. The electrical conductivity at 10°C is 0.068 to 2.947 $\Omega^{-1} \cdot cm^{-1}$, at 25°C 0.098 to 4.930 $\Omega^{-1} \cdot cm^{-1}$, and at 90°C 0.261 to 16.925 $\Omega^{-1} \cdot cm^{-1}$. The specific electrical conductivity has maximum values at a concentration of 3 mol/L for all temperatures between 10 to 98.2°C. In this temperature range the maximum values of conductivity rise monotonically from 5.273 to 22.275 $\Omega^{-1} \cdot cm^{-1}$; the maximum value at 25°C is 7.668 $\Omega^{-1} \cdot cm^{-1}$. Also given in the same ranges of temperatures and concentrations are the viscosity (0.287×10^{-3} to 17.06×10^{-3} $N \cdot s \cdot m^{-2}$) and the ionic strength (0.014 to 10.526). All the properties are shown in graphs; so are the values for molar volume, equivalent and reduced conductivity. Additionally, values are also tabulated for 10, 20, 30, 40, 50, 60, 70, 80, and 98.2°C. **Fig. 35** shows the pH values and the indices of refraction, n_D, of the Li_2MoO_4 solutions at 25°C. For the density, viscosity, and electrical conductivity of saturated solutions at temperatures from −2.5 to 100°C see the paper [2].

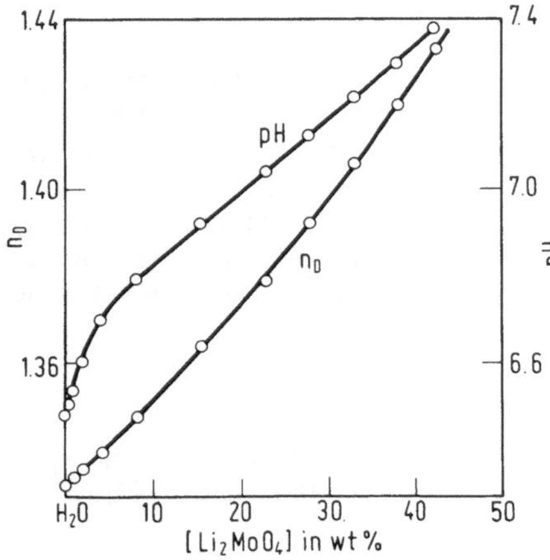

Fig. 35. Index of refraction, n_D, and pH of aqueous Li_2MoO_4 solutions at 25°C [2].

References:

[1] P. A. G. O'Hare, K. J. Jensen, H. R. Hoekstra (J. Chem. Thermodyn. **6** [1974] 681/91). – [2] Z. G. Karov, N. I. Tkhashokov, T. I. Oranova (Khim. Tekhnol. Molibdena Vol'frama No. 3 [1976] 105/25; C.A. **87** [1977] No. 157795).

2.2.2.5 $Li_6[Mo_7O_{24}] \cdot nH_2O$ (= $3Li_2O \cdot 7MoO_3 \cdot nH_2O$), n = 10, 22; Lithium Paramolybdates

$3Li_2O \cdot 7MoO_3 \cdot 22H_2O$. This compound crystallizes from solutions of lithium molybdate, acidified with HCl to pH 3 to 5.6, after 3 to 4 months in white agglomerates. It crystallizes also from solutions acidified with acetic acid to pH 3.4 to 11(?) after more than 12 months [1, 2].

The salt is soluble in water and alcohol, but insoluble in ether [1, 2].

$Li_6[Mo_7O_{24}] \cdot 10H_2O$. This compound was prepared from LiOH and MoO_3 in aqueous solution [4] according to an older description [3] given for the preparation of sodium paramolybdate.

Specimens with various contents of water of hydration are obtained by isothermal dehydration at room temperature over P_2O_5 and H_2SO_4, or by keeping specimens in air at certain temperatures. At 85°C 3.5 H_2O molecules are lost. Further dehydration of the 6.5-hydrate starts at 180°C and takes place in three steps, at 220, 240, and 260°C. For analytical determination of the water content the substance is heated at 500 to 550°C [4].

All crystalline hydrates are free of any form of "constitutional" water and the structures of the polyanions are assumed to be the same as in other paramolybdates [4] (see Fig. 1, p. 9).

According to ^1H NMR spectroscopic investigations there are two groups of water molecules with greatly different bond energies. Approximately 6.5 H_2O molecules are bonded firmly, the remaining 3.5 H_2O molecules form weak bonds and intermingle relatively freely in the structure. The 6.5 H_2O molecules are assumed to enter into the coordination shell of the cation in a definite manner [4].

References:

[1] H. Guiter (Bull. Soc. Chim. France [5] **11** [1944] 537/8). – [2] H. Guiter (Compt. Rend. **218** [1944] 406/8). – [3] A. Rosenheim (Z. Anorg. Allgem. Chem. **96** [1916] 139/81, 143). – [4] V. F. Chuvaev, V. N. Nagovitsyn, V. I. Spitsyn (Zh. Neorgan. Khim. **20** [1975] 1556/60; Russ. J. Inorg. Chem. **20** [1975] 870/3).

2.2.2.6 $[Li_4Mo_6O_{20} \cdot 2nH_2O]_\infty$ (= $Li_2O \cdot 3MoO_3 \cdot nH_2O$), n = 5.7, 7; Lithium "Trimolybdates"

$Li_2O \cdot 3MoO_3 \cdot 7H_2O$. This lithium "trimolybdate" heptahydrate crystallizes from solutions of lithium molybdate, acidified with concentrated acetic acid to pH 2.8 to 3.4, after 2 to 3 weeks [1, 2].

The salt forms white microcrystals isomorphous with the corresponding sodium compound. It is soluble in alcohol, insoluble in ether, and relatively sparingly soluble in H_2O and acetic acid [1, 2].

$[Li_4Mo_6O_{20} \cdot 11.4H_2O]_\infty$ (= $Li_2O \cdot 3MoO_3 \cdot 5.7H_2O$). This hydrate was prepared from a sodium "trimolybdate" solution in water-acetone medium (c_{Mo} = 0.25 M; vol ratio water : acetone = 4 : 1) by ion exchange. Dowex 50 W resin was first used in the hydrogen form, then the resulting solution was passed through the resin in the Li form. The solution was evaporated at 80°C and then kept in a desiccator over P_2O_5 until, in a few days time, the white crystalline product was obtained. The fibrillar crystals were filtered, washed first with ice-cold water, then with a mixture of water with acetone (vol ratio 1:1), and finally with pure acetone. The product was stored in a desiccator over $CaCl_2$ [3]. The compound was characterized by its infrared spectrum [5] and X-ray powder diffraction diagram [3].

Most crystals are multi-twinned and form aggregates of parallel fibers. They are monoclinic, space group P2/n-C^4_{2h} (No. 13) or Pn-C^2_s (No. 7), lattice parameters $a = 20.17 \pm 0.02$, $b = 7.60 \pm 0.01$, $c = 9.56 \pm 0.01$ Å, $\beta = 95.9° \pm 0.1°$; Z = 4. $D_m = 2.56 \pm 0.01$, $D_x = 2.56$ g/cm³. From the Patterson projection P(u0w) a chain structure was deduced. Each link of the chain consists of three MoO_6 octahedra [4].

The DTA and TG curves show endothermal effects at 116, 170, 230, 286, and 350°C corresponding to the loss of 3.0 H_2O at 110°C, 4.7 H_2O at 160°C, 5.3 H_2O at 220°C, and 5.7 H_2O at 280°C (values of n). For analytical determination of the water content the substance was heated at 550°C (the anhydrous compound melts at 559°C). From the X-ray patterns registered during the thermal dehydration (line diagrams are given in the paper) it was concluded that the loss of 4.7 H_2O molecules does not bring about essential changes in the crystal structure. With further dehydration at 220 to 280°C the substance becomes amorphous. At temperatures higher than 350°C a new diffraction pattern appears [3]. Infrared spectra of differently dehydrated samples have been recorded in the 3700 to 400 cm^{-1} region (for plots see the paper) [5].

From the DTA, TG, and X-ray diffraction experiments the formula $Li_2H_2Mo_3O_{11} \cdot 4.7\,H_2O$ with one molecule of constitutional water and 4.7 molecules of water of crystallization was deduced. According to the two steps of dehydration at 220 and 280°C, some of the constitutional water is assumed to exist in the form of coordinated water molecules and another portion in the form of OH groups [3]. The presence of OH groups was confirmed by infrared [5] and ^{1}H NMR [5, 6] spectral methods and the anion was described as $[MoO_{1-\alpha}(OH)_{2\alpha}(H_2O)_{1-\alpha}O(MoO_4)_2^{2-}]_\infty$ ($0 < \alpha < 1$). α is temperature dependent according to ^{1}H NMR experiments in the range −225 to 25°C and decreases with decreasing temperature [5, 6]. At 220°C $\alpha = 0.30$ (by thermogravimetric analysis), at −195°C $\alpha = 0.15$ (from NMR measurements) [5].

In contrast to some of the above structural conclusions, the nearly identical infrared spectra [5, 7, 9] and Raman spectra [7, 8], as well as the nearly identical length of the b axis (which is parallel to the extension of the chain) [7] of the water-containing "trimolybdates" with different cations have been interpreted as showing the same chain-like polymolybdate anion in all "trimolybdates" of the alkali metals [7, 8]. Hence this lithium "trimolybdate" should contain no constitutional water and should have the same chain structure as the rubidium "trimolybdate" monohydrate [7] (see Fig. 7, p. 13).

References:

[1] H. Guiter (Bull. Soc. Chim. France [5] **11** [1944] 537/8). – [2] H. Guiter (Compt. Rend. **218** [1944] 406/8). – [3] S. Hodorowicz, W. Surga (Roczniki Chem. **51** [1977] 411/5). – [4] S. Hodorowicz (Krist. Tech. **13** [1978] K4/K6). – [5] S. Hodorowicz, S. Sagnowski (Acta Phys. Polon. A **50** [1976] 817/21).

[6] S. Hodorowicz, S. Sagnowski (J. Inorg. Nucl. Chem. **40** [1978] 49/51). – [7] H. U. Kreusler, A. Förster, J. Fuchs (Z. Naturforsch. **35b** [1980] 242/4). – [8] K. H. Tytko, B. Schönfeld (Z. Naturforsch. **30b** [1975] 471/84). – [9] B. Schönfeld (Diss. Göttingen, West Germany, 1973, p. 47).

2.2.2.7 $Li_2O \cdot 4MoO_3 \cdot 7H_2O$, Lithium "Tetramolybdate"

The compound crystallizes from solutions of lithium molybdate, acidified with HCl to pH 0.3 to 2, after 4 weeks. It precipitates at pH > 1 in colorless, efflorescent needles, at lower pH as a white mass which turns blue on exposure to light. It is soluble in water and alcohol, and insoluble in ether.

H. Guiter (Bull. Soc. Chim. France [5] **11** [1944] 537/8; Compt. Rend. **218** [1944] 406/8).

2.2.2.8 Lithium Phase C Polymolybdate[1)]

The lithium phase C compound, in admixture with MoO_3, can be obtained from the reaction of hot, concentrated HNO_3 on solid normal or isopolymolybdates of lithium, or by reacting 1N HNO_3 with 0.5M Li_2MoO_4 solution at 100°C for a few minutes. The lithium phase C is very unstable and is entirely converted to MoO_3 by boiling in 1N HCl for 10 min [1]. The lithium phase C polymolybdate could not be obtained by [2].

References:

[1] M. L. Freedman (J. Chem. Eng. Data **8** [1963] 113/6). – [2] H. Peters, L. Till, K. H. Radeke (Z. Anorg. Allgem. Chem. **365** [1969] 14/21).

2.3 Sodium Molybdate Hydrates

Older data are given in "Molybdän", 1935, pp. 214/26.

2.3.1 Sodium Molybdate Hydrates with Mo of Oxidation State < 6

Compounds of this type are not reported.

2.3.2 Sodium Molybdate(VI) Hydrates

2.3.2.1 Na_3HMoO_5 ($=3Na_2O \cdot 2MoO_3 \cdot H_2O$) and $Na_8H_6MoO_{10}$ ($=4Na_2O \cdot MoO_3 \cdot 3H_2O$)

Investigations of fused mixtures of NaOH and Na_2MoO_4 by thermographic and visual-polythermic methods show that the composition $NaOH \cdot Na_2MoO_4$ melts congruently at 516°C and $6NaOH \cdot Na_2MoO_4$ melts incongruently at 325°C. The compositions are considered to be acid salts and described by the formulas Na_3HMoO_5 and $Na_8H_6MoO_{10}$, respectively [1, 2].

References:

[1] V. A. Khitrov, N. N. Khitrova (Dokl. Akad. Nauk SSSR **88** [1953] 853/4; C.A. **1953** 12082). – [2] V. A. Khitrov (Izv. Voronezhsk. Gos. Ped. Inst. **16** [1955] 13/8; C.A. **1959** 3964).

---

1) See the notes added in proof on pp. 7 and 23/4.

2.3.2.2 The Na_2O-MoO_3-H_2O and Na_2MoO_4-NaOH-H_2O Systems

The **Na_2O-MoO_3-H_2O** system (see **Figs. 36** and **37**) has been investigated at 24.1°C. The curve A to D in Fig. 36 passes through the following values:

point	A	B	C	D
Na_2O in wt%	40.6	40.3	32.6	11.82
MoO_3 in wt%	0	0.206	0.465	27.50

For a delineation in an enlarged scale see [1].

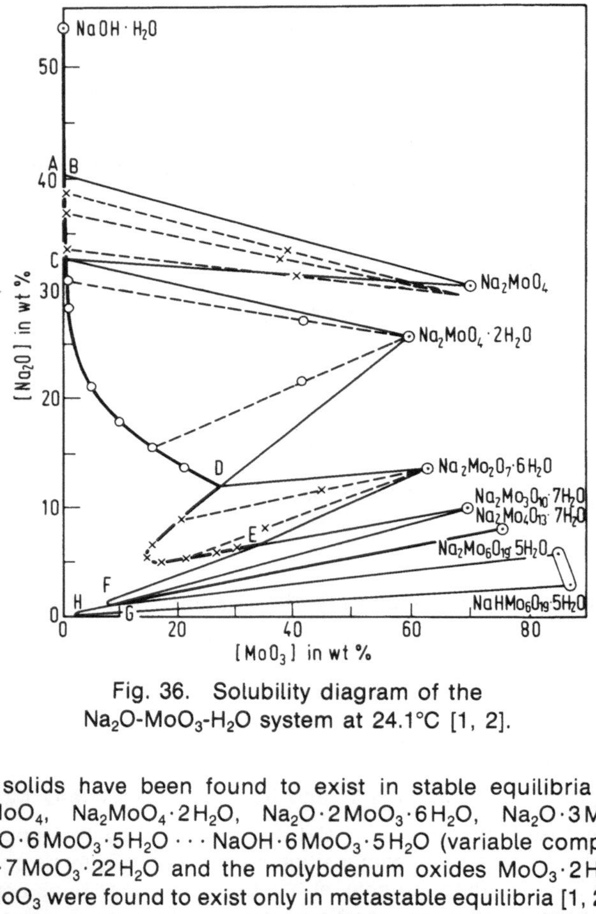

Fig. 36. Solubility diagram of the Na_2O-MoO_3-H_2O system at 24.1°C [1, 2].

The following solids have been found to exist in stable equilibria with the solution: NaOH · H_2O, Na_2MoO_4, $Na_2MoO_4 \cdot 2H_2O$, $Na_2O \cdot 2MoO_3 \cdot 6H_2O$, $Na_2O \cdot 3MoO_3 \cdot 7H_2O$, $Na_2O \cdot 4MoO_3 \cdot 7H_2O$, $Na_2O \cdot 6MoO_3 \cdot 5H_2O$ ··· NaOH · $6MoO_3 \cdot 5H_2O$ (variable composition). The paramolybdate $3Na_2O \cdot 7MoO_3 \cdot 22H_2O$ and the molybdenum oxides $MoO_3 \cdot 2H_2O$ (yellow), $MoO_3 \cdot H_2O$ (white), and MoO_3 were found to exist only in metastable equilibria [1, 2]. $Na_2MoO_4 \cdot 10H_2O$ exists below 10°C, above which temperature $Na_2MoO_4 \cdot 2H_2O$ crystallizes [3]. The transition temperature for the phases $Na_2MoO_4 \cdot 10H_2O$-$Na_2MoO_4 \cdot 2H_2O$ is between 10 and 11°C [4]. This system has to be corrected at least with respect to the identity of the following compounds: $Na_2O \cdot 2MoO_3 \cdot 6H_2O$ has been shown in many cases to be a mixture of $Na_2[MoO_4] \cdot 2H_2O$ and $Na_6[Mo_7O_{24}] \cdot 22H_2O$ [5]; $Na_2O \cdot 6MoO_3 \cdot 5H_2O$ ··· NaOH · $6MoO_3 \cdot 5H_2O$ is most probably $[NaMo_5O_{15}(OH)(H_2O) \cdot H_2O]_\infty$ ($Na_2O \cdot 10MoO_3 \cdot 5H_2O$) (cf. p. 69)[1]. Additionally, it would seem that at least one of the molybdenum oxides must have some stable range of existence [6].

[1] Compare the notes added in proof on pp. 7 and 23/4.

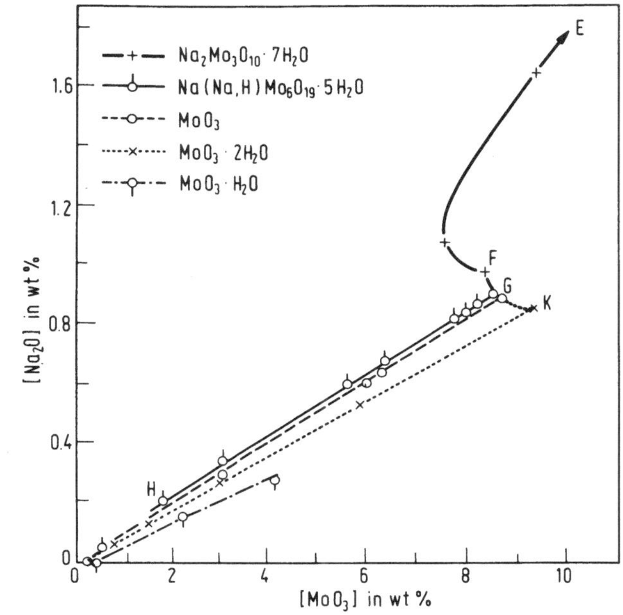

Fig. 37. Solubility diagram of the Na_2O-MoO_3-H_2O system at low concentrations and 24.1°C [1, 2].

There are two largely consistent investigations showing that acidified sodium molybdate solutions spontaneously deposit just five polymolybdates: the paramolybdate $Na_6[Mo_7O_{24}] \cdot 22H_2O$, the "trimolybdate" $Na_2O \cdot 3MoO_3 \cdot nH_2O$, the "tetramolybdate" $Na_2O \cdot 4MoO_3 \cdot 6H_2O$ (yellowish), the 36-molybdate $Na_8[Mo_{36}O_{112}(H_2O)_{16}] \cdot \sim 64H_2O$, and the "decamolybdate" $Na_2O \cdot 10MoO_3 \cdot 5H_2O$ [6, 7][1)]. However, in [6] the 36-molybdate was assumed to be a 19-molybdate (cf. p. 67), and the "decamolybdate" was thought to be $MoO_3 \cdot H_2O$ (cf. p. 69). The value of n for the "trimolybdate" was found to be 4 [8] (in [6] no determination of n was undertaken).

The **Na_2MoO_4-NaOH-H_2O** system has been investigated by measuring the density, viscosity, refractive index, electrical conductivity, and pH at 25°C. Chemical compounds or solid solutions are not formed at this temperature. The transition of $Na_2MoO_4 \cdot 2H_2O$ into Na_2MoO_4 is not clearly expressed by the solubility curve; it seems that the two branches do not intersect, but that they directly converge. The change of the two solids in equilibrium with the solution (corresponding to point C mentioned above) is found to be between the solution compositions 1.54 wt% Na_2MoO_4, 41.06 wt% NaOH and 1.53 wt% Na_2MoO_4, 41.84 wt% NaOH. The other invariant point (point B mentioned above), at which Na_2MoO_4 and $NaOH \cdot H_2O$ are in equilibrium with the solution at 25°C, is determined at 1.35 wt% Na_2MoO_4 and 52.41 wt% NaOH in the solution. (The solubility limit of pure NaOH is 53.43 wt%) [9].

References:

[1] J. Byé (Ann. Chim. [Paris] [11] **20** [1945] 463/550, 516/29). – [2] J. Byé (Bull. Soc. Chim. France [5] **10** [1943] 239/44). – [3] R. G. Samuseva, T. F. Zhatkina, V. E. Plyushchev (Zh. Neorgan. Khim. **14** [1969] 270/6; Russ. J. Inorg. Chem. **14** [1969] 141/3). – [4] J. E. Ricci, L. Doppelt (J. Am. Chem. Soc. **66** [1944] 1985/7). – [5] I. Lindqvist (Nova Acta Regiae Soc. Sci. Upsaliensis [4] **15** No. 1 [1950] 1/22, 6).

1) Compare the notes added in proof on pp. 7 and 23/4.

[6] Y. Sasaki, L. G. Sillén (Arkiv Kemi **29** [1968/69] 253/77, 275). – [7] K. H. Tytko, B. Schönfeld (Z. Naturforsch. **30b** [1975] 471/84). – [8] B. Schönfeld (Diss. Göttingen, West Germany, 1973, p. 46). – [9] M. J. Shavaev, Z. G. Karov, K. S. Sulaimankulov (Izv. Akad. Nauk Kirg. SSR **1982** No. 1, pp. 44/50; C.A. **96** [1982] No. 206275).

2.3.2.3 $Na_2MoO_4 \cdot 10H_2O$ (= $Na_2O \cdot MoO_3 \cdot 10H_2O$)

For the occurrence of this decahydrate in the Na_2O-MoO_3-H_2O system see above. The heat of formation of solid $Na_2MoO_4 \cdot 10H_2O$ is estimated as $\Delta H^\circ_{f,298} = -1078$ kcal/mol (maximal error given as 9%) [1]. On crystallization from aqueous solutions $Na_2MoO_4 \cdot 10H_2O$ may take up a small amount of $Na_2SO_4 \cdot 10H_2O$ in solid solution [2, 3].

References:

[1] P. G. Maslov (Zh. Fiz. Khim. **33** [1959] 1461/6; Russ. J. Phys. Chem. **33** No. 7 [1959] 6/9). – [2] W. E. Cadbury (J. Phys. Chem. **59** [1955] 257/60). – [3] J. E. Ricci, W. F. Linke (J. Am. Chem. Soc. **73** [1951] 3607/12).

2.3.2.4 $Na_2MoO_4 \cdot 2H_2O$ (= $Na_2O \cdot MoO_3 \cdot 2H_2O$)

Preparation. This compound crystallizes from aqueous hydrochloric acid solutions at pH > 1.8; from solutions containing acetic acid at pH > 5.5. Optimum pH for crystallization is 13 to 13.3 (NaOH). Big triclinic crystals, with faces on the order of cm^2, are obtained in crystallization times of up to ten months at 10 to 20°C [1, 2]. According to a patent specification $Na_2MoO_4 \cdot 2H_2O$ is prepared by adding $NaHCO_3$ to an aqueous suspension of MoO_3 until pH 7 to 8 is reached while during this addition the mixture is heated to boiling in order to eliminate CO_2. The solution is filtered, and the compound crystallized at 10 to 20°C and dried at 30 to 60°C [3], see also [4].

From the ore molybdenite (MoS_2) the compound can be prepared by electrolysis. The ore is mixed with NaCl and water and electrolyzed with carbon electrodes at 5 V and 25 to 30°C for 36 h. The resulting black paste is diluted with water, filtered, and the solution is concentrated and $Na_2MoO_4 \cdot 2H_2O$ crystallized. For further details see the patent specification [5].

The enthalpy of formation in the solid state is estimated to be $\Delta H^\circ_{f,298} = -510$ kcal/mol (maximal error 9%) [6]. The free energy of formation is calculated for the crystallized compound as $\Delta G^\circ_{f,298} = -437.6 \pm 0.3$ kcal/mol [7].

Crystallographic Properties. Bonding. From aqueous solutions $Na_2MoO_4 \cdot 2H_2O$ crystallizes in the form of thin transparent plates [8] or rhombic plates with smallest dimension in the [110] direction [9]. Frequent forms are {111} bipyramids and {001} pinacoids; {100} and {010} are also found [9].

$Na_2MoO_4 \cdot 2H_2O$ crystallizes in the orthorhombic system [9, 10]. The following values of the lattice parameters have been determined:

a in Å	8.463(3)	8.458±0.005	8.514±0.002	8.453±0.005	8.57
b in Å	10.552(3)	10.582±0.005	10.562±0.003	10.537±0.005	10.59
c in Å	13.827(6)	13.904±0.005	13.823±0.004	13.825±0.005	13.89
Ref.	[11]	[8]	[12]*)	[10]	[9]

*) Axes designation according to Pbca.

Z = 8 [11]. For d values see tables in [9, 13]. The space group is Pbca-D_{2h}^{15} (No. 61) [8 to 10]. The alternative orientation Pcab was also given [5].

All atoms occupy the general position 8c. The atomic coordinates of the anisotropically refined structure (R = 0.037) are [11]:

atom	x	y	z
Na(1)	0.6565(4)	0.4952(3)	0.4145(2)
Na(2)	0.7424(4)	0.4492(3)	0.6482(2)
Mo	0.5149(1)	0.1982(1)	0.5233(0)
O(1)	0.5573(6)	0.3602(5)	0.5431(4)
O(2)	0.4505(7)	0.1759(5)	0.4013(4)
O(3)	0.6877(7)	0.1090(6)	0.5393(4)
O(4)	0.3716(6)	0.1486(5)	0.6097(4)
$H_2O(1)$	0.2288(7)	0.1416(6)	0.2011(4)
$H_2O(2)$	0.4617(8)	0.4077(6)	0.2996(4)

The crystal structure as shown in **Fig. 38**, p. 56, is considered to consist of alternate layers (parallel to (001)) of MoO_4^{2-} tetrahedra and of water molecules. The MoO_4^{2-} tetrahedra within a layer are connected by the interlinking Na(1) and Na(2) atoms and by the hydrogen bonding O(2)-H_2O(1)-O(4). Adjacent layers are bridged by interlinking cations and by the hydrogen bonding O(2)-H_2O(2)-O(4) linkages. The MoO_4^{2-} anions form slightly deformed tetrahedra, Mo-O distances being between 1.752(6) and 1.788(6) Å; for a detailed listing of distances and angles see the tables in the paper. The O(2) atom, which is coordinated to the Na(2) cation, forms two hydrogen bonds with two water molecules, and O(4) forms two hydrogen bonds and is attached to Na(1). The Na(2)-OH_2(1) distance of 2.297(7) Å is one of the shortest Na-O distances reported. Oxygen atoms from two MoO_4^{2-} anions and two sodium cations form a distorted tetrahedron around H_2O(1) and H_2O(2). This suggests that the two lone pairs of a water oxygen atom point to the two sodium ions, and the O-H···O hydrogen bonding probably is not linear in the case of H_2O(1); it is almost linear in the case of H_2O(2), however. The O-H···O distances (2.78 and 2.87 Å) are in accordance with the O-H stretching frequencies (3250 and 3450 cm^{-1} [14]) reported. The Na-O bond lengths where O belongs to tetrahedral MoO_4^{2-} lie between 2.304(6) and 2.481(7) Å. The oxygen coordinations around the two crystallographically independent Na atoms are remarkably different. Na(1) is surrounded by two water oxygen atoms and four MoO_4^{2-} oxygen atoms, each of which belongs to a different MoO_4^{2-} group. Na(2) has the coordination number 5, being surrounded by two water oxygen atoms and three MoO_4^{2-} oxygen atoms [11]. Earlier, less precise structure investigations were performed by [8, 12].

From measurements of the broad ^{1}H NMR lines of $Na_2MoO_4 \cdot 2H_2O$ (as well as $Na_2WO_4 \cdot 2H_2O$) it is concluded that there are strong hydrogen bonds with a small amplitude of oscillation for the H_2O molecules in the structure. The H-H distance of the water molecules is calculated to be 1.62 Å according to the experimental second moment. This value is somewhat larger than the usual values for hydrate water (1.54 to 1.60 Å) [15].

With the ^{22}Na isotope the exchange between the saturated solution and the crystals of $Na_2MoO_4 \cdot 2H_2O$ has been measured (equilibrium is established within 10 to 15 min). A high value of the rate constant is obtained, which characterizes the bonding force among ions in the crystal structure, and is thought to be primarily due to the highly labile character of the water of crystallization during exchange with the saturated solution [16].

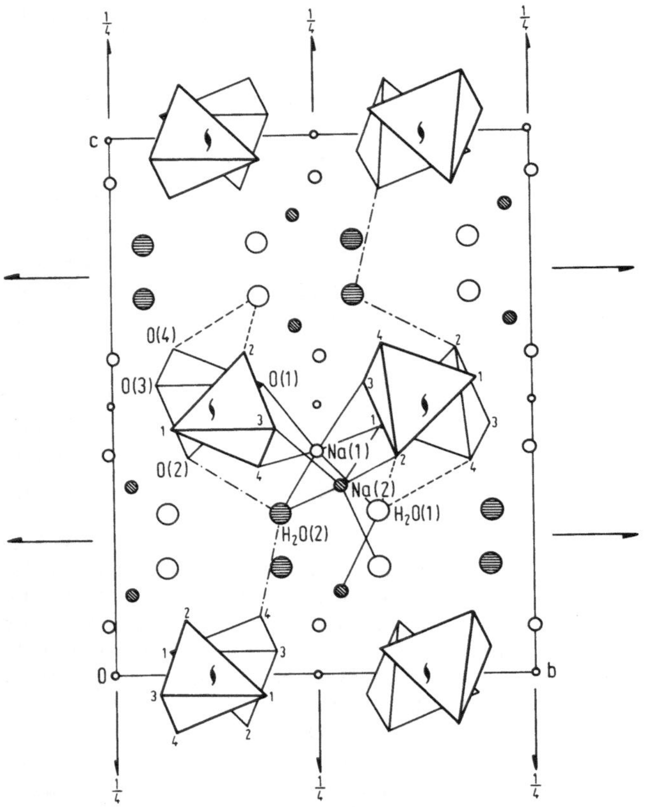

Fig. 38. Crystal structure of $Na_2MoO_4 \cdot 2H_2O$ viewed along [100] [11].

X-ray photoelectron measurements give an Mo $3d_{5/2}$ energy of 231.5 eV (standard Au $4f_{7/2}$ = 83.0 eV) [17]. The following energies of Mo ionization are determined (in eV): $3p_{1/2}$ = 421.8, $3p_{3/2}$ = 404.3, $3d_{3/2}$ = 241.9, $3d_{5/2}$ = 238.85, 4p = 47.4 (standard C1s = 290.0 eV) [18].

Physical Properties. Crystal plates of $Na_2MoO_4 \cdot 2H_2O$ are very brittle [8]. The measured density is 2.551 [8], 2.56 [12], 2.51 at 25°C [9, 11], 2.28 g/cm³ [19]; the calculated density is 2.55 [9], 2.566 [10], 2.60 g/cm³ [11].

The electrical conductivity of pellets (made by pressing the powder at 5.1×10^8 Pa) measured by ac of 1 kHz has considerable magnitude only within a narrow temperature range (at about 370 K), see figure in the paper. In that temperature range the conductivity increases and decreases abruptly, forming only a sharp peak. It is assumed that this happens because the compound dehydrates and dissolves in the water of crystallization liberated at these temperatures, and because dissolved sample ions (in addition to the H^+ and OH^- ions from the water) are present [20]. Also measurements of the dc electrical conductivity and of the dielectric constant of polycrystalline pellets in the temperature ranges 40 to 280°C and 30 to 130°C, respectively, show a peak at 98°C in their temperature variation. On cooling from about 280°C a peak in the dc conductivity is not observed. The conductivity and dielectric constant peaks produced upon heating are also explained by the presence of H^+ and OH^- ions dissociated from H_2O molecules which are released from their lattice sites. The dissociation energy in the solid is found to have decreased to 0.46 eV as compared to its free space value of 1.2 eV [21].

The ^1H NMR spectrum at 80 and 300 K shows the usual doublet of fixed water molecules. Up to 50 to 60 K the lines are readily saturated and spin-lattice relaxation time is expected to be high as in the case of strong bonds. The width of the lines and the second moment are $\delta H = 11$ Oe and $\Delta H^2 = 20$ Oe^2. Both values are equal with those of $Na_2WO_4 \cdot 2H_2O$. In $Na_2MoO_4 \cdot 2H_2O$ the line width is unchanged up to 80 to 110°C (contrary to the case of $Na_2WO_4 \cdot 2H_2O$ where the broad line of fixed water is converted to a narrow line) [15]. NMR absorption spectra of the ^{23}Na central resonance in powder samples were investigated with the hydrous and anhydrous compound [22].

The EPR spectrum of irradiated polycrystalline $Na_2MoO_4 \cdot 2H_2O$ indicates a paramagnetic center, which is attributed to an MoO_4 radical, and is resistant to temperatures up to 140°C. The centers appear after doses of 10^8 R of γ-rays [23].

Crystals of $Na_2MoO_4 \cdot 2H_2O$ are glass-like, white, and show weak birefringence. Indices of refraction are $n_\gamma = 1.578$, n_β (estimated) = 1.576, 2V(−) = 14° to 16° [9].

The Raman and IR absorption spectra of $Na_2MoO_4 \cdot 2H_2O$ differ considerably from those of anhydrous Na_2MoO_4. Absorption at 3300 cm^{-1} is accounted for as an asymmetric stretch and that at 1680 cm^{-1} as water bending vibrations [14]. Absorption bands observed at 3260 and 2220 cm^{-1} are characteristic of the symmetric and antisymmetric vibrations ν(O-H) and δ(H-O-H) of the lattice water; bands at 1800, 1692, and 1680 cm^{-1} are probably water overtone and combination bands, whereas those at 900 and 800 are due to ν(Mo-O) of normal molybdate groups [24]. A band at 2218 cm^{-1} may be accounted for as a combination of 1675 (H_2O bending) and 550 cm^{-1}. The band at 550 cm^{-1} is in the overtone-combination region for the bands below 400 cm^{-1}. A definite analysis cannot be made until data below 300 cm^{-1} are available [14]. Frequencies between 1667 and 817 cm^{-1} are also listed by [25], between 3280 and 820 cm^{-1} by [26], and 645 and 314 cm^{-1} by [27]. For the assignment of the IR bands in the region 800 to 900 cm^{-1} see the discussion of the Raman spectrum below.

From $Na_2MoO_4 \cdot 2H_2O$ powder the following Raman spectrum is observed (in cm^{-1}; w = weak, m = medium, s = strong, vs = very strong, sh = shoulder): 905 m, sh, 897 vs, 843 m, 836 m, 805 w, 325 m, 285, 120, 83 [14]. Also, crystalline state Raman lines at 897 (strong, sharp) and at 839 cm^{-1} (less strong, more diffuse) are reported [28, 29]. The strong line at 897 cm^{-1} is practically coincident in the dihydrate, in aqueous solution (MoO_4^{2-} aq), and in the anhydrous compound (892 cm^{-1}) and corresponds to the $\nu_1(A_1)$ symmetrical stretch vibration of the tetrahedral molybdate ion. The three lines at 843, 836, and 805 cm^{-1} correspond to the $\nu_3(F_2)$, consistent with C_1 or C_2 site symmetry. The line at 905 cm^{-1}, barely detectable in the Raman spectrum but resolved in the infrared, is not accounted for by a tetrahedral ion in C_1 or C_2 site symmetry. It probably is a coordinated-water rocking frequency, which are known to occur in this spectral region. The very asymmetric line centered at 325 cm^{-1} consists of at least two unresolved lines (compare also the IR spectrum in this region; see above) [14].

Chemical Reactions. By heating at 110°C for up to 10 h $Na_2MoO_4 \cdot 2H_2O$ is converted to the anhydrous state, as is shown by the X-ray powder diffraction pattern and by the absence of any proton resonance [22]. According to another work the compound begins to lose water at 90°C and is dehydrated on heating to 146°C [25]. The total weight loss between 90 and 360°C in thermal analysis is determined as 15.1%, and as 14.3% between 10 and 160°C, which corresponds to the removal of two molecules of water [24]. From the results of ^1H NMR measurements $Na_2MoO_4 \cdot 2H_2O$ is dehydrated on heating in the interval 110 to 135°C. Water leaves the structure immediately after cleavage of the corresponding bonds [15]. The peaks at 98°C, observed for dc electric conductivity or dielectric constant of polycrystalline $Na_2MoO_4 \cdot 2H_2O$

pellets versus temperature, are explained by assuming that H_2O molecules released from their lattice sites are still randomly trapped in the structure up to 98°C and simultaneously are partly dissociated into H^+ and OH^- [21]. After heating $Na_2MoO_4 \cdot 2H_2O$ powder at 350°C for 30 min in a stream of He, the white color of the material is maintained and no EPR signal is observed [30].

For the reaction $Na_2MoO_4 \cdot 2H_2O(c) \rightarrow Na_2MoO_4(c) + 2H_2O(g)$ $\Delta G^\circ_{298} = 4.46 \pm 0.02$ kcal/mol is determined by the isopiestic method [7].

Solubility. At 24°C the solubility is determined as 39.4 wt% Na_2MoO_4 [31], 39.44 wt% at 25°C [32]. Also at this temperature 39.42 wt% solubility is obtained as an average value, both from undersaturation and from supersaturation [33]. At 100°C the solubility in H_2O is determined as 45.47 wt%, at 104°C as 45.90 wt%, and at the boiling point of 107°C as 46.32 wt% [34].

Investigations with high concentrations of sodium molybdate at 25°C show that addition of any molybdic oxide to Na_2MoO_4-H_2O compositions causes the formation of a second solid phase, a hydrated dimolybdate ($Na_2Mo_2O_7 \cdot 6H_2O$), and the solubility curve for solutions saturated with the normal molybdate practically does not exist in such cases at 25°C. The solution in equilibrium with both solid phases is nearly coincident with that at the point representing the aqueous solubility of the pure normal molybdate. It contains no detectable concentration of molybdic oxide, but 39.48 wt% Na_2MoO_4 [33]. Later the dimolybdate could not be confirmed, see p. 52.

References:

[1] H. Guiter (Bull. Soc. Chim. France [5] **9** [1942] 622/5). – [2] H. Guiter (Compt. Rend. **216** [1943] 587/9). – [3] O. G. S. Comerzan (Rom. 44239 [1964/66]; C.A. **68** [1968] No. 31643). – [4] O. Comerzan (Rev. Chim. [Bucharest] **18** [1967] 281/2; C.A. **68** [1968] No. 31618). – [5] T. P. Petrovici (Rom. 51889 [1967/69]; C.A. **72** [1970] No. 27828).

[6] P. G. Maslov (Zh. Fiz. Khim. **33** [1959] 1461/6; Russ. J. Phys. Chem. **33** No. 7 [1959] 6/9). – [7] A. P. Zhidikova, I. L. Khodakovskii, M. A. Urusova, V. M. Valyashko (Zh. Neorgan. Khim. **18** [1973] 1160/5; Russ. J. Inorg. Chem. **18** [1973] 612/5). – [8] R. M. Mitra, H. K. L. Verma (Indian J. Chem. **7** [1969] 598/602). – [9] R. J. Poljak, L. N. Becka (Anales Asoc. Quim. Arg. **46** [1958] 199/203). – [10] C. W. F. T. Pistorius (Z. Krist. **114** [1960] 154/5).

[11] K. Matsumoto, A. Kobayashi, Y. Sasaki (Bull. Chem. Soc. Japan **48** [1975] 1009/13). – [12] L. O. Atovmyan, O. A. D'yachenko (Zh. Strukt. Khim. **10** [1969] 504/7; J. Struct. Chem. [USSR] **10** [1969] 416/8). – [13] M. Théodoresco (Compt. Rend. **218** [1944] 233/4). – [14] R. H. Busea, O. L. Keller (J. Chem. Phys. **41** [1964] 215/25). – [15] V. F. Chuvaev, R. A. Gazarov, V. I. Spitsyn (Izv. Akad. Nauk SSSR Ser. Khim. **1974** 1679/83; Bull. Acad. Sci. USSR Div. Chem. Sci. **1974** 1605/9).

[16] S. G. Babayan, S. S. Isakhanyan, L. P. Medvedeva (Arm. Khim. Zh. **23** [1970] 1078/84; C.A. **74** [1971] No. 116668). – [17] S. O. Grim, L. J. Matienco (Inorg. Chem. **14** [1975] 1014/8). – [18] A. Mueller, Ch. K. Joergensen, E. Diemann (Z. Anorg. Allgem. Chem. **391** [1972] 38/53). – [19] R. G. Samuseva, T. F. Zhatkina, V. E. Plyushchev (Zh. Neorgan. Khim. **14** [1969] 270/6; Russ. J. Inorg. Chem. **14** [1969] 141/3). – [20] M. V. Susic, D. M. Minic (Glasnik Hem. Drustva Beograd **46** [1981] 485/91; C.A. **96** [1982] No. 153444).

[21] R. Singh, S. D. Pandey (Proc. Natl. Acad. Sci. India A **50** [1980] 145/52). – [22] G. F. Lynch, S. L. Segel (Can. J. Phys. **50** [1972] 567/72). – [23] M. Constantinescu, I. Pascaru, O. Constantinescu (Rev. Roumaine Phys. **11** [1966] 97/8 from C.A. **64** [1966] 18743). – [24] P. K. Sinhamahapatra, S. Sinhamahapatra, S. K. Bhattacharya (Indian J. Chem. A **15** [1977] 741/2). – [25] C. Duval (Mikrochim. Acta **1962** 947/53).

[26] F. A. Miller, Ch. H. Wilkins (Anal. Chem. **24** [1952] 1253/91, 1258). – [27] F. A. Miller, G. L. Carlson, F. F. Bentley, W. H. Jones (Spectrochim. Acta **16** [1960] 135/235, 155). – [28] E. Darmois, M. Théodoresco (Compt. Rend. **208** [1939] 1308/9). – [29] M. Théodoresco (Compt. Rend. **206** [1938] 753/4). – [30] M. Otake, Y. Komiyama, T. Otaki (J. Phys. Chem. **77** [1973] 2896/903).

[31] J. Byé (Bull. Soc. Chim. France [5] **10** [1943] 239/44). – [32] Z. G. Karov, I. N. Lepeshkov, E. I. Kukulieva (Zh. Neorgan. Khim. **17** [1972] 509/13; Russ. J. Inorg. Chem. **17** [1972] 267/9). – [33] J. E. Ricci, L. Doppelt (J. Am. Chem. Soc. **66** [1944] 1985/7). – [34] W. F. Linke, I. A. Cooper (J. Phys. Chem. **60** [1956] 1662/3).

2.3.2.5 Aqueous Na_2MoO_4 Solution

Aqueous solutions of Na_2MoO_4 are formed by the reaction of Na_2CO_3 solution with $CaMoO_4$, $CuMoO_4$, or $FeMoO_4$, but the reactions proceed slowly owing to the formation of insoluble films [1].

The vapor pressure p over Na_2MoO_4 solutions of various molality has been measured at 294.3°C; p in kg/cm², molality m in mol/kg [2]:

m	0.0	0.25	0.5	1.2	2.7	3.4	4.6	5.7	6.6	7.2
p	81.0	80.5	80.0	77.0	75.0	73.5	69.5	64.5	62.5	59.5

The density of the saturated solution at 25°C is 1.430 g/cm³ [3]. For pH, density, viscosity, index of refraction, and specific electrical conductivity of saturated solutions in the Na_2MoO_4-NaOH-H_2O system at 25°C, see [4].

At 25 ± 0.1°C for a saturated solution (3.16 molal) the activity coefficient $\gamma_\pm = 0.500$ is determined by the isopiestic method. The variation of $\gamma_\pm$ versus molal Na_2MoO_4 concentration is shown in **Fig. 39** [2].

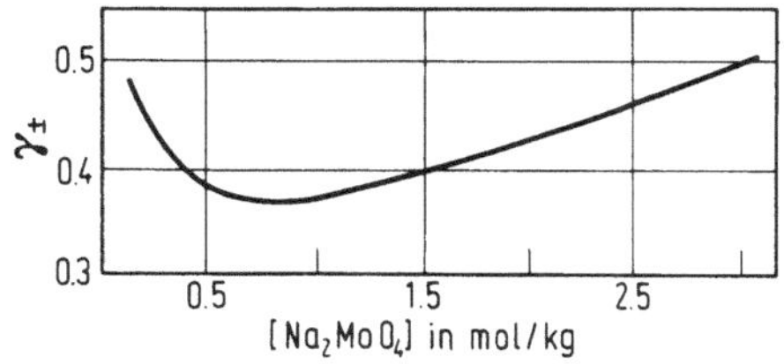

Fig. 39. Activity coefficient $\gamma_\pm$ of aqueous Na_2MoO_4 solutions at 25°C [2].

Lowering the pH of Na_2MoO_4 solutions by adding HCl leads to the formation of polymolybdates [5]. From aqueous solutions with HCl, compounds may be formed as follows: $Na_2MoO_4 \cdot 2HCl$ and $Na_2MoO_4 \cdot 2HCl \cdot 4H_2O$ at 20°C, $Na_2MoO_4 \cdot 2HCl \cdot 2H_2O$ and $Na_2MoO_4 \cdot 6HCl$ at 50°C, see the solubility diagrams of the Na_2MoO_4-HCl-H_2O system at 20 and 50°C in the paper [6].

No other compounds are formed from aqueous solutions of $Na_2MoO_4 \cdot 2H_2O$ with NaF (at 25°C) [7, 8], NaCl (25 and 40°C) [9], Na_2CO_3 (25°C) [10], or $NaNO_3$ [3, 11]. For the adsorption of Na^+ and Cl^- ions on $Na_2MoO_4 \cdot 2H_2O$ dispersed in aqueous solutions see [12]. With aqueous solutions containing Na_2SO_4, solubility relations of sodium molybdate are investigated at 0, 8.5, 12.1, 20.0, and 28.0°C [13], as well as at 35 and 100°C [14]. $Na_2MoO_4 \cdot 2H_2O$ does not dissolve detectable amounts of Na_2SO_4, in contrast to the decahydrate (compare p. 54) [13, 15]. NaSCN in aqueous solutions at 25°C has a pronounced salting-out effect on sodium molybdate. For the Na_2MoO_4-NaSCN-H_2O system at this temperature see [16]. Also salting out takes place for $Na_2MoO_4 \cdot 2H_2O$ in solutions containing carbamide, and the formation of anhydrous Na_2MoO_4 is suppressed in such solutions. From investigations of the Na_2MoO_4-urea-H_2O system reaction in solution cannot be fully ruled out [17]. For the decomposition of H_2O_2 in aqueous solutions of $Na_2MoO_4 \cdot 2H_2O$ see [19].

Investigations of the corrosion inhibition of water-based, metal-working nitrite-amine fluids with additions of $Na_2MoO_4 \cdot 2H_2O$ on cast iron, steel, copper, and brass show that the molybdate may significantly reduce the nitrite requirement of such fluids, and improves the corrosion protection by amine borate and carboxylate salt fluids [18].

References:

[1] A. N. Zelikman, L. V. Belyaevskaya (Zh. Prikl. Khim. **29** [1956] 11/7; J. Appl. Chem. [USSR] **29** [1956] 9/15). – [2] A. P. Zhidikova, I. L. Khodakovskii, M. A. Urusova, V. M. Valyashko (Zh. Neorgan. Khim. **18** [1973] 1160/5; Russ. J. Inorg. Chem. **18** [1973] 612/5). – [3] J. E. Ricci, L. Doppelt (J. Am. Chem. Soc. **66** [1944] 1985/7). – [4] M. I. Shavaev, Z. G. Karov, K. S. Suleimankulov (Izv. Akad. Nauk Kirg.SSR **1982** No. 1, pp. 44/50; C.A. **96** [1982] No. 206275). – [5] G. S. Rao, S. N. Banerji (Proc. Natl. Acad. Sci. India A **21** [1952] 22/8; C.A. **1955** 6715).

[6] G. G. Babayan, A. A. Danielyan, E. A. Kapantsyan (Uch. Zap. Erevan. Gos. Univ. **1977** No. 3, pp. 63/8; C.A. **89** [1978] No. 66189). – [7] Z. G. Karov, R. Kh. Urusova (Khim. Tekhnol. Molibdena Vol'frama **1974** No. 2, pp. 228/37; C.A. **83** [1975] No. 121676). – [8] Z. G. Karov, I. A. Kharkovskii, N. A. Kul'bashnaya (Zh. Neorgan. Khim. **20** [1975] 2821/7; Russ. J. Inorg. Chem. **20** [1975] 1561/4). – [9] Z. G. Karov, V. I. Timchenko (Khim. Tekhnol. Molibdena Vol'frama **1971** No. 1, pp. 158/62; C.A. **81** [1974] No. 159668). – [10] Z. G. Karov, S. B. Semenova, Zh. I. Shorova (Khim. Tekhnol. Molibdena Vol'frama **1974** No. 2, pp. 238/47; C.A. **83** [1975] No. 137711).

[11] Z. G. Karov, F. M. Perel'man (Uch. Zap. Kabardino-Balkarsk. Gos. Univ. **1962** No. 12, pp. 261/75 from C.A. **59** [1963] 2215). – [12] U. V. Seshaiah, S. N. Banerji (Madhya Bharati II A **8** No. 8 [1959] 17/20; C.A. **57** [1962] 10555). – [13] W. E. Cadbury (J. Phys. Chem. **59** [1955] 257/60). – [14] W. F. Linke, J. A. Cooper (J. Phys. Chem. **60** [1956] 1662/3). – [15] J. E. Ricci, W. F. Linke (J. Am. Chem. Soc. **73** [1951] 3607/12).

[16] Z. G. Karov, I. N. Lepeshkov, E. I. Kukulieva (Zh. Neorgan. Khim. **17** [1972] 509/13; Russ. J. Inorg. Chem. **17** [1972] 267/9). – [17] T. E. Kutsenko, L. I. Zhukova, A. L. Rapoport (Ukr. Khim. Zh. **43** [1977] 813/7; Soviet Progr. Chem. **43** No. 8 [1977] 27/9). – [18] M. S. Vukasovich (Lubrication Eng. **36** [1980] 708/12). – [19] M. Sahin, A. R. Berkem (Chim. Acta Turc. **5** [1977] 311/31; C.A. **89** [1978] No. 153361).

2.3.2.6 $2Na_2O \cdot 3MoO_3 \cdot 10H_2O$?

The compound has been obtained from a 0.42 M solution of sodium molybdate, acidified with acetic acid to pH 4.1 to 6.4, after about 10 d as a white powder. The product was merely characterized by analysis.

This compound has never been mentioned by other authors. Moreover, its composition is very unusual. It is, besides the compound $3K_2O \cdot 4MoO_3 \cdot 2H_2O$ described by the same author, the only polymolybdate requiring less than 1, namely 0.67 H^+/MoO_4^{2-}, for its formation.

H. Guiter (Compt. Rend. **216** [1943] 587/9).

2.3.2.7 $Na_2O \cdot 2MoO_3 \cdot nH_2O$, n = 4, 5, 6; Sodium "Dimolybdates"

The hydrates $Na_2O \cdot 2MoO_3 \cdot 6H_2O$ [1 to 3], $Na_2O \cdot 2MoO_3 \cdot 5H_2O$ [4, 5], and $Na_2O \cdot 2MoO_3 \cdot 4H_2O$ [6] have been reported. Lindqvist [7] has repeated the preparations following the descriptions given by [5, 6, 8]. In all cases the analyses gave the correct mole ratio $Na_2O \cdot 2MoO_3$, but X-ray powder diagrams indicated the existence of mixtures of normal and paramolybdate [7]. Investigations of recent date [9, 10] do not report "dimolybdates". However, in the case of the potassium polymolybdates a compound $K_2O \cdot 2MoO_3 \cdot H_2O$ has been obtained apparently by accident (compare p. 78).

The crystallization of the sodium "dimolybdate" is very difficult and requires rather concentrated solutions with $P \approx 0.3$ (calculated from some informations given in [2]). Crystals appear spontaneously only after vigorous shaking for a week and only if the solution is not too acidic [2]. Very small crystals are obtained [1 to 3].

References:

[1] J. Byé (Bull. Soc. Chim. France [5] **10** [1943] 239/44). – [2] J. Byé (Ann. Chim. [Paris] [11] **20** [1945] 463/550, 518/27). – [3] J. E. Ricci, L. Doppelt (J. Am. Chem. Soc. **66** [1944] 1985/7). – [4] Z. Soubarew-Châtelain (Compt. Rend. **208** [1939] 1153/4). – [5] H. Frey (Ann. Chim. [Paris] [11] **18** [1943] 5/60, 26).

[6] H. Guiter (Bull. Soc. Chim. France [5] **9** [1942] 622/5). – [7] I. Lindqvist (Nova Acta Regiae Soc. Sci. Upsaliensis [4] **15** No. 1 [1950] 1/22, 6). – [8] Gmelin Handbuch "Molybdän", 1935, p. 220. – [9] Y. Sasaki, L. G. Sillén (Arkiv Kemi **29** [1968/69] 253/77). – [10] K. H. Tytko, B. Schönfeld (Z. Naturforsch. **30b** [1975] 471/84).

2.3.2.8 $Na_6[Mo_7O_{24}] \cdot nH_2O$ (= $3Na_2O \cdot 7MoO_3 \cdot nH_2O$), n = 4, 10, 14, ~22; Sodium Paramolybdates

The compounds of this type are reported in [1] as $5Na_2O \cdot 12MoO_3 \cdot nH_2O$.

$Na_6[Mo_7O_{24}] \cdot 21$ to $23H_2O$. The compound crystallizes from 1 to 2 M solutions of Na_2MoO_4 acidified with hydrochloric acid or nitric acid to 1.1 to 1.0 H^+/MoO_4^{2-}, pH ≈ 5 to 6 (no pH control necessary), within some hours after acidification [2, 3]. According to [4] the compound is formed by isothermal evaporation at 20°C, the other conditions of the solution being approximately the same. The salt crystallizes with 22 [4 to 6, 8], 21 to 23 [7], 23 [3], or 22 to 23 (recalculated from [1]) H_2O.

The compound is characterized by its Raman spectrum [2] (see Fig. 2, p. 10). The infrared spectrum reported by [8] is identical with that of the 14-hydrate [3] probably due to the loss of water of crystallization on sample preparation (KBr tablet [8]).

The compound is monoclinic, space group P2/c-C_{2h}^4 (No. 13) or Pc-C_s^2 (No. 7). Lattice parameters a = 12.91, b = 10.07, c = 20.14 Å, β = 126.8°; Z = 2, $D_m \approx 2.52$ g/cm³ [7]. From the similarity of the Raman spectra of this 22-hydrate and those of $Na_6[Mo_7O_{24}] \cdot 14H_2O$, $K_6[Mo_7O_{24}] \cdot 4H_2O$, and $(NH_4)_6[Mo_7O_{24}] \cdot 4H_2O$ it is concluded that the polyanion structure of the 22-hydrate is the same as in the other compounds [2] (see Fig. 1, p. 9).

The crystals readily effloresce with formation of the 14-hydrate. The compound is soluble in water without decomposition of the polyanion [2]. Solubility in water: 46.9 g $Na_6[Mo_7O_{24}]$/100 g saturated solution [5].

$Na_6[Mo_7O_{24}] \cdot 14H_2O$. The compound crystallizes from 1 to 2 M solutions of Na_2MoO_4 acidified with hydrochloric acid or nitric acid to 1.1 to 1.0 H^+/MoO_4^{2-}, pH ≈ 5 to 6 (no pH control necessary), on slow evaporation at room temperature after about 7 d [2, 3]. Solutions of the composition 2.04 M MoO_4^{2-} (total) and 2.33 M $HClO_4$ (total) were subjected to slow evaporation at room temperature. Two colorless crystalline phases were obtained within a few days [7]. A 1 M solution of sodium molybdate is acidified with 1.5 N HNO_3 to ~1 H^+/MoO_4^{2-}, diluted with the same volume of ethanol and vigorously shaken for 48 h. The precipitate is collected on a glass filter, washed with ethanol/water (1:1), then with ethanol and dried in the air. For further procedures see the paper [9]. The 14-hydrate is also formed by efflorescence of the 22-hydrate [2, 3].

The compound is characterized by its Raman [2, 12] (see Fig. 2, p. 10) and infrared spectra [3] (see Fig. 3, p. 10). The latter is identical with that reported for the 22-hydrate [8] (see above).

The 14-hydrate crystallizes in two forms: orthorhombic with a long prismatic habit and monoclinic with a short prismatic habit. Crystal data of the orthorhombic form: space group $P2_1ab$-C_{2v}^5 (No. 29); lattice parameters a = 15.626(1), b = 21.130(1), c = 10.377(1) Å; Z = 4. D_m = 2.75, D_x = 2.80 g/cm³ [7].

The crystal structure has been determined from three-dimensional X-ray diffraction data; the atomic coordinates are listed in the paper. The structure of the polyanion consists of $Mo_7O_{24}^{6-}$ blocks composed of seven (distorted) MoO_6 octahedra sharing edges (see Fig. 1, p. 9). The Mo-O distances fall in three ranges, 1.67 to 1.76, 1.88 to 2.01, and 2.11 to 2.33 Å, depending on the coordination number of the oxygen atoms. All Na^+ ions are directly coordinated to $Mo_7O_{24}^{6-}$ blocks, and most of them act as links in O-Na-O bridges in the x, y, and z directions thus forming a three-dimensional network (see **Fig. 40**). Each Na^+ ion is (approximately) octahedrally surrounded by O atoms donated by both H_2O molecules and $Mo_7O_{24}^{6-}$ groups. The NaO_6 octahedra (Na-O distances in the range 2.29 to 2.71 Å) are all coupled together through common edges and corners forming a double-chain arrangement in which the $Mo_7O_{24}^{6-}$ groups are embedded, see figure in the paper. The double chains are approximately parallel with the (100) plane and stretch along the y axis. In the structure there are also O-Na-H_2O-Na-O linkages as well as numerous hydrogen bonds indicated by short H_2O-O and H_2O-H_2O distances. One of the fourteen water molecules is a "free" water molecule not coordinated to any Na^+ ion. It binds to other atoms through hydrogen bonds. For other bond lengths and angles between Mo and O and Na and O see the paper [7].

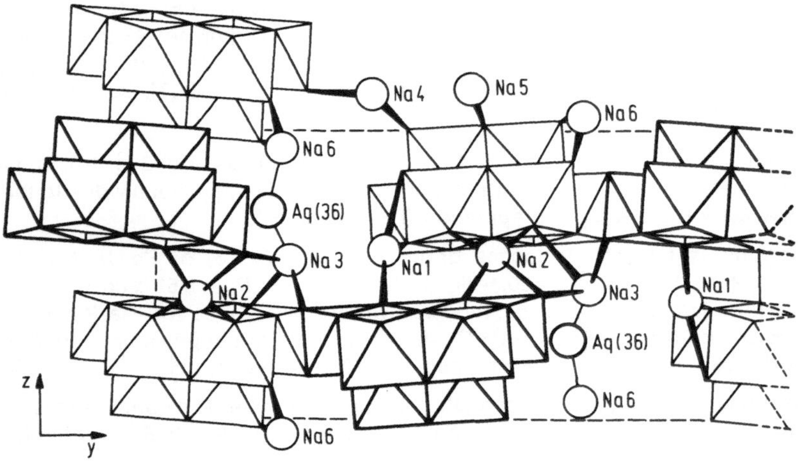

Fig. 40. The O-Na-O connections between the $[Mo_7O_{24}]^{6-}$ anions in the crystal structure of $Na_6[Mo_7O_{24}] \cdot 14H_2O$. Anions at x≈0.7 are drawn with thick lines, anions at x≈0.2 with thin lines [7].

Both the crystal forms are unstable in air [7]. The compound is soluble in water without decomposition of the polyanion [2].

$3Na_2O\cdot 7MoO_3\cdot 10H_2O$. This hydrate has been obtained from a 0.42 M solution of sodium molybdate acidified with conc. acetic acid and adjusted to pH≈4.0 using acetic acid/sodium acetate. The solution was placed at 10 to 20°C in a desiccator over KOH. After 15 d the decahydrate appeared in the form of a white powder [10]. The decahydrate has also been obtained from solutions acidified with hydrochloric acid to pH 5.6 [11]. It is insoluble in alcohol and ether [10].

$Na_6[Mo_7O_{24}]\cdot 4H_2O$. The compound was obtained by adding 10.0 g of MoO_3 to a solution of 2.4 g of NaOH in 50 mL of water. The solution volume was reduced to ca. 10 mL at 50°C, and crystals formed upon cooling the solution to 25°C. The crystals were filtered off and dried in vacuo over P_2O_5 for 24 h [13, 14].

A ^{17}O-enriched compound suited for ^{17}O FT (Fourier transform) NMR investigations was obtained by dissolving the unenriched compound in ^{17}O-enriched water [13].

References:

[1] Gmelin Handbuch "Molybdän", 1935, p. 220. – [2] K. H. Tytko, B. Schönfeld (Z. Naturforsch. **30b** [1975] 471/84). – [3] B. Schönfeld (Diss. Göttingen, West Germany, 1973, pp. 41/5). – [4] I. Lindqvist (Acta Chem. Scand. **2** [1948] 88/9). – [5] J. Byé (Bull. Soc. Chim. France [5] **10** [1943] 239/44).

[6] Y. Sasaki, L. G. Sillén (Arkiv Kemi **29** [1968/69] 253/77, 275). – [7] K. Sjöbom, B. Hedman (Acta Chem. Scand. **27** [1973] 3673/91). – [8] M. J. Schwing-Weill, F. Arnaud-Neu (Bull. Soc. Chim. France **1970** 853/60). – [9] J. Byé (Ann. Chim. [Paris] [11] **20** [1945] 463/550, 529/34). – [10] H. Guiter (Compt. Rend. **216** [1943] 587/9).

[11] H. Guiter (Bull. Soc. Chim. France [5] **9** [1942] 622/5). – [12] G. Johansson, L. Pettersson, N. Ingri (Acta Chem. Scand. A **33** [1979] 305/12). – [13] M. Filowitz, R. K. C. Ho, W. G. Klemperer, W. Shum (Inorg. Chem. **18** [1979] 93/103, 95). – [14] B. Grüttner, G. Jander (in: G. Brauer, Handbook of Preparative Inorganic Chemistry, Vol. 2, Academic, New York 1965, pp. 1710/1 from [13]).

2.3.2.9 $[Na_4Mo_6O_{20}\cdot 2nH_2O]_\infty$ (=$Na_2O\cdot 3MoO_3\cdot nH_2O$), n=1.5, 3, 4, 7; Sodium "Trimolybdates"

$Na_2O\cdot 3MoO_3\cdot 7H_2O$. This heptahydrate occurs in the ternary Na_2O-MoO_3-H_2O system according to [1, 2] as one of the stable solid phases (compare p. 52). To prepare the compound a 1 M solution of the neutral (normal) sodium molybdate was acidified with the theoretical amount of hydrochloric acid (1.33 H^+/MoO_4^{2-}). The mixture was shaken for 48 h, and the crystals were separated from the mother liquor by centrifugation. They consist of very fine needles [1, 2]. This hydrate has also been obtained from a 0.42 M solution of sodium molybdate acidified with conc. acetic acid and adjusted to pH≈3.8 using acetic acid/sodium acetate. The solution was placed at 10 to 20°C in a desiccator over KOH for 4 months where white, star-like crystals precipitated [3, 4]. They are isomorphous with the corresponding lithium compound (see p. 49).

The solubility in water is 8.6 g $Na_2Mo_3O_{10}$/100 g saturated solution [1]. The compound is insoluble in alcohol and ether [3].

$[Na_4Mo_6O_{20} \cdot 8H_2O]_\infty$ ($= Na_2O \cdot 3MoO_3 \cdot 4H_2O$). This "trimolybdate" 4-hydrate was obtained by [5] using the method described for the preparation of the 7-hydrate [1]. The 4-hydrate [6] crystallizes from 1 M sodium molybdate solutions acidified to ~1.3 H^+/MoO_4^{2-} (pH ≈ 4) after several days in the form of a white, voluminous cotton-like product consisting of very fine needles often growing as a bundle radiating from a center [6, 7].

The compound was characterized, apart from analysis, by infrared [5, 6] (see Fig. 9, p. 14) and Raman spectra [6, 7] (see Fig. 8, p. 14). Both spectra are typical "trimolybdate" spectra suggesting this trimolybdate to have the same chain-like polymolybdate anion as the rubidium "trimolybdate" 1-hydrate [16] (see Fig. 7, p. 13).

The compound is slightly soluble in water [7].

$[Na_4Mo_6O_{20} \cdot 6H_2O]_\infty$ ($= Na_2O \cdot 3MoO_3 \cdot 3H_2O$). This hydrate is obtained from a 0.25 M solution of sodium molybdate acidified with HNO_3 to 1.33 H^+/MoO_4^{2-} (pH ≈ 3.3) in the presence of 3 M $NaNO_3$. The first crystals appear after 3 d; crystallization is terminated after 10 to 20 d. The Mo concentration in the mother liquor is 0.076 M. The crystals are washed with an ice-cold solution of 1 M $NaNO_3$ containing some drops of HNO_3 (pH ≈ 5), then with a mixture of 1 M $NaNO_3$ and acetone and finally with acetone/water (3:1) mixture and pure acetone. The downy crystals were stored in weighing bottles in a desiccator over $CaCl_2$ [8].

The crystallization process of the "whiskers" has been studied in dependence of the initial concentration of molybdenum and of the ratio H^+/MoO_4^{2-} at constant ionic strength (3 M $NaNO_3$) and temperature. The changes of pH and of molybdenum concentration in the mother liquor with time were observed. There are three parts in the crystallization curves: an induction period, a period of rapid crystallization, and a period of crystallization restrained by the decrease of concentration [8]. The crystallization proceeds from the surface layer of the solutions or from bubbles of air. The kinetics of the growth of the crystals obey the equation $-dc/dt = k_1c_t(c_t - c_s)$ (k_1 = specific rate constant; c_t, c_s = concentration of molybdenum in the actual solution and the saturated solution, respectively), the solution of which is given by $t = (1/k_1c_s)\{\ln[c_t/(c_t - c_s)] + \ln[(c_0 - c_s)/c_0]\}$. The growth of the "trimolybdate" can be treated as one-dimensional and therefore the area of crystallization surface is constant. The "trimolybdate" crystallizes at the expense of other polymolybdate forms in the solution [9].

The compound was characterized, apart from analysis, by its infrared spectrum and X-ray powder diffraction diagram [10] (for line diagrams see the paper). The former is similar to that of the ("trimolybdate") tetrahydrate [5, 6]. The infrared spectrum of the deuterated substance is also reported [10].

The crystals are fibers a few millimeters in length and 0.1 to 0.3 μm in thickness [8]. They are monoclinic, space group P2/n-C_{2h}^4 (No. 13) or Pn-C_s^2 (No. 7); lattice parameters $a = 19.60 \pm 0.05$, $b = 7.44 \pm 0.04$, $c = 8.55 \pm 0.04$ Å, $\beta = 91.9° \pm 0.1°$; Z = 4. (The data were obtained from a package of fibers of parallel orientation.) $D_m = 2.94 \pm 0.01$, $D_x = 2.93$ g/cm³ [11]. Morphological arguments [10] and the Patterson projection P(u0w) [11] hint that the polyanion has a chain structure, fitting well with the theoretical models [12, 13] dealing with the polymerization process of molybdates [10].

Thermal dehydration processes were studied by TG and DTA measurements, X-ray powder diffraction diagrams, and infrared spectra as a function of temperature. Two water molecules are lost within the range 120 to 150°C, the third at 287°C, accompanied by endothermic effects. The powder diagrams show the formation of a new phase precisely when complete dehydration has occurred. The changes in the infrared spectra are taken as a sign that most of the observed OH groups are bound in the crystal lattice in the form of water molecules [10]. ^1H NMR spectra taken at different temperatures in the range −196 to 27°C and the above dehydration experiments were interpreted in terms of an equilibrium between $Na_2[Mo_3O_{10}(H_2O)] \cdot 2H_2O$ (at low temperatures) and $Na_2[Mo_3O_9(OH)_2] \cdot 2H_2O$ (at normal temperatures) [10]. Further, more detailed broad line NMR investigations led to the conclusion that at −195°C the formula of the "trimolybdate" anion should be written $[MoO_{0.81}(OH)_{0.38}(H_2O)_{0.81}O(MoO_4)_2^{2-}]_n$ [14]. In contrast to some of the above structural conclusions, the typical "trimolybdate" infrared spectrum (see above) suggests this "trimolybdate" to have the same chain-like polymolybdate anion as the rubidium "trimolybdate" 1-hydrate [16] (see Fig. 7, p. 13).

$Na_2O \cdot 3MoO_3 \cdot 1.5H_2O$. The isolated and investigated crystals were obtained from a 0.25 M sodium molybdate solution, acidified with nitric acid to 1.20 H^+/MoO_4^{2-} (pH≈4 to 5) in the presence of 3 M $NaNO_3$, after some days. Crystals of the same appearance occur, however, also in the range 1.15 to 1.3 H^+/MoO_4^{2-} and at lower molybdate concentrations. The crystals were washed thoroughly with water to remove $NaNO_3$ [15]. According to another description an 1.5-hydrate was obtained from a 0.42 M solution of sodium molybdate acidified with 10 N hydrochloric acid to pH 0.8 to 4.5. First, after about 20 d of storage in a desiccator over KOH at 10 to 20°C, colorless, transparent small square rods, soluble in alcohol, crystallized ($Na_2O \cdot 4MoO_3 \cdot >2.5H_2O$). Thereupon a white powder formed which was collected and washed with alcohol and ether [3, 4].

The crystal fibers are a few millimeters long and 0.2 to 2 μm thick. They are orthorhombic; lattice parameters a=15.0, b=9.6, c=7.6 Å; Z=4. Supposed density 3.0 to 3.3 g/cm³ [15].

The solubility in water is ~0.01 M Mo [15]. According to Guiter [3] the compound is soluble in water, insoluble in alcohol and ether [3]. The description of the crystals as a white powder soluble in water is not in accordance with the typical appearance of "trimolybdates".

References:

[1] J. Byé (Bull. Soc. Chim. France [5] **10** [1943] 239/44). – [2] J. Byé (Ann. Chim. [Paris] [11] **20** [1945] 463/550, 516/29, 533/4). – [3] H. Guiter (Compt. Rend. **216** [1943] 587/9). – [4] H. Guiter (Bull. Soc. Chim. France [5] **9** [1942] 622/5). – [5] M. J. Schwing-Weill, F. Arnaud-Neu (Bull. Soc. Chim. France **1970** 853/60).

[6] B. Schönfeld (Diss. Göttingen, West Germany, 1973, pp. 46/8). – [7] K. H. Tytko, B. Schönfeld (Z. Naturforsch. **30b** [1975] 471/84). – [8] J. Chojnacki, S. Hodorowicz (Krist. Tech. **8** [1973] 589/94). – [9] S. Hodorowicz (Krist. Tech. **12** [1977] 431/4). – [10] J. Chojnacki, S. Hodorowicz (Roczniki Chem. **47** [1973] 2213/9).

[11] J. Chojnacki, S. Hodorowicz (Roczniki Chem. **48** [1974] 1399/400). – [12] J. Chojnacki (Proceedings of the Symposium on Theory and Structure of Complex Compounds, London 1964, p. 415). – [13] J. Chojnacki (Bull. Acad. Polon. Sci. Ser. Sci. Chim. **11** [1963] 365/8, 369/74). – [14] S. Hodorowicz, S. Sagnowski (Acta Phys. Polon. A **53** [1978] 185/8). – [15] J. Chojnacka, M. Madejska (Roczniki Chem. **46** [1972] 553/63).

[16] H. U. Kreusler, A. Förster, J. Fuchs (Z. Naturforsch. **35b** [1980] 242/4).

2.3.2.10 $2Na_2O \cdot 7MoO_3 \cdot xH_2O$

This ill-defined compound is formed only in a very small acidity range between the "tri-" and "tetramolybdate" and thus cannot be prepared entirely pure. Analysis: $2Na_2O \cdot 6.44$ to 7.16 $MoO_3 \cdot xH_2O$. The substance was characterized by powder photographs (not reproduced) showing the presence of impurities.

This type of compound has never been described by other authors.

I. Lindqvist (Nova Acta Regiae Soc. Sci. Upsaliensis [4] **15** No. 1 [1950] 1/22, 7).

2.3.2.11 $Na_2O \cdot 4MoO_3 \cdot nH_2O$, n = 2.5, 5, 6, 7; Sodium "Tetramolybdates"

$Na_2O \cdot 4MoO_3 \cdot 7H_2O$. This heptahydrate has been described as one of the stable solid phases in the ternary Na_2O-MoO_3-H_2O system [1, 2] (see p. 52).

The "tetramolybdate" heptahydrate was obtained from a sodium molybdate solution, acidified with hydrochloric acid to 1.5 H^+/MoO_4^{2-}, by vigorous shaking with a 2 N sodium chloride solution [1].

The compound was merely characterized by analysis. It forms very small crystals [1, 2]. The solubility in water is 9.41 g $Na_2Mo_4O_{13}$/100 g saturated solution [2].

$Na_2O \cdot 4MoO_3 \cdot 6H_2O$ (yellowish). For preparation of the hexahydrate 50 g of sodium "trimolybdate" 7-hydrate are dissolved in 50 mL of hot 1 N nitric acid. The solution is boiled for some minutes, filtered, cooled, and subsequently diluted with 200 mL of alcohol and vigorously shaken. After about 15 min crusts are formed which are extracted and crushed with a part of the mother liquor. The crushed crusts are restored to the mother liquor and the whole is again vigorously shaken for about 2 h. The precipitate is collected on a filter and washed with alcohol. Yellowish crystals are obtained this way [1].

Another method starts from a 1 M solution of sodium molybdate which is acidified to 1.55 to 1.6 H^+/MoO_4^{2-} (pH ≈ 2.5); use of the theoretical amount of 1.5 H^+/MoO_4^{2-} yields products mixed with the trimolybdate. After some days fine yellowish crystals precipitate. No other "tetramolybdates" (7-hydrate, white 6-hydrate) were observed [3]. Yellowish crystals of composition $Na_2O \cdot 4MoO_3 \cdot 6H_2O$ have also been obtained from solutions of sodium molybdate with pH 2.0 to 2.9 by [4], but they were assumed to be $Na_3H_3Mo_7O_{24} \cdot xH_2O$.

The compound was characterized, apart from analysis, by its infrared [5] (see Fig. 12, p. 16) and Raman spectra [3, 5] (see Fig. 11, p. 16). These spectra are different from the spectra of "tetramolybdates" formed with other cations, thus showing the presence of different structures of the polymetalate ions [3, 5]. The compound is slightly soluble in water [3].

$Na_2O \cdot 4MoO_3 \cdot 6H_2O$ (white). A further sodium "tetramolybdate" hexahydrate is described as forming colorless crystals. This compound was obtained as follows: 1 mL of 12 N HCl was added to 50 mL of boiling 1 M sodium polymolybdate solution (its composition corresponding to "$Na_4Mo_8O_{26}$") to establish the "octamolybdate" composition ("$Na_2Mo_8O_{25}$"). The solution was kept hot for 1 h and the precipitate ($Na_2O \cdot 10.4MoO_3 \cdot 5.3H_2O$) was then filtered. The filtrate was evaporated to 20 mL and kept at room temperature for 4 d to obtain 1.4 g (15%) of colorless crystals which were identified by X-ray diffraction as $Na_4Mo_8O_{26} \cdot 12H_2O$. (No source for the X-ray diffraction reference is reported.) This colorless hexahydrate is completely dehydrated at 100°C, where its X-ray pattern altered [6].

$Na_2O \cdot 4MoO_3 \cdot 5H_2O$(?). Deposits obtained from sodium molybdate solutions with hydrochloric, nitric, or sulphuric acid at higher acid concentrations, upon aging, were described as mixtures of sodium "tetramolybdate" 5-hydrate and molybdic acid [8]. According to the conditions (acid/molybdate ratio 2.5 to 5:1) used, however, the "decamolybdate" respectively phase C compound (cf. pp. 69/70) would be the appropriate polymolybdate.

$Na_2O \cdot 4MoO_3 \cdot nH_2O$ (n > 2.5). A 0.42 M solution of sodium molybdate is acidified with 10 N HCl to pH 0.8 to 2.0. After about 20 d storage in a desiccator over KOH at 10 to 20°C colorless, transparent small square rods crystallize [7].

The crystals readily effloresce. The number of water molecules has been determined on a product partly effloresced. The compound is soluble in water and alcohol, insoluble in ether. It is the only sodium polymolybdate soluble in alcohol [7].

References:

[1] J. Byé (Ann. Chim. [Paris] [11] **20** [1945] 463/550, 516/34). – [2] J. Byé (Bull. Soc. Chim. France [5] **10** [1943] 239/44). – [3] K. H. Tytko, B. Schönfeld (Z. Naturforsch. **30b** [1975] 471/84). – [4] Y. Sasaki, L. G. Sillén (Arkiv Kemi **29** [1968/69] 253/77, 275). – [5] B. Schönfeld (Diss. Göttingen, West Germany, 1973, p. 51).

[6] M. L. Freedman (J. Chem. Eng. Data **8** [1963] 113/6). – [7] H. Guiter (Bull. Soc. Chim. France [5] **9** [1942] 622/5). – [8] G. S. Rao, S. N. Banerji (Proc. Natl. Acad. Sci. India A **28** Pt. 4 [1959] 182/9; C.A. **1960** 24078).

2.3.2.12 $Na_2[Mo_6O_{19}] \cdot 5H_2O$ (= $Na_2O \cdot 6MoO_3 \cdot 5H_2O$), $Na_2[Mo_6O_{19}] \cdot xH_2O \cdot yC_{12}H_{24}O_6$ ($C_{12}H_{24}O_6$ = 18-crown-6), Sodium Hexamolybdates

$Na_2[Mo_6O_{19}] \cdot 5H_2O$. The method of synthesis is based on ion exchange between polymolybdates and ion-exchange resins in acetone as the solvent, and it consists of the following steps: preparation of a solution of "hexamolybdic acid", preparation of the hexamolybdate solution, and separation of the salt as a crystalline precipitate. The mixture of 5 g of $(NH_4)_6[Mo_7O_{24}] \cdot 4H_2O$, 15 g of air-dried KU-23 ion-exchange resin in the H^+ form, and 2 L of acetone is stirred for 2 h at room temperature. The sorbent and the solid, unreacted ammonium paramolybdate are removed by filtration. 20 g of air-dried cation exchanger in the Na^+ form are added to the solution and allowed to stand in contact with the solution for a day. After separation of the sorbent, the solution is concentrated to a volume of 250 mL by evaporation of the acetone at room temperature under vacuum (5 to 10 mm Hg). After addition of 1 L of ethyl ether to the solution the mixture is shaken several times, and is then allowed to stand for 3 to 4 h. The precipitated salt is separated, immediately washed with ether, and dried in air. Yellow-green crystals are obtained [1].

The compound was characterized by analysis and infrared spectrum (see Fig. 18, p. 19) [1]. From the similarity of the infrared spectra the structure is inferred to be the same as that of the $R_2[Mo_6O_{19}]$ compounds (R = organic cation) [2, 3] (see Fig. 16, p. 18). Assignments of the infrared bands are made [1] (compare pp. 33/4).

$Na_2[Mo_6O_{19}] \cdot xH_2O \cdot yC_{12}H_{24}O_6$. The compound has been prepared by [4], apparently analogously to the compound $(C_{12}H_{24}O_6)_2 \cdot K_2[Mo_6O_{19}] \cdot H_2O$ (see pp. 82/3). No other data are reported.

References:

[1] A. M. Golubev, E. A. Torchenkova, V. I. Spitsyn (Dokl. Akad. Nauk SSSR **217** [1974] 345/7; Dokl. Chem. Proc. Acad. Sci. USSR **214/219** [1974] 495/7). – [2] R. Mattes, H. Bierbüsse, J. Fuchs (Z. Anorg. Allgem. Chem. **385** [1971] 230/42, 236). – [3] H. R. Allcock, E. S. Bissell, E. T. Shawl (J. Am. Chem. Soc. **94** [1972] 8603/4). – [4] O. Nagano, Y. Sasaki (Acta Cryst. B **35** [1979] 2387/9).

2.3.2.13 $Na_8[Mo_{36}O_{112}(H_2O)_{16}] \cdot \sim 64H_2O$ (= $Na_2O \cdot 9MoO_3 \cdot \sim 20H_2O$), Sodium 36-Molybdate

Colorless, transparent crystals of the compound are obtained from a 1 M solution of sodium molybdate acidified with HCl or HNO_3 to ~1.9 H^+/MoO_4^{2-} (pH ≈ 1) [1, 2]. Elevated temperatures, long periods of crystallization (e.g., in case of small concentrations), and slightly increased acidities (> 2.0 H^+/MoO_4^{2-}) favor the formation of the "decamolybdate". The "decamolybdate" is also formed on long standing of the crystals under the mother liquor [4].

The compound was characterized by analysis [2], infrared [3] (see Fig. 21, p. 21) and Raman spectra [1, 2] (see Fig. 20, p. 21). The analytical ratio Na_2O/MoO_3 shows (for reasons open to question) deviations from the theoretical value down to −10%, corresponding to a composition $Na_2O \cdot 9$ to $10\,MoO_3$ (similar observations have been made for the potassium and ammonium compounds). Different compositions of the compound do not, however, affect the spectra [1, 2]. Similar observations have been made on powder photographs of substances [5, 6] obviously identical to this 36-molybdate. The presence of H_3O^+ cations (in addition to Na^+) [5] has been assumed to explain the differences in composition.

The compound crystallizes in monoclinic prisms with space group Cc-C_s^4 (No. 9) or C2/c-C_{2h}^6 (No. 15) and the following lattice parameters: a = 40.2, b = 19.4, c = 33.4 Å, β = 139.2°; Z = 4. D_m = 2.72, D_x = 2.77 g/cm³ [2].

The degree of aggregation of the discrete polyanion was deduced from a determination of the molecular weight of the polymolybdate anion in aqueous solution in combination with the crystal data and the Raman finger-prints of the solid and solution and found to be 36, the compound thus having the formula $Na_8Mo_{36}O_{112} \cdot \sim 80\,H_2O$ [2]. In the light of the X-ray structure determination of the potassium 36-molybdate 36 to 40-hydrate [7, 8] (see p. 84) and considering the identity of the Raman spectra of these compounds, the polymolybdate must now be formulated as $Na_8[Mo_{36}O_{112}(H_2O)_{16}] \cdot \sim 64\,H_2O$ and obviously has the same structure as the potassium 36-molybdate (see Fig. 19, p. 20).

The crystals readily effloresce [1, 2] with conversion into the sodium "decamolybdate" [1]. The compound is readily soluble in water (if it is insoluble, conversion into the "decamolybdate" has occurred) [1].

The literature on this subject prior to 1973 is extremely confusing [1, 5, 6]. There is described a number of different sodium polymolybdates obviously identical to this 36-molybdate [1]. The criteria for identifying these polymolybdates as 36-molybdates are: (1) the variable composition (ratio $M_2O:MoO_3$), often reported [1, 2, 5, 6, 13] to have no influence on the properties (e.g., powder photographs, Raman spectra) of the compounds; (2) the large content of water, the ratio $H_2O:MoO_3$ being ~2; (3) the rapid efflorescence of the crystals; (4) the considerable solubility of the compound and limited solubility of its product of efflorescence; (5) in singular cases finger-prints (e.g., infrared spectra, powder photographs) [4]. In the following some examples are given: $Na_2O \cdot 8\,MoO_3 \cdot 17\,H_2O$, $Na_2O \cdot 10\,MoO_3 \cdot 21(22)\,H_2O$ [9], $Na_2O \cdot 6.96$ to $8.64\,MoO_3 \cdot x\,H_2O$ [5, 6], $Na_2O \cdot 7.6\,MoO_3 \cdot 15.3\,H_2O$ [10], $Na_4Mo_{19}O_{59} \cdot x\,H_2O$ [6]. Some of these identities have been recognized before [1, 5, 6, 8]. However, the "hexamolybdate" of variable composition $Na_{1+\alpha}H_{1-\alpha}Mo_6O_{19} \cdot \sim 5\,H_2O$ [11, 12] has erroneously [1, 5, 6] been placed in this group since (a) it is slightly soluble in water and (b) its water content is too small (thus showing no efflorescence) [4].

References:

[1] K. H. Tytko, B. Schönfeld (Z. Naturforsch. **30b** [1975] 471/84). – [2] K. H. Tytko, B. Schönfeld, B. Buss, O. Glemser (Angew. Chem. **85** [1973] 305/7; Angew. Chem. Intern. Ed. Engl. **12** [1973] 330/2). – [3] B. Schönfeld (Diss. Göttingen, West Germany, 1973, p. 55). – [4] Remark of the reviewer. – [5] I. Lindqvist (Nova Acta Regiae Soc. Sci. Upsaliensis [4] **15** No. 1 [1950] 1/22, 8).

[6] Y. Sasaki, L. G. Sillén (Arkiv Kemi **29** [1968/69] 253/77, 275). – [7] I. Paulat-Böschen (J. Chem. Soc. Chem. Commun. **1979** 780/2). – [8] B. Krebs, I. Paulat-Böschen (Acta Cryst. B **38** [1982] 1710/8). – [9] Gmelin Handbuch "Molybdän", 1935, p. 224. – [10] M. L. Freedman (J. Chem. Eng. Data **8** [1963] 113/6).

[11] J. Byé (Bull. Soc. Chim. France [5] **10** [1943] 239/44). – [12] J. Byé (Ann. Chim. [Paris] [11] **20** [1945] 463/550, 518/28). – [13] M. J. Schwing-Weill, F. Arnaud-Neu (Bull. Soc. Chim. France **1970** 853/60).

2.3.2.14 $[NaMo_5O_{15}(OH)(H_2O)\cdot H_2O]_\infty$ (= $Na_2O\cdot 10MoO_3\cdot 5H_2O$), Sodium "Decamolybdate"[1)]

Crystals of the compound are obtained from a 1 M solution of Na_2MoO_4, acidified with HCl or HNO_3 to 2.0 to 4 H^+/MoO_4^{2-}, at room temperature within some days [1]. The compound crystallizes from solutions of Na_2MoO_4 acidified to 1.9 to 5.7 H^+/MoO_4^{2-} at room temperature or ca. 40°C [2]. The compound is also formed instead of the 36-molybdate when there are some (small) deviations from the procedure described for the preparation of the 36-molybdate (see above) and on efflorescence of the sodium 36-molybdate [1].

The compound was characterized by analysis [1, 2], powder diagram (see the paper) [3], infrared [3] (see Fig. 24, p. 23) and Raman spectra [1, 3] (see Fig. 23, p. 23). The powder diagram and infrared spectrum of the "hexagonal hydrate of molybdenum trioxide" [4] are in reality those of "decamolybdates" [2].

The crystals form very small, colorless, hexagonal columns [1, 2]. The sodium and potassium salts are isomorphous [2]. Hexagonal lattice parameters are a = 10.55, c = 3.68 Å [5] (erroneously stated for a molybdate $Na_2O\cdot 12.20$ to $12.84MoO_3\cdot xH_2O$ [2]).

From the similarity of the powder diagrams [3], infrared [3] and Raman spectra [1, 3] the formula and structure of the sodium "decamolybdate" were derived to be the same as those of the potassium "decamolybdate" (see Fig. 22, p. 22). A direct determination of the crystal structure has also been performed [6]. The compound described as hexagonal hydrate of molybdenum trioxide [4] has been studied, apart from X-ray diffraction analysis and infrared spectroscopy, by NMR spectroscopy, thermogravimetric, and differential thermal analysis in air, hydrogen, and a hydrogen-thiophene mixture [4]. The compound is slightly soluble in water [1].

The literature on this compound prior to 1975 is extremely confusing [1 to 3, 5, 8]. There are described a number of different sodium polymolybdates obviously identical with this "decamolybdate" [1 to 3]. The criteria for identifying these polymolybdates as "decamolybdates" are in particular: (1) the description of the crystal form [1 to 3]; (2) the slight solubility of the "decamolybdate" [1]; (3) the easy conversion of the 36-molybdate into the "decamolybdate" [1] (bearing in mind that the 36-molybdate itself is erroneously described); (4) the ratio H_2O : $MoO_3 \approx 0.5$; (5) in some cases finger-prints (e.g., infrared spectra, powder diagrams [3, 4]). In the following some examples are given: $Na_2O\cdot 8MoO_3\cdot 4H_2O$, $Na_2O\cdot 10MoO_3\cdot 6H_2O$ [9], $Na_2O\cdot 16MoO_3\cdot 9H_2O$ [5, 9], $Na_2Mo_6O_{19}\cdot 5H_2O$ to $NaHMo_6O_{19}\cdot 5H_2O$ [10, 11], $Na_2O\cdot 12.20$ to $12.84 MoO_3\cdot xH_2O$ [5], $Na_2O\cdot 10.4MoO_3\cdot 5.3H_2O$ (termed "phase C") [12], $yNa_2O\cdot MoO_3\cdot xH_2O$ (termed "hexagonal hydrate of molybdenum trioxide") [4], $MoO_3\cdot H_2O$ [8, 13]. Some of these identities have been recognized before [1 to 3, 5, 8, 12].

References:

[1] K. H. Tytko, B. Schönfeld (Z. Naturforsch. **30b** [1975] 471/84). – [2] B. Krebs, I. Paulat-Böschen (Acta Cryst. B **32** [1976] 1697/704). – [3] I. Paulat-Böschen (Diss. Kiel, West Germany, 1974, pp. 92/4, 174, 176). – [4] N. Sotani (Bull. Chem. Soc. Japan **48** [1975] 1820/5). – [5] I. Lindqvist (Nova Acta Regiae Soc. Sci. Upsaliensis [4] **15** No. 1 [1950] 1/22, 8).

[1)] See the notes added in proof on pp. 7 and 23/4.

[6] B. Hedman, R. Strandberg (from [7]). – [7] B. Krebs, I. Paulat-Böschen (Acta Cryst. B **38** [1982] 1710/8). – [8] Y. Sasaki, L. G. Sillén (Arkiv Kemi **29** [1968/69] 253/77, 255/6, 275). – [9] Gmelin Handbuch "Molybdän", 1935, pp. 224/5. – [10] J. Byé (Bull. Soc. Chim. France [5] **10** [1943] 239/44).

[11] J. Byé (Ann. Chim. [Paris] [11] **20** [1945] 463/550, 518/29). – [12] M. L. Freedman (J. Chem. Eng. Data **8** [1963] 113/6). – [13] B. Schönfeld (Diss. Göttingen, West Germany, 1973, p. 56).

2.3.2.15 0.04 to 0.10 $Na_2O \cdot MoO_3 \cdot$ 0.3 to 0.6 H_2O, Sodium Phase C Polymolybdate[1)]

The conditions of preparation may vary within wide limits [1, 2]: concentration of Na_2MoO_4 0.4 to 1.5 N, excess concentration of HNO_3 1.2 to 6.0 N, temperature of precipitation 100°C, time of formation 5 to 10 min [1]; 0.4 to 1 M Na_2MoO_4, concentration of HNO_3 1 to 2.5 N, temperature of precipitation 100°C, time of formation 5 min; solution of $MoO_3 \cdot 2H_2O$ in the presence of 1 M NaCl, 100°C; solution or suspension of $MoO_3 \cdot 2H_2O$ or $MoO_3 \cdot H_2O$ in the presence of 1 M NaCl, room temperature, a few days; slurries of $MoO_3 \cdot 2H_2O$ in 1 M solution of NaCl at 50°C, a few hours; reaction of hot, concentrated HNO_3 on solid sodium isopoly or normal molybdates, 30 min. Yields are 60 to 98%, depending on the conditions. HCl can also be used, but at HCl concentrations greater than 2 N no precipitates are formed due to the high solubility of Mo^{VI} in HCl [2]. High molybdate concentrations and high temperatures favor the formation of the phase C polymolybdate. At lower temperatures and correspondingly longer reaction times $MoO_3 \cdot H_2O$ (white) [1, 2] or $MoO_3 \cdot 2H_2O$ [2] is obtained.

The white products are dried at 40°C and stored away from air and light, since humid products more or less rapidly become blue-colored depending on the conditions of preparation [1]. Other products were dried at 25 or 100°C [2].

The range of homogeneity of the sodium phase C was found to be 0.044 to 0.098 $Na_2O \cdot MoO_3 \cdot$ 0.32 to 0.42 H_2O for a product dried at 100°C to constant weight [2]. Products dried at 40°C showed the formula $0.074Na_2O \cdot MoO_3 \cdot 0.54H_2O$ [1].

The phase C compound was characterized by X-ray powder diagrams (see the papers) [1, 3]. It is bcc, lattice parameter $a = 13.03 \pm 0.02$ [1], 13.01 ± 0.01 Å [1, 3]. The amounts of sodium and water may vary within wide limits without change in the X-ray diffraction pattern [1, 3] thus showing the water in the lattice of the C phase to be unimportant for the structure [1]. However, the cation is an integral part of the phase C structure since it does not exist in its absence [2].

The dehydration diagram of the product $0.074\ Na_2O \cdot MoO_3 \cdot 0.54H_2O$ shows a break at ~130°C where about 50% of the water content has been liberated, indicating loosely adsorbed and firmly bound water to be present in nearly equal amounts. This is confirmed by ^1H NMR investigations. The water is quantitatively lost at ~310°C [1].

Refluxing with 1 N HCl for 6 h reduced the Na_2O content of a sample with 3.27% Na_2O to 0.87% Na_2O and produced a mixture of MoO_3 and phase C, whereas refluxing with 3.5 N HNO_3 reduced the Na_2O content only to 3.01% and left the X-ray pattern unchanged [2]. The compound is readily soluble in alkalies [1].

References:

[1] H. Peters, L. Till, K. H. Radeke (Z. Anorg. Allgem. Chem. **365** [1969] 14/21). – [2] M. L. Freedman (J. Chem. Eng. Data **8** [1963] 113/6). – [3] M. L. Freedman, S. Leber (J. Less-Common Metals **7** [1964] 427/32).

[1)] See the notes added in proof on pp. 7 and 23/4.

2.3.3 Sodium Peroxomolybdate Hydrates

Older data are given in "Molybdän", 1935, pp. 131/2, 225/6.

2.3.3.1 $Na_2MoO_5 \cdot H_2O$?

A yellow compound of this formula (or $Na_2MoO_4 \cdot H_2O_2$), which would be analogous to a corresponding tungsten compound [1], is reported. It is formed in aqueous solution at pH 5.0 from Na_2MoO_4 and H_2O_2 [2].

References:

[1] G. P. Aleeva, L. A. Kotorlenko, V. A. Lunenok-Burmakina (Zh. Strukt. Khim. **13** [1972] 632/6; J. Struct. Chem. [USSR] **13** [1972] 590/3). – [2] F. M. Perel'man, A. K. Verkovskaya (Kinetika Kataliz Akad. Nauk SSSR Sb. Statei **1960** 75/80; C.A. **57** [1962] 10568).

2.3.3.2 $Na_2MoO_6 \cdot H_2O$

A compound $Na_2MoO_6 \cdot nH_2O$ is obtained from Na_2MoO_4 and H_2O_2 at pH 6 [1]. The investigations in the following sections are accomplished with the monohydrate.

The intensely yellow sodium peroxomolybdate has a mole ratio of active oxygen to dry residue of 1 and the water content is ~7 wt% [2]. From its IR spectrum in analogy with peroxotungstates the compound is a true peroxide and not a product of molecular attachment of hydrogen peroxide. IR absorption bands are observed at 910, 860, 835, and 770 cm^{-1}; for a detailed discussion see the paper [3]. Intensive bands in the IR spectrum in the range 860 to 830 cm^{-1} are assigned to the O-O bond. This is confirmed by the shift of bands observed after introducing ^{18}O [4]. For the bond energies of the active oxygen atom see [5].

The heat of decomposition according to the reaction $Na_2MoO_6 \cdot H_2O(s) \rightarrow Na_2MoO_4(aq) + H_2O(g) + O_2(g)$ is 40.00 kcal/mol, measured calorimetrically by decomposition at 95°C [5], while 37.70 kcal/mol was obtained indirectly by reaction with $KMnO_4$ solution [6]. For the reaction $Na_2MoO_6 \cdot H_2O(s) \rightarrow Na_2MoO_4(s) + H_2O(g) + O_2(g)$ 15.40 kcal/mol is calculated by the direct method [5]. It is supposed that decomposition takes place in two successive steps [7]. From investigations with isotopic methods, IR spectroscopy, and EPR it is concluded that oxygen is separated with a 1:1 ratio of broken and preserved peroxide groups. At decomposition in aqueous solution reaction with H_2O_2 also may occur along with the formation of oxygen without rupture of the peroxo bond [4].

References:

[1] F. M. Perel'man, A. K. Verkovskaya (Kinetika Kataliz Akad. Nauk SSSR Sb. Statei **1960** 75/80; C.A. **57** [1962] 10568). – [2] G. V. Kosmodem'yanskaya, K. G. Khomyakov (Zh. Neorgan. Khim. **4** [1959] 2242/3; Russ. J. Inorg. Chem. **4** [1959] 1022). – [3] G. P. Aleeva, L. A. Kotorlenko, V. A. Lunenok-Burmakina (Zh. Strukt. Khim. **13** [1972] 632/6; J. Struct. Chem. [USSR] **13** [1972] 590/3). – [4] G. P. Aleeva, V. A. Lunenok-Burmakina (Tezisy Dokl. Vses. Soveshch. Khim. Neorgan. Perekisnykh Soedin., Riga 1973, pp. 91/2; C.A. **83** [1975] No. 70845). – [5] G. V. Kosmodem'yanskaya, K. G. Khomyakov (Zh. Neorgan. Khim. **4** [1959] 2432/5; Russ. J. Inorg. Chem. **4** [1959] 1117/9).

[6] G. V. Kosmodem'yanskaya, K. G. Khomyakov (Zh. Neorgan. Khim. **4** [1959] 2428/31; Russ. J. Inorg. Chem. **4** [1959] 1115/6). – [7] G. V. Kosmodem'yanskaya, K. G. Khomyakov (Nauchn. Dokl. Vyssh. Shk. Khim. Khim. Tekhnol. **1958** No. 3, pp. 426/9; C.A. **1959** 831).

2.3.3.3 $Na_2MoO_8 \cdot 4H_2O$

This red peroxomolybdate cannot be prepared in a pure form [1]. $Na_2MoO_8 \cdot nH_2O$ (or $Na_2MoO_4 \cdot 4H_2O_2 \cdot nH_2O$) is reported to be formed in aqueous solution from Na_2MoO_4 and H_2O_2 at pH 7.0 [2]. It is a peroxohydrate rather than a true peroxo compound [3]. The mole ratio of active oxygen to dry residue should be 2 for the pure compound, but varies for practical preparations between 1.6 and 1.96. In the latter the peroxomolybdate, which is too unstable to be isolated pure, is mixed with the normal molybdate $Na_2MoO_4 \cdot 2H_2O$, and not with the yellow peroxomolybdate $Na_2MoO_6 \cdot H_2O$ [1]. The water content of the preparations is between 20 and 23 wt% [3].

Two water molecules can be removed by dehydration at 28°C without loss of active oxygen. $Na_2MoO_8 \cdot 4H_2O \rightarrow Na_2MoO_8 \cdot 2H_2O + 2H_2O$ is a slightly endothermal reaction (2.4 kcal/mol). At higher temperatures $Na_2MoO_8 \cdot 4H_2O$ decomposes [3]. The heat of decomposition according to $Na_2MoO_8 \cdot 4H_2O(s) \rightarrow Na_2MoO_4(aq) + 4H_2O(l) + 2O_2(g)$, determined indirectly by reaction with acidified $KMnO_4$ solution, is 77.9 kcal/mol. In the mixtures with $Na_2MoO_4 \cdot 2H_2O$ the heat of decomposition is proportional to the amount of active oxygen they contain [1]. For a mixture with the mole ratio of active oxygen to dry residue of 1.96 and the reaction equation $Na_2MoO_8 \cdot 4H_2O(s) \rightarrow Na_2MoO_4(aq) + 4H_2O(g) + 2O_2(g)$ the heat of decomposition from calorimetric measurement at about 55°C is 75.40 kcal/mol; the corresponding value from the indirect method with $KMnO_4$ solution is 75.80 kcal/mol. For the reaction $Na_2MoO_8 \cdot 4H_2O(s) \rightarrow Na_2MoO_4(s) + 4H_2O(g) + 2O_2(g)$ the heat determined from the indirect method is given as 63.00 kcal/mol. For the bond energies of the active oxygen atoms see the paper [4].

The kinetics of decomposition depend on the water content. For material with molar ratio of oxygen to dry residue of about 1.8 and a water content of 19 wt% at 28.8°C the initial period, in which decomposition is very slow, lasts 1.5 h, whereas for a water content of 22.3 wt% the initial period is decreased, and still more so for a water content of 36 wt% (see the kinetic curves in the paper) [5].

References:

[1] G. V. Kosmodem'yanskaya, K. G. Khomyakov (Zh. Neorgan. Khim. **4** [1959] 2428/31; Russ. J. Inorg. Chem. **4** [1959] 1115/6). – [2] F. M. Perel'man, A. K. Verkovskaya (Kinetika Kataliz Akad. Nauk SSSR Sb. Statei **1960** 75/80; C.A. **57** [1962] 10568). – [3] G. V. Kosmodem'yanskaya, K. G. Khomyakov (Zh. Neorgan. Khim. **4** [1959] 2242/3; Russ. J. Inorg. Chem. **4** [1959] 1022). – [4] G. V. Kosmodem'yanskaya, K. G. Khomyakov (Zh. Neorgan. Khim. **4** [1959] 2432/5; Russ. J. Inorg. Chem. **4** [1959] 1117/9). – [5] G. V. Kosmodem'yanskaya, K. G. Khomyakov (Nauchn. Dokl. Vyssh. Shk. Khim. Khim. Tekhnol. **1958** No. 3, pp. 426/9; C.A. **1959** 831).

2.3.3.4 $Na_2MoO_8 \cdot 2H_2O$

A compound of this composition is prepared from $Na_2MoO_8 \cdot 4H_2O$ by dehydration under reduced pressure at 28°C. For $Na_2MoO_8 \cdot 2H_2O$ the mole ratio of active oxygen to dry residue is 2 (whereas for the tetrahydrate preparations it is <2 because of the instability of this compound, see above). The dihydrate (like the tetrahydrate) is a peroxohydrate rather than a true peroxo compound [1].

$Na_2MoO_8 \cdot 2H_2O$ loses active oxygen at temperatures above about 28°C and decomposes explosively when heated [1]. At 55°C it is not entirely decomposed; however, it decomposes explosively under vacuum [2]. The heat of decomposition according to $Na_2MoO_8 \cdot 2H_2O(s) \rightarrow Na_2MoO_4 \cdot H_2O(s) + H_2O(g) + 2O_2(g)$ is 54.70 kcal/mol (or 57.70 kcal/mol according to the summary in the paper [3]) measured directly at the decomposition temperature of ~75°C. Since a composition $Na_2MoO_4 \cdot H_2O$, as found in the dry residue, is not known as a compound, the residue probably is partially dehydrated $Na_2MoO_4 \cdot 2H_2O$. For the bond energies of the active oxygen atoms in this compound see the paper [3].

On hydrolysis, in contrast to the tetrahydrate, $Na_2MoO_8 \cdot 2H_2O$ does not yield the yellow sodium peroxomolybdate $Na_2MoO_6 \cdot H_2O$ [1].

References:

[1] G. V. Kosmodem'yanskaya, K. G. Khomyakov (Zh. Neorgan. Khim. **4** [1959] 2242/3; Russ. J. Inorg. Chem. **4** [1959] 1022). – [2] G. V. Kosmodem'yanskaya, K. G. Khomyakov (Nauchn. Dokl. Vyssh. Shk. Khim. Khim. Tekhnol. **1958** No. 3, pp. 426/9; C.A. **1959** 831). – [3] G. V. Kosmodem'yanskaya, K. G. Khomyakov (Zh. Neorgan. Khim. **4** [1959] 2432/5; Russ. J. Inorg. Chem. **4** [1959] 1117/9).

2.3.3.5 $Na_2MoO_8 \cdot H_2O$

The formula $Na_2MoO_8 \cdot H_2O$ is stated without any proof, perhaps by analogy with the corresponding tungsten compound [1].

IR absorption bands are given at 915 and 860 cm^{-1}. By analogy with peroxotungstates (see absorption bands and detailed discussion in the paper) it is said to be a true peroxide and not a product of molecular attachment of hydrogen peroxide [1]. Likewise intensive bands in the IR spectrum are reported, which are assigned to O-O bonding. This is confirmed by the shift of bands observed after introducing ^{18}O. Thermal decomposition separates oxygen with a 1:1 ratio of broken and preserved peroxide groups [2].

References:

[1] G. P. Aleeva, L. A. Kotorlenko, V. A. Lunenok-Burmakina (Zh. Strukt. Khim. **13** [1972] 632/6; J. Struct. Chem. [USSR] **13** [1972] 590/3). – [2] G. P. Aleeva, V. A. Lunenok-Burmakina (Tezisy Dokl. Vses. Soveshch. Khim. Neorgan. Perekisnykh Soedin., Riga 1973, pp. 91/2; C.A. **83** [1975] No. 70845).

2.3.3.6 $Na_2Mo_2O_{11} \cdot 2H_2O$

Presumably the compound has been prepared analogously to the corresponding K compound (see p. 87).

Raman bands of this compound, for which the formula $Na_2[Mo_2O_3(O_2)_4(H_2O)_2]$ is given, are measured and assigned as follows:

$\tilde{\nu}$ in cm^{-1}	relative intensity	assignment
958	10	ν(Mo=O)
865	10	ν(O–O), A_1
584	7	ν(MoO_2), B_2, asym. Mo-peroxide stretch
535	7	ν(MoO_2), A_1, sym. Mo-peroxide stretch
353	7	—
324	7	δ(Mo=O)
220	weak	—

The complex involves metal-peroxide isosceles triangles of local symmetry C_{2v}. For a discussion of the assignment see the paper [1], see also [2].

References:

[1] W. P. Griffith, T. D. Wickins (J. Chem. Soc. A **1968** 397/400). – [2] W. P. Griffith (J. Chem. Soc. **1964** 5248/53).

2.3.4 Sodium Lithium Molybdate Hydrate

2.3.4.1 $Na_3Li(MoO_4)_2 \cdot 6H_2O$

A material of this composition and with the point group C_{3v} is reported to be piezoelectric.

E. E. Flint (Trans. All-Union Sci. Res. Inst. Econ. Mineral. [USSR] No. 142 [1939] 50/102, 103 [Engl.] from P. H. Egli, Am. Mineralogist **33** [1948] 622/33, 630; C.A. **1940** 7688).

2.4 Potassium Molybdate Hydrates

Older data are given in "Molybdän", 1935, pp. 229/36.

2.4.1 Potassium Molybdate Hydrates with Mo of Oxidation State <6

Compounds of this type are not reported.

2.4.2 Potassium Molybdate(VI) Hydrates

2.4.2.1 $K_4H_2MoO_6$ ($=2K_2O \cdot MoO_3 \cdot H_2O$)

Investigations of fused mixtures of KOH and K_2MoO_4 by thermographic and visual-polythermic methods show that the composition $2KOH \cdot K_2MoO_4$ melts incongruently at 580°C. It is considered to be an acid salt and is described by the formula $K_4H_2MoO_6$.

V. A. Khitrov, N. N. Khitrova (Dokl. Akad. Nauk SSSR **88** [1953] 853/4; C.A. **1953** 12082).

2.4.2.2 The K_2O-MoO_3-H_2O and K_2MoO_4-KOH-H_2O Systems

The solid phase (in addition to ice) that can be crystallized from aqueous solutions of K_2MoO_4 is found to be anhydrous when it is dried in the open air at room temperature. The solid is evidently anhydrous throughout since the solubility curve is quite smooth. Relations in the K_2MoO_4-H_2O system are shown in **Fig. 41**. The system has a eutectic point at −38°C and 62.7 wt% K_2MoO_4. The boiling point of the solution saturated with 67.6 wt% K_2MoO_4 is 119°C. The solubility (S) in wt% in the steep branch of Fig. 41 is $S=63.82+0.0300\,t$, temperature t in °C. Values selected from a table in the paper are [1]:

t in °C	−18.1	+9.00	17.94	25.00	40.41	64.46	89.96
S in wt%	63.15	64.22	64.46	64.57	65.03	65.72	66.54

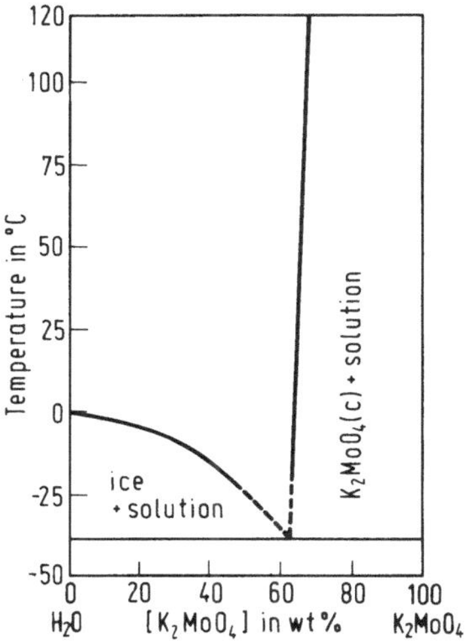

Fig. 41. The K_2MoO_4-H_2O system [1].

Increasingly acidified solutions of potassium molybdate spontaneously deposit the following compounds at room temperature:

compound	P	Z	pH	updated formula	Ref.
K_2MoO_4	0	0	$\gtrsim$8		[2]
$K_6[Mo_7O_{24}]\cdot 4H_2O$	1.0 to 1.1	1.0 to 1.1	~5.5		[2,3]
$K_2O\cdot 3MoO_3\cdot 3H_2O$	1.1 to 1.5	1.1 to 1.5	5.6 to 3.0	$[K_4Mo_6O_{20}\cdot 6H_2O]_\infty$	[2,3]
	1.1 to 1.4				[4]
$K_8Mo_{36}O_{112}\cdot$ ~$80H_2O$	~1.9	~1.78	~1.5	$K_8[Mo_{36}O_{112}(H_2O)_{16}]\cdot$ ~64 or 36 to $40H_2O$	[2,3]
	1.8 to 2.0				[5,6]
$[KMo_5O_{15}(OH)(H_2O)\cdot H_2O]_\infty$	2.0 to 4	~1.8	1.5 to 0.4		[2,3]
	1.9 to 5.7				[7]

See also the note added in proof on p. 82.

In the K_2MoO_4-KOH-H_2O system at 25°C no hydrate of K_2MoO_4 is detected. KOH is neither expected nor observed to form either compounds or solid solutions with the molybdate. Results of experiments with selected compositions (in wt%) between about 72 K_2MoO_4, 1.7 KOH and 16 K_2MoO_4, 33.5 KOH are shown in **Fig. 42**, p.76. Only anhydrous solid K_2MoO_4 is formed, according to algebraic extrapolation of the tie lines. Preliminary experiments, carried out with solid KOH (not carbonate-free), show that the isothermally invariant solution for saturation with KOH and K_2MoO_4 at 25°C is close to the solubility of KOH alone (point (a) in Fig. 42) within about 1% [1].

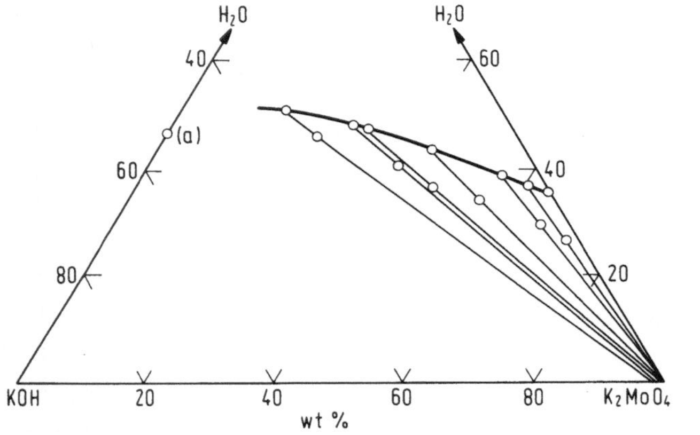

Fig. 42. Part of the K_2MoO_4-KOH-H_2O system at 25°C. Point (a) represents the solubility of KOH alone [1].

References:

[1] J. E. Ricci, F. J. Loprest (J. Am. Chem. Soc. **75** [1953] 1224/6). – [2] K. H. Tytko, B. Schönfeld (Z. Naturforsch. **30b** [1975] 471/84). – [3] B. Schönfeld (Diss. Göttingen, West Germany, 1973, pp. 32, 38/55). – [4] J. Chojnacki, S. Hodorowicz (Roczniki Chem. **49** [1975] 679/82). – [5] I. Paulat-Böschen (J. Chem. Soc. Chem. Commun. **1979** 780/2).

[6] B. Krebs, I. Paulat-Böschen (Acta Cryst. B **38** [1982] 1710/8). – [7] B. Krebs, I. Paulat-Böschen (Acta Cryst. B **32** [1976] 1697/704).

2.4.2.3 $K_2MoO_4 \cdot 5H_2O$ (= $K_2O \cdot MoO_3 \cdot 5H_2O$)?

From studies of the K_2MoO_4-H_2O system (see above) the existence of any hydrated potassium molybdate is doubtful [1]. In spite of this later literature refers to the formula $K_2MoO_4 \cdot 5H_2O$ and experimental data are given with respect to such a compound. $K_2MoO_4 \cdot 5H_2O$ on application of pressure at 25°C shows no degree of dehydration but undergoes a first-order reversible volume transition at 10.3 kbar [2]. IR absorption bands are given as 900 (m) and 825 cm^{-1} (s, br) [3].

References:

[1] J. E. Ricci, F. J. Loprest (J. Am. Chem. Soc. **75** [1953] 1224/6). – [2] R. R. Sood, R. A. Stager (Science **154** [1966] 388/90). – [3] W. P. Griffith, J. Lewis, G. Wilkinson (J. Chem. Soc. **1959** 872/5).

2.4.2.4 Aqueous K_2MoO_4 Solution

The density, viscosity, and electrical conductivity of aqueous solutions of K_2MoO_4 were measured at 15 to 98.2°C and concentrations of 0.119 to 62.951 wt%. The results are represented in detailed tables and diagrams together with the values of molar volume, equivalent and reduced electrical conductivity in the paper [1]. In these ranges of temperatures and concentrations the density values are between 0.9994 and 1.7969 g/cm^3, respectively, at 15°C and between 0.9610 and 1.7525 g/cm^3, respectively, at 98.2°C. The viscosity values are between 1.114×10^{-3} and 7.059×10^{-3} N·s·m^{-2} at 15°C and between 0.287×10^{-3} and 1.496×10^{-3} N·s·m^{-2} at 98.2°C, at the same concentrations, respectively, as above. The electrical conductivity has maximum values at about 47 wt% K_2MoO_4 for all temperatures investigated. The specific conductivity over all is between 15.64 and 60.49 $\Omega^{-1} \cdot cm^{-1}$. The equivalent electrical conductivity shows a change of slope in the concentration dependence for all temperatures investigated at about 5.7 wt% K_2MoO_4 [1].

From aqueous solutions of K_2MoO_4 with K_2SO_4 added no double salt crystallizes at 25°C [2]. Also together with KNO_3 or KIO_3 at 25°C only the separate salts form as solid phases [3].

References:

[1] Z. G. Karov, S. B. Semenova, V. G. Ashkhotov (Khim. Tekhnol. Molibdena Vol'frama **1971** No. 1, pp. 114/28; C.A. **81** [1974] No. 159894). – [2] Z. G. Karov, F. M. Perel'man, R. F. Vasechko (Uch. Zap. Karbadino-Balkarsk. Gos. Univ. Ser. Biol. **1961** No. 10, pp. 237/46 from C.A. **57** [1962] 2920). – [3] J. E. Ricci, F. J. Loprest (J. Am. Chem. Soc. **75** [1953] 1224/6).

2.4.2.5 $(C_{12}H_{24}O_6)_2 \cdot K_2MoO_4 \cdot 5H_2O$ $(= K_2O \cdot MoO_3 \cdot 5H_2O \cdot 2C_{12}H_{24}O_6)$

Colorless crystals were obtained from an aqueous solution of 18-crown-6 and K_2MoO_4 (pH 7.0 to 7.5, 18-crown-6: $K_2MoO_4 = 2:1$).

The crystals are triclinic, space group P1-C_1^1 (No. 1), lattice parameters a = 12.089(2), b = 10.547(2), c = 8.472(2) Å, $\alpha = 113.14(2)°$, $\beta = 77.88(2)°$, $\gamma = 100.39(2)°$; Z = 1. $D_m = 1.50$, $D_x = 1.47$ g/cm^3. The atomic coordinates are listed in the paper, R = 0.053.

The molybdate anion is approximately tetrahedral: Mo-O distances 1.748 to 1.773 Å, O-Mo-O angles 106.5° to 112.5°. Each of the four O atoms of the molybdate anion is linked by a hydrogen bond to at least one H_2O molecule. The water of crystallization is not zeolitic. The H_2O molecules interact with one another as well as with K^+ ions and the MoO_4^{2-} ion. One K^+ ion is coordinated to an 18-crown-6 molecule and 2 H_2O molecules. The other K^+ ion is coordinated to another 18-crown-6 molecule, 1 H_2O molecule, and the MoO_4^{2-} anion. Both K^+ ions are displaced from the mean oxygen planes of the corresponding 18-crown-6 molecules by 0.92 and 0.78 Å, respectively. Distances involving the K^+ ions and the H_2O molecules are listed. For figures of the crystal structure see the paper.

The 18-crown-6 · K^+ cations are considered to be too large, compared with the MoO_4^{2-} anion, to form anhydrous crystals. The crystals are soluble in polar solvents such as CH_3OH and CH_3CN.

O. Nagano (Acta Cryst. B **35** [1979] 465/7).

2.4.2.6 $KHMoO_4$ $(= K_2O \cdot 2MoO_3 \cdot H_2O)$

From experiments with fuel-rich laminar H_2-O_2-N_2 flames at 1815 to 2475 K with additions of K and Mo (10^{11} to 5×10^{13} molecules/mL and mole fractions of 3×10^{-6} to 1.5×10^{-3} $Mo(CO)_6$ vapor or 3×10^{-9} to 3×10^{-6} K_2MoO_4), the formation of the stable species $KHMoO_4$ in the flames is concluded. When only an Mo compound is supplied to a flame without K, no significant quantities of ions are observed. Additions of Mo to K-seeded flames which contain K^+ ions apparently lead to compound formation accompanied by a large reduction of the K concentration. A homogeneous reaction mechanism is assumed, and the enthalpy change and equilibrium constant for the reaction $K + H_2MoO_4 \rightleftharpoons KHMoO_4 + H$ are calculated to be $\Delta H_0^\circ = -46 \pm 40$ kJ/mol and $K = 3.6\ \exp(3500/T)$ [1]. Gaseous $KHMoO_4$ is also formed in H_2-O_2 flames at atmospheric pressure when gaseous MoF_6 and K vapor are added to the gas mixture. From equilibrium intensities of the species at an average flame temperature of 2300 K, a ΔG of −7.2 kcal/mol for the reaction $K(g) + H_2MoO_4(g) \rightleftharpoons KHMoO_4(g) + H(g)$ is obtained. With free energy functions calculated by statistical mechanical methods $\Delta H_{f,298} = -243.0 \pm 5$ kcal/mol is estimated for $KHMoO_4(g)$ [2].

References:

[1] D. E. Jensen, W. J. Miller (13th Symp. Combust., Salt Lake City 1970 [1971], pp. 363/70; C.A. **75** [1971] No. 153480). – [2] M. Farber, R. D. Srivastava (Combust. Flame **20** [1973] 33/42; AD-743514 [1972] 1/29; C.A. **78** [1973] No. 99956).

2.4.2.7 $3K_2O \cdot 4MoO_3 \cdot 2H_2O$?

The compound is reported to crystallize as a white product from 0.5 M solutions of potassium molybdate, acidified with >1 M acetic acid to pH 6.3 (to 11.6?), after standing 2 months in a desiccator at 10 to 20°C over KOH [1, 2].

The product was characterized only by analysis. Its composition, requiring 0.5 H^+/MoO_4^{2-} for its formation, is rather uncommon and is possibly due to a mixture of substances. In a systematic study of the K_2O-MoO_3-H_2O system [3] it was not found.

References:

[1] H. Guiter (Bull. Soc. Chim. France [5] **10** [1943] 261/3). – [2] H. Guiter (Compt. Rend. **216** [1943] 796/8). – [3] K. H. Tytko, B. Schönfeld (Z. Naturforsch. **30b** [1975] 471/84).

2.4.2.8 $[K_4Mo_4O_{14} \cdot 2H_2O]_\infty$ (= $K_2Mo_2O_7 \cdot H_2O$, $K_2O \cdot 2MoO_3 \cdot H_2O$), Potassium "Dimolybdate"

Crystals of the compound were obtained during an attempt to crystallize potassium heptamolybdate.

The crystals are triclinic, space group $P\bar{1}$-C_i^1 (No. 2), lattice parameters $a = 7.635(4)$, $b = 8.906(4)$, $c = 7.647(4)$ Å, $\alpha = 109.44(7)°$, $\beta = 95.73(6)°$, $\gamma = 119.20(7)°$; $Z = 2$. $D_m = 3.24(4)$, $D_x = 3.27$ g/cm^3.

The structure of the polyanion consists of a chain of edge-shared MoO_6 octahedra and MoO_5 square pyramids approximately parallel to the a axis of the crystal. The chain is shown in idealized form in Fig. 25, p. 25. The Mo-Mo distances are Mo(1)-Mo(2) = 3.263(1), Mo(1)-Mo(1′) = 3.360(1), Mo(2)-Mo(2′) = 3.225(1), and Mo(2′)-Mo(1) = 3.826(1) Å. The Mo-O distances in the octahedron range from 1.734 to 2.252 Å, averaging 1.973 Å; in the square pyramid the range is 1.727 to 2.057 Å, averaging 1.871 Å. The potassium ions are situated in interchain positions with closest contacts to a water molecule (2.7 Å) and an oxygen anion (2.5 Å). The water molecule is 2.6 Å from the nearest oxygen anion.

B. M. Gatehouse (J. Less-Common Metals **54** [1977] 283/8).

2.4.2.9 $K_6[Mo_7O_{24}] \cdot nH_2O$ ($= 3K_2O \cdot 7MoO_3 \cdot nH_2O$), n = 4, 22; Potassium Paramolybdates, Heptamolybdates

The compounds of this type are reported in [1] as $5K_2O \cdot 12MoO_3 \cdot nH_2O$.

$3K_2O \cdot 7MoO_3 \cdot 22H_2O$. The compound crystallizes as a white product from 0.5 M solutions of potassium molybdate, acidified with 1 to 10 M hydrochloric acid or >1 M acetic acid to pH 2.8 to 7, after standing 3 weeks in a desiccator at 10 to 20°C over KOH. It was only characterized by analysis [2, 3].

$K_6[Mo_7O_{24}] \cdot 4H_2O$. The crystalline compound is prepared by dissolving molybdenum trioxide in aqueous potassium hydroxide solution, the pH of which is adjusted to 6, and slowly evaporating until the product crystallized [4]. The compound crystallizes from 0.2 M solutions of potassium molybdate acidified with hydrochloric or nitric acid to (at most [7, p. 44]) 1.1 H^+/MoO_4^{2-}, pH ≈ 5 (no pH control necessary) [5]. To prepare the compound, a suspension of 4.23 g of MoO_3 in 800 mL of hot water is added to 3.64 g of K_2CO_3 in 50 mL of water and heated until all the solid has dissolved. The solution is slowly concentrated to 200 mL, cooled, and then the heptamolybdate is precipitated by addition of an equal volume of alcohol [9].

The compound was characterized by its infrared [7, p. 44], [9] (see Fig. 3, p. 10) and Raman spectra [5] (see Fig. 2, p. 10) and X-ray powder diagram [7, p. 84].

The crystals show perfect cleavage parallel to (010), the plane along which the anions form double layers. The heptamolybdate is monoclinic, space group $P2_1/c-C_{2h}^5$ (No. 14), lattice parameters a = 8.1318(4), b = 35.6097(16), c = 10.3376(6) Å, β = 115.397(5)°; Z = 4. D_m = 3.23(1), D_x = 3.41 g/cm³ [4]. (Note: The values given in the text differ markedly from those in the summary.) For short preliminary data see [8]. The atomic coordinates are listed, R = 0.120. No attempt was made to locate hydrogen atoms [4].

The structure of the polyanion consists of $Mo_7O_{24}^{6-}$ blocks composed of seven MoO_6 octahedra sharing edges (see Fig. 1, p. 9, and **Fig. 43**). The $Mo_7O_{24}^{6-}$ blocks are held together by the K^+ ions. The O coordination around the K^+ ions is irregular; it is 8- to 10-fold, where the water oxygens also form part of the coordination polyhedron. Each water molecule bridges a pair of K^+ ions. There are two types of cations and two types of crystal water. For the different bond lengths (1.71 to 2.42 Å) and bond angles between Mo and O, for the different types of Mo-O bonds (including π bonds) and for the cation-oxygen distances (2.62 to 3.49 Å) see the papers [4, 8]. The dihedral Mo-Mo-Mo angle is 163.1° [8].

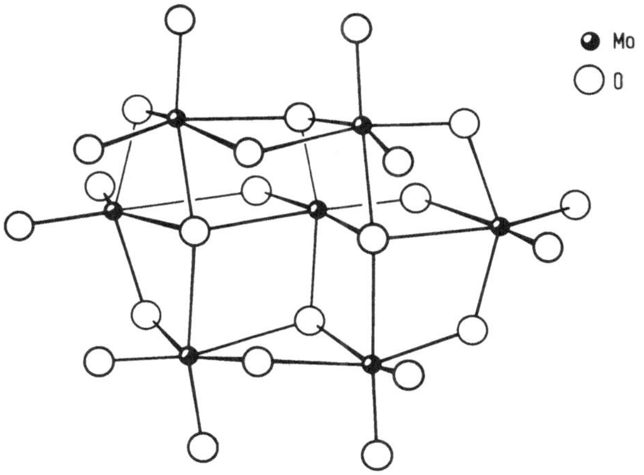

Fig. 43. Structure of the heptamolybdate ion $[Mo_7O_{24}]^{6-}$ showing bonding between atoms and strong distortion of the MoO_6 octahedra [4].

Thermogravimetric analysis indicates that 2 H_2O are lost at ~115°C and 2 at ~150°C [4, 8]. Other investigations showed that all water of crystallization is lost between 140 and 190°C. At 250°C a weak exothermic reaction indicates decomposition to potassium "di-" and "trimolybdates" [9]. According to yet another investigation (TG, DTA, and ^1H NMR spectroscopy) all the water molecules form bonds of equal strength and the anhydrous compound is formed in a single step at 160 to 170°C. The 4-hydrate melts at 170°C. (For analytical determination of the water content the substance is heated at 500 to 550°C.) The molecules of water of hydration are isolated and rigidly fixed in the structure. The bond energy of the water molecules is mainly determined by their interaction with the polyanion and is independent of their position relative to the anion [6].

The compound is soluble in water without decomposition of the polymolybdate ion [5].

References:

[1] Gmelin Handbuch "Molybdän", 1935, pp. 232/3. – [2] H. Guiter (Bull. Soc. Chim. France [5] **10** [1943] 261/3). – [3] H. Guiter (Compt. Rend. **216** [1943] 796/8). – [4] H. T. Evans Jr., B. M. Gatehouse, P. Leverett (J. Chem. Soc. Dalton Trans. **1975** 505/14). – [5] K. H. Tytko, B. Schönfeld (Z. Naturforsch. **30b** [1975] 471/84).

[6] V. F. Chuvaev, V. N. Nagovitsyn, V. I. Spitsyn (Zh. Neorgan. Khim. **20** [1975] 1556/60; Russ. J. Inorg. Chem. **20** [1975] 870/3). – [7] B. Schönfeld (Diss. Göttingen, West Germany, 1973, pp. 44, 84). – [8] B. M. Gatehouse, P. Leverett (Chem. Commun. **1968** 901/2). – [9] A. La Ginestra, G. Rubino (Atti Accad. Nazl. Lincei Classe Sci. Fis. Mat. Nat. Rend. [8] **41** [1966] 510/20).

2.4.2.10 $[K_4Mo_6O_{20}\cdot 6H_2O]_\infty$ (= $K_2O\cdot 3MoO_3\cdot 3H_2O$), Potassium "Trimolybdate"

The compound crystallizes from 0.2 to 0.25 M solutions of potassium molybdate acidified with hydrochloric or nitric acid to 1.33 H^+/MoO_4^{2-} [1, 2]. The pH of the solution is ~4 [2]. Crystallization is accelerated by working in a nearly saturated solution of KNO_3 [1]. The degree of acidification is not critical and may range from 1.1 to 1.4 [1], or 1.1 to 1.5 [2, 4] H^+/MoO_4^{2-}. The crystals are filtered from the mother liquor, washed with an ice-cold solution of 1 M KNO_3 of pH ≈ 5, then with a mixture of 1 M KNO_3 and acetone, and finally with pure acetone [1]. The final product is a fluffy powder [1, 2] and is stored in a desiccator over $CaCl_2$ [1]. According to another specification, a suspension of 13.8 g of MoO_3 in 1 L of hot water is added to 4.6 g of K_2CO_3 in 59 mL of water and heated until all solid has dissolved. The compound precipitates on cooling [3].

Microscopic observations show fibrillar crystals, best developed in the solution of Z = 1.33 [1] (a few millimeters in length and 0.1 to 4 μm in thickness [1, 4]). Most crystals are multi-twinned and form aggregates of parallel-oriented fibers [4]. (Moreover, in solutions of Z = 1.40 plate-like crystals are also formed [1].) The mechanism of crystal growth is assumed to have the diffusion-dislocation character [4].

The compound was characterized by analysis [1 to 3, 7], infrared spectrum [3, 5 to 7, 14] (see Fig. 9, p. 14), Raman spectrum [2] (see Fig. 8, p. 14), and X-ray powder diffraction diagram (see the paper) [1]. It is presumably identical with the compound described in the Gmelin Handbook [8] although its morphology is quite different [1].

The crystals are orthorhombic, space group Pmnn-D_{2h}^{12} (No. 58) or P2nn-C_{2v}^{10} (No. 34), lattice parameters a = 13.72 ± 0.01, b = 7.66 ± 0.02, c = 12.00 ± 0.01 Å; Z = 4. D_m = 3.06 ± 0.01, D_x = 3.05 g/cm³ [4, 5]. From the P(u0w) Patterson projection a chain structure perpendicular to the projection is deduced. Each link of the chain consists of three MoO_6 octahedra [4].

Thermogravimetric measurements, DTA investigations, and X-ray powder diffraction diagrams show at 150°C [1], or between 115 to 160°C [3], the loss of the three water molecules (formation of $K_2O\cdot 3MoO_3$) [1, 3] accompanied by an endothermic effect and structural changes, and at 270 and 335°C two additional endothermic effects related to structural changes. This behavior indicates that the three water molecules are crystallographically equivalent to one another and that the initial structure is completely disrupted upon dehydration. In this respect the potassium compound is not a true homologue of the corresponding sodium compound but is assumed to be a hydrated homologue of the chain potassium polymolybdate $K_2Mo_3O_{10}$ (cf. [9]) [1]. The compound contains a hydrogen bond system. The bonding of the water molecules is assumed to be of zeolitic character [5].

From ^1H NMR spectra of a polycrystalline sample in the temperature range 77 to 370 K the presence of two kinds of hydrogen (OH groups and H_2O molecules) is deduced and the coexistence of the two anions $[Mo(OH)_2O(MoO_4)_2^{2-}]_n$ and $[MoO(H_2O)O(MoO_4)_2^{2-}]_n$ assumed. At 77 K the formula of the "trimolybdate" anion is written as $[MoO_{0.84}(OH)_{0.32}(H_2O)_{0.84}O(MoO_4)_2^{2-}]_n$ [10, 11]. At 323 K the structure of the polymolybdate anion changes completely according to $K_2Mo_3O_{10-\alpha}(OH)_{2\alpha}(H_2O)_{1-\alpha}\cdot 2H_2O \rightleftharpoons K_2Mo_3O_7(OH)_6$. This type of transition is not observed in the other chain trimolybdates due to the different ability of cations to polarize water molecules in $A^-\cdots aq\cdots M^+\cdots aq\cdots A^-$ bond systems in the crystal structure [10].

In contrast to some of the above structural conclusions, the nearly identical infrared [6, 12] and Raman [2, 12] spectra of the water-containing "trimolybdates" with different cations have been interpreted as showing the same chain-like polymolybdate anion in all "trimolybdates" of the alkali metals [2, 12] (see Fig. 7, p. 13). This is confirmed by the identical length of the b axis (which is parallel to the extension of the chain) of the potassium, lithium, and rubidium compounds [12].

The compound is sparingly soluble in water, with decomposition. The solution contains the polymolybdate ions usually occurring in aqueous polymolybdate solutions at P=1.33, i.e., predominantly protonated $[Mo_7O_{24}]^{6-}$ ions [2] (compare "Molybdenum" Suppl. Vol. B3, to be published).

Potassium "trimolybdate" reacts with diazonium fluoroborates in aqueous solution (suspension?) at pH 4.3 to 4.4 to give diazonium "trimolybdates" $[XC_6H_4N_2]HMo_3O_{10}$ (X = p-CH_3, p-OCH_3, p-Cl, p-NO_2) and with hydrochlorides of aromatic amines at pH 4.2 to 4.7 to give arylammonium "trimolybdates" $[XC_6H_4NH_3]_2Mo_3O_{10}$ (X = H, p-CH_3, p-OCH_3, p-Cl) in yields of 50 to 75% [13]. (If there is really a solution, these reactions are not reactions of the "trimolybdate" ion, as such, since in solution such an ion does not exist [2], see above; compare also p. 45.)

References:

[1] J. Chojnacki, S. Hodorowicz (Roczniki Chem. **49** [1975] 679/82). – [2] K. H. Tytko, B. Schönfeld (Z. Naturforsch. **30b** [1975] 471/84). – [3] A. La Ginestra, G. Rubino (Atti Accad. Nazl. Lincei Classe Sci. Fis. Mat. Nat. Rend. [8] **41** [1966] 510/20). – [4] S. Hodorowicz (Krist. Tech. **13** [1978] K4/K6). – [5] S. Hodorowicz (Roczniki Chem. **50** [1976] 1031/3).

[6] B. Schönfeld (Diss. Göttingen, West Germany, 1973, p. 47). – [7] M. J. Schwing-Weill, F. Arnaud-Neu (Bull. Soc. Chim. France **1970** 853/60). – [8] Gmelin Handbuch "Molybdän", 1935, pp. 233/4. – [9] Gmelin Handbuch "Molybdän" Erg.-Bd. B 1, 1975, pp. 220/1. – [10] S. Hodorowicz, S. Sagnowski (Inorg. Chim. Acta **27** [1978] L69/L70).

[11] S. Hodorowicz, S. Sagnowski (Acta Phys. Polon. A **53** [1978] 185/8). – [12] H. U. Kreusler, A. Förster, J. Fuchs (Z. Naturforsch. **35b** [1980] 242/4). – [13] V. G. Smirnova, V. V. Kozlov (Zh. Obshch. Khim. **46** [1976] 105/8; J. Gen. Chem. [USSR] **46** [1976] 107/9). – [14] F. Arnaud-Neu, M. J. Schwing-Weill (Bull. Soc. Chim. France **1973** 3225/32).

2.4.2.11 $K_4[Mo_8O_{26}]\cdot nH_2O$ (= $K_2O\cdot 4MoO_3\cdot 0.5nH_2O$), n = 6, x; Potassium Octamolybdates, "Tetramolybdates"

β-$K_4[Mo_8O_{26}]\cdot xH_2O$. The compound is formed by mixing a solution of α-$[(n\text{-}C_4H_9)_4N]_4[Mo_8O_{26}]$ in acetonitrile with an aqueous solution of KCl. Heating should be avoided since the solution reacts to form $[Mo_6O_{19}]^{2-}$ when heated.

The substance has been characterized by analysis, infrared and Raman spectroscopy; however, no data or spectra were reported. The value of x in the formula is also not reported.

The structure of the polymolybdate ion (see Fig. 10, p. 16) has been deduced by (fingerprint) Raman and infrared spectroscopy in comparison with other compounds containing the β-$[Mo_8O_{26}]^{4-}$ ion [1].

$K_2O\cdot 4MoO_3\cdot 3H_2O$ (β-$K_4[Mo_8O_{26}]\cdot 6H_2O$?). To prepare the compound, HCl is added to a suspension containing 40 g MoO_3/L dissolved in KOH until pH = 2.2. The tetramolybdate precipitates immediately [2].

The compound was characterized by analysis and infrared spectrum (see the paper) [2]. The IR spectrum [2] shows some similarities to that of the compound $(NH_4)_4[Mo_8O_{26}]\cdot 4(5)H_2O$ [3] which is known to have the β-$[Mo_8O_{26}]^{4-}$ (Lindqvist) structure (cf. p. 110 and Fig. 10, p. 16). Accordingly the potassium "tetramolybdate" can also be assumed to be a β-octamolybdate.

Thermogravimetric measurements and DTA investigations indicate between 60 and 110°C the loss of $2.5H_2O$ and between 150 and 190°C the loss of $0.5H_2O$. The anhydrous salt melts at 559°C. Over a period of 6 months, $1.5H_2O$ are lost spontaneously at room temperature [2].

This compound is the only well-established example of the preparation of a potassium "tetramolybdate" from an aqueous solution. Other authors [4, 5] report of unsuccessful attempts to prepare potassium "tetramolybdates" since under the appropriate conditions the "trimolybdate" $K_2O\cdot 3MoO_3\cdot 3H_2O$ is always formed.

Note added in proof: The preparation and identity of this compound have recently [6] been confirmed using the method described by [2]. Raman and infrared spectra clearly show the presence of a β-octamolybdate. The decisive factor in the preparation procedure apparently is the high molybdate concentration used by [2]. This compound is the only β-octamolybdate with a cation other than ammonium or organo-ammonium prepared from (pure) aqueous solution [6].

References:

[1] W. G. Klemperer, W. Shum (J. Am. Chem. Soc. **98** [1976] 8291/3). – [2] A. La Ginestra, G. Rubino (Atti Accad. Nazl. Lincei Classe Sci. Fis. Mat. Nat. Rend. [8] **41** [1966] 510/20). – [3] B. Schönfeld (Diss. Göttingen, West Germany, 1973, p. 51). – [4] K. H. Tytko, B. Schönfeld (Z. Naturforsch. **30b** [1975] 471/84). – [5] J. Chojnacki, S. Hodorowicz (Roczniki Chem. **49** [1975] 679/82).

[6] G. Baethe (Diss. Göttingen, West Germany, 1985, pp. 94/6, 154/5).

2.4.2.12 $(C_{12}H_{24}O_6)_2\cdot K_2[Mo_6O_{19}]\cdot H_2O$ (= $K_2O\cdot 6MoO_3\cdot H_2O\cdot 2C_{12}H_{24}O_6$), Hexaoxacyclooctadecane (18-Crown-6) Complex of Potassium Hexamolybdate Monohydrate

The compound is prepared from an aqueous solution of 18-crown-6 and K_2MoO_4 (the ratio 18-crown-6 : K_2MoO_4 = 1.3), acidified to pH 1.0 and heated on a steam-bath for 1 h. Repeated recrystallization of the precipitate from CH_3CN gives lemon-yellow crystals.

The crystals are orthorhombic, space group $P2_12_12_1$-D_2^4 (No. 19); lattice parameters a = 11.678(3), b = 11.361(3), c = 35.84(1) Å; Z = 4. D_x = 2.101 g/cm³. The atomic coordinates are listed in the paper; R = 0.045.

In the structure a $[Mo_6O_{19}]^{2-}$ anion is located between two crown complex cations, forming a sandwich structure, see **Fig. 44**. The sandwich structures connect with each other through a water molecule. Thus, the water molecules are strongly linked to two K^+ cations by an ion-dipole interaction. The $[Mo_6O_{19}]^{2-}$ anion structure consists of six MoO_6 octahedra sharing edges (see Fig. 16, p. 18). Average bond lengths (in Å): Mo–O(terminal) = 1.674, Mo–O (bridge) = 1.926, Mo–O(center) = 2.320; Mo–Mo = 3.281. The two 18-crown-6 molecules form a "garland" structure with approximate D_{3d} symmetry. Each K^+ cation is surrounded by a nearly planar hexagon of O atoms (K–O distances 2.72 to 2.88 Å) of a corresponding 18-crown-6 molecule, but displaced from the mean oxygen planes by about 0.26 Å. Cation coordination is completed by a terminal O atom of the $[Mo_6O_{19}]^{2-}$ anion (K–O distances 2.70 and 2.72 Å) and a water molecule (K–O distances 2.89 and 2.93 Å), giving a hexagonal-bipyramidal coordination of K^+. For further bond lengths and angles see the paper.

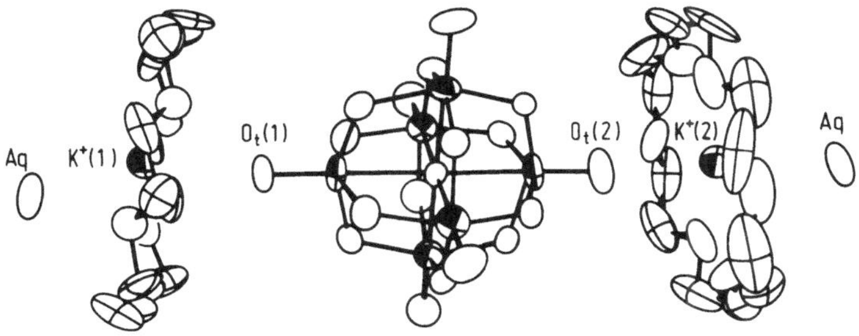

Fig. 44. The sandwich structure of $(C_{12}H_{24}O_6)_2 \cdot K_2[Mo_6O_{19}] \cdot H_2O$.

The water molecules are released from the crystals at 90°C, indicating that they are not zeolitic.

O. Nagano, Y. Sasaki (Acta Cryst. B **35** [1979] 2387/9).

2.4.2.13 $K_8[Mo_{36}O_{112}(H_2O)_{16}] \cdot nH_2O$ (= $K_2O \cdot 9MoO_3 \cdot (4 + 0.25\,n)H_2O$), n = 36 to 64; Potassium 36-Molybdates

$K_8[Mo_{36}O_{112}(H_2O)_{16}] \cdot {\sim}64H_2O$. Colorless transparent crystals of the compound are obtained from a 0.2 M solution of potassium molybdate acidified to Z = 1.8 (~1.9 H^+ added/MoO_4^{2-}; pH ≈ 1) [1]; see also [2].

The compound was characterized by analysis [2] and Raman spectrum [1, 2] (see Fig. 20, p. 21). The analytical ratio $K_2O:MoO_3$ shows, for reasons open to question, deviations up to +10% of the theoretical value, corresponding to a composition $K_2O \cdot 8$ to $9MoO_3$ (similar observations have been made for the sodium and ammonium compounds). Different compositions of the compound do not, however, affect the Raman spectrum [1, 2]. The presence of small amounts of the salt of the acid used for acidification has been assumed to explain the differences in composition [1].

From the identity of the Raman spectra the formula $[Mo_{36}O_{112}]^{8-}$ was proposed for the polymolybdate anion, in analogy to the corresponding sodium compound [1, 2]. In the light of the X-ray structure determination of the compound with 36 to 40H_2O [3, 4] the polymolybdate anion is now formulated $[Mo_{36}O_{112}(H_2O)_{16}]^{8-}$ (see Fig. 19, p. 20). This compound is possibly identical with the 36 to 40-hydrate.

The crystals readily effloresce [1, 2] with conversion into the potassium "decamolybdate" [1]. The compound is soluble in water (if it is insoluble, conversion into the "decamolybdate" has occurred) [1].

$K_8[Mo_{36}O_{112}(H_2O)_{16}]\cdot$36 to 40$H_2O$. Colorless, transparent crystals of the compound are obtained from a mixture of equal volumes of 0.2 M K_2MoO_4 solution and 0.4 M HNO_3 at room temperature after 3 to 6 d. On addition of a little more acid or by raising the temperature to 40°C the "decamolybdate" crystallizes [3, 4].

The compound was characterized by analysis, infrared and Raman spectra [5]. The latter are identical with those of the other [1, 2] 36-molybdates.

The crystals are monoclinic, space group $P2_1/c$-C^5_{2h} (No. 14); lattice parameters a = 16.436(4), b = 18.739(3), c = 27.651(5) Å, β = 117.00(2)° (for the 36-hydrate) [3], a = 16.550(4), b = 18.810(3), c = 27.730(5) Å, β = 116.82(2)° [4]; Z = 2. D_m = 2.81(1), D_x = 2.81 g/cm^3 (for the 37-hydrate) [4]. The unit cell dimensions vary significantly from crystal to crystal owing to differences in the number of molecules of water of crystallization [3, 4]. Atomic parameters are listed in [4].

The crystal structure of the compound has been determined from three-dimensional X-ray diffraction data, final R value 0.057. The structure of the $[Mo_{36}O_{112}(H_2O)_{16}]^{8-}$ polyanion consists of two 18-molybdate sub-units which combine *via* four common oxygen atoms to form a "ring" around the b axis (see Fig. 19, p. 20). The anion contains in each half a compact group made of five edge-sharing MoO_6 octahedra and of two MoO_7 pentagonal bipyramids and several groups of mainly corner-sharing octahedra, forming three- and four-membered rings in most of which octahedral coordination of the Mo atoms is completed by H_2O molecules. Hence, this unusual structure contains features typical of isopolymolybdates precipitating from less acidified solutions as well as those of polymeric compounds obtained from strongly acidic solutions. Features otherwise unknown in the chemistry of isopolymolybdates (and molybdenum oxide hydrates) are the MoO_7 polyhedra and the H_2O molecules which are simultaneously coordinated to two Mo atoms. Some of the cations and the molecules of water of crystallization are in the middle of the "ring" formed by the two sub-units of the 36-molybdate ion, the others separate the large anions in the crystal and are partly disordered. The coordination of the K^+ ions is normal (8- to 10-fold). The geometry of the $H_2O\cdots H_2O$ and $H_2O\cdots O$ systems indicates some degree of hydrogen bonding between the anions and the cation-hydrate arrangement. As a whole, the packing of the hydrate water system appears quite loose and the bonding forces very weak, in accordance with the easy loss of water. The Mo–O bond distances for the different kinds of O atoms vary in quite wide limits (in Å): from 1.664 to 1.715 (average 1.693) for Mo–O (terminal) bonds, from 1.719 to 2.362 (average 1.952) for Mo–O bonds with O connected to 2 Mo, 1.869 to 2.311 (average 2.063) for Mo–O(triply bridging) bonds, 2.057 to 2.377 (average 2.170) for Mo–O(quadruply bridging) bonds, 2.320 to 2.402 (average 2.367) for Mo–OH_2(terminal) and 2.387 to 2.441 (average 2.415) for Mo–OH_2(bridging) bonds. For other bond lengths see the paper [4]; see also the earlier structure investigation (R = 0.076) in [3].

The crystals are unstable in dry air [3, 4].

This potassium 36-molybdate is described in [6] as "octamolybdate" $K_2O\cdot8MoO_3\cdot13H_2O$ and "decamolybdate" $K_2O\cdot10MoO_3\cdot15H_2O$ [1] and in [7] as "octamolybdate" $K_2O\cdot8MoO_3\cdot12H_2O$. The criteria for these assignments are given on p. 68. In the latter case [7] the IR spectrum of the compound is identical with those [5, 8] of the 36-molybdates.

TG and DTA investigations of the compound $K_2O\cdot 8MoO_3\cdot 12H_2O$ showing the successive loss of 1.5, 3.5, 3, and 2.5 H_2O between 60 and 130°C and 1.5 H_2O between 140 and 200°C [7] thus must be interpreted as successive loss of about 6, 15, 13, 11, and 6 H_2O with regard to the compound $K_8[Mo_{36}O_{112}(H_2O)_{16}]\cdot 36$ to 40 H_2O.

References:

[1] K. H. Tytko, B. Schönfeld (Z. Naturforsch. **30b** [1975] 471/84). – [2] K. H. Tytko, B. Schönfeld, B. Buss, O. Glemser (Angew. Chem. **85** [1973] 305/7; Angew. Chem. Intern. Ed. Engl. **12** [1973] 330/2). – [3] I. Paulat-Böschen (J. Chem. Soc. Chem. Commun. **1979** 780/2). – [4] B. Krebs, I. Paulat-Böschen (Acta Cryst. B **38** [1982] 1710/8). – [5] I. Paulat-Böschen (Diss. Kiel, West Germany, 1974, pp. 76/7, 87/8, 168/70).

[6] Gmelin Handbuch "Molybdän", 1935, p. 234. – [7] A. La Ginestra, G. Rubino (Atti Acad. Nazl. Lincei Classe Sci. Fis. Mat. Nat. Rend. [8] **41** [1966] 510/20). – [8] B. Schönfeld (Diss. Göttingen, West Germany, 1973, p. 55).

2.4.2.14 $[KMo_5O_{15}(OH)(H_2O)\cdot H_2O]_\infty$ ($=K_2O\cdot 10MoO_3\cdot 5H_2O$), Potassium "Decamolybdate"[1)]

Crystals of the compound are obtained from 0.1 to 0.2 M K_2MoO_4 solutions, acidified with HCl or HNO_3 to 2.0 to 4 H^+/MoO_4^{2-}, at room temperature [1] or from 0.2 M K_2MoO_4 solutions, acidified with HCl to 1.9 to 5.7 H^+/MoO_4^{2-}, at 40°C [2] after some days in the form of small hexagonal columns [1, 2]. The compound is also formed on efflorescence of the potassium 36-molybdate [1].

The compound was characterized by analysis [1, 2], X-ray powder diagram (see the paper) [3], infrared spectrum [3] (see Fig. 24, p. 23), and Raman spectrum [1, 3] (see Fig. 23, p. 23).

The compound crystallizes with hexagonal symmetry, space group $P6_3/m\text{-}C_{6h}^2$ (No. 176), lattice parameters a = 10.550(8), c = 3.727(3) Å; Z = 1. $D_m = 3.73(3)$, $D_x = 3.75$ g/cm³. The substance is isomorphous with the sodium compound. The atomic coordinates are listed in the paper [2].

The crystal structure was determined from single crystal diffractometer data. The structure was solved from the Patterson function and chemical considerations and refined to R = 0.042. The structure (see Fig. 22, p. 22) consists of double chains made of edge-sharing, strongly distorted MoO_6 octahedra. The double chains are linked by common corners of octahedra to form a three-dimensional network. The double chains are not perfect due to statistical non-occupation of one of the six equivalent Mo positions in the unit cell. Along the short c axis the polymeric structure forms tunnels (minimal diameter 2.9 Å) in which the octahedrally coordinated potassium ions [K···O distances 2.92(3) (3×) and 3.08(5) Å (3×)] are located. Mo–O bond lengths in the MoO_6 octahedra are 1.700 (2×), 1.955 (2×), 2.190, and 2.373 Å. In the neighborhood of a vacant Mo position the O atoms are partially replaced by coordinated H_2O molecules, isolated H_2O molecules (crystal water), and OH groups. The statistical distribution of the Mo vacancies can be interpreted as resulting in smaller sub-units of edge-sharing octahedra, linked in the direction of the double chains only by common corners. The structure is only distantly related to the structures of the other typical polymolybdates crystallizing from aqueous solutions, but more closely related to the structures of $\alpha\text{-}MoO_3\cdot H_2O$ and of MoO_3. For other interatomic distances and angles see the paper [2].

1) This polymolybdate type is identical with the phase C polymolybdate type, see the notes added in proof on pp. 7 and 23/4.

The compound is slightly soluble in water [1].

This potassium "decamolybdate" is described in [4, p. 235] as $K_2O \cdot 10MoO_3 \cdot 9H_2O$, in [4, p. 109] as $MoO_3 \cdot H_2O$ [1, 2].

References:

[1] K. H. Tytko, B. Schönfeld (Z. Naturforsch. **30b** [1975] 471/84). – [2] B. Krebs, I. Paulat-Böschen (Acta Cryst. B **32** [1976] 1697/704). – [3] I. Paulat-Böschen (Diss. Kiel, West Germany, 1974, pp. 94, 174, 176). – [4] Gmelin Handbuch "Molybdän", 1935, pp. 109, 235.

2.4.2.15 0.058 to 0.067 $K_2O \cdot MoO_3 \cdot$0.33 to 0.65 H_2O, Potassium Phase C Polymolybdate[1)]

The conditions of preparation may vary within wide limits (concentration of K_2MoO_4 0.4 to 1.5 N; excess concentration of HNO_3 0.8 to 2.5 N; temperature of precipitation 25 to 100°C, corresponding to a time of formation of 3 weeks to 5 min). The white products are dried at 40°C and stored away from air and light, since humid products more or less rapidly turn blue-colored (depending on the conditions of preparation) [1].

This phase C compound was characterized by its X-ray powder diagram [1, 2] which is in accordance with that of the sodium phase C (p. 70). It is bcc, lattice parameter $a = 12.98 \pm 0.02$ Å. The amounts of potassium and water may vary within wide limits without change in the X-ray diffraction pattern, thus showing the water in the C phase to be unimportant for the structure [1].

The compound is readily soluble in alkalies [1].

The dehydration diagram shows a break at 130 to 140°C where about 50% of the water content has been lost, indicating loosely adsorbed and firmly bound water to be present in nearly equal amounts. This is confirmed by ^1H NMR investigations. The water is quantitatively removed at ~340°C [1].

The ion-exchange properties of a sample prepared by acidification with 2.5 N (excess) HNO_3 [1] and dried at 50°C have been studied with a batch technique: 5 g of the phase C compound (~0.06 $K_2O \cdot MoO_3 \cdot$ ~0.6 H_2O [1]) were heated with 100 mL of 1N HNO_3 and 50 mL of 1M NH_4NO_3 for 1 h to reflux. A sample containing 1.060 mval K/g contained after the exchange procedure 0.302 mval K and 0.776 mval NH_4 (1.078 mval $(K+NH_4)$/g) [3]. See also p. 46.

References:

[1] H. Peters, L. Till, K. H. Radeke (Z. Anorg. Allgem. Chem. **365** [1969] 14/21). – [2] M. L. Freedman (J. Chem. Eng. Data **8** [1963] 113/6). – [3] H. Peters, K. H. Radeke (Monatsber. Deut. Akad. Wiss. Berlin **11** [1969] 351/4).

2.4.3 Potassium Peroxomolybdate Hydrates

2.4.3.1 $K_2MoO_6 \cdot K_2MoO_5 \cdot nH_2O$, $n \approx 0.5, 1, 1.5, 2, 3$

$K_2MoO_6 \cdot K_2MoO_5 \cdot 3H_2O$ is prepared from saturated potassium molybdate solution and H_2O_2 (15% solution) in a neutral medium at −5°C. Products with lower water content are obtained by dehydration in vacuum at 20°C. After 25 min $K_2MoO_6 \cdot K_2MoO_5 \cdot 1.5H_2O$ results, and it is believed

[1)] This polymolybdate type is identical with the "decamolybdate" type, see the notes added in proof on pp. 7 and 23/4.

that a water content exceeding this is absorbed water only. After about 7 h a product with 1 H_2O is obtained. Prolonged dehydration finally leads to the loss of all water. With less than 0.5 H_2O the products are unstable [1]. The EPR spectrum of $K_2MoO_6 \cdot K_2MoO_5 \cdot 2H_2O$ is a superposition of an anisotropic signal ($g_\perp = 1.986$, $g_\parallel = 2.059$) and an asymmetric singlet (g = 2.008). A few seconds of thermal treatment raise the intensity of the anisotropic signal. At heating for more than 10 min the intensities of both the anisotropic signal and the asymmetric singlet are decreased [2].

On heating $K_2MoO_6 \cdot K_2MoO_5 \cdot 3H_2O$ at 90°C water (80%) and oxygen (66.5%) are lost. At 175°C the rest of the water and oxygen are given off and K_2MoO_4 is formed, according to X-ray diffraction and IR spectroscopy results [1].

References:

[1] G. L. Smorgonskaya, G. A. Bogdanov, G. L. Petrova, Z. K. Alekhina, L. N. Belous (Izv. Vysshikh Uchebn. Zavedenii Khim. Khim. Tekhnol. **17** [1974] 1730/1; C.A. **82** [1975] No. 77567). – [2] G. L. Smorgonskaya, G. A. Bogdanov, G. L. Petrova, M. V. Savina (Zh. Obshch. Khim. **45** [1975] 2745; J. Gen. Chem. [USSR] **45** [1975] 2706).

2.4.3.2 $2K_2MoO_6 \cdot 3H_2O$ and $K_2MoO_6 \cdot K_2MoO_5 \cdot H_2O_2 \cdot 2H_2O$

These compounds have the same overall composition. They are reported to be prepared in a neutral medium [1].

$2K_2MoO_6 \cdot 3H_2O$ is rose colored and $K_2MoO_6 \cdot K_2MoO_5 \cdot H_2O_2 \cdot 2H_2O$ is dark yellow colored. The formulas have been concluded from the different dehydration kinetics of the two compounds and their decomposition kinetics in water at 15 to 45°C, see figures in the paper [1]. The EPR spectrum of $2K_2MoO_6 \cdot 3H_2O$ consists of an anisotropic signal ($g_\perp = 1.986$, $g_\parallel = 2.059$) analogous to that of $K_2Mo_2O_{11} \cdot 4H_2O$. After thermal treatment an asymmetric singlet (g = 2.008) also appears in the spectrum. It is suggested that upon thermal treatment part of the peroxide oxygen is converted into a molecular adduct of H_2O_2 [2].

References:

[1] G. L. Smorgonskaya, G. L. Petrova, T. M. Kurokhtina, G. A. Bogdanov (Izv. Vysshikh Uchebn. Zavedenii Khim. Khim. Tekhnol. **18** [1975] 1183/6; C.A. **83** [1975] No. 201300). – [2] G. L. Smorgonskaya, G. A. Bogdanov, G. L. Petrova, M. V. Savina (Zh. Obshch. Khim. **45** [1975] 2745; J. Gen. Chem. [USSR] **45** [1975] 2706).

2.4.3.3 $K_2Mo_2O_{11} \cdot 4H_2O$

For the preparation, an H_2O_2 solution (30%, 10 mL) is mixed in the cold with a solution of K_2MoO_4 (5 g/100 mL). To this dark red colored alkaline solution concentrated HCl is added dropwise until the solution becomes light red. Then 1 M HCl is added until the red color of the mixture is changed to light yellow. From this solution needle-shaped crystals of the compound are separated [1].

$K_2Mo_2O_{11} \cdot 4H_2O$ is triclinic; the lattice parameters are a = 10.086 ± 0.004, b = 11.444 ± 0.005, c = 6.257 ± 0.004 Å, α = 108.14° ± 0.03°, β = 109.52° ± 0.03°, γ = 88.49° ± 0.04°; Z = 2. For the dimensions of the Delaunay reduced cell see the paper. Space group $P\bar{1}$-C_i^1 (No. 2). For a list of indexed d values see the paper. The structure was determined by single crystal film method and isotropically refined to R = 0.104, the atomic parameters are listed.

As **Fig. 45** [1] shows, the structure consists of dinuclear complex ions, K^+ ions, and coordinated and lattice H_2O molecules. The anions contain a nonlinear oxygen bridge, the Mo–O–Mo angle being 136.1°. For detail drawings of the mean configuration around the Mo atom see the paper. There are one short Mo–O_{term} (1.66(1) Å) and five Mo–O_{bridge} bondings (one from the Mo–O–Mo bridge and four from the peroxo groups with distances between 1.92(1) and 1.98(1) Å). A seventh coordination site is occupied by a water molecule (2.420(9) and 2.470(9) Å). Thus the compound is a μ-oxo-bis(oxodiperoxoaquomolybdenum(VI)) dihydrate, $K_2[(H_2O)(O_2)_2OMoOMoO(O_2)_2(H_2O)](H_2O)_2$. For a discussion, also with respect to the structure of $K_2W_2O_{11}\cdot4H_2O$, see the paper [1].

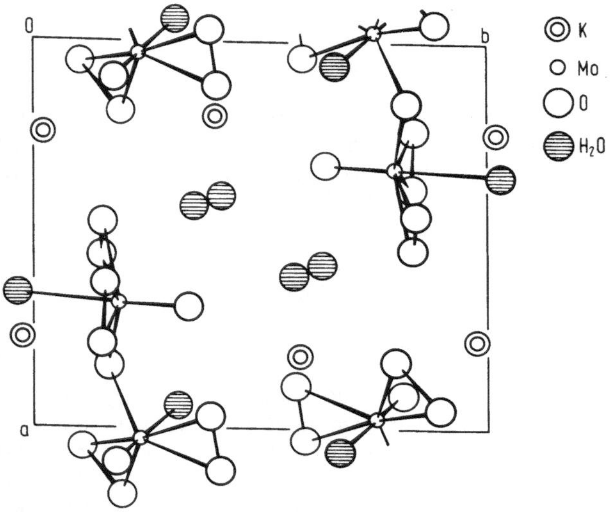

Fig. 45. The crystal structure of $K_2Mo_2O_{11}\cdot4H_2O$ viewed along [001] [1].

The measured density is 2.58, the calculated density 2.66 g/cm³ [1]. The EPR spectrum of the compound shows an anisotropic signal with $g_\perp = 1.987$ and $g_\parallel = 2.056$. On heating the signal increases in intensity but the spectrum does not change [2].

References:

[1] R. Stomberg (Acta Chem. Scand. **22** [1968] 1076/90). – [2] G. L. Smorgonskaya, G. A. Bogdanov, G. L. Petrova, M. V. Savina (Zh. Obshch. Khim. **45** [1975] 2745; J. Gen. Chem. [USSR] **45** [1975] 2706).

2.5 Ammonium Molybdate Hydrates and Solvates

Older data are given in "Molybdän", 1935, pp. 253/61.

2.5.1 Ammonium Molybdate Hydrates with Mo of Oxidation State <6

2.5.1.1 $(NH_4)_2Mo^{V}_4Mo^{VI}_2O_{17}\cdot7H_2O$ (=$(NH_4)_2O\cdot2$"Mo_2O_5"$\cdot2MoO_3\cdot7H_2O$)?

The rust brown product described by [1] to be $(NH_4)_2Mo^{V}_4Mo^{VI}_2O_{17}\cdot7H_2O$ (see "Molybdän", 1935, p. 145) was shown by polarographic determination of the oxidation state to contain far less (the amount was found to be "twice weaker") molybdenum of oxidation state V than indicated by the above formula. Additionally, the product was found to be a mixture with a hydroxylamine-molybdate complex, the hydroxylamine resulting from a molybdenum-catalyzed oxidation of the hydrazine which was used for the reduction of Mo^{VI}. The proportion of hydroxylamine was the reason for the erroneous determination of the Mo^V content [2].

References:

[1] W. F. Jakób, M. R. Reznar (Collection Czechoslov. Chem. Commun. **5** [1933] 93/102; Chem. Listy **26** [1933] 461/6). – [2] S. Ostrowetsky (Bull. Soc. Chim. France **1964** 1003/11, 1009/10).

2.5.1.2 $(NH_4)_6[Mo^V_nMo^{VI}_{7-n}O_{24-n/2}]\cdot4H_2O$ or $(NH_4)_{6-n}[Mo^V_nMo^{VI}_{7-n}O_{24-n}]\cdot4H_2O$

The γ irradiation of polycrystalline $(NH_4)_6[Mo_7O_{24}]\cdot4H_2O$ (p. 96) produces paramagnetic centers detectable by EPR. A 600 Ci ^{60}Co source was used, and irradiation doses of 1 to 140 Mr were applied. (Similar results are obtained after 10 h irradiation in a $10^{13}\,n\cdot cm^{-2}\cdot s^{-1}$ neutron flux.) Samples irradiated at room temperature became gray and exhibited EPR spectra with feebly resolved hyperfine structure; samples irradiated at liquid nitrogen temperature produced complex spectra having a large number of hyperfine structure components. These spectra are attributed to $\bullet NH_3^+$ ($g\approx2.00$) and to Mo^V ($g=1.919$; $g(MoOCl_3)=1.949$, $g(Mo_2O_3(SO_4)_3)=1.925$). Mo^V is formed by the reduction of Mo^{VI} during irradiation (see below). The $\bullet NH_3^+$ is formed according to

$$NH_4^+ \xrightarrow{h\nu} \bullet NH_3^+ + H. \qquad (1)$$

However, it is not observable at room temperature due to rapid recombination reactions (see below). The concentration of Mo^V increases with the radiation dose and decreases with time. Samples irradiated at room temperature and then heated to temperatures of 25 to 200°C showed an increase in the Mo^V concentration up to about 125°C followed by gradual decrease. This post-irradiation increase in the Mo^V concentration is attributed to the reduction of Mo^{VI} by hydrazine or by other reducing agents produced by the radiation. The subsequent decrease in the Mo^V concentration with increasing temperature and with increasing time is attributed to air oxidation of Mo^V to Mo^{VI}. Hydrazine is assumed to form by recombination of the radical ions $\bullet NH_3^+$ according to

$$\bullet NH_3^+ + \bullet NH_3^+ \rightarrow N_2H_4 + 2H^+ \qquad (2)$$

$$\bullet NH_3^+ + NH_3 \rightarrow NH_4^+ + \bullet NH_2;\ 2\bullet NH_2 \rightarrow N_2H_4 \qquad (3)$$

[1]. To find out whether Mo^V is possibly formed by a direct radiolytic reduction mechanism, a compound of Mo^{VI} not containing the ammonium ion was studied under the same conditions. The fact that Mo^V does not form with irradiated $Na_2MoO_4\cdot2H_2O$ was taken to strengthen the hypothesis according to which Mo^V would be formed from Mo^{VI} by reduction with hydrazine or with other radiolytically formed reducing groups [2].

A further EPR study was performed on γ-irradiated single crystals of ammonium paramolybdate. Irradiation was effected at room temperature using a ^{60}Co source, dosage absorbed was 1 Mr. The sample was light brown. The spectra consisted of a set of two lines, but when the crystal was at certain orientations in the magnetic field, a set of four lines was observed. The corresponding spectra of deuterated ammonium paramolybdate consisted of three lines and six lines, respectively. The two resonance lines were contributed from the magnetic interaction of an odd electron and a proton. The unchanging coupling constant of 10 ± 0.7 G showed that the coupling was isotropic. The four lines have been contributed from two chemically equivalent but differently oriented (i.e., magnetically unequal) paramagnetic centers. The principal values of g were found to be 1.900, 1.923, and 1.937. Based on the g values and the analysis of the spectra, the authors assume in accordance with [1] the occurrence of reactions (1) and (2) and the presence of Mo^V, but the path of the reduction of Mo^{VI} to Mo^V is assumed to be somewhat different. As the central Mo atom in the $[MoO_6]^{6-}$ units of the Mo–O skeleton of $[Mo_7O_{24}]^{6-}$ possesses an entirely vacant 4 d orbital and is often a strong electron acceptor, the radiatively-generated H will then transfer its electron to the 4 d orbital of this central Mo atom, thus reducing the $[MoO_6]^{6-}$ ion according to

$$[MoO_6]^{6-} + H \rightarrow [MoO_6]^{7-} + H^+. \qquad (4)$$

The vanishingly small H^+ cation will then be joined to the highly electronegative oxygen atom through a pseudo-hydrogen bond in order to attain charge compensation so as to make the whole crystal electrically neutral. The chemical structure of the paramagnetic center in question is thus represented as

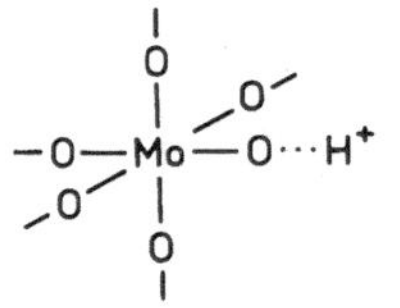

The H^+ ($I = 1/2$) joined to the oxygen atom then interacts with the unpaired electron and thus induces the so-called superhyperfine interaction. EPR spectra of the γ-irradiated deuterated ammonium paramolybdate single crystal confirm this suggestion on the origin of the superhyperfine structure, because when H^+ is substituted by D^+ ($I = 1$) the resonance lines split further. The small superhyperfine interaction coupling constant observed (10 ± 0.7 G) for the magnetic interaction between the H^+ and the unpaired electron, and its almost isotropic nature mean, however, that such magnetic interaction is weak. It is further suggested that the odd electron located in the 4d orbital of the Mo atom transfers a part of its electron cloud to its nearest ligand neighbor so that the spin of this odd electron interacts with the nuclear magnetic moment of the oxygen atom joined to the H^+. Thus, the magnetic moment of this electron spin-interacted H^+ is the cause of the superhyperfine interaction observed [3].

In a third investigation the paramagnetic species produced by 50 kV X-ray irradiation of single crystals of $(NH_4)_6[Mo_7O_{24}] \cdot 4H_2O$ at room temperature has been studied by EPR spectroscopy. The crystals became brown on irradiation. The spectra show that the hyperfine splitting is due primarily to only one of the seven molybdenum nuclei, together with a nearly isotropic splitting (~7 G) due to a nearby hydrogen nucleus. This indicates that most of the spin density of the radical resides on one molybdenum atom. The experimental results are explained by assuming that the $[Mo_7O_{24}]^{6-}$ ion captures a hydrogen atom. This H atom presumably attaches itself to an unshared oxygen leading to the ion $[Mo_7O_{24}H]^{6-}$. The electronic structure of the paramagnetic species is discussed in terms of molecular orbital theory in a semi-quantitative manner. The bonding coefficient, α, for the odd electron orbital and the contact spin density, χ, were estimated. Values of the g tensor and hyperfine interaction tensor for $[Mo_7O_{24}H]^{6-}$:

$g_{\parallel}$	$g_{\perp}$	$T_{\parallel}$ in 10^{-4} cm^{-1}	$T_{\perp}$ in 10^{-4} cm^{-1}	α^2	χ in a.u.
1.90	1.90[a)]	78.0	33.7	0.68	−5.25
1.891	1.915[b)]	78.0	33.7	0.67	−5.35

[a)] g values in T-axes frame. – [b)] g tensor not parallel to T tensor.

The large value of χ indicates that there is little or no direct contribution from the 5s orbital of molybdenum to the contact term and hence this is due primarily to core polarization [4].

References:

[1] I. Pascaru, O. Constantinescu, M. Constantinescu, D. Arizan (J. Chim. Phys. **62** [1964/65] 1283/8). – [2] M. Constantinescu, I. Pascaru, O. Constantinescu (Rev. Roumaine Phys. **11** [1966] 97/8). – [3] H. C. Wang, Y. Yang, Y. W. Chen (Sci. Sinica **17** [1974] 17/25). – [4] C. R. Byfleet, F. G. Herring, W. C. Lin, C. A. McDowell, D. J. Ward (Mol. Phys. **15** [1968] 239/47).

2.5.2 Ammonium Molybdate(VI) Hydrates and Solvates

2.5.2.1 The NH_3-MoO_3-H_2O System

There are several investigations [1 to 6] concerning the nature of the solid phases occurring in equilibrium with the solution and the course of the solubility curves. The results are rather controversial [4 to 6] as can be seen from Table 5 (p. 93) and **Figs. 46** and **47** (p. 92). According to [5, 6], the differences in the course of the solubility curves are simply due to the fact that certain authors made the measurements relative to domains of the metastable paramolybdate equilibrium. Considering the fact that, according to [7], in dependence on the time under otherwise identical conditions three different solids (the paramolybdate, the "trimolybdate", and the (3:8)-molybdate) can deposit from an ammonium paramolybdate solution (compare pp. 105 and 108), the different results in this region are explained. On the other side, there is a region in which in a similar manner the paramolybdate and the "dimolybdate" can deposit [6].

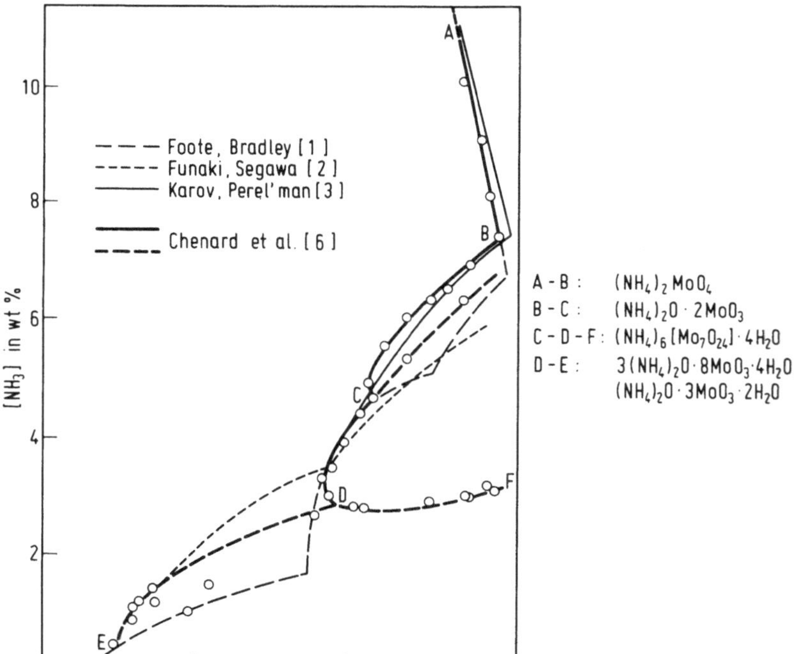

Fig. 46. Summary of the solubility data reported by [1 to 3, 5, 6] on the ternary NH_3-MoO_3-H_2O system at 25°C according to [6].

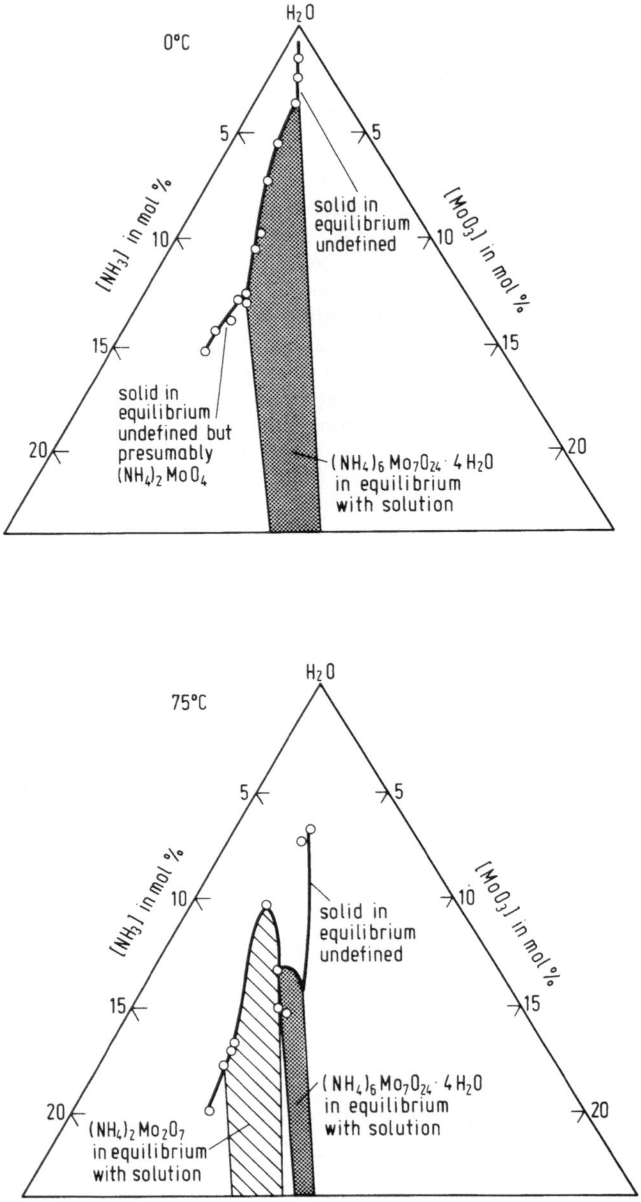

Fig. 47. Solubility isotherms of the NH_3-MoO_3-H_2O system at 0 and 75°C; equilibration period one week [4].

The results obtained by [1 to 3] at 25°C were summarized by [4] (see the paper) and [6] (Fig. 46). Solubility isotherms determined by [4] at 0 and 75°C are shown in Fig. 47 (diagrams for 25 and 50°C can be found in the paper). In this case the time of equilibration was about one week [4]. Isotherms determined at 25 and 85°C applying several months for equilibration are given by [5, 6]. Four incongruently soluble compounds are observed. In accordance with [7], the (3:8)-molybdate occurs only in those investigations [1, 5, 6] in which a very long equilibration period (several months) was applied. With respect to the "dimolybdate" there is no clear picture. It was observed after a short equilibration period according to [3], and after a long period [5, 6], but was not observed after a short equilibration period [4] at 25°C.

Table 5

Compounds Occurring as Solid Phases in Equilibrium with the Solution in the NH_3-MoO_3-H_2O System.

temperature	equilibration time	solid phases (updated formulae)	Ref.
25, 85°C	several months	$(NH_4)_2MoO_4$ [c)] $[(NH_4)_4Mo_4O_{14}]_\infty$ [c)] $(NH_4)_6[Mo_7O_{24}]\cdot 4H_2O$ [c)] $[(NH_4)_6Mo_8O_{27}\cdot 4H_2O]_\infty$ [a), c)]	[5, 6]
0, 25, 50, 75°C	one week	$(NH_4)_2MoO_4$ [b), c)] $(NH_4)_6[Mo_7O_{24}]\cdot 4H_2O$ [c)]	[4]
25°C	2 to 3 d	$(NH_4)_2MoO_4$ [d)] $[(NH_4)_4Mo_4O_{14}]_\infty$ [d)] $(NH_4)_6[Mo_7O_{24}]\cdot 4H_2O$ [d)]	[3]
25, 40°C	(?)	$(NH_4)_2MoO_4$ [d)] $(NH_4)_2Mo_2O_7\cdot H_2O$ [d)] $(NH_4)_6[Mo_7O_{24}]\cdot 4H_2O$ [d)] $(NH_4)_2O\cdot 3MoO_3\cdot 2H_2O$ [d)]	[2]
25°C	several months	$(NH_4)_6[Mo_7O_{24}]\cdot 4H_2O$ [d)] $[(NH_4)_6Mo_8O_{27}\cdot 4H_2O]_\infty$ [d)]	[1]

[a)] At 85°C, the solid phase $[(NH_4)_6Mo_8O_{27}\cdot 4H_2O]_\infty$ was not observed. – [b)] At 75°C, the solid phase $[(NH_4)_4Mo_4O_{14}]_\infty$ was also observed. – [c)] Identified by X-ray diffraction, thermal analysis, and chemical analysis. – [d)] Identified only by chemical analysis.

Depending on the degree of acidification and the time, ammonium molybdate solutions spontaneously deposit seven polymolybdates at room temperature: The paramolybdate $(NH_4)_6[Mo_7O_{24}]\cdot 4H_2O$, the (3:8)-molybdate $[(NH_4)_6Mo_8O_{27}\cdot 4H_2O]_\infty$, the "trimolybdate" $[(NH_4)_4Mo_6O_{20}\cdot 18(?)H_2O]_\infty$, the "tetramolybdates" $(NH_4)_2O\cdot 4MoO_3$ and $(NH_4)_4[Mo_8O_{26}]\cdot 4(5)H_2O$, the 36-molybdate $(NH_4)_8[Mo_{36}O_{112}(H_2O)_{16}]\cdot \sim 64H_2O$, and the "decamolybdate" $[NH_4Mo_5O_{15}(OH)(H_2O)\cdot H_2O]_\infty$ [1)] [7]. At temperatures above 60°C the "dimolybdate" $[(NH_4)_4Mo_4O_{14}]_\infty$ [7, 8] (see "Molybdän" Erg.-Bd. B1, 1975, p. 227), the decamolybdate

[1)] The "decamolybdate" and phase C polymolybdate types have been shown to be identical, see the notes added in proof on pp. 7 and 23/4.

$(NH_4)_8[Mo_{10}O_{34}]$ [9, 10] (see "Molybdän" Erg.-Bd. B1, 1975, p. 228), and the phase C polymolybdate[1)] [11, 12] crystallize.

It is thought that for the temperatures investigated (0 to 75°C) the normal molybdate is present at all $NH_3:MoO_3$ mole ratios greater than two, i.e., above those ratios where ammonium paramolybdate or ammonium "dimolybdate" exist. But the normal molybdate decomposes unless stored under NH_3, and the method of drying the solids that was employed in the study (air drying several days) led to decomposition of any $(NH_4)_2MoO_4$ that was present as an equilibrium solid [4]. Other authors [3] report the presence of a transition from one equilibrium solid to another, of $(NH_4)_2Mo_2O_7$ to $(NH_4)_2MoO_4$, at a solution composition of about 5 mol% MoO_3 and 10 mol% NH_3 at 25°C, compare figure 1 in [4] and table in [3]; $(NH_4)_2MoO_4$ crystallizes at the higher ammonia content side of this transition point. The existence of a transition point at this composition is confirmed [4]. Other data [2] agree quite well with the solubility isotherm at 25°C, see a figure in [4]; but "trimolybdate", paramolybdate, and "dimolybdate" are found as equilibrium solids in addition to the monomolybdate at 25 and 40°C [2], whereas only the paramolybdate is present as a well-defined solid at 25°C besides probably $(NH_4)_2MoO_4$ according to [4].

Solubility results obtained in the alkaline region of the NH_3-MoO_3-H_2O system, namely in the $(NH_4)_2MoO_4$-NH_3-H_2O subsystem at 25°C, are represented in **Fig. 48**, for special values see table in the paper [14]. On successive increase of the concentration of ammonia in the saturated solution of $(NH_4)_2MoO_4$, the solubility of this compound decreases steadily up to the point E′, where the solution becomes saturated with ammonia and a new phase, gaseous NH_3, appears in the closed system. On the other hand, when an increasing amount of $(NH_4)_2MoO_4$ is added to a saturated aqueous ammonia solution (the gas pressure over the solution being kept constant) the solubility of ammonia also decreases up to point E′, where the solution becomes saturated with $(NH_4)_2MoO_4$. The system is thus quite analogous in structure to solubility diagrams for ternary water-salt systems of corresponding type; the point E′ corresponds to a eutonic point E (E′ is called a eutonoid or quasieutonic point) [14].

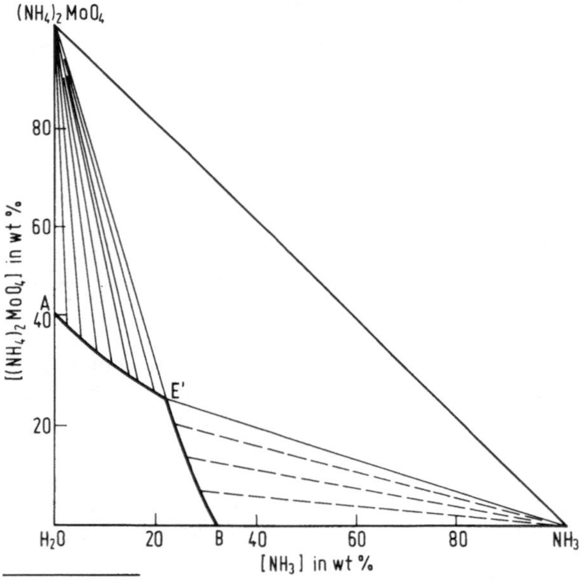

Fig. 48. Solubility in the $(NH_4)_2MoO_4$-NH_3-H_2O subsystem at 25°C and 720±5 Torr [14].

[1)] The "decamolybdate" and phase C polymolybdate types have been shown to be identical, see the notes added in proof on pp. 7 and 23/4.

For investigations studying the influence of NH_4Cl see [3] (investigation of the ammonium molybdate-ammonium chloride-water system and improvement of the conditions to prepare the normal ammonium molybdate) and [8, 13] (improvement of the conditions to prepare ammonium "dimolybdate" and ammonium (3:8)-molybdate, compare p. 105).

The knowledge of the equilibrium system NH_3–MoO_3–H_2O is important in any theoretical development of the extracting process of the molybdenum with ammonia from its roasted ores [2] and for the preparation of compounds for applications demanding high-purity levels of molybdenum or molybdenum trioxide, since the production of ammonium molybdates has been used as the purification step in producing pure feed material for metal powder. Solubility is furthermore a paramount consideration in the application of ammonium molybdates when these are used as a means by which molybdenum is deposited on alumina or silica carriers for catalytic applications [4].

References:

[1] H. W. Foote, W. M. Bradley (J. Am. Chem. Soc. **58** [1936] 930/1). – [2] K. Funaki, T. Segawa (J. Electrochem. Soc. Japan **18** [1950] 152/4). – [3] Z. G. Karov, F. M. Perel'man (Zh. Neorgan. Khim. **5** [1960] 713/9; Russ. J. Inorg. Chem. **5** [1960] 343/6). – [4] C. J. Hallada (J. Less-Common Metals **36** [1974] 103/10). – [5] A. Chenard, R. Cohen-Adad (Compt. Rend. C **280** [1975] 1371/3).

[6] A. Chenard, R. Tenu, R. Cohen-Adad (Bull. Soc. Chim. France **1976** 1797/801). – [7] K. H. Tytko, B. Schönfeld (Z. Naturforsch. **30b** [1975] 471/84). – [8] W. D. Hunnius (Z. Naturforsch. **29b** [1974] 599/602). – [9] W. D. Hunnius (Z. Naturforsch. **30b** [1975] 63/5). – [10] J. Fuchs, H. Hartl, W. D. Hunnius, S. Mahjour (Angew. Chem. **87** [1975] 634/5; Angew. Chem. Internat. Ed. Engl. **14** [1975] 644).

[11] M. L. Freedman (J. Chem. Eng. Data **8** [1963] 113/6). – [12] H. Peters, L. Till, K. H. Radeke (Z. Anorg. Allgem. Chem. **365** [1969] 14/21). – [13] W. D. Hunnius (Diss. F.U. Berlin 1970, pp. 54/6). – [14] E. G. Karov (Zh. Neorgan. Khim. **16** [1971] 3110/4; Russ. J. Inorg. Chem. **16** [1971] 1651/3).

2.5.2.2 $(NH_4)_2MoO_4 \cdot 2H_2O$(?), Aqueous $(NH_4)_2MoO_4$ Solutions

The normal ammonium molybdate as prepared by cooling saturated solutions with $NH_3:MoO_3$ ratios greater than 2:1 [1] has the composition $(NH_4)_2MoO_4$; a hydrate has not been obtained, cf. "Molybdän" Erg.-Bd. B1, 1975, p. 225. However, Raman spectra of a compound $(NH_4)_2MoO_4 \cdot 2H_2O$ and of the saturated aqueous $(NH_4)_2MoO_4$ solution are given by [2] (wavenumbers $\tilde{\nu}$ in cm^{-1}):

$\tilde{\nu}$ of $(NH_4)_2MoO_4 \cdot 2H_2O$ (cryst.)	936	915, 893, 874	360	218
$\tilde{\nu}$ of $(NH_4)_2MoO_4$ (sat. solution)	944	896(br)	367	218
assignment	ν_1	ν_2	ν_4	ν_2

When ammonium nitrate is added to the saturated aqueous $(NH_4)_2MoO_4$ solution at 25°C, the anhydrous compound separates in the solid phase; for the $(NH_4)_2MoO_4$–NH_4NO_3–H_2O system at 25°C see [3]. In aqueous solutions with NH_4Cl at 25°C no double salt with the molybdate is formed [4]. The cathodic reduction of HF-containing $(NH_4)_2MoO_4$ solutions has been investigated to obtain Mo coatings [5], see also [6].

References:

[1] C. J. Hallada (J. Less-Common Metals **36** [1974] 103/10, 106). – [2] C. S. Venkateswaran (Proc. Indian Acad. Sci. A **7** [1938] 144/55, 145). – [3] Z. G. Karov, F. M. Perel'man (Zh. Neorgan. Khim. **7** [1962] 2450/8; Russ. J. Inorg. Chem. **7** [1962] 1270/4). – [4] Z. G. Karov, F. M. Perel'man (Zh. Neorgan. Khim. **5** [1960] 713/20; Russ. J. Inorg. Chem. **5** [1960] 343/6). – [5] N. D. Ivanova (Ukr. Khim. Zh. **44** [1978] 252/6; Soviet Progr. Chem. **44** No. 3 [1978] 29/32).

[6] A. V. Gorodyskii, N. D. Ivanova, K. B. Kladnitskaya (Ukr. Khim. Zh. **48** [1982] 377/9; Soviet Progr. Chem. **48** No. 4 [1982] 44/6).

2.5.2.3 $(NH_4)_6[Mo_7O_{24}]\cdot nSolv$ ($=3(NH_4)_2O\cdot7MoO_3\cdot nSolv$), Ammonium Paramolybdates

2.5.2.3.1 $(NH_4)_6[Mo_7O_{24}]\cdot4H_2O$ ($=3(NH_4)_2O\cdot7MoO_3\cdot4H_2O$)

2.5.2.3.1.1 Formula

The compound is formulated in [1] like the paramolybdates of some other cations as $5(NH_4)_2O\cdot12MoO_3\cdot7H_2O$. The difficulty in distinguishing between the formulations $3(NH_4)_2O\cdot7MoO_3\cdot4H_2O$ and $5(NH_4)_2O\cdot12MoO_3\cdot7H_2O$ arises from the nearly identical analytical data for both the formulations [1 to 4]. Analysis of mixed paramolybdates of ammonia and triethanolamine, in which case the possible ratios of the cations show greater deviations, reaffirmed the older (3:7) formulation [2]. This (3:7) formulation was unambigously confirmed by the X-ray measurements of the lattice parameters and the determination of the density of crystalline ammonium paramolybdate [3]. Since the paramolybdates represent a polymolybdate type [1], all the paramolybdates have to be formulated as (3:7)-molybdates. Crystal structure determinations have shown discrete heptamolybdate ions of formula $[Mo_7O_{24}]^{6-}$ to be present in ammonium paramolybdate [5 to 10] and in the paramolybdates of other cations (see p. 9).

References:

[1] Gmelin Handbuch "Molybdän", 1935, pp. 113/9, 210/98. – [2] F. Garelli, A. Tettamanzi (Atti Accad. Sci. Torino Classe Sci. Fis. Mat. Nat. **70** [1934/35] 382/90; Gazz. Chim. Ital. **65** [1935] 1009/15). – [3] J. H. Sturdivant (J. Am. Chem. Soc. **59** [1937] 630/1). – [4] I. Lindqvist (Acta Chem. Scand. **2** [1948] 88/9). – [5] I. Lindqvist (Acta Cryst. **3** [1950] 159/60).

[6] I. Lindqvist (Arkiv Kemi **2** [1950/51] 325/41). – [7] E. Shimao (Nature **214** [1967] 170/1). – [8] E. Shimao (Bull. Chem. Soc. Japan **40** [1967] 1609/13). – [9] H. T. Evans Jr. (J. Am. Chem. Soc. **90** [1968] 3275/6). – [10] H. T. Evans Jr., B. M. Gatehouse, P. Leverett (J. Chem. Soc. Dalton Trans. **1975** 505/14).

2.5.2.3.1.2 Preparation

The general method of preparing ammonium paramolybdate is, as for other paramolybdates, adding of the necessary amount of a mineral acid or of molybdenum trioxide to the aqueous solution of the normal molybdate [1]. The pH of the solution is about 5 [2]. Commercial products often contain the "dimolybdate" $(NH_4)_2O\cdot2MoO_3$ [1 to 3, 11] in amounts up to 10% [2]. For purification the impure products are recrystallized from aqueous solution (cf. [4 to 6]), for instance by cooling of a saturated solution at 25°C to 0°C [7, 10] or by isothermal evaporation [3]. On very long standing, concentrated ammonium paramolybdate solutions deposit first (after several days) ammonium "trimolybdate" and later (after several weeks or months) ammonium (3:8)-molybdate [2] (see pp. 105 and 108).

To recover molybdenum from filtrates of P determinations, molybdenum is precipitated as ammonium dodecamolybdophosphate by controlled addition of ammonia and $NH_4H_2PO_4$. The precipitate is washed and dissolved in ammonia solution, and the PO_4^{3-} ion is removed by precipitation of $MgNH_4PO_4$. The filtrate is brought to pH 5 with acetic acid and stored at 0°C to recover ammonium paramolybdate by crystallization in yields of 85 to 90%. For further details see the paper [8]. For a method developed and studied to recover high-purity ammonium paramolybdate from impure sodium molybdate solutions of low and high concentrations on an industrial scale using tri-*n*-caprylylamine to extract molybdenum, see [9]. To produce pure ammonium paramolybdate from native molybdenite, the MoS_2 enriched by flotation is only partially (60 to 70%) roasted to avoid the formation of the sparingly soluble MoO_3. Oxidation is completed with nitric acid. $MoO_3\cdot H_2O$, obtained from the ammoniacal solution by precipitation with conc. HNO_3, is treated with gaseous NH_3 to give beautiful, absolutely white crystals of $(NH_4)_6[Mo_7O_{24}]\cdot nH_2O$ within 10 h in a yield of 98 to 99% [13]. For studies of the crystallization process of ammonium paramolybdate on a technical scale see [14, 15].

$(ND_4)_6[Mo_7O_{24}]\cdot 4D_2O$ was prepared by repeating crystallization in D_2O (heavy water). After this exchange process, the hydrogen atoms in NH_4^+ and water of crystallization were completely substituted by deuterium [12]. An ^{17}O-enriched sample was obtained by dissolving the recrystallized compound in ^{17}O-enriched water [7].

References:

[1] I. Lindqvist (Nova Acta Regiae Soc. Sci. Upsaliensis **15** No. 1 [1950] pp. 1/22, 5/6). – [2] K. H. Tytko, B. Schönfeld (Z. Naturforsch. **30b** [1975] 471/84). – [3] A. Hansson, I. Lindqvist (Acta Chem. Scand. **3** [1949] 1430/6). – [4] I. Lindqvist (Acta Chem. Scand. **2** [1948] 88/9). – [5] I. Lindqvist (Arkiv Kemi **2** [1950/51] 325/41, 325).

[6] H. T. Evans Jr., B. M. Gatehouse, P. Leverett (J. Chem. Soc. Dalton Trans. **1975** 505/14, 506). – [7] M. Filowitz, R. K. C. Ho, W. G. Klemperer, W. Shum (Inorg. Chem. **18** [1979] 93/103, 94). – [8] H. Dietrich, P. von Polheim (Z. Anal. Chem. **166** [1959] 18/23). – [9] M. B. MacInnis, T. K. Kim, J. M. Laferty (J. Less-Common Metals **36** [1974] 111/6; Chem. Uses Molybdenum Proc. 1st Conf., Reading, Engl., 1973, pp. 56/8). – [10] J. H. Sturdivant (J. Am. Chem. Soc. **59** [1937] 630/1).

[11] W. D. Hunnius (Z. Naturforsch. **30b** [1975] 63/5). – [12] H. C. Wang, Y. Yang, Y. W. Chen (Sci. Sinica **17** [1974] 17/25). – [13] R. Ripan, A. Duca (Acad. Repub. Populare Romine Bul. Stiint. Ser. Mat. Fiz. Chim. **2** [1950] 381/6). – [14] V. A. Postnikov, V. N. Morozov, V. A. Ryabkov (Tsvetn. Metal. **41** No. 5 [1968] 70/3; Soviet J. Non-Ferrous Metals **9** No. 5 [1968] 90/4). – [15] V. A. Postnikov (U.S.S.R. 216677 [1965/69] from C.A. **71** [1969] No. 72458).

2.5.2.3.1.3 "Finger-Prints", Spectra

X-ray powder patterns are referred to in [1, 2], [3, p. 84].

Raman spectra are reported by [4 to 6] (see Fig. 2, p. 10), infrared spectra by [3, p. 44], [7 to 13] (see Fig. 3, p. 10). The spectra resemble those of other paramolybdates.

The diffuse reflectance spectrum of a finely ground sample against blanks of magnesium oxide, lithium fluoride, and alumina shows UV peaks assigned to charge-transfer transitions $O^{2-}\rightarrow Mo^{6+}$ at 44350, 36450, and 32950 cm^{-1}. The peak at 32950 cm^{-1} is regarded as characteristic of MoO_6 octahedra, since in the spectrum of normal molybdates this peak is not present [15] and the effect of increasing the coordination number of molybdenum (VI) from four to six in aqueous solution by protonation and polymerization is to produce additional absorption peaks below 45000 cm^{-1} [14, 15].

References:

[1] E. Ma (Bull. Chem. Soc. Japan **37** [1964] 171/5, 648/53). – [2] M. J. Schwing-Weill (Bull. Soc. Chim. France **1967** 3795/8). – [3] B. Schönfeld (Diss. Göttingen, West Germany, 1973, pp. 44, 84). – [4] J. Gupta (Indian J. Phys. **12** [1938] 223/32). – [5] J. Aveston, E. W. Anacker, J. S. Johnson (Inorg. Chem. **3** [1964] 735/46).

[6] K. H. Tytko, B. Schönfeld (Z. Naturforsch. **30b** [1975] 471/84). – [7] F. A. Miller, C. H. Wilkins (Anal. Chem. **24** [1952] 1253/94, 1258, 1288). – [8] M. Hacskaylo (Anal. Chem. **26** [1954] 1410/2). – [9] F. A. Miller, G. L. Carlson, F. F. Bentley, W. H. Jones (Spectrochim. Acta **16** [1960] 135/235, 141, 156, 226). – [10] M. J. Schwing-Weill, F. Arnaud-Neu (Bull. Soc. Chim. France **1970** 853/60).

[11] A. Kiss, S. Holly, E. Hild (Magy. Kem. Folyoirat **77** [1971] 418/26; C.A. **75** [1971] No. 135415). – [12] A. W. Armour, M. G. B. Drew, P. C. H. Mitchell (J. Chem. Soc. Dalton Trans. **1975** 1493/6). – [13] T. Yamase, T. Ikawa (Bull. Chem. Soc. Japan **50** [1977] 746/9). – [14] A. Bartecki, D. Dembicka (J. Inorg. Nucl. Chem. **29** [1967] 2907/16). – [15] J. H. Ashley, P. C. H. Mitchell (J. Chem. Soc. A **1968** 2821/7).

2.5.2.3.1.4 Crystal Data, Structure

The compound crystallizes in the form of colorless rhomboidal plates [1 to 3], but also crystallizes in a more needle-shaped form elongated along the a axis [2, 3], and some having a stubby prismatic habit [4], were found. There is a perfect cleavage parallel to (010) [4].

The compound is monoclinic, space group $P2_1/c-C_{2h}^5$ (No. 14), lattice parameters a = 8.3934(8), b = 36.1703(45), c = 10.4715(11) Å, β = 115.958(8)°; Z = 4 [4]; see also [1 to 3]. D_m = 2.871 ± 0.003 [1], D_x = 2.872 g/cm³ [4]. Atomic coordinates (R = 0.076) are listed in [4], for older values (R > 0.2) see [2, 3].

The compound is isomorphous with $K_6[Mo_7O_{24}] \cdot 4H_2O$ [4] and $Rb_6[Mo_7O_{24}] \cdot 4H_2O$ [2]. The structure of the heptamolybdate anion was first deduced by [2, 5] from approximate coordinates of the molybdenum atoms. The positions of the oxygen atoms were inferred from assumed octahedral coordination with molybdenum and appropriate edge sharing (see Fig. 1, p. 9). Later, more complete X-ray studies, in which all non-hydrogen atoms were located, confirmed this structure and showed that the MoO_6 octahedra in the $Mo_7O_{24}^{6-}$ blocks are strongly distorted [3, 4, 6, 7] (see Fig. 43, p. 79). The structure has the point symmetry 2mm-C_{2v} and the molecular ion lies in a general position in the unit cell. There are three different kinds of MoO_6 octahedra. The Mo–O bond lengths span a considerable range (1.71 to 2.42 Å). Of the oxygen atoms there are twelve bonded to a single Mo atom (mean Mo–O bond length 1.72 Å), eight bonded to two Mo each (mean Mo–O bond length 2.00 Å), two bonded to three Mo (mean Mo–O length 2,04 Å), and two bonded to four Mo (mean Mo–O bond length 2.21 Å). The terminal Mo–O bonds occur in pairs for the exterior molybdenum atoms, forming V-shaped MoO_2 groups, where the mean angle is 105.2°. This configuration is also found in other isopoly and heteropoly complexes; the consistency of this characteristic configuration suggests a rather rigid π-bond system in which the bond number is ca. 1.5. Such a configuration is assumed to increase the stability of the poly complexes. The oxygen atoms of the MoO_2 groups are involved in further, more distant, bonding with cations. The central Mo atom of the heptamolybdate ion also conforms to the V-shaped configuration with two short bonds. The need for the central Mo atom to adopt this distorted coordination could explain why this polyion has the unusual bent configuration rather than the much more symmetrical Anderson [8] structure [4].

The ammonium ions occupy positions between the polymolybdate anions in irregular eightfold to tenfold coordination with oxygen (oxygen atoms of the polyanions and of water).

The mean N···O distances for the NH_4-oxygen polyhedra are in the rather narrow range of 3.04 to 3.11 Å, overall mean 3.06 Å (the individual values are in the range of 2.74 to 3.67 Å). For other bond lengths and angles between Mo and O and NH_4^+ and O see the paper [4]. The Mo–O bond lengths reported by [3] differ markedly from those given by other authors for the heptamolybdate ion [4, 9].

The heptamolybdate anions are assembled in layers extended normal to the *y* axis. In these layers the cations and water molecules serve to bind the molybdate anions together by a complex system of ionic and hydrogen bonds, see a figure in the paper [4].

For a system of the structures of ternary oxides in which ammonium paramolybdate 4-hydrate was included see [10].

References:

[1] J. H. Sturdivant (J. Am. Chem. Soc. **59** [1937] 630/1). – [2] I. Lindqvist (Arkiv Kemi **2** [1950/51] 325/41). – [3] E. Shimao (Bull. Chem. Soc. Japan **40** [1967] 1609/13). – [4] H. T. Evans Jr., B. M. Gatehouse, P. Leverett (J. Chem. Soc. Dalton Trans. **1975** 505/14). – [5] I. Lindqvist (Acta Cryst. **3** [1950] 159/60).

[6] E. Shimao (Nature **214** [1967] 170/1). – [7] H. T. Evans Jr. (J. Am. Chem. Soc. **90** [1968] 3275/6). – [8] J. S. Anderson (Nature **140** [1937] 850). – [9] B. M. Gatehouse, P. Leverett (Chem. Commun. **1968** 901/2). – [10] R. Hoppe (Z. Anorg. Allgem. Chem. **294** [1958] 135/45, 144).

2.5.2.3.1.5 Other Physical Properties

The diamagnetic susceptibility of the compound in the solid state was found to be $\chi_{mol} = (-163.1 \pm 0.84) \times 10^{-6}$ cm³/mol at room temperature [1].

A compound named "ammonium molybdate 4aq", presumably being ammonium paramolybdate 4-hydrate, was found to show triboluminescence [2].

References:

[1] M. E. Bedwell, J. E. Spencer, V. C. G. Trew (Trans. Faraday Soc. **45** [1949] 217/23). – [2] G. Wolff, G. Gross, I. N. Stranski (Z. Elektrochem. **56** [1952] 420/8, 424).

2.5.2.3.1.6 Thermal Behavior

Oxides of molybdenum are catalysts of industrial importance and may be obtained as the end products of thermal decomposition of the ammonium salts. The knowledge of the process of thermal decomposition is important in obtaining these oxides at a temperature at which no appreciable sintering occurs [1]. Molybdenum oxide supported on alumina (prepared by thermal decomposition of the ammonium paramolybdate impregnated on alumina) is also widely used as a catalyst [2].

In Air at Normal Pressure. Thermal decomposition of ammonium heptamolybdate 4-hydrate proceeds, according to the majority of the authors [1 to 3, 5 to 7, 11 to 13, 18, 27], in three steps first proposed by [11, 12] *via* ammonium (2:5)-molybdate, $2(NH_4)_2O \cdot 5MoO_3$, and ammonium (1:4)-molybdate, $(NH_4)_2O \cdot 4MoO_3$, to give molybdenum trioxide:

$$5(NH_4)_6[Mo_7O_{24}] \cdot 4H_2O \rightarrow 7(2(NH_4)_2O \cdot 5MoO_3) + 2NH_3 + 21H_2O \quad (1)$$

$$4(2(NH_4)_2O \cdot 5MoO_3) \rightarrow 5((NH_4)_2O \cdot 4MoO_3) + 6NH_3 + 3H_2O \quad (2)$$

$$(NH_4)_2O \cdot 4MoO_3 \rightarrow 4MoO_3 + 2NH_3 + H_2O \quad (3)$$

The first step occurs at 126°C (maximum of decomposition rate) [2], 110°C [5], 150 to 200°C [1], 115 to 130°C [7] ([13] arrives at similar values), 110 to 120°C [6], 103 to 113°C [18], 109 to 158°C [3, 27] (this temperature range supersedes those stated in [11, 12]), and is accompanied by an endothermic effect [1, 7]. The compound $2(NH_4)_2O \cdot 5MoO_3$ was shown to be identical with a product obtained from aqueous solution [5] (see "Molybdän" Erg.-Bd. B1, 1975, p. 228), being a discrete decamolybdate $(NH_4)_8[Mo_{10}O_{34}]$ [10]. The second step occurs at 243°C (maximum of decomposition rate) [2], 165 to 215°C [5], 245 to 310°C [1], 225 to 245°C [7] ([13] reported similar values), 220°C [6], 191 to 210°C [18], 208 to 256°C [3, 27] (for [11, 12] see above), and is also accompanied by an endothermic effect [1, 7]. This decomposition product is often formulated $(NH_4)_4Mo_8O_{26}$ and named octamolybdate (see "Molybdän" Erg.-Bd. B1, 1975, pp. 228/9); however, there is no experimental justification for this formulation. The third step occurs at 354°C (maximum of decomposition rate) [2], 325 to 390°C [1], 320 to 355°C [7] ([13] arrives at similar values), 300°C [6], 306 to 350°C [18], 304 to 393°C [3, 27] (for [11, 12], see above), and is accompanied by an endothermic peak when an alumina cup [1, 7] is used or by an exothermic peak when the material of the cup is platinum [1] (platinum tends to oxidize ammonia exothermically [9]). The foregoing thermogravimetric investigations were completed by differential thermal analyses [1, 6, 7], infrared spectra [1], X-ray powder patterns [1, 7, 15], mass spectrometric investigations of the gaseous reaction products [15, 16], the determination of nitrogen in the decomposition products [1, 15], and isothermal decomposition experiments [1].

Four research groups [4, 8, 20, 28] postulated, in addition to the (2:5)- and (1:4)-molybdates, additional species as intermediates. According to [4], some additional but rather small inflections in the TG, DTA, and other curves and some other investigations (see below) were interpreted to show the formation of a (1:4)-molybdate of a slightly different structure, of a (1:14)- and (1:22)-molybdate, and of $MoO_3 \cdot <0.15H_2O$. (In a somewhat older investigation [17] the authors arrived at somewhat different conclusions.) The (2:5)-molybdate is formed at ~100°C, the (1:4)-molybdate at ~210°C, MoO_3 above 300°C; the additional compounds are formed at ~160°C (1:4), ~260°C (1:14), ~275°C (1:22), and ~280°C ($MoO_3 \cdot <0.15H_2O$). In this study, joint thermogravimetry and infrared spectroscopic measurements of the gaseous decomposition products H_2O and NH_3, differential thermogravimetry and differential thermal analysis, X-ray and infrared investigations of the solids, and some other methods were applied. The composition of the (1:14)- and (1:22)-molybdates, which exist in a very narrow temperature range and therefore could not be prepared as pure compounds in this way, was obtained from compounds with the same infrared and X-ray diffraction characteristics as those of the compounds in question that could be separated from solutions as the 3.3-hydrate and 2.1-hydrate, respectively. The (1:14)-molybdate is characterized by a cubic unit cell with a = 12.90 Å, the (1:22)-molybdate by a hexagonal unit cell with a = 10.52 and c = 3.725 Å. The dimensions of the cubic unit cell of the (1:14)-molybdate and those of the hexagonal unit cell of the (1:22)-molybdate are not quite independent from each other. A structural relationship between the two compounds is suggested also by their very similar diffraction patterns and infrared spectra [4]. The unit cell dimensions and the diffraction patterns of the (1:14)- and (1:22)-molybdates are those of phase C polymolybdates (see pp. 70, 86, 116) and "decamolybdates" (see pp. 69; 85, 115), respectively[1)] [25]. (According to [24] the d(hkl) values given for the "14-molybdate" in [4] are incorrect.) The second group [8] postulated from additional, but again rather small inflections in the DTA curve the occurrence of small amounts of a precursor of the (1:4)-molybdate (assumed to be a (1:3)-molybdate) and of a (1:12)-molybdate. The (2:5)-molybdate is formed at 80 to 85°C, the (1:4)-molybdate at 160 to 185°C, MoO_3 at 225 to 270°C; the additional compounds are formed at 200°C(?) (1:3) and 315°C(?) (1:12). In this study thermogravimetry, differential thermal analysis, and X-ray powder diffraction investigations were applied. The

[1)] The "decamolybdate" and phase C polymolybdate types have been shown to be identical, see the notes added in proof on pp. 7 and 23/4.

(1:12)-molybdate was identified by its powder diagram as described by [18]. This "dodeca-molybdate", however, has been shown to be a "decamolybdate" (see pp. 114 and 115)[1]. The third group postulated from isothermal studies in connection with X-ray analysis and infrared absorption, which were carried out in addition to thermogravimetric analysis, the occurrence of a (1:12)- or (1:14)-molybdate, of a (1:22)-molybdate, and of $MoO_3\cdot xH_2O$. The (2:5)-molybdate is formed at 80 to 120°C, the (1:4)-molybdate at 180 to 210°C, and MoO_3 at 310 to 350°C. The additional compounds are formed at 240°C (1:2), 260°C (1:22), and 280°C ($MoO_3\cdot xH_2O$) [28]. Hence, these three groups [4, 8, 28] are, in the main, in agreement. The fourth group [20] postulated from an endothermic effect at 123 to 125°C the occurrence of the anhydrous paramolybdate. The (2:5)-molybdate is formed at about 220°C, the (1:4)-molybdate at 250 to 265°C, and MoO_3 at 320 to 370°C. In this investigation the authors still use the incorrect (5:12) formulation for the paramolybdate (and an incorrect content of water).

Two additional research groups are more or less in agreement with the above results. In [19] $3(NH_4)_2O\cdot 7MoO_3$ is postulated instead of $2(NH_4)_2O\cdot 5MoO_3$; the (1:4)-molybdate is formed at ~200°C and MoO_3 at 280°C. In [22] the author states the occurrence of "$5(NH_4)_2O\cdot 14MoO_3$" (which is a formula between $2(NH_4)_2O\cdot 5MoO_3$ and $(NH_4)_2O\cdot 3MoO_3$) instead of $2(NH_4)_2O\cdot 5MoO_3$, and $(NH_4)_2O\cdot 28MoO_3$ was found additionally. The (5:14)-molybdate is formed at 90 to 130°C, the (1:4)-molybdate at 175 to 210°C, the (1:28)-molybdate at 285 to 315°C, and MoO_3 at 360 to 395°C. In this study thermogravimetry, differential thermogravimetry, X-ray powder analysis [22] and mass spectroscopic investigations of the gaseous decomposition products [23] were applied. The latter showed that above 300°C N_2 instead of NH_3 is increasingly evolved due to oxidation of the NH_3 by Mo^{VI} oxide [23].

In two further investigations the authors [21, 9] disagree completely with the above results with respect to the formulae of the decomposition products (with the exception of MoO_3). In these investigations of a larger number of ammonium salts the conclusions were obviously drawn without special knowledge (see reference list) on polymolybdates. Postulated formulae of intermediates are $(NH_4)_6Mo_7O_{24}$ [21, 9], $6NH_3\cdot 7MoO_3$ [21], $2(NH_4)_2O\cdot 7MoO_3$ [9]. These investigations were carried out using thermogravimetry, differential thermogravimetry, differential thermal analysis [21, 9], and a special method for thermoanalysis of evolved gases [9].

For the investigation of a sample that was stored in Egypt at room temperature (which may reach up to 45°C in summer) for many years and assumed to be the monohydrate see [28].

Effect of Various Flowing Atmospheres. The methods in this study were thermogravimetry, differential thermogravimetry, differential thermal analysis, X-ray powder diffraction, and high resolution mass spectrometry.

Atmospheres of dry air, nitrogen, or helium: Compared with atmospheres of static air, the same intermediates were formed, i.e., $2(NH_4)_2O\cdot 5MoO_3$ and $(NH_4)_2O\cdot 4MoO_3$. However, the steps were less sharp, and the changes began at lower and terminated at higher temperatures.

Atmospheres of wet air (= H_2O-air mixtures): Before heating, the weight of the sample increased very rapidly as the wet air flow was introduced. Accordingly an additional step was found at the beginning of the curves.

Atmospheres of NH_3–N_2 mixtures: Before heating, the weight of the sample increased by 7 to 12%, depending on the flow rate. Accordingly an additional step was found at the beginning of the curves. A second additional step was observed between $2(NH_4)_2O\cdot 5MoO_3$ and $(NH_4)_2O\cdot 4MoO_3$. The NH_3 evolution appeared at temperatures 20 to 30 K higher than in the other cases [14].

---

[1] The "decamolybdate" and phase C polymolybdate types have been shown to be identical, see the notes added in proof on pp. 7 and 23/4.

In Air at Reduced Pressures. Thermolysis was carried out at 580, 130, 70, 20, 5, 10^{-3} Torr [22], and at 10^{-2} Torr; however, the gaseous compounds evolved led to a pressure of several Torr [8]. Generally, at pressures ≦20 Torr the curves showed gentle slopes with very small plateaus or only inflection points for the different compounds [8, 22]. Both the authors agree that at reduced pressures partially dehydrated ammonium heptamolybdates occur, $3(NH_4)_2O \cdot 7MoO_3 \cdot H_2O$ at 130 Torr and ~105°C [22], $3(NH_4)_2O \cdot 7MoO_3 \cdot 2H_2O$ at 20 Torr and ~85°C [22] and at 10^{-2} Torr and 50 to 65°C [8]. In the latter case the 2-hydrate was characterized by its X-ray powder pattern which is similar to that of the 4-hydrate. There is also agreement with respect to the non-occurrence of the (2:5)-compound, of a decreasing importance of the (1:4)-compound with decreasing pressure, and of an increasing importance of a fairly stable intermediate between the "tetramolybdate" and molybdenum trioxide. This latter is assumed to be $(NH_4)_2O \cdot 15MoO_3$ occurring at 20 to 130 Torr and 300 to 360°C [22] or $(NH_4)_2O \cdot 12MoO_3$ occurring at 10^{-2} Torr and 235 to 315°C [8]. $(NH_4)_2O \cdot 12MoO_3$ was identified by its X-ray powder pattern [18] (cf. pp. 114 and 115). The final product was MoO_3 formed at 360 to 400°C at 130 Torr, 360 to 380°C at 20 Torr [22], and above 340°C at 10^{-2} Torr [8]. The methods applied in these studies were thermogravimetry [8, 22], differential thermogravimetry [22], differential thermal analysis [8], and X-ray powder analysis [8, 22].

Kinetic Studies. Decomposition in the pure state: The activation energies of the decomposition of ammonium paramolybdate under 760 Torr are 24.6, 27.1, and 31.9 kcal/mol, respectively, for the reactions in which $2(NH_4)_2O \cdot 5MoO_3$, $(NH_4)_2O \cdot 4MoO_3$, and MoO_3 are formed (see above) [8]; 38.6, 194.8, 81.9, and 151.2 kcal/mol, respectively, for the steps leading to $5(NH_4)_2O \cdot 14MoO_3$, $(NH_4)_2O \cdot 4MoO_3$, $(NH_4)_2O \cdot 28MoO_3$, and MoO_3 (see above). For details see the paper [22].

Decomposition of alumina-supported paramolybdate: Ammonium paramolybdate supported on alumina decomposes giving the same intermediates and end product as pure ammonium paramolybdate, i.e., $2(NH_4)_2O \cdot 5MoO_3$, $(NH_4)_2O \cdot 4MoO_3$, and MoO_3, but with much smaller stability intervals. For the unsupported as well as for the supported molybdate the value of the reaction order has been found to be 2 for the reactions (1) and (2) and 1 for the reaction (3). The values of the activation parameters of the processes (1) to (3) for different concentrations of ammonium paramolybdates are given in the following table:

process	concentration in %	activation energy in kcal/mol	preexponential factor in s^{-1}
(1)	14	15	2.5×10^{6}
	20	18	1×10^{8}
	29	20	1×10^{9}
	40	23	5.7×10^{10}
	100	31.5	2.4×10^{15}
(2)	14	37.4	7×10^{13}
	20	45.5	3×10^{17}
	29	48.2	4×10^{18}
	40	54.3	3×10^{21}
	100	63	1.4×10^{25}
(3)	14	45.5	7×10^{13}
	20	50.6	7×10^{15}
	29	51.8	1.7×10^{16}
	40	54.2	1×10^{17}
	100	56	3×10^{17}

The dependences of both activation energy and preexponential factor upon concentration are assumed to be due to the support effect. For a more detailed discussion of the mode of operation of the support see the paper [2].

Low Temperature Dehydration. $(NH_4)_6Mo_7O_{24}$ was obtained by rapid freezing to −10°C (liquid nitrogen) of a solution of 150 g $(NH_4)_6Mo_7O_{24}\cdot 4H_2O$/L followed by sublimation at −10°C and 10^{-2} Torr and degassing at 30°C and 10^{-4} Torr. The X-ray powder pattern differs from that of the 4-hydrate. For further investigations see [8, 26].

References:

[1] I. K. Bhatnagar, D. K. Chakrabarty, A. B. Biswas (Indian J. Chem. **10** [1972] 1025/8). – [2] I. Sălăgeanu, D. Trestianu, E. Segal (Rev. Roumaine Chim. **18** [1973] 1537/46). – [3] C. Duval (Anal. Chim. Acta **15** [1956] 223/5). – [4] A. B. Kiss, P. Gadó, I. Asztalos, A. J. Hegedüs (Acta Chim. [Budapest] **66** [1970] 235/49). – [5] W. D. Hunnius (Z. Naturforsch. **30b** [1975] 63/5).

[6] W. Romanowski (Chem. Stosowana **7** [1963] 105/23, 122; C.A. **59** [1963] 14623). – [7] E. Ma (Bull. Chem. Soc. Japan **37** [1964] 648/53). – [8] A. Louisy, J. M. Dunoyer (J. Chim. Phys. **67** [1970] 1390/4). – [9] M. Berényi (Talanta **16** [1969] 101/6). – [10] J. Fuchs, H. Hartl, W. D. Hunnius, S. Mahjour (Angew. Chem. **87** [1975] 634/5; Angew. Chem. Intern. Ed. Engl. **14** [1975] 644).

[11] T. Dupuis (Mikrochemie **35** [1950] 449/65, 451/2). – [12] T. Dupuis, C. Duval (Anal. Chim. Acta **4** [1950] 173/9, 173/4). – [13] C. J. Hallada (J. Less-Common Metals **36** [1974] 103/10, 109). – [14] K. Isa, Y. Hirai, H. Ishimura (Therm. Anal. Proc. 5th Intern. Conf., Kyoto 1977, pp. 348/51). – [15] E. Ma (Bull. Chem. Soc. Japan **37** [1964] 171/5).

[16] M. Onchi, E. Ma (J. Phys. Chem. **67** [1963] 2240/1). – [17] P. Gadó, A. B. Kiss (Hiradastech. Ipari Kut. Intez. Kozl. **8** [1968] 113/24 from C.A. **70** [1969] No. 120682). – [18] M. J. Schwing-Weill (Bull. Soc. Chim. France **1967** 3795/8). – [19] K. Funaki, T. Segawa (J. Electrochem. Soc. Japan **18** [1950] 152/4). – [20] E. Ya. Rode, V. N. Tverdokhlebov (Zh. Neorgan. Khim. **3** [1958] 2343/6; Russ. J. Inorg. Chem. **3** No. 10 [1958] 157/62).

[21] L. Erdey, S. Gál, G. Liptay (Talanta **11** [1964] 913/40, 926/7). – [22] I. H. Park (Bull. Chem. Soc. Japan **45** [1972] 2739/44). – [23] I. H. Park (Bull. Chem. Soc. Japan **45** [1972] 2745/8). – [24] F. A. Schröder, J. Scherle (Z. Naturforsch. **28b** [1973] 46/55). – [25] Remark of the reviewer.

[26] J. G. Batteux, A. F. Louisy, J. M. Dunoyer (Progr. Refrig. Sci. Technol. Proc. 13th Intern. Congr. Refrig., Washington, D.C. 1971 [1973], Vol. 3, pp. 611/6 from C.A. **81** [1974] No. 9204). – [27] C. Duval (Inorganic Thermogravimetric Analysis, 2nd Ed., Elsevier, Amsterdam – London – New York 1963, pp. 96/7, 462). – [28] Z. M. Hanafi, M. A. Khilla, M. H. Askar (Thermochim. Acta **45** [1981] 221/32).

2.5.2.3.1.7 Chemical Reactions. Solubility

The reaction of $(NH_4)_6[Mo_7O_{24}]\cdot 4H_2O$ with NH_3 in the solid state gives $(NH_4)_2MoO_4$ [1].

The interaction of $Ce(NO_3)_3\cdot 6H_2O$ and $(NH_4)_6[Mo_7O_{24}]\cdot 4H_2O$ was studied at 500 to 850°C in air. By variation of the Ce/Mo ratio and the oxidation state of Ce, the following compounds were obtained: β-$Ce_2Mo_3O_{13}$, α- and β-$Ce_2Mo_4O_{15}$, and $Ce_8Mo_{12}O_{49}$. The formation of the cerium molybdates passes through several stages [2].

By evaporation of a solution containing 10 g of $(NH_4)_6[Mo_7O_{24}]\cdot 4H_2O$ and 15 g of H_3BO_3 in 50 mL of water to a 10 to 15 mL volume of water, $H_3[B(Mo_2O_7)_6]\cdot 28H_2O$ was formed. The complex was used as a dry powder as a catalyst for the epoxidation of cyclohexene with cumene hydroperoxide or *t*-butyl hydroperoxide [3].

$(NH_4)_6[Mo_7O_{24}]\cdot 4H_2O$ is readily soluble in water [4 to 9], see also p. 91. The solution of the solid [10, 11] as well as the solution from which the compound crystallizes [11] have been shown unambiguously by Raman "finger-prints" to contain the $[Mo_7O_{24}]^{6-}$ ion. Solubility S in dependence on the temperature t [9]:

t in °C	−3.0	−1.4	3.6	8.7	19.8	24.6	30.1
S in g/100 g solution[a)]	10.5	11.4	13.9	18.4	29.5	33.5	40
[b)]	—	—	—	—	31.7	37.1	42.8
t in °C	40.2	49.9	59.8	70.0	80.2	89.9	
S in g/100 g solution[a)]	50	57	62	64.5	68	72	
[b)]	51.5	—	—	—	—	—	

[a)] Visual observation of the disappearance of the last crystal. – [b)] Conductometric measurement.

In trifluoroacetic acid the compound dissolves in an amount from 1 to 10 g salt per 100 g CF_3COOH. The polymolybdate ion reacts with the solvent but products were not identified [12].

References:

[1] F. Rémy, R. Mercier, K. Keowkamnerd, J. J. Hantzpergue (Compt. Rend. C **275** [1972] 733/6). – [2] J. C. J. Bart, N. Giordano (Proc. 11th Rare Earth Res. Conf., Traverse City, Mich., 1974, Vol. 1, pp. 157/65 from C.A. **83** [1975] No. 104038). – [3] L. Cerveny, A. Marhoul, V. Ruzicka, A. Hora, J. Novak (Czech. 156839 [1972/75] from C.A. **83** [1975] No. 131434). – [4] H. W. Foote, W. M. Bradley (J. Am. Chem. Soc. **58** [1936] 930/1). – [5] K. Funaki, T. Segawa (J. Electrochem. Soc. Japan **18** [1950] 152/4).

[6] Z. G. Karov, F. M. Perel'man (Zh. Neorgan. Khim. **5** [1960] 713/9; Russ. J. Inorg. Chem. **5** [1960] 343/6). – [7] C. J. Hallada (J. Less-Common Metals **36** [1974] 103/10; Chem. Uses Molybdenum Proc. 1st Conf., Reading, Engl., 1973, pp. 52/5). – [8] A. Chenard, R. Cohen-Adad (Compt. Rend. C **280** [1975] 1371/3). – [9] A. Chenard, R. Tenu, R. Cohen-Adad (Bull. Soc. Chim. France **1976** 1797/801). – [10] J. Aveston, E. W. Anacker, J. S. Johnson (Inorg. Chem. **3** [1964] 735/46).

[11] K. H. Tytko, B. Schönfeld (Z. Naturforsch. **30b** [1975] 471/84). – [12] R. Hara, G. H. Cady (J. Am. Chem. Soc. **76** [1954] 4285/7).

2.5.2.3.2 $3(NH_4)_2O\cdot 7MoO_3\cdot 7H_2O$?

A compound $5(NH_4)_2O\cdot 12MoO_3\cdot 12H_2O$ was obtained from ammonium molybdate solutions acidified with acetic acid to pH 6.2. The "3:7" formulation is $3(NH_4)_2O\cdot 7MoO_3\cdot 7H_2O$.

H. Guiter (Bull. Soc. Chim. France [5] **12** [1945] 74/5).

2.5.2.3.3 $(NH_4)_6Mo_7O_{24}\cdot 3C_2H_5OH$ (=$3(NH_4)_2O\cdot 7MoO_3\cdot 3C_2H_5OH$)

The compound was prepared by controlled hydrolysis of $3MoO_2(OC_2H_5)_2\cdot 2NH_3$ (I) in absolute ethanol-ether in the presence of ammonia according to $7MoO_2(OC_2H_5)_2 + 10H_2O + 6NH_3 \rightarrow (NH_4)_6Mo_7O_{24} + 14C_2H_5OH$. A ca. 0.7 M solution of I in absolute ethanol was diluted with a twofold volume of absolute ether, saturated with gaseous ammonia, and precipitated by adding half the volume of a solution of 3% H_2O in ether. All the reactions were performed at room temperature in special equipment excluding atmospheric humidity. The precipitate instantaneously formed was filtered off and dried in a stream of dry nitrogen.

The composition of the product is not sensitive to the amount of H_2O applied, indicating a uniform compound. The colorless substance is X-ray amorphous.

J. Fuchs, K. F. Jahr, A. Nebelung (Chem. Ber. **100** [1967] 2415/20).

2.5.2.4 $[(NH_4)_6Mo_8O_{27}\cdot 4H_2O]_\infty$ ($=3(NH_4)_2O\cdot 8MoO_3\cdot 4H_2O$), Ammonium Polyoctamolybdate

The compound was first obtained by [1] when studying the NH_3-MoO_3-H_2O system and later several times [2, 3, 7] rediscovered. It precipitates from concentrated [2], saturated [4], and 0.2 to 0.25 M [3, 5, 6] solutions of ammonium heptamolybdate after several weeks to months. Precipitation is accelerated by acidification to $1.2H^+/MoO_4^{2-}$ and addition of NH_4Cl [7]. In [2] and [3] the content of water of crystallization is erroneously stated to be $3H_2O$. Furthermore, in [3] the formula is reported to be $(NH_4)_6Mo_8O_{26}(OH)_2\cdot 2H_2O$. The compound was also obtained by [8] when studying the NH_3-MoO_3-H_2O system; however, other authors [9 to 11] failed to find this compound in the system. The interrelations between the solid polymolybdates occurring in this range of acidification have been described as follows: a saturated solution of ammonium heptamolybdate usually deposits first (after some days) the cotton-like "trimolybdate", then much later (after several weeks) the "trimolybdate" is converted into the polyoctamolybdate [4]. The formation of the solid compound ceases when about 60 to 70% of the heptamolybdate has reacted. The formation of the compound is associated with the release of NH_3: $8[Mo_7O_{24}]^{6-}+48NH_4^+ + 25H_2O \rightarrow 7[(NH_4)_6Mo_8O_{27}\cdot 4H_2O]_\infty + 6NH_3$ (updated equation). Thus, in case of the alkali metal salts the formation of OH^- instead of NH_3 should prevent the formation of the corresponding alkali metal compounds [3].

The compound forms colorless needles [3, 5, 6] and was characterized by analyses [1 to 3, 8, 12 to 14], infrared [3, 12 to 14] (see Fig. 6, p. 12) and Raman spectra [3, 4, 13, 14] (see Fig. 5, p. 12), and X-ray powder patterns (see the papers) [2, 7].

The crystals are monoclinic, space group $P2_1/c$-C_{2h}^5 (No. 14), lattice parameters $a=9.519\pm 0.005$, $b=11.309\pm 0.006$, $c=15.049\pm 0.008$ Å, $\beta=109.50°\pm 0.03°$; $Z=2$. $D_m=3.01\pm 0.02$, $D_x=3.00$ g/cm^3 [5, 6] (see also [3]). Atomic parameters are given in [6].

The crystal structure has been determined from three-dimensional X-ray diffractometer data and refined to $R=0.064$. The structure of the polymolybdate ion (see Fig. 4, p. 12) consists of octamolybdate units which are connected through one common oxygen atom (free corner of the octamolybdate unit, the only atom on a special atomic position) to form infinite chains running parallel to the a axis. (The chain structure is similar to that of the compound $K_4Mo_8O_{26}$ obtained from melts; in that case, however, the octameric units are connected through two common oxygen atoms (free edge of the octamolybdate unit) [15, 16], see "Molybdän" Erg.-Bd. B1, 1975, p. 221). The octamolybdate units are built up of distorted MoO_6 octahedra sharing edges (and corners). Mo-O bond lengths vary from 1.697 to 2.401 Å, with values of 1.697 to 1.732 Å in the terminal bonds due to π-bonding. Mo···Mo distances are 3.258 to 3.472 Å (edge-sharing) and 3.778 to 3.980 Å (corner-sharing). The short distance of the corner-sharing octahedra is that between octahedra of different octameric units due to the coordination number two of the connecting oxygen atom, whereas within the octameric units the corner-sharing oxygen atoms have coordination numbers three or four. O···O distances of the common edges between MoO_6 octahedra (2.447 to 2.723 Å) are shorter than those of the free edges (2.691 to 3.088 Å). The connection of the anionic chains in the b and c directions is determined mainly by an extended hydrogen-bridge bond system between predominantly terminal oxygen atoms of the polyanion, all the ammonium ions, and molecules of water of crystallization (see a figure in [6]). The distances N···O and O···O within the bridges vary from 2.800 to 3.145 Å and 2.718 to 3.307 Å, respectively. There are also bridges between the octameric units (i.e., in the a direction). Many of the hydrogen bridges occur as bifurcated bridges. The oxygen atoms of the polymolybdate ions ($r(O^{2-})=1.40$ Å), of the water molecules, and the ammonium cations ($r(NH_4^+)=1.43$ Å) form a distorted cubic closest spherical packing in which the molybdenum atoms occupy some of the octahedral gaps. For the individual bond lengths and bond angles within the polymolybdate ion and within the hydrogen bond system see the paper [6] (see also [5]).

The structure of the polyoctamolybdate ion has also been derived from some simple experimental data (such as analytical results, lattice parameters) and assumptions (presence of MoO_6 octahedra, chain structure to explain the insolubility) in combination with considerations on the mechanism of formation (see the paper). A mechanism comprised of the following stages was proposed: formation of tetramolybdate ions according to an addition mechanism (compare "Molybdenum" Suppl. Vol. B 3, to be published); formation of octamolybdate ions from the tetramolybdate ions by further aggregation according to a condensation mechanism (compare "Molybdenum" Suppl. Vol. B 3) (**Fig. 49a**); and polycondensation of the octameric units (**Fig. 49b**). The formation of the polyoctamolybdate ion is thermodynamically unfavorable in comparison with that of the parent compound, the heptamolybdate ion (see the paper), and is only possible since the polyoctamolybdate ion is continuously removed from the equilibrium by precipitation. The very slow formation of the compound is a consequence of the low thermodynamic and kinetic stability (and hence small concentrations) of the intermediate stages compared to the heptamolybdate ion. An analogous planar octamolybdate ion is considered to be incapable of crystallization (cf. section 2.1.9.11, p. 38) [17]. Considerations on the formation mechanism of the polyoctamolybdate ion have also been published by [19] (cf. "Molybdenum" Suppl. Vol. B 3).

The compound dissolves slowly in water with hydrolysis to form a solution with P = 1.25 [4], i.e., $[Mo_7O_{24}]^{6-}$ and its protonated forms [18] (compare "Molybdenum" Suppl. Vol. B3). Thus, $[Mo_8O_{26}]^{4-}$ as stated in [3] cannot occur as a product of hydrolysis. For the solubility see p. 91.

The compound is stable in air [3]. Thermogravimetric analysis shows decomposition in three steps. Between 115 and 130°C the anhydrous "trimolybdate" $(NH_4)_2O\cdot3MoO_3$ is formed, between 202 and 226°C the anhydrous "tetramolybdate" $(NH_4)_2O\cdot4MoO_3$, and between 290 and 330°C MoO_3 [2].

References:

[1] H. W. Foote, W. M. Bradley (J. Am. Chem. Soc. **58** [1936] 930/1). – [2] M. J. Schwing-Weill (Bull. Soc. Chim. France **1967** 3795/8). – [3] O. Glemser, G. Wagner, B. Krebs (Angew. Chem. **82** [1970] 639; Angew. Chem. Intern. Ed. Engl. **9** [1970] 639). – [4] K. H. Tytko, B. Schönfeld (Z. Naturforsch. **30b** [1975] 471/84). – [5] I. Böschen, B. Buss, B. Krebs, O. Glemser (Angew. Chem. **85** [1973] 409; Angew. Chem. Intern. Ed. Engl. **12** [1973] 409).

[6] I. Böschen, B. Buss, B. Krebs (Acta Cryst. B **30** [1974] 48/56). – [7] W. D. Hunnius (Diss. F.U. Berlin 1970, pp. 52a/53). – [8] A. Chenard, R. Cohen-Adad (Compt. Rend. C **280** [1975] 1371/3). – [9] K. Funaki, T. Segawa (J. Electrochem. Soc. Japan **18** [1950] 152/4). – [10] C. J. Hallada (J. Less-Common Metals **36** [1974] 103/10).

[11] Z. G. Karov, F. M. Perel'mann (Zh. Neorgan. Khim. **5** [1960] 713/9; Russ. J. Inorg. Chem. **5** [1960] 343/6). – [12] B. Schönfeld (Diss. Göttingen, West Germany, 1973, p. 49). – [13] W. D.

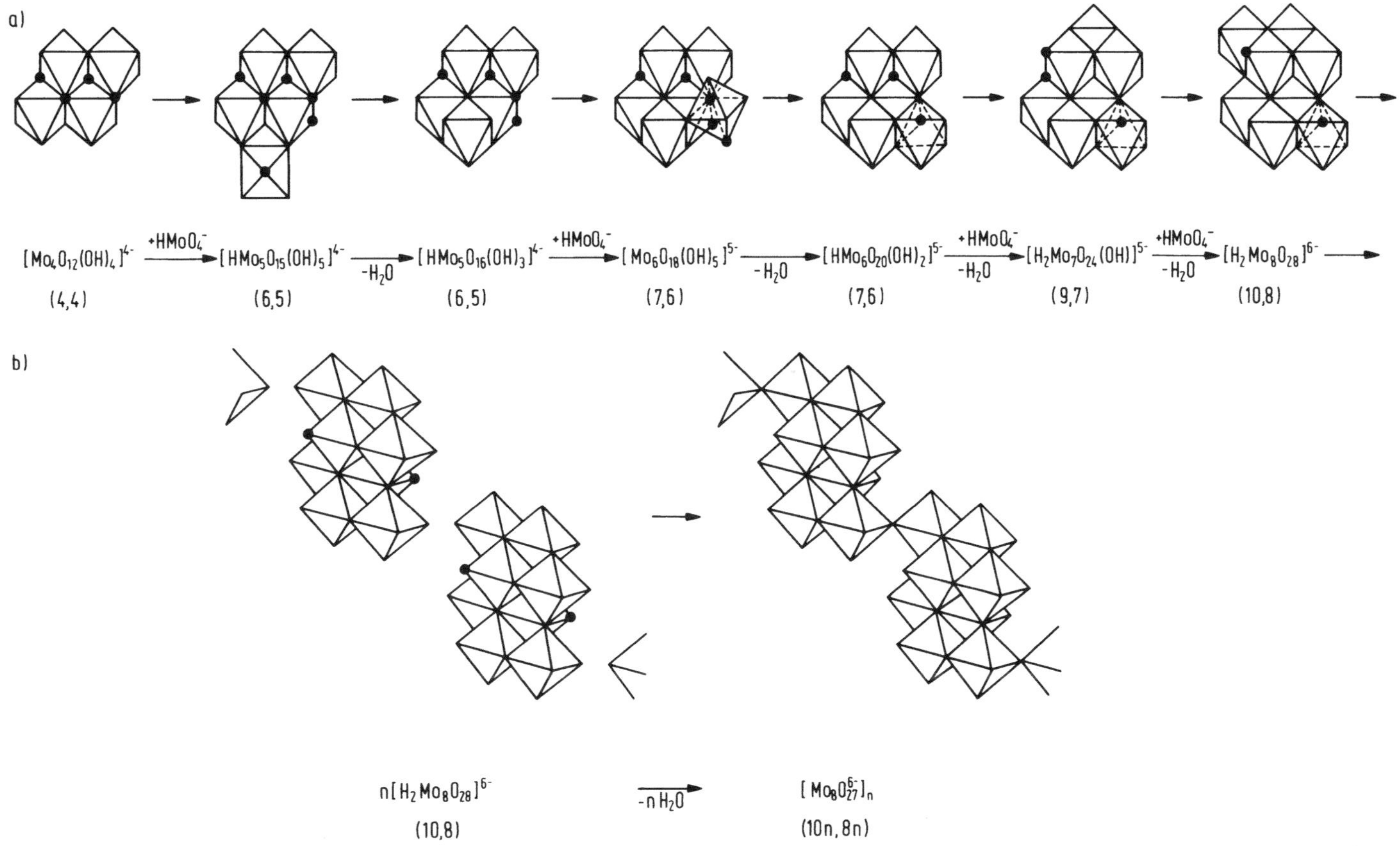

Fig. 49. Reaction mechanism proposed for the formation of the polyoctamolybdate chains. (a) Formation of the octameric units from tetramolybdate ions according to a condensation mechanism. (b) Polycondensation of the octameric units [17].

Hunnius (Z. Naturforsch. **30b** [1975] 63/5). – [14] I. Böschen (Diss. Kiel, West Germany, 1974, p. 167). – [15] B. M. Gatehouse, P. Leverett (Chem. Commun. **1970** 740/1).

[16] B. M. Gatehouse, P. Leverett (J. Chem. Soc. A **1971** 2107/12). – [17] K. H. Tytko (Z. Naturforsch. **28b** [1973] 272/5). – [18] K. H. Tytko, G. Baethe, E. R. Hirschfeld, K. Mehmke, D. Stellhorn (Z. Anorg. Allgem. Chem. **503** [1983] 43/66). – [19] A. Goiffon, B. Spinner (Bull. Soc. Chim. France **1977** 1081/8, 1086).

2.5.2.5 $[(NH_4)_4Mo_6O_{20}\cdot 2nH_2O]_\infty$ (= $(NH_4)_2O\cdot 3MoO_3\cdot nH_2O$), n = 0.5, 1, 2, 3, 9, x; Ammonium "Trimolybdates"

$[(NH_4)_4Mo_6O_{20}\cdot 18H_2O]_\infty$ (= $(NH_4)_2O\cdot 3MoO_3\cdot 9H_2O$). This compound crystallizes from 0.1 to 0.2 M solutions of ammonium molybdate acidified with hydrochloric or nitric acid to ~1.3 H^+/MoO_4^{2-} (pH ≈ 4) after several days in the form of a white, voluminous, cotton-like product consisting of very fine needles. Starting material is ammonium paramolybdate [1]. For an account of an unsuccessful attempt to prepare ammonium trimolybdate see [5].

The compound was characterized by analysis (the value of n = 9 [2] appears to be rather high), infrared [2] (see Fig. 9, p. 14) and Raman spectra [1] (see Fig. 8, p. 14). Both the spectra are typical "trimolybdate" spectra suggesting that this "trimolybdate" contains the same chain-like polymolybdate anion as the rubidium "trimolybdate" monohydrate [3] (see Fig. 7, p. 13).

The compound is sparingly soluble in water, accompanied by decomposition. The solution contains the polymolybdate ions usually occurring in aqueous polymolybdate solutions at P = 1.33 [1], i.e., predominantly protonated $[Mo_7O_{24}]^{6-}$ ions [4] (compare "Molybdenum" Suppl. Vol. B 3, to be published).

$[(NH_4)_4Mo_6O_{20}\cdot 6H_2O]_\infty$ (= $(NH_4)_2O\cdot 3MoO_3\cdot 3H_2O$). This hydrate crystallizes, like all the "trimolybdates", as very small needles which adhere to each other. The compound is characterized by chemical analysis and X-ray powder photographs (no data are given). The latter indicate a very large unit cell [8].

$(NH_4)_2O\cdot 3MoO_3\cdot 2H_2O$. This hydrate was reported to deposit from ammonium molybdate solutions of pH 3 to 5.9 (if acidified with HCl) or 4.6 to 5.9 (if acidified with CH_3COOH) after a month as a downy substance [12]. This hydrate was also described by [7] as one of the solid phases in the ternary NH_3–MoO_3–H_2O system. Other authors observed the (3:8)-molybdate $[(NH_4)_6Mo_8O_{27}\cdot 4H_2O]_\infty$ instead or only the paramolybdate $(NH_4)_6[Mo_7O_{24}]\cdot 4H_2O$ (see p. 91). The compound described in [6] is, according to its infrared spectrum, no "trimolybdate".

Thermal decomposition studies showed the formation of $(NH_4)_2O\cdot 3MoO_3$ at ca. 110°C, the formation of $(NH_4)_2O\cdot 4MoO_3$ at ca. 200°C, and the formation of MoO_3 at ca. 280°C [7].

$[(NH_4)_4Mo_6O_{20}\cdot 2H_2O]_\infty$ (= $(NH_4)_2O\cdot 3MoO_3\cdot H_2O$). To prepare this hydrate 100 mL of a 0.15 M solution of ammonium paramolybdate were acidified with 6.3 mL of 1 N HCl to 1.20 H^+/MoO_4^{2-}. After approximately a week 10 mL of a 5.1 M solution of NH_4Cl and some crystals of the trimolybdate were added to initiate crystallization. After 3 d the voluminous precipitate was filtered off and washed with water. The crystals form fine needles which stick firmly together [11].

The compound was characterized by analysis, infrared spectrum, and X-ray powder diagram [11]. The spectrum is a typical "trimolybdate" spectrum suggesting the same polymolybdate anion to be present as in other cases (see Fig. 7, p. 13).

$(NH_4)_4H_2Mo_6O_{21}$ (= $(NH_4)_2O\cdot 3\,MoO_3\cdot 0.5\,H_2O$). This "trimolybdate" is obtained by a controlled hydrolysis of the compound $3\,MoO_2(OC_2H_5)_2\cdot 2\,NH_3$ in an ethanol-ether mixture as the solvent in the presence of NH_4Cl. A special apparatus is used. To 10 to 15 mL of a solution of $3\,MoO_2(OC_2H_5)_2\cdot 2\,NH_3$ in absolute ethanol (0.7 M Mo) are added 20 to 30 mL of absolute ether, 2 mL of a saturated solution of NH_4Cl in absolute ethanol, and 5 to 6 mL of a solution of water (3%) in ether. The precipitated compound is admixed with NH_4Cl. The use of less ether in the solvent prevents coprecipitation of NH_4Cl, but gives smaller yields.

The X-ray amorphous compound was characterized by analysis. The conditions of preparation are assumed to prevent the formation of hydrates; hence the compound is formulated as an acid polymolybdate. The structure of the polyanion is assumed to be a six-membered ring consisting of MoO_6 octahedra and MoO_4 tetrahedra [13].

$(NH_4)_2O\cdot 3\,MoO_3\cdot x\,H_2O$. The preparation of the compound is not reported. The thermal behavior has been studied by thermogravimetric investigation combined with a determination of the NH_3 evolved. Dehydration of this hydrate starts at about 60°C and is complete around 150°C. Above 200°C deammoniation occurs [9] and the crystal structure changes into that of the "tetramolybdate" [9, 10] (by the author assumed to be an octamolybdate). At 255°C the "tetramolybdate" crystallizes. Above 360°C molybdenum trioxide is produced, at about 380°C the trioxide becomes predominant and at 425°C crystallization of the trioxide occurs [10].

References:

[1] K. H. Tytko, B. Schönfeld (Z. Naturforsch. **30b** [1975] 471/84). – [2] B. Schönfeld (Diss. Göttingen, West Germany, 1973, p. 47). – [3] H. U. Kreusler, A. Förster, J. Fuchs (Z. Naturforsch. **35b** [1980] 242/4). – [4] K. H. Tytko, G. Baethe, E. R. Hirschfeld, K. Mehmke, D. Stellhorn (Z. Anorg. Allgem. Chem. **503** [1983] 43/66). – [5] M. J. Schwing-Weill (Bull. Soc. Chim. France **1967** 3795/8).

[6] F. Arnaud-Neu, M. J. Schwing-Weill (Bull. Soc. Chim. France **1973** 3225/32). – [7] K. Funaki, T. Segawa (J. Electrochem. Soc. Japan **18** [1950] 152/4). – [8] I. Lindqvist (Nova Acta Regiae Soc. Sci. Upsaliensis [4] **15** No. 1 [1950] 1/22, 7). – [9] E. Ma (Bull. Chem. Soc. Japan **37** [1964] 171/5). – [10] E. Ma (Bull. Chem. Soc. Japan **37** [1964] 648/53).

[11] W. D. Hunnius (Diss. F.U. Berlin 1970, pp. 50/3, 56/7). – [12] H. Guiter (Bull. Soc. Chim. France [5] **12** [1945] 74/5). – [13] J. Fuchs, K. F. Jahr, A. Nebelung (Chem. Ber. **100** [1967] 2415/20).

2.5.2.6 $2(NH_4)_2O\cdot 7\,MoO_3\cdot x\,H_2O$ (?)

This ill-defined compound is formed only in a very narrow acidity range between the "tri-" and "tetramolybdates" and thus cannot be prepared entirely pure. Analysis: $2(NH_4)_2O\cdot 7.04$ to $7.20\,MoO_3\cdot x\,H_2O$. The substance was characterized by X-ray powder photographs (not reproduced) showing the presence of impurities.

This "2:7" type compound has never been described by other authors.

I. Lindqvist (Nova Acta Regiae Soc. Sci. Upsaliensis [4] **15** No. 1 [1950] 1/22, 7).

2.5.2.7 $(NH_4)_4[Mo_8O_{26}]\cdot 2nH_2O$ (= $(NH_4)_2O\cdot 4MoO_3\cdot nH_2O$), n = 0, 2, 2.5, 3, 7; Ammonium "Tetramolybdates"

$(NH_4)_2O\cdot 4MoO_3\cdot nH_2O$, n = 7, 3, 2, 0. These hydrates are reported to deposit from ammonium molybdate solutions acidified with hydrochloric or acetic acid as white crystalline precipitates [1]:

n	7	3	2	2	0
acid	HCl	HCl	HCl	CH_3COOH	CH_3COOH
pH	2.6	3	2.6	2.2 to 3.5	2.2 to 3.5
time	1 d	15 d	30 d	some days rapid (?) crystallization	some days slow (?) crystallization

Some statements regarding the time necessary for formation or crystallization are confused. For the anhydrous compound see "Molybdän" Erg.-Bd. B 1, 1975, pp. 228/9.

β-$(NH_4)_4[Mo_8O_{26}]\cdot 4(5)H_2O$ (= $(NH_4)_2O\cdot 4MoO_3\cdot 2(2.5)H_2O$), **Ammonium β-Octamolybdate.** Large transparent crystals of the 4- and 5-hydrate were obtained from 0.1 to 0.2 M ammonium molybdate solutions acidified with hydrochloric or nitric acid to 1.5 H^+/MoO_4^{2-} (pH ≈ 2.5). Usually the finely crystalline compound $(NH_4)_2O\cdot 4MoO_3$ precipitates first; the large crystals of the 4- and 5-hydrate appear later after several days. Recrystallization always yielded $(NH_4)_2O\cdot 4MoO_3$ [2]. The 4-hydrate was prepared by adding the calculated amount of "molybdic acid" to ammonium paramolybdate solution, followed by crystallization [3]. The 4-hydrate was also obtained from a saturated solution of ammonium paramolybdate (21 g/30 mL H_2O) acidified by dropwise addition of 1 N HNO_3 to pH ≈ 2.5. The abundant precipitate formed was left undisturbed. After 2 d the octamolybdate crystals appeared [4]. The 5-hydrate was obtained from a suspension of 10.0 g of $MoO_3\cdot 2H_2O$ in 80 mL of hot (70°C) water on the addition of small amounts (3.5 g) of ammonium citrate. The solution (pH 3.1) was treated while hot with an equal volume of ethanol. The crystals separated rapidly on cooling or overnight at 20°C if no ethanol had been added. After drying in air, the compound is a 4-hydrate. Attempts to recrystallize the 5-hydrate from warm water always afforded the 4-hydrate [5]. A solution of ammonium paramolybdate acidified with HNO_3 to pH 2 to 2.5 at 40 to 55°C yields $(NH_4)_2O\cdot 4MoO_3\cdot 2H_2O$ [19].

To produce $(NH_4)_2O\cdot 4MoO_3\cdot 2H_2O$ on a technical scale from solutions with large sodium sulfate content, the solution is acidified to pH 2.0 to 3.0 and precipitated with an amount of NH_4Cl which is 60 to 80% of that required to form $(NH_4)_2MoO_4$. Recovery is 99.8% at Mo concentrations >60 g/L, Na content 0.4 to 0.5% of the Mo content. Purification to 0.01% is obtained by agitating the precipitate for 1 h at 100°C with a 7% solution of NH_4Cl at pH 2 and a solid:liquid ratio of 1:5 to 1:7.5 [6]. See also [7].

The compounds were characterized by analyses [2 to 5, 8], infrared [9] (see Fig. 12, p. 16) and Raman spectra [2, 3] (see Fig. 11, p. 16). The IR spectrum of $(ND_4)_4[Mo_8O_{26}]\cdot 4D_2O$ in D_2O is given in [14]. The problem of the number of molecules of water of crystallization was investigated by [8] and [5]. Both the authors state that the dimensions of the unit cells are definitely different, compare Table 6 (the infrared and Raman spectra do not differ [2, 9]). It was postulated that the 5-hydrate occurs in the presence of excess ammonium ions during crystallization. It was further postulated that the loss of one molecule of water on exposure to air occurs with partial disordering of the remaining lattice water but without other loss of crystallinity [5]. The reported stability of the crystals of both the hydrates in air of 30% relative humidity [8] refers possibly only to the dimensions of the unit cell.

Table 6

Crystal Data of Ammonium β-Octamolybdate 5-Hydrate and 4-Hydrate.

H_2O content	$5H_2O$	$4H_2O$[a)]	$4H_2O$	$4H_2O$	$4H_2O$	$4H_2O$
lattice parameters in Å	a=7.76	c=7.848(11)	—	a= 7.31(3)	c= 7.882(4)	c= 7.881
	b=9.75	a=9.769(16)	a=10.0	b=10.21(4)	b=10.047(4)	a=10.051
	c=9.78	b=9.832(13)	—	c=10.62(4)	a=10.595(5)	b=10.603
interaxial angles in °	α= 97.2	γ= 97.40(4)	—	α=112.5(5)	γ=113.40(5)	γ=113.41
	β=100.5	β=101.03(11)	—	β=100.8(5)	β=101.08(7)	α=101.06
	γ= 98.4	α= 99.11(4)	—	γ=105.8(5)	α=105.66(5)	β=105.67
R	—	0.080	—	0.096	0.08	0.0638
V in $Å^3$	707	—	686	718.9	697.6	698.4
D_m in g/cm^3	3.1	3.02	3.1	2.94	3.13(5)	—
D_x in g/cm^3	3.14[b)]	3.06	—	3.08	3.16	3.15
Ref.	[8, 10]	[5]	[8]	[11]	[12]	[4]

a) The loss of one molecule of water occurred on exposure to air for drying. – b) From J. D. H. Donnay, H. M. Ondik (Crystal Data, Determinative Tables, Vol. 2, Inorganic Compounds, Washington, D.C., 1973, p. A-42).

The crystal data of different authors and preparations [4, 5, 8, 10 to 12] are summarized in Table 6. All the authors find the crystals to be triclinic, space group $P\bar{1}$-C_i^1 (No. 2); Z=1. Combination of the results of [8] with the results of a molecular weight determination by [13] (giving hexamolybdate ions) led to the postulation of octamolybdate ions [8]. Atomic coordinates are given in [4, 10].

The structure of the octamolybdate anion was first deduced by [10] for the 5-hydrate from approximate coordinates of the molybdenum atoms as given by the three-dimensional Patterson function. The oxygen atoms were not directly located. The proposed anion was based on eight edge-sharing MoO_6 octahedra (see Fig. 10, p. 16). Later, more complete X-ray studies in which all non-hydrogen atoms were located confirmed this structure [5] and showed that this anion is also present in the 4-hydrate [4, 11, 12] (and in some salts with other cations) and that the MoO_6 octahedra in the $Mo_8O_{26}^{4-}$ blocks are strongly distorted (see **Fig. 50**, p. 112). The bonds around each Mo atom fall into three groups: a short *cis*-pair (1.69 to 1.75 Å), a *trans*-pair of medium length (1.88 to 2.01 Å), and a long *cis*-pair (2.18 to 2.48 Å). Mo–Mo distances: 3.22 to 3.59 Å (edge-sharing), 4.53 to 4.62 Å (corner-sharing) [4, 5]. O···O distances of common edges between MoO_6 octahedra (2.54 to 2.80 Å) are somewhat shorter than those of free edges (2.66 to 3.12 Å) [4]. The $[Mo_8O_{26}]^{4-}$ anion can be described alternatively as two cyclic Mo_4O_{12} units formed from distorted MoO_4 tetrahedra, cross-linked by long Mo–O bonds and by additional long bonds from Mo to two extra O^{2-} ions, O(12), O(12′) [4, 5, 14]. The packing of the anions in the 5-hydrate is a somewhat distorted version of the packing in the 4-hydrate. The orientations of the anions in the projection along [001] are similar, but the cell of the 5-hydrate is 4.6% larger; although its a and b axes are shorter, the interaxial angles are closer to 90° [5]. Polymolybdate anions, ammonium cations, and molecules of water of crystallization are held together by an extended three-dimensional hydrogen-bridge bond system (see a figure in the paper) in which many bifurcated bridges exist. Two of the four water molecules form bridges to two, and two to only one NH_4^+ cation. The distances N···O and O···O within the bridges vary from 2.784 to 3.194 Å and 2.726 to 3.319 Å, respectively. The oxygen atoms and the ammonium cations form a cubic spherical packing [4]. For the individual bond lengths and bond angles within the polymolybdate ion [4, 5] and within the hydrogen bond system [4] see the papers. Short descriptions of the structure and some interatomic distances are also given in [11, 12].

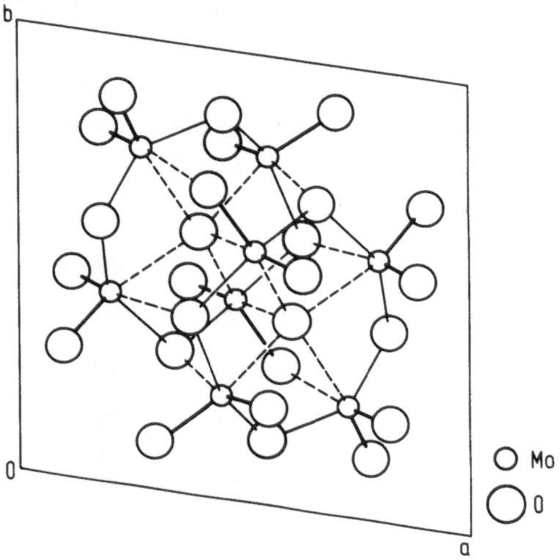

Fig. 50. The bondings within the $[Mo_8O_{26}]^{4-}$ anion. Heavy, light, and dashed lines denote short, medium, and long bonds [5].

The compound dissolves slowly in water to form a solution with P = 1.50 [2] which contains mainly $[Mo_8O_{26}]^{4-}$, $[H_2Mo_7O_{24}]^{4-}$, $[HMo_8O_{26}]^{3-}$, and $[H_3Mo_7O_{24}]^{3-}$ [15].

According to [8] both the hydrates are stable in air of 30% relative humidity. The 5-hydrate is described to be unstable at room temperature and to form a 4-hydrate having cell dimensions somewhat different from those of the 4-hydrate directly obtained from the solution (see above) [5]. According to [4] the 4-hydrate loses water of crystallization at room temperature even in sealed tubes. Thermogravimetric investigations of the 4-hydrate indicate the loss of two molecules of water at 80°C (130°C in an atmosphere of NH_3) [4], 106 to 120°C [18] (in [18] the authors still use the obsolete "1:4" formulation). The anhydrous compound $(NH_4)_2O \cdot 4\,MoO_3$ is formed at 145 to 155°C [18], under NH_3 only at 245°C [4]. This different behavior of the water molecules is explained with the differently strong hydrogen bondings in which the water molecules are involved (see above) [4]. MoO_3 is formed at 320 to 370°C [18]. Another thermogravimetric investigation (combined with a determination of the NH_3 evolved) of a dihydrate (the origin of this product is not reported) shows continuous loss of the two water molecules between 100 and about 280°C. Deammoniation starts at about 250°C and is complete at about 360°C (formation of MoO_3) [16]. DTA investigations and X-ray diffraction patterns confirm these results [17]. NH_3 is partially decomposed to N_2 and H_2. This was interpreted as meaning that ammonia is oxidized to nitrogen by the molybdenum oxide whereas the latter is reduced to a lower oxide. An alternative explanation is that ammonia is decomposed through the catalytic action of the molybdenum oxide [16].

The compound is converted in boiling water (100 g MoO_3 per 250 to 270 g H_2O) at a constant volume within 30 to 60 min into β-$(NH_4)_2O \cdot 4\,MoO_3$ [19].

References:

[1] H. Guiter (Bull. Soc. Chim. France [5] **12** [1945] 74/5). – [2] K. H. Tytko, B. Schönfeld (Z. Naturforsch. **30b** [1975] 471/84). – [3] J. Aveston, E. W. Anacker, J. S. Johnson (Inorg. Chem. **3** [1964] 735/46, 737). – [4] H. Vivier, J. Bernard, H. Djomaa (Rev. Chim. Minerale **14** [1977] 584/604). – [5] T. J. R. Weakley (Polyhedron **1** [1982] 17/9).

[6] K. Ya. Shapiro (Tsvetn. Metal. **34** No. 7 [1961] 57/60; Soviet J.Non-Ferrous Metals **34** No. 7 [1961] 57/60). – [7] K. Ya. Shapiro (Zh. Prikl. Khim. **35** [1962] 486/91; J. Appl. Chem. [USSR] **35** [1962] 466/70). – [8] I. Lindqvist (Acta Chem. Scand. **4** [1950] 551/2). – [9] B. Schönfeld (Diss. Göttingen, West Germany, 1973, p. 51). – [10] I. Lindqvist (Arkiv Kemi **2** [1950/51] 349/55).

[11] L. O. Atovmyan, O. N. Krasochka (Zh. Strukt. Khim. **13** [1972] 342/3; J. Struct. Chem. [USSR] **13** [1972] 319/20). – [12] B. M. Gatehouse (J. Less-Common Metals **54** [1977] 283/8). – [13] G. Jander, K. F. Jahr, W. Heukeshoven (Z. Anorg. Allgem. Chem. **194** [1930] 383/428, 391/402). – [14] W. G. Klemperer, W. Shum (J. Am. Chem. Soc. **98** [1976] 8291/3). – [15] K. H. Tytko, G. Baethe, E. R. Hirschfeld, K. Mehmke, D. Stellhorn (Z. Anorg. Allgem. Chem. **503** [1983] 43/66).

[16] E. Ma (Bull. Chem. Soc. Japan **37** [1964] 171/5). – [17] E. Ma (Bull. Chem. Soc. Japan **37** [1964] 648/53). – [18] E. Ya. Rode, V. N. Tverdokhlebov (Zh. Neorgan. Khim. **3** [1958] 2343/6; Russ. J. Inorg. Chem. **3** No. 10 [1958] 157/62). – [19] G. Robert, F. Fauvarque, H. Cugnardey, R. Durand, Manufactures de Produit Chimique du Nord Etablissement Kuhlmann (Fr. 1268595 [1960/61]; C. **1966** No. 31-2091).

2.5.2.8 $(NH_4)_2[Mo_6O_{19}]\cdot H_2O$ $(=(NH_4)_2O\cdot 6MoO_3\cdot H_2O)$, $(NH_4)_2[Mo_6O_{19}]\cdot xH_2O\cdot yC_{12}H_{24}O_6$ $(C_{12}H_{24}O_6 = 18\text{-crown-6})$, Ammonium Hexamolybdates

$(NH_4)_2[Mo_6O_{19}]\cdot H_2O$. The synthesis of this compound is based on ion exchange between polymolybdates and ion-exchange resins in acetone as the solvent, and it consists of the following steps: preparation of a solution of "hexamolybdic acid", preparation of the hexamolybdate solution, and separation of the salt as a crystalline precipitate. The mixture of 5 g of $(NH_4)_6[Mo_7O_{24}]\cdot 4H_2O$, 15 g of air-dried KU-23 ion exchange resin in the H^+ form, and 2 L of acetone is stirred for 2 h at room temperature. The sorbent and the solid unreacted ammonium paramolybdate are removed by filtration. 20 g of air-dried cation exchanger in the NH_4^+ form are added to the solution and allowed to stand in contact with the solution for a day. After separation of the sorbent, the solution is concentrated to a volume of 250 mL by evaporation of the acetone at room temperature under vacuum (5 to 10 Torr). After addition of 1 L of ethyl ether to the solution the mixture is shaken several times, and is then allowed to stand for 3 to 4 h. The precipitated yellow-green crystals are separated, immediately washed with ether, and dried in air [1].

The compound was characterized by chemical analysis, infrared spectrum (see Fig. 18, p. 19), and UV spectrum of its acetone solution [1]. From the similarity of the infrared spectra it was derived that the structure of the polymolybdate ion present in this compound is the same as that in the compounds $R_2[Mo_6O_{19}]$ (R = organic cation) [2, 3] (see Fig. 16, p. 18). Assignments of the infrared (compare section 2.1.9.6, p. 33) and UV bands are made [1].

$(NH_4)_2[Mo_6O_{19}]\cdot xH_2O\cdot yC_{12}H_{24}O_6$. The compound has been prepared by [4], apparently analogously to the compound $(C_{12}H_{24}O_6)_2\cdot K_2[Mo_6O_{19}]\cdot H_2O$ (see p. 82). No other data are reported.

References:

[1] A. M. Golubev, E. A. Torchenkova, V. I. Spitsyn (Dokl. Akad. Nauk SSSR **217** [1974] 345/7; Dokl. Chem. Proc. Acad. Sci. USSR **214/219** [1974] 495/7). – [2] R. Mattes, H. Bierbüsse, J. Fuchs (Z. Anorg. Allgem. Chem. **385** [1971] 230/42, 236). – [3] H. R. Allcock, E. S. Bissel, E. T. Shawl (J. Am. Chem. Soc. **94** [1972] 8603/4). – [4] O. Nagano, Y. Sasaki (Acta Cryst. B **35** [1979] 2387/9).

2.5.2.9 $(NH_4)_8[Mo_{36}O_{112}(H_2O)_{16}]\cdot{\sim}64\,H_2O$ $(=(NH_4)_2O\cdot 9\,MoO_3\cdot{\sim}20\,H_2O)$, Ammonium 36-Molybdate

Colorless crystals of the compound are obtained from 0.1 to 0.2 M solutions of ammonium molybdate acidified with hydrochloric or nitric acid to Z = 1.8 (~1.9 H^+ added/MoO_4^{2-}; pH ≈ 1) [1], see also [2].

The compound was characterized by analysis [2] and Raman spectrum [1, 2] (see Fig. 20, p. 21). The analytical ratio $(NH_4)_2O:MoO_3$ shows for reasons open to question deviations from the theoretical value up to +26%, corresponding to a composition $(NH_4)_2O\cdot 7$ to $9\,MoO_3$ (similar observations have been made for the sodium and potassium compounds). Different compositions of the compound do not, however, affect the Raman spectrum [1, 2]. The presence of small amounts of the ammonium salt of the acid used for acidification has been assumed to explain the differences in composition [1].

From the identity of the Raman spectra the formula of the polymolybdate anion was proposed to be $[Mo_{36}O_{112}]^{8-}$ in analogy to the corresponding sodium compound [1, 2]. In the light of the X-ray structure determination of the potassium compound [3, 4] the polymolybdate anion is now formulated as $[Mo_{36}O_{112}(H_2O)_{16}]^{8-}$ (see Fig. 19, p. 20).

Crystal data of a monoclinic compound according to these data obviously being the ammonium isomorph of $K_8[Mo_{36}O_{112}(H_2O)_{16}]\cdot 36\,H_2O$ have been reported: lattice parameters a = 16.65(2), b = 18.69(1), c·sin β = 24.19(2) Å, space group $P2_1/c\text{-}C_{2h}^5$ (No. 14). This compound was sometimes obtained in place of the ammonium octamolybdate 5-hydrate when crystallization was performed in the absence of ethanol [5] (see p. 110).

The elongated prismatic crystals crumble rapidly on exposure to air [5]. They readily effloresce [1, 2] with conversion into the "decamolybdate" [1]. The compound is soluble in water without decomposition [1].

This ammonium 36-molybdate is described as "octamolybdate" $(NH_4)_2O\cdot 8\,MoO_3\cdot 19\,H_2O$ [6], as "octamolybdate" $(NH_4)_2O\cdot 8\,MoO_3\cdot x\,H_2O$ (x = 13, 16), as "decamolybdate" $(NH_4)_2O\cdot 10\,MoO_3\cdot 19\,H_2O$ [7], and as $(NH_4)_2O\cdot 6.02$ to $8.74\,MoO_3\cdot x\,H_2O$ [8]. The criteria for these assignments are given on p. 68. A further "octamolybdate" of formula $(NH_4)_2O\cdot 8.1$ to $8.5\,MoO_3\cdot x\,H_2O$ [9] (x = 5 [10]) shows an infrared spectrum [9] very similar to that of other 36-molybdates [11, 12] (see Fig. 21, p. 21); however, the content of water and the conditions of the preparation (working at 70°C) [10] are not in accordance with those for 36-molybdates.

Thermal investigation of the compound $(NH_4)_2O\cdot 8.1$ to $8.5\,MoO_3\cdot 5\,H_2O$ up to 275°C shows the loss of all the water but none of the ammonia. Above ~300°C the product shows nearly the same properties and behavior [10] as that of the ammonium "decamolybdate" (in [10] described as "dodecamolybdate"), see below.

References:

[1] K. H. Tytko, B. Schönfeld (Z. Naturforsch. **30b** [1975] 471/84). – [2] K. H. Tytko, B. Schönfeld, B. Buss, O. Glemser (Angew. Chem. **85** [1973] 305/7; Angew. Chem. Intern. Ed.

Engl. **12** [1973] 330/2). – [3] I. Paulat-Böschen (J. Chem. Soc. Chem. Commun. **1979** 780/2). – [4] B. Krebs, I. Paulat-Böschen (Acta Cryst. B **38** [1982] 1710/8). – [5] T. J. R. Weakley (Polyhedron **1** [1982] 17/9).

[6] H. Guiter (Bull. Soc. Chim. France [5] **12** [1945] 74/5). – [7] Gmelin Handbook "Molybdän", 1935, pp. 259/60). – [8] I. Lindqvist (Nova Acta Regiae Soc. Sci. Upsaliensis [4] **15** No. 1 [1950] 1/22, 8). – [9] M. J. Schwing-Weill, F. Arnaud-Neu (Bull. Soc. Chim. France **1970** 853/60). – [10] M. J. Schwing-Weill (Bull. Soc. Chim. France **1967** 3795/8).

[11] B. Schönfeld (Diss. Göttingen, West Germany, 1973, p. 55). – [12] I. Böschen (Diss. Kiel, West Germany, 1974, p. 169).

2.5.2.10 $[NH_4Mo_5O_{15}(OH)(H_2O) \cdot H_2O]_\infty$ (= $(NH_4)_2O \cdot 10 MoO_3 \cdot 5 H_2O$), Ammonium "Decamolybdate"[1)]

Crystals of the compound are obtained from 0.1 to 0.2 M ammonium molybdate solutions acidified with HCl or HNO_3 to 2.0 to 4 H^+/MoO_4^{2-} at room temperature within some days in the form of small hexagonal columns [1]. A "decamolybdate" 4-hydrate was obtained from an ammonium molybdate solution of pH 0.9 left standing for a month [2]. The compound is also formed on efflorescence of the ammonium 36-molybdate [1].

The compound was characterized by analysis and Raman spectrum [1] (see Fig. 23, p. 23). The X-ray powder diagram and infrared spectrum of a "hexagonal hydrate of molybdenum trioxide" [3] are in reality those of "decamolybdates" [4]. The analytical data of this "hexagonal hydrate of molybdenum trioxide" correspond approximately to the formula $0.75(NH_4)_2O \cdot 10 MoO_3 \cdot 4.6 H_2O$. The infrared spectrum of a compound described to be ammonium "dodecamolybdate", $(NH_4)_2O \cdot 10.6$ to $12 MoO_3 \cdot x H_2O$ [5], $(NH_4)_2O \cdot 12 MoO_3 \cdot 4 H_2O$ [6], is actually that of a "decamolybdate", and the X-ray diffraction data of this "dodecamolybdate" [6], the „hexagonal hydrate of molybdenum trioxide" [3], and sodium and potassium "decamolybdates" [7] are also identical. The compounds described as "dodecamolybdate" and "reactive molybdic acid" [8] are obviously also "decamolybdates". For these compounds an isomorphic replacement of NH_4^+ and H_3O^+ in the crystal structure was assumed to explain the differences in the compositions [8] (see also below). The X-ray diffraction data and lattice parameters of the "22-molybdate" described by [9] are also those of "decamolybdates" (compare pp. 69, 85, and 100). Of the two ammonium "decamolybdates" described in [10], the 3-hydrate is in fact a "decamolybdate"; the 19-hydrate, however, is a 36-molybdate. For the criteria for these assignments see pp. 68 and 69.

From the identity of the Raman spectra [1], infrared spectra, and X-ray powder diagrams [3, 4] the formula and structure of the ammonium "decamolybdate" were derived to be the same as those of the potassium "decamolybdate" [1, 4] (see Fig. 22, p. 22).

The compound is insoluble in water [1].

The compound described as "hexagonal hydrate of molybdenum trioxide" has been studied, apart from infrared spectroscopy, by NMR spectroscopy, electron microscopic observations, thermogravimetric and differential thermal analysis, and other techniques. Heat treatment of this compound in air shows continuous liberation of water up to ~400°C. The content of ammonia shows only small (positive and negative) changes in the range 20 to 350°C and decreases in the range 350 to 450°C. Exothermic effects occur at ~315°C (small) and 420 to 430°C (large; solid phase transformation accompanied by endothermic deammoniation). From

---

[1)] This polymolybdate type is identical with the phase C polymolybdate type described in the next section, see the notes added in proof on pp. 7 and 23/4.

the continuous liberation of the water, the high hygroscopicity of the dehydrated product, and from the changes in the densitiy without any change in the crystal structure (X-ray powder diagrams) it was concluded that the dehydrated phase has a structure with vacancies in which water molecules can be occluded. At 315 and 420 to 430°C(?) the irreversible formation of a stable phase (rhombic MoO_3) occurs and hence the first dehydrated phase is a metastable one [3]. The thermal investigation of the compound described to be a "dodecamolybdate" also shows a rather continuous loss of weight, at about 300°C the formation of very hygroscopic "$(NH_4)_2O \cdot 12MoO_3$" and at 420°C the formation of MoO_3. In this case there was also observed a constancy of the powder diffraction patterns thus indicating the possibility of an isomorphous exchange of H_3O^+ and NH_4^+ in the crystal structure without any other change. This could explain the differing analytical ratios $(NH_4)_2O:MoO_3$ usually observed [6].

References:

[1] K. H. Tytko, B. Schönfeld (Z. Naturforsch. **30b** [1975] 471/84). – [2] H. Guiter (Bull. Soc. Chim. France [5] **12** [1945] 74/5). – [3] N. Sotani (Bull. Chem. Soc. Japan **48** [1975] 1820/5). – [4] B. Krebs, I. Paulat-Böschen (Acta Cryst. B **32** [1976] 1697/704, 1697). – [5] M. J. Schwing-Weill, F. Arnaud-Neu (Bull. Soc. Chim. France **1970** 853/60).

[6] M. J. Schwing-Weill (Bull. Soc. Chim. France **1967** 3795/8). – [7] I. Böschen (Diss. Kiel, West Germany, 1974, p. 94). – [8] J. Byé, M. J. Weill (Bull. Soc. Chim. France **1960** 1130/3). – [9] A. B. Kiss, P. Gadó, I. Asztalos, A. J. Hegedüs (Acta Chim. [Budapest] **66** [1970] 235/49). – [10] Gmelin Handbook "Molybdän", 1935, p. 260.

2.5.2.11 0.030 to 0.085$(NH_4)_2O \cdot MoO_3 \cdot$0.30 to 0.51$H_2O$, Ammonium Phase C Polymolybdate[1)]

The conditions of preparation may vary within wide limits: ammonium molybdate concentration 0.1 to 1 N, excess concentration of HNO_3 1 to 2.5 N, temperature of precipitation 25 to 100°C, corresponding to a time of formation of 3 weeks to 10 min [1]; ammonium paramolybdate concentration 0.1 to 1.3 M, concentration of HCl 0(?) to 1 N, of HNO_3 2.5 N, temperature of precipitation 70 to 100°C, corresponding time of formation 30 to 2 min; solid ammonium paramolybdate, 14.7 N HNO_3, 100°C, 30 min; yields 50 to 99% according to the conditions [2]. High molybdate concentrations and high temperatures favor the formation of the C phase [1, 2].

The white (prepared at 100°C) or yellowish (prepared at 25 or 70°C) products are dried at 40°C, and stored away from air and light since humid products more or less rapidly turn blue, depending on the conditions of preparation [1]. Other products were dried at 25 or 100°C [2].

The composition of the phase C compounds was found to vary in the range 0.033 to 0.083$(NH_4)_2O \cdot MoO_3 \cdot$0.3 to 0.4$H_2O$ for products dried at 100°C [2]. Products dried at 40°C showed the formula 0.075 to 0.085$(NH_4)_2O \cdot MoO_3 \cdot$0.32 to 0.42$H_2O$ [1].

These ammonium phase C compounds were characterized by X-ray powder patterns [1, 2] which are found to be identical with those of the sodium phase C described by [2]. The compound has a bcc crystal structure like the sodium and potassium compounds, lattice parameter $a = 12.98 \pm 0.02$ Å [1]; Z = 2 [3]. The amounts of ammonium and water may vary within wide limits without change in the X-ray powder pattern [1, 2] thus showing the water in the structure of the C phase to be unimportant for the structure type [1].

[1)] This polymolybdate is identical with the "decamolybdate" type described in the preceding section, see the notes added in proof on pp. 7 and 23/4.

The compound described as $(NH_4)_2O \cdot 14MoO_3 \cdot 3.3H_2O$, obtained from acidified solutions, or anhydrous by thermal treatment of ammonium paramolybdate 4-hydrate at 260°C [5] (see p. 100), shows the X-ray powder pattern and lattice parameter (a = 12.90 Å) of a phase C compound. An additional compound having a very similar X-ray powder pattern was described by [6] as "blue product". A compound prepared according to the description given for "molybdic acid", "$H_2MoO_4 \cdot H_2O$" [7], also shows the same X-ray powder pattern and, accordingly, has to be considered as a phase C compound [1].

The pycnometric density is 3.10 ± 0.04 g/cm^3 [3].

Some NH_3 as well as H_2O is lost on drying the phase C compounds at 100°C [2]. TG, DTA, and other experiments between 50 and 340°C show only dehydration. At 260°C two thirds of the water are lost. Between 340 and 380°C small amounts of water and ammonia are simultaneously lost, and between 380 and 420°C ammonia is rapidly liberated. The residue is MoO_3. Isothermal degradation experiments at different temperatures lead to the same conclusions. X-ray diagrams of products decomposed stepwise at different temperatures also show complete decomposition at 410°C. This thermal behavior is quite different from that of the compound $MoO_3 \cdot H_2O$ (white) [1].

The compound is readily soluble in alkaline solutions [1]. The ammonium phase C compounds are more stable in acid solution than the sodium phase C compounds. A 5 g sample was recovered in nearly quantitative yield after refluxing for 6 h with 50 mL 1 N HCl; however, the composition changed from $0.084(NH_4)_2O \cdot MoO_3 \cdot 0.51H_2O$ to $0.030(NH_4)_2O \cdot MoO_3 \cdot 0.42H_2O$ [2]. Refluxing in conc. HNO_3 for 2 d gives pure MoO_3 [3].

The ion exchange properties of a sample prepared by acidification with 2.5 N (excess) HNO_3 [1] and dried at 50°C have been studied using a batch technique: 5 g of the phase C compound ($\sim 0.08(NH_4)_2O \cdot MoO_3 \cdot \sim 0.4H_2O$ [1]) were heated with 100 mL of 1N HNO_3 and 50 mL of 1M M^INO_3 or $M^{II}(NO_3)_2$ (M^I = Li, Na, K; M^{II} = Ca, Sr, Ba) for 1 h to reflux. A sample containing 1.09 mval NH_4/g contained after the exchange procedure (in mval/g) 0.01 Li, 0.21 Na, 0.46 K, 0.05 Ca, 0.27 Sr, 0.64 Ba; total of M^I or $M^{II} + NH_4$ approximately 1.1 mval/g. The exchange process is reversible. Concentrated HNO_3 hampers the exchange process by formation of MoO_3 and leads to a decrease of the total cation content together with an increase of the water content of approximately 2 mol H_2O/1 val cation [3]. See also p. 4.

The dimensional behavior of a pressed (10 to 15 kg/cm^2) sample of the compound, activated at 280°C for 12 h, on the adsorption of n-heptane was found to be similar to that of zeolite X13 (characterized by a complete desorption of the n-heptane in vacuum), while the adsorption of the water led to an immediate strong expansion followed by a slow contraction without reaching the zero line on desorption [4].

References:

[1] H. Peters, L. Till, K. H. Radeke (Z. Anorg. Allgem. Chem. **365** [1969] 14/21). – [2] M. L. Freedman (J. Chem. Eng. Data **8** [1963] 113/6). – [3] H. Peters, K. H. Radeke (Monatsber. Deut. Akad. Wiss. Berlin **11** [1969] 351/4). – [4] K. H. Radeke, H. Peters (Z. Chem. [Leipzig] **15** [1975] 414/5). – [5] A. B. Kiss, P. Gadó, I. Asztalos, A. J. Hegedüs (Acta Chim. [Budapest] **66** [1970] 235/49).

[6] F. A. Schröder, J. Scherle (Z. Naturforsch. **28b** [1973] 46/55). – [7] G. Brauer (Handbuch der Präparativen Anorganischen Chemie, 2nd Ed., Vol. 2, Enke, Stuttgart 1962, pp. 1233/4).

2.5.3 Ammonium Peroxomolybdate Hydrates

2.5.3.1 $(NH_4)_2Mo_2O_{11}\cdot nH_2O$, n = 1 or 2

From diammonium "dimolybdate", "trimolybdate", "tetramolybdate", or ammonium paramolybdate and H_2O_2 in aqueous solution the same peroxide compound is formed when the amount of H_2O_2 is two or more moles per mole of MoO_3. This compound is diammonium tetraperoxodimolybdate, formulated as $(NH_4)_2O\cdot 2MoO_3\cdot 4O\cdot nH_2O$, n = 1 or 2. It is prepared, e.g., from ammonium paramolybdate (12 g), which is dissolved in concentrated hydrogen peroxide (12 mL, 9 M). The compound is obtained as a yellowish-white microcrystalline precipitate from absolute alcohol. It is repeatedly washed with alcohol and ether and dried in air at room temperature. By the action of excess peroxide on the above-mentioned ammonium molybdates and evaporation in a stream of air at 40 to 50°C, on the other hand, only peroxocompounds with the same $MoO_3:(NH_4)_2O$ ratio as the starting molybdates and with only one atom of active oxygen per Mo atom are obtained.

J. Beltran Martinez (Anales Real Soc. Espan. Fis. Quim. B **45** [1949] 665/96, C.A. **1950** 3828).

2.6 Rubidium Molybdate Hydrates

Older data are given in "Molybdän", 1935, pp. 281/3.

2.6.1 Rubidium Molybdate Hydrates with Mo of Oxidation State <6

2.6.1.1 $Rb_2Mo^{V}_2Mo^{VI}_4O_{18}\cdot 10H_2O$ or $Rb_2[H_2Mo^{V}_2Mo^{VI}_4O_{19}]\cdot 9H_2O$ (= $Rb_2O\cdot Mo_2O_5\cdot 4MoO_3\cdot 10H_2O$)

The compound is obtained from a solution of 2 Mo^{VI} and 1 Mo^{V} at pH 1.3 to 1.5. Fifty milliliters of a 0.5 M solution of Na_2MoO_4, 11 mL H_2O, 41.5 mL of a 0.3 M solution of Mo^{V} in 3 N HCl (obtained by reduction with Hg), and 22.5 mL 3 N NaOH are mixed in an atmosphere of N_2. The solution is slightly heated and precipitation is induced by addition of 10 g RbCl. After 24 h (under N_2) the crystals are rapidly filtered, successively washed with a buffer solution (chloracetic acid/chloracetate, pH 1.3 to 1.5), ethanol, and petroleum ether and subsequently dried at room temperature under vacuum [1, 5].

The blue-colored compound was characterized by analysis [1, 2] and UV-visible spectrum (see the paper) of its aqueous solution [1].

From the molar ratio of the components and from the polyacidity the degree of aggregation is deduced to be at least six [1, 2]. The polymolybdate ion most probably has the Mo_6O_{19} structure (see Fig. 16, p. 18) and, therefore, the compound should be formulated $Rb_2[H_2Mo^{V}_2Mo^{VI}_4O_{19}]\cdot 9H_2O$ [3]. For a relation between the UV-visible band maximum (of the dissolved species) and the reciprocal of the number of reduced molybdenum atoms in the species using the free electron model, and for the postulation of direct metal-metal bonds, see [4].

At higher acidities, no precipitate forms, but the complex decomposes [2].

References:

[1] S. Ostrowetsky (Bull. Soc. Chim. France **1964** 1003/11). – [2] P. Souchay, S. Ostrowetsky (Compt. Rend. **250** [1960] 4168/70). – [3] M. T. Pope (Inorg. Chem. **11** [1972] 1973/4). – [4] E. E. Kriss, V. K. Rudenko, K. B. Yatsimirskii (Zh. Neorgan. Khim. **16** [1971] 2147/53; Russ. J. Inorg. Chem. **16** [1971] 1146/50). – [5] S. Ostrowetsky, P. Souchay (Compt. Rend. **251** [1960] 373/5).

2.6.1.2 $Rb_2HMo_3^VMo_3^{VI}O_{18}\cdot 12H_2O$ or $Rb_2[H_3Mo_3^VMo_3^{VI}O_{19}]\cdot 11H_2O$ ($=2Rb_2O\cdot 3Mo_2O_5\cdot 6MoO_3\cdot 25H_2O$)

The compound is obtained from a solution of Mo^{VI} and Mo^V in equal amounts at pH 2.7. To 100 mL of 1 M chloracetic/chloracetate (1:9) buffer solution are added in an atmosphere of N_2 50 mL of a 0.5 M solution of Na_2MoO_4, and 83 mL of a 0.3 M solution of Mo^V in 3 N HCl (neutralized with 17.5 mL 10 N NaOH). The solution is slightly heated and precipitated by addition of 16 g RbCl. After 5 to 6 d the crystals are rapidly filtered, successively washed with the buffer solution, ethanol, and petroleum ether and subsequently dried at room temperature under vacuum [1, 2].

The brown compound was characterized by analysis [1, 2] and UV-visible spectrum (see the paper) of its aqueous solution [1].

From the molar ratio of the components and from the polyacidity the degree of aggregation is derived to be at least six [1, 2]. The polymolybdate ion has, by analogy with the preceding compound, most probably the Mo_6O_{19} structure and, therefore, the compound should be formulated $Rb_2[H_3Mo_3^VMo_3^{VI}O_{19}]\cdot 11H_2O$.

For a relation between the UV-visible band maximum (of the dissolved species) and the reciprocal of the number of reduced molybdenum atoms in the species using the free electron model, and for the postulation of direct metal-metal bonds, see [3].

References:

[1] S. Ostrowetsky (Bull. Soc. Chim. France **1964** 1003/11). – [2] S. Ostrowetsky, P. Souchay (Compt. Rend. **251** [1960] 373/5). – [3] E. E. Kriss, V. K. Rudenko, K. B. Yatsimirskii (Zh. Neorgan. Khim. **16** [1971] 2147/53; Russ. J. Inorg. Chem. **16** [1971] 1146/50).

2.6.2 Rubidium Molybdate(VI) Hydrates

2.6.2.1 The Rb_2MoO_4–$RbOH$–H_2O System

This system was investigated at 25°C, equilibrium being reached in 24 to 30 h. Three crystallization branches are determined, which belong to $Rb_2MoO_4\cdot 2H_2O$, to the anhydrous Rb_2MoO_4, and the very short one to $RbOH\cdot H_2O$, see **Fig. 51**, p. 120. The invariant points are at 11.04 wt% RbOH, 64.16 wt% Rb_2MoO_4 and 67.82 wt% RbOH, 1.98 wt% Rb_2MoO_4. The density, refractive index, viscosity, and electrical conductivity of the solutions at various concentrations have been measured and tabulated [1]. The solubility of Rb_2MoO_4 in water at 18°C is 67.88 wt% or 6.4 mol/L [2].

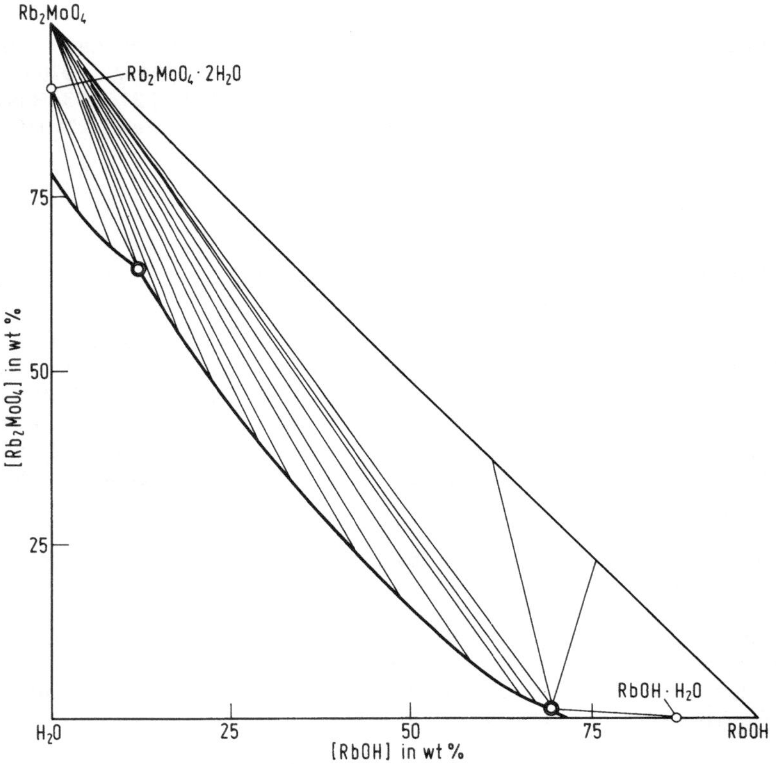

Fig. 51. Solubility isotherm in the Rb_2MoO_4–RbOH–H_2O system at 25°C [1].

References:

[1] M. I. Shavaev, Z. G. Karov, K. Sulaimankulov (Zh. Neorgan. Khim. **27** [1982] 1311/3; Russ. J. Inorg. Chem. **27** [1982] 737/9). – [2] V. I. Spitsyn, I. M. Kuleshov (Zh. Obshch. Khim. **21** [1952] 408/12; J. Gen. Chem. [USSR] **21** [1952] 453/6).

2.6.2.2 $Rb_2MoO_4 \cdot 2H_2O$

This compound has been found in the Rb_2MoO_4–RbOH–H_2O system (see above) but not investigated in detail.

M. I. Shavaev, Z. G. Karov, K. Sulaimankulov (Zh. Neorgan. Khim. **27** [1982] 1311/3; Russ. J. Inorg. Chem. **27** [1982] 737/9).

2.6.2.3 $[Rb_4Mo_6O_{20} \cdot 2nH_2O]_\infty$ (= $Rb_2O \cdot 3MoO_3 \cdot nH_2O$), n = 1, 3; Rubidium "Trimolybdates"

$Rb_2O \cdot 3MoO_3 \cdot 3H_2O$ or **$[Rb_4Mo_6O_{20} \cdot 6H_2O]_\infty$**. The compound crystallizes from solutions of acidity 1.33 H^+/MoO_4^{2-} at room temperature. Two molecules of water per formula weight are lost over silica gel. Due to the similarity of the vibrational spectra of the hydrous "trimolybdates" the same structure of the polyanion is assumed as for the monohydrate [1] (see below).

$[Rb_4Mo_6O_{20} \cdot 2H_2O]_\infty$ ($= Rb_2Mo_3O_{10} \cdot H_2O$). Crystals suitable for X-ray structure analysis (0.03 mm in diameter) are obtained from a 0.67 M solution of rubidium molybdate of acidity $0.71\,H^+/MoO_4^{2-}$ (which is far away from the theoretical value of 1.33) by slow crystallization at 70°C: 1.57 g $Rb_6Mo_7O_{24} \cdot 4H_2O$ are added to a solution of 0.41 g $RbOH \cdot 2H_2O$ in 10 mL H_2O at 70°C and stored for 12 d at this temperature in a stoppered vessel; the precipitate is dried over silica gel. On longer storage under the mother liquor the monoclinic $Rb_2Mo_3O_{10}$ (without water of crystallization) is formed [1]. According to the following method the usual, fibrillar form of the "trimolybdates" is obtained. A solution of 10 g of $RbOH \cdot 2H_2O$ in 10 mL of water was heated to its boiling point and then portions of solid MoO_3 were added to saturation. The yellow-green solution was filtered. After cooling crystals of Rb_2MoO_4 precipitated. The crystals were filtered and the colorless filtrate was acidified to pH 4.5 using 1M HNO_3. After about 24 h the fibrillar crystals appeared. They were filtered, washed first with small portions of water, then with aqueous acetone (vol ratio 1:1), and finally with pure acetone. The fluffy crystals were stored in a desiccator over $CaCl_2$. Microscopic observation showed fiber-shaped monocrystals about 1 mm long and 0.1 to 5 μm thick with smooth side faces [2].

The compound was characterized by analysis, infrared [1, 2] and Raman spectra [1] (see Figs. 9 and 8, p. 14), and interplanar distances [2].

The crystals are orthorhombic [1, 2], space group Pnma-D_{2h}^{16} (No. 62); lattice parameters a = 9.868(5), b = 7.607(6), c = 14.940(5) Å [1], a = 14.962(5), b = 7.580(5), c = 9.940(9) Å [2]; Z = 4. $D_m(20°C) = 3.72 \pm 0.2(?)$, $D_x = 3.74\ g/cm^3$ [2]. Atomic coordinates are listed, R = 0.056. The length of the b axis is characteristic of the "trimolybdates" [1].

The polymolybdate anion forms chains of MoO_6 octahedra running parallel to the (shortest) b axis and has a hexameric period of identity (see Fig. 7, p. 13). The structure of a $Mo_3O_{10}^{2-}$ unit is shown in **Fig. 52**. The Mo–O distances range from 1.68 Å (Mo(1)–O(5)) to 2.82 Å (Mo(1)–O(1)) and from 1.69 Å (Mo(2)–O(6)) to 2.24 Å (Mo(2)–O(3)). The O coordination around the Rb^+ cation is 7-fold, Rb–O distances 2.86 to 3.19 Å. The position of the water of crystallization is half-occupied. (A similar arrangement of the chains has been found in $Ag_6Mo_{10}O_{33}$ where the chains are connected by groups of four MoO_6 octahedra, see "Silber" B4, 1974, pp. 338/9) [1].

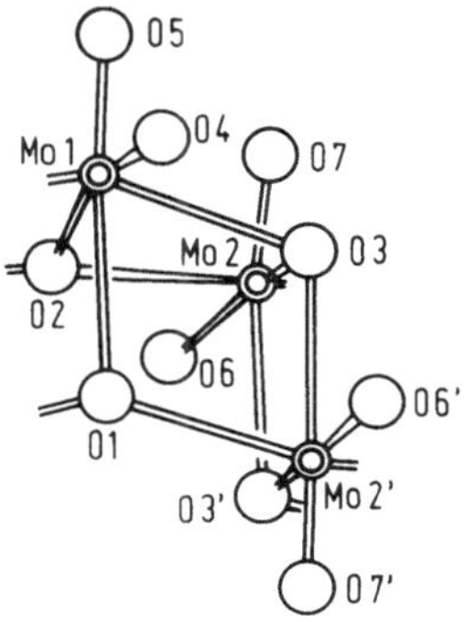

Fig. 52. Structure of an $Mo_3O_{10}^{2-}$ unit in $[Rb_4Mo_6O_{20} \cdot 2H_2O]_\infty$ [1].

The absence of Mo–O vibrational bands in the range 670 to 875 cm^{-1} of the infrared and Raman spectra indicates the absence of twofold bonded O atoms in the structure in accordance with the results of the X-ray crystal structure determination [1]. The frequency 1120 cm^{-1} is regarded as an indication that the structure of the compound contains OH groups [2] which is in disagreement with the structure described above.

The monohydrate is continuously dehydrated above 100°C. At 205°C the monoclinic (anhydrous) phase having a quite different structure of the polyanion (see "Molybdän" Erg.-Bd. B 1,

1975, p. 233) is formed [1]. The dehydration is accompanied by an endothermic peak in the DTA diagram at 185°C and, as has been concluded from the X-ray powder pattern, is connected with the total destruction of the initial structure. From this it has been assumed that the water is strongly bound and constitutional [2] which is in disagreement with the structure described above.

From ^1H NMR spectra of a polycrystalline sample in the temperature range 25 to −196°C, taken on a broad line spectrometer, the presence of two kinds of hydrogen (OH groups and H_2O molecules) has been deduced, the total number of protons existing in the structure being 1.752 at −196°C. At room temperature a more substantial portion of the protons should exist in the form of hydroxyl groups coordinated to the Mo atoms due to a growing movement of the water molecules [2]. Again there is a disagreement with the structure described above.

References:

[1] H. U. Kreusler, A. Förster, J. Fuchs (Z. Naturforsch. **35b** [1980] 242/4). – [2] W. Surga, S. Sagnowski, S. Hodorowicz (J. Inorg. Nucl. Chem. **43** [1981] 1821/5).

2.6.2.4 β-$Rb_4[Mo_8O_{26}]\cdot xH_2O$ (= $Rb_2O\cdot 4MoO_3\cdot 0.5xH_2O$), Rubidium β-Octamolybdate

The compound is formed by mixing a solution of α-$[(n\text{-}C_4H_9)_4N]_4[Mo_8O_{26}]$ in acetonitrile with an aqueous solution of RbCl. Heating should be avoided since the solution reacts to form $[Mo_6O_{19}]^{2-}$ when heated.

The substance has been characterized by analysis, infrared and Raman spectroscopy; however, the latter are not reproduced and the value of x in the formula is not reported. The structure of the polymolybdate ion (see Fig. 10, p. 16) has been assigned by (fingerprint) Raman and infrared spectroscopy in comparison with other compounds containing the β-$[Mo_8O_{26}]^{4-}$ ion.

W. G. Klemperer, W. Shum (J. Am. Chem. Soc. **98** [1976] 8291/3).

2.6.2.5 $Rb_2[Mo_6O_{19}]\cdot xH_2O\cdot yC_{12}H_{24}O_6$ (= $Rb_2O\cdot 6MoO_3\cdot xH_2O\cdot yC_{12}H_{24}O_6$) ($C_{12}H_{24}O_6$ = 18-crown-6), Rubidium Hexamolybdate

The compound has been prepared apparently analogously to the compound $(C_{12}H_{24}O_6)_2\cdot K_2[Mo_6O_{19}]\cdot H_2O$ (see p. 82). No other data are reported.

O. Nagano, Y. Sasaki (Acta Cryst. B **35** [1979] 2387/9).

2.6.3 Rubidium Sodium Molybdate Hydrates

2.6.3.1 The Rb_2MoO_4–Na_2MoO_4–H_2O System

Solubility in this ternary system is studied by the isothermal method, equilibrium being established within three weeks at 25°C and two weeks at 50°C. The solubility isotherm at 25°C consists of four crystallization branches: $Na_2MoO_4\cdot 2H_2O$, the double salts $Rb_2MoO_4\cdot 3Na_2MoO_4\cdot 18H_2O$ and $2Rb_2MoO_4\cdot Na_2MoO_4\cdot 2H_2O$, and the anhydrous rubidium molybdate, see **Fig. 53**. The points P_1 and P_2 are peritonic, E is a eutonic point.

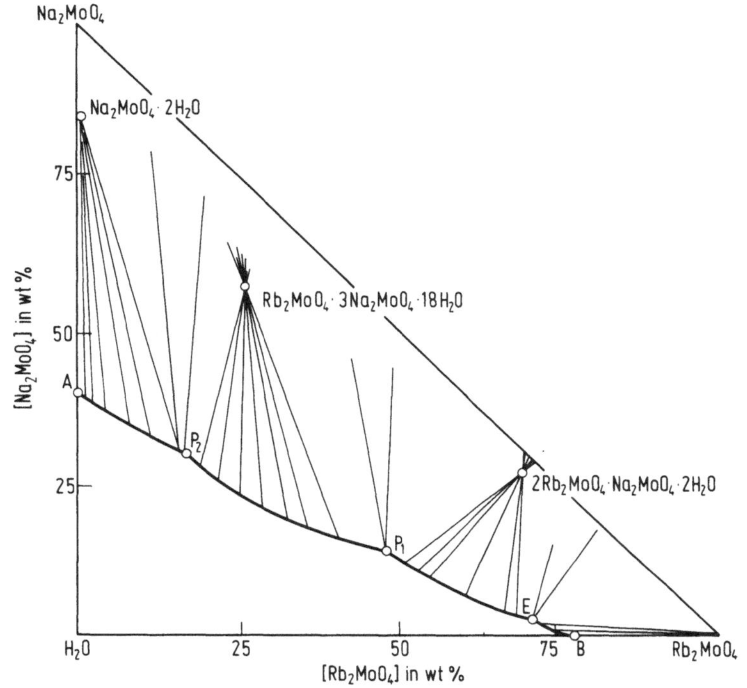

Fig. 53. Solubility isotherm in the Rb_2MoO_4–Na_2MoO_4–H_2O system at 25°C.

At 50°C the formation of only one double molybdate, $2Rb_2MoO_4 \cdot Na_2MoO_4 \cdot 2H_2O$, is observed; cf. the solubility isotherm in the paper.

R. G. Samuseva, T. F. Zhatkina, V. E. Plyushchev (Zh. Neorgan. Khim. **14** [1969] 270/6; Russ. J. Inorg. Chem. **4** [1969] 141/4).

2.6.3.2 $RbNa_3(MoO_4)_2 \cdot 9H_2O$ (= $Rb_2MoO_4 \cdot 3Na_2MoO_4 \cdot 18H_2O$)

The compound is observed in the Rb_2MoO_4–Na_2MoO_4–H_2O system, see above. Crystals are classified as hexagonal and are prisms with an isotropic section in the form of a hexagon. For lattice spacings and the X-ray powder line diagram see the paper. The pycnometric density is 2.14 g/cm³ at 25°C. In polarized light the crystals are characterized by a weakly colored interference picture. They are uniaxial positive. Refractive indices are $n_e = 1.5895 \pm 0.003$ and $n_o = 1.5865 \pm 0.003$.

Whether in the air or in a desiccator over sulfuric acid the compound loses its crystallization water completely, but in steps. The last 5 water molecules are removed after standing for a month. On heating, the thermogravimetric curves show that the crystallization water is also lost in steps, which correspond to endothermic effects at 100, 120, and 130°C. In isothermal drying experiments the greatest loss of water is at 40°C (10 mol H_2O), whereas at 50 and 60°C 3 and 5 mol H_2O, respectively, are removed.

R. G. Samuseva, T. F. Zhatkina, V. E. Plyushchev (Zh. Neorgan. Khim. **14** [1969] 270/6; Russ. J. Inorg. Chem. **14** [1969] 141/4).

2.6.3.3 $Rb_4Na_2(MoO_4)_3 \cdot 2H_2O$ ($= 2Rb_2MoO_4 \cdot Na_2MoO_4 \cdot 2H_2O$)

The compound is observed in the Rb_2MoO_4–Na_2MoO_4–H_2O system, see above. The crystals can have a columnar or an acicular form. The surfaces of the crystals are smooth and they have sharply expressed relief in the mother liquor. For lattice spacings and the X-ray powder line diagram see the paper. The pycnometric density is 3.78 g/cm³ at 25°C. Only one refractive index, 1.6015 ± 0.003, could be determined by the immersion method. The thermogravimetric heating curve shows an effect at 120°C whereas under isothermal drying conditions the water of crystallization is already lost at 40°C (see graphs in the paper).

R. G. Samuseva, T. F. Zhatkina, V. E. Plyushchev (Zh. Neorgan. Khim. **14** [1969] 270/6; Russ. J. Inorg. Chem. **14** [1969] 141/4).

2.7 Caesium Molybdate Hydrates

Older data are given in "Molybdän", 1935, pp. 285/6.

2.7.1 Caesium Molybdate Hydrates with Mo of Oxidation State <6

Compounds of this type have not been reported.

2.7.2 Caesium Molybdate(VI) Hydrates

2.7.2.1 The Cs_2MoO_4–$CsOH$–H_2O System

This system is studied at 25°C and equilibrium is attained in 24 to 30 h as in the corresponding system with Rb, see p. 119. Likewise three crystallization branches are found belonging to $Cs_2MoO_4 \cdot 2H_2O$, the anhydrous Cs_2MoO_4, and $CsOH \cdot H_2O$. The solubility isotherm resembles very closely that of the corresponding Rb system; it is shown in the paper. The invariant points are at 9.64 wt% CsOH, 69.27 wt% Cs_2MoO_4 and 77.59 wt% CsOH, 3.40 wt% Cs_2MoO_4. The density, refractive index, viscosity, and electrical conductivity of the solutions at various concentrations have been measured and tabulated [1]. The solubility of Cs_2MoO_4 in water at 18°C is 67.07 wt% or 4.8 mol/L [2].

References:

[1] M. I. Shavaev, Z. G. Karov, K. Sulaimankulov (Zh. Neorgan. Khim. **27** [1982] 1311/3; Russ. J. Inorg. Chem. **27** [1982] 737/9). – [2] V. I. Spitsyn, I. M. Kuleshov (Zh. Obshch. Khim. **21** [1952] 408/12; J. Gen. Chem. [USSR] **21** [1952] 453/6).

2.7.2.2 $Cs_2MoO_4 \cdot 2H_2O$

This compound has been found in the Cs_2MoO_4–$CsOH$–H_2O system (see above) but not investigated in detail.

M. I. Shavaev, Z. G. Karov, K. Sulaimankulov (Zh. Neorgan. Khim. **27** [1982] 1311/3; Russ. J. Inorg. Chem. **27** [1982] 737/9).

2.7.2.3 $Cs_6[Mo_7O_{24}]\cdot nH_2O$ (= $3Cs_2O\cdot 7MoO_3\cdot nH_2O$), Caesium Paramolybdate

Since the paramolybdates have been shown to be (3:7)- and not (5:12)-molybdates [1], the compound $5Cs_2O\cdot 12MoO_3\cdot 11H_2O$ [2] has to be reformulated $3Cs_2O\cdot 7MoO_3\cdot 6.5H_2O$ or $Cs_6[Mo_7O_{24}]\cdot 6.5H_2O$.

The caesium salt is the least soluble alkali polymolybdate which is precipitated from solutions of pH 4.5 to 5 (presumably as the paramolybdate). Hence it can be used for the preparative separation of caesium from other alkali metals, especially rubidium [3].

References:

[1] J. H. Sturdivant (J. Am. Chem. Soc. **59** [1937] 630/1). – [2] Gmelin Handbook "Molybdän", 1935, p. 285. – [3] G. Kraft (Ger. 1065391 [1959] from C.A. **1961** 11781).

2.7.2.4 $[Cs_4Mo_6O_{20}\cdot 2H_2O]_\infty$ (= $Cs_2O\cdot 3MoO_3\cdot H_2O$), Caesium "Trimolybdate"

To prepare this compound a 0.5 M solution of caesium carbonate was heated to its boiling point and then portions of solid MoO_3 were added until its saturation. After cooling crystals of Cs_2MoO_4 were precipitated; the mother liquor was heated until the Cs_2MoO_4 precipitate had dissolved completely and then it was left for a few days to crystallize. The crystals were filtered, washed first with small portions of water, then with a mixture of water with acetone (vol ratio 1:1), and finally with pure acetone. The white, fluffy crystals were stored in a desiccator over P_2O_5. Microscopic observation showed fiber-shaped crystals [1].

The compound is characterized by chemical analysis, the X-ray powder pattern and interplanar distances (see the paper [1]). It is presumably identical with a compound described by [2] although the form of the precipitate is quite different [1].

The crystals are orthorhombic, lattice parameters a = 15.50(1), b = 7.61(6), c = 10.23(4) Å; Z = 4. D_m(20°C) = 4.00 ± 0.03, D_x = 4.02 g/cm³ [1].

Thermogravimetric measurements showed the loss of the water at approximately 100 to 180°C, DTA investigations endothermic effects at 160 (loss of the water), 298 (structural transformations, see the X-ray data), and 535°C (melting point). The X-ray powder pattern of a sample heated at 100°C for 20 min was unchanged; those of samples heated for 20 min at 200 and 350°C were quite different showing that the crystal structure is completely reorganized by the loss of the water, and structural changes without loss of water, respectively. The strong changes accompanied with the loss of the water are assumed to indicate that water is bound strongly as constitutional water [1].

The fibrillar shape of the single crystals is mainly the result of their chain structure. The dimension of the links of the chain is practically equal to that of the other "trimolybdates" (b = 7.60, 7.44, 7.66 Å for the lithium, sodium, and potassium "trimolybdates", respectively) [1].

From ^1H NMR spectra of a polycrystalline sample in the temperature range 25 to −196°C, taken on a broad line spectrometer, the presence of two kinds of hydrogen (OH groups and H_2O molecules) is deduced. At −196°C, 1.88 protons are present as H_2O and 0.12 as OH per one molecule of the compound. Accordingly, the formula of the compound is written $Cs_2[Mo_3O_{0.94}(OH)_{0.12}(H_2O)_{0.94}O_9]$. This formula is in accordance with the conceptions of [3, 7] on the mechanism of the polymerization process according to which the anion of the chain-like "trimolybdates" should comprise two OH groups or one water molecule bound directly to the central molybdenum atom. At room temperature a more considerable portion of protons exists in the form of OH groups as has been found for the lithium and potassium "trimolybdates" [1].

In contrast to some of the above structural conclusions the nearly identical infrared [4, 5] and Raman [5, 6] spectra of the water-containing "trimolybdates" with different cations have been interpreted as showing the same chain-like polymolybdate anion in all "trimolybdates" of the alkali metals [5, 6] (see Fig. 7, p. 13). This is confirmed by the identical length of the b axis (which is parallel to the extension of the chain) of the lithium, potassium, rubidium [5], and caesium compounds.

References:

[1] S. Hodorowicz, E. Hodorowicz, S. Sagnowski, W. Surga (Pol. J. Chem. **54** [1980] 1859/64). – [2] F. Ephraim, H. Herschfinkel (Z. Anorg. Allgem. Chem. **64** [1909] 263/72, 270/2). – [3] S. Hodorowicz (Krist. Tech. **12** [1977] 431/4). – [4] B. Schönfeld (Diss. Göttingen, West Germany, 1973, p. 47). – [5] H. U. Kreusler, A. Förster, J. Fuchs (Z. Naturforsch. **35b** [1980] 242/4).

[6] K. H. Tytko, B. Schönfeld (Z. Naturforsch. **30b** [1975] 471/84). – [7] W. Surga, S. Sagnowski, S. Hodorowicz (J. Inorg. Nucl. Chem. **43** [1981] 1821/5).

2.7.3 Caesium Peroxomolybdate Hydrates

2.7.3.1 $Cs_4Mo_2O_{11} \cdot H_2O$, $Cs_2MoO_6 \cdot H_2O$, $Cs_2MoO_7 \cdot 3H_2O$, $Cs_2MoO_8 \cdot nH_2O$

When a saturated aqueous Cs_2MoO_4 solution is titrated with hydrogen peroxide solution (2.5, 5, or 10%) definite caesium peroxomolybdates are formed, as is revealed by the potentiometric titration curves. If a 10% H_2O_2 solution is used, four compounds are observed. Crystals of these compounds are precipitated at the equivalence points or they may be precipitated with ethanol. The existence of the following compounds and their formations at the given pH values were stated [1]:

composition of compounds formed	molar ratio $MoO_4^{2-}:O_2$	pH
$Cs_4Mo_2O_{11} \cdot H_2O$, triperoxodimolybdate	1:0.75	11.8
$Cs_2MoO_6 \cdot H_2O$, diperoxomolybdate	1:1	11.4
$Cs_2MoO_7 \cdot 3H_2O$, triperoxomolybdate	1:1.5	10.7
$Cs_2MoO_8 \cdot nH_2O$, tetraperoxomolybdate	1:2	9.4

For preparation, 2N Cs_2MoO_4 solution is used and the peroxomolybdates are precipitated with ethanol cooled to −5°C [2]. The anhydrous compounds $Cs_4Mo_2O_{11}$, Cs_2MoO_6, and Cs_2MoO_7 are readily obtained from the corresponding hydrates by removing the water of crystallization in a desiccator over P_2O_5. Only $Cs_2MoO_8 \cdot nH_2O$ is not dehydrated because it is unstable, and so its water content remains undetermined [1].

The compounds are considered as true peroxomolybdates [1, 2]. For line diagrams of the X-ray powder patterns (except $Cs_2MoO_8 \cdot nH_2O$) see [1].

Some properties of $Cs_2MoO_6 \cdot H_2O$, $Cs_2MoO_7 \cdot 3H_2O$, and of the corresponding anhydrous compounds are given in [2]:

compound	color	stability during storage	products of the phase transformation during storage
$Cs_4Mo_2O_{11}$	white	a year at 25°C	Cs_2MoO_4
Cs_2MoO_6	pink	8 to 10 months at −5 to 0°C, 4 to 5 months at 25°C	$Cs_4Mo_2O_{11}$
$Cs_2MoO_6 \cdot H_2O$	yellow	3 to 4 months at −5 to 0°C, 1 to 2 months at 25°C	$Cs_4Mo_2O_{11} \cdot H_2O$
Cs_2MoO_7	red	3 to 4 h at −5 to 0°C, 0.5 to 1 h at 25°C	Cs_2MoO_6
$Cs_2MoO_7 \cdot 3H_2O$	red	7 to 8 h at −5 to 0°C, 1 to 2 h at 25°C	$Cs_2MoO_6 \cdot H_2O$

For some IR frequencies see the paper [2].

Except for $Cs_2MoO_8 \cdot nH_2O$ the compounds are stable in air [1]. On thermal analysis with 2.5 K/min heating, $Cs_2MoO_6 \cdot H_2O$ is dehydrated at 90°C and loses one oxygen molecule at about 135°C (exothermic effect). $Cs_2MoO_7 \cdot 3H_2O$ shows an exothermic loss of one oxygen atom at 35°C and probably has an overlap at about 120°C of the exothermic effect due to the loss of peroxide oxygen with an endothermic effect of dehydration. The anhydrous $Cs_4Mo_2O_{11}$ and Cs_2MoO_6 lose oxygen at about 135°C whereas anhydrous Cs_2MoO_7 loses oxygen in two steps, at ~35 and ~100°C. $Cs_2MoO_8 \cdot nH_2O$ has not been investigated by thermal analysis because of the danger of explosion [2].

All the compounds are soluble in water and almost insoluble in organic solvents [1].

References:

[1] A. V. Arkhipov, T. M. Kurokhtina, G. L. Petrova, G. A. Bogdanov (Zh. Neorgan. Khim. **24** [1979] 1688/91; Russ. J. Inorg. Chem. **24** [1979] 936/8). – [2] A. V. Arkhipov, T. M. Kurokhtina, G. L. Petrova, G. A. Bogdanov, M. O. Bronnikova (Zh. Neorgan. Khim. **25** [1980] 2707/10; Russ. J. Inorg. Chem. **25** [1980] 1493/5).

2.8 Molybdate Hydrates and Solvates of Organic Cations

Older data are given in "Molybdän", 1935, pp. 271/3.

2.8.1 Molybdates with Molybdenum of Oxidation State <6

2.8.1.1 $[(C_2H_5)_4N]_2[H_2Mo^{V}_2Mo^{VI}_4O_{19}]$ (= $[(C_2H_5)_4N]_2O \cdot Mo_2O_5 \cdot 4MoO_3 \cdot H_2O$), Tetraethylammonium Hexamolybdate(V, VI)

A 0.5 M solution of hexavalent molybdenum in tributylphosphate was prepared by extraction of Na_2MoO_4 in 6 N HCl into tributylphosphate. A 0.5 M solution of pentavalent molybdenum in tributylphosphate was prepared by reduction of MoO_3 in concentrated hydrochloric acid by hydroiodic acid followed by extraction of the Mo^V formed into tributylphosphate. Stoichiometric quantities of the above solutions were mixed and shaken with an equal volume of water until a dark blue color was obtained. After the two phases were separated, the organic phase was washed twice with equal volumes of an aqueous 2N LiCl solution and finally with a saturated NaCl solution. The compound was precipitated by addition of an equal volume of aqueous 1N $(C_2H_5)_4NCl$ solution to the tributylphosphate solution and shaking. The precipitate was filtered off after addition of ethanol, washed with absolute ethanol and petroleum ether, and dried overnight in a vacuum desiccator. The compound was also obtained by partial oxidation of pentavalent molybdenum in tributylphosphate by an aqueous solution of iron(III) sulfate [1].

The compound was characterized by analysis and infrared spectrum (see the paper). The latter shows, compared with that of the hexamolybdate(VI) (see Fig. 18, p. 19), a large number of additional bands in the range 890 to 970 cm^{-1} due to a variety of different Mo-O bonds existing in the mixed valence compound [1].

The electronic spectrum of the solution in dimethyl sulfoxide shows three bands which were assigned by comparison with the spectrum of $MoOCl_5^{2-}$ as follows:

molybdenum blue E in cm^{-1}	ε in $L \cdot mol^{-1} \cdot cm^{-1}$	$MoOCl_5^{2-}$ [2] E in cm^{-1}	ε in $L \cdot mol^{-1} \cdot cm^{-1}$	assignments
900	134	—	—	intervalence CT transition
13400	512	14050	16	$^2B_2 \rightarrow {}^2E(I)$
—	—	22500	14	$^2B_2 \rightarrow {}^2B_1$
—	—	28200	500	$^2B_2 \rightarrow {}^2E(II)$
31750	7660	32200	4400	$^2B_2 \rightarrow {}^2B_2(I)$

For a detailed discussion see the paper [1].

EPR measurements were made on an undiluted powder and in dimethyl sulfoxide solution. The g values found were $g_\perp = 1.938$ and $g_\parallel = 1.890$. For a more detailed discussion see the paper [1].

Differential thermal analysis shows no reaction up to 280°C, and from this it is concluded that the compound is anhydrous. At 280°C a violent pyrolysis of the organic cation takes place [1].

References:

[1] C. Heitner-Wirguin, D. Hall (J. Inorg. Nucl. Chem. **36** [1974] 3870/1). – [2] C. K. Jœrgensen (Acta Chem. Scand. **11** [1957] 73/85, 82/3).

2.8.1.2 $(i\text{-}C_3H_7NH_3^+)_4[Mo_{13}O_{40}^{4-}]_{0.33}[H_4Mo_{12}^{VI}O_{40}^{4-}]_{0.67}$

The blue compound was obtained from an aqueous solution of the UV photoreduced isopropylammonium heptamolybdate, $(i\text{-}C_3H_7NH_3)_6[Mo_7O_{24}]$ [1].

The crystals are tetragonal, space group $I\bar{4}\text{-}S_4^2$ (No. 82), lattice parameters a = 14.908(2), c = 10.323(2) Å; Z = 2. $D_m = 2.97$, $D_x = 2.99$ g/cm³ [1].

The structure was solved by the heavy atom method and refined with anisotropic temperature factors, R = 0.037. The disordered crystals contain two anionic species, $[Mo_{13}O_{40}]^{4-}$ (Mo atoms having mixed-valencies, Mo^V and Mo^{VI}) and $[H_4Mo_{12}O_{40}]^{4-}$, which are randomly distributed over the anion sites in a ratio of ca. 1:2. Both anions have configurations similar to that of the Keggin type [2] structure. The occupancy factor of the central Mo atom is 0.33. An attempt to resolve the two anionic structures was unsuccessful. The positions of the hydrogen atoms were obscure in the difference map owing to the disordered structure. The average structure of the anions having S_4 symmetry is shown in **Fig. 54**. For the anion $[H_4Mo^{VI}_{12}O_{40}]^{4-}$ the central Mo atom, Mo(1), is lost. In $[Mo_{13}O_{40}]^{4-}$, Mo(1) is tetrahedrally surrounded by four O(10) atoms; the Mo(1)-O(10) distance is 1.628(7) Å and the O(10)-Mo-O(10') angle is 109.2(4)°. The other three Mo atoms are octahedrally coordinated by O atoms. The MoO_6 octahedra are joined together by sharing edges and corners to form an $Mo_{12}O_{40}$ group and connected to the central tetrahedron by sharing corners to form the $Mo_{13}O_{40}$ anion. The Mo atom in each octahedron is shifted from its center towards the unshared O atom owing to repulsion between the neighboring Mo atoms. The Mo-O distances and O-Mo-O angles in these octahedra are in agreement with the corresponding ones [3, 4] found in the octahedra of $[Mo_7O_{24}]^{6-}$ and $[H_2Mo_8O_{28}]^{6-}$ [1].

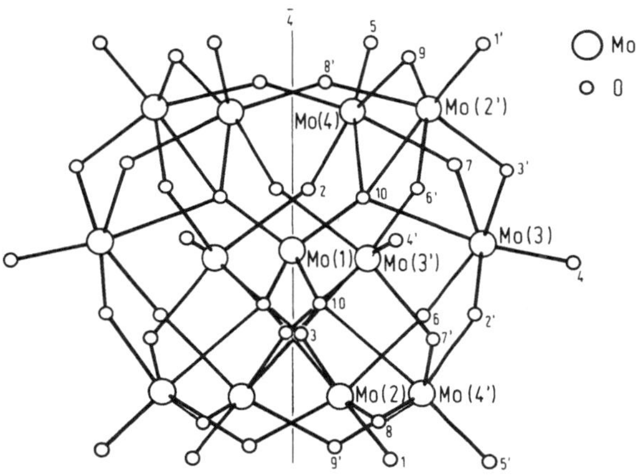

Fig. 54. Perspective view of the mixed anions $[Mo_{13}O_{40}]^{4-}$ and $[H_4Mo_{12}O_{40}]^{4-}$; in the latter the central Mo atom is lost [1].

An EPR study (one broad singlet, ΔH_{ms} ca. 150 G, with an isotropic g factor of 1.947) indicates that Mo^V species are present in the crystal. Since neither hyperfine nor superhyperfine structures were observed, the paramagnetic electrons due to Mo^V species may not be present in the 12-molybdate anion but only in the H-free 13-molybdate anion. In comparison with EPR studies of two-electron heteropoly blue complexes [5 to 7] it is concluded that both the paramagnetic electrons are not localized on the central Mo atom, one of them being delocalized on the surrounding Mo atoms in the octahedron network [1].

Note added in proof: According to [8] the Mo_{12}-Mo_{13} complex is an ordinary $SiMo_{12}$ Keggin type molecule in a reduced state. There are two arguments that confirm this view: (1) The central tetrahedral bond length of 1.628 Å is typical of SiO_4 tetrahedra (e.g., in isolated silicates, olivines, and dodecamolybdosilicates the Si-O bond lengths vary from 1.61 to 1.66 Å with an average close to 1.63 Å), while the typical mean Mo-O bond length in MoO_4 tetrahedra is 1.767 Å. (2) Si has just one third the number of electrons that Mo has, and the scattering curve for Si is very close to that of ⅓ Mo which allows an equally satisfactory refinement. This alternative interpretation is also suggested by [9].

References:

[1] T. Yamase, T. Ikawa, Y. Ohashi, Y. Sasada (J. Chem. Soc. Chem. Commun. **1979** 697/8). – [2] J. F. Keggin (Proc. Roy. Soc. [London] A **144** [1934] 75/100). – [3] H. T. Evans Jr., B. M. Gatehouse, P. Leverett (J. Chem. Soc. Dalton Trans. **1975** 505/14). – [4] M. Isobe, F. Marumo, T. Yamase, T. Ikawa (Acta Cryst. B **34** [1978] 2728/31). – [5] R. A. Prados, P. T. Meiklejohn, M. T. Pope (J. Am. Chem. Soc. **96** [1974] 1261/3).

[6] G. M. Varga Jr., E. Papaconstantinou, M. T. Pope (Inorg. Chem. **9** [1970] 662/7). – [7] J. M. Fruchart, P. Souchay (Compt. Rend. C **266** [1968] 1571/4). – [8] H. T. Evans Jr., M. T. Pope (Inorg. Chem. **23** [1984] 501/4). – [9] V. I. Spitsyn, L. P. Kazanskii, E. A. Torchenkova (Sov. Sci. Rev. B **3** [1981] 111/96, 183).

2.8.1.3 Compounds Containing Mo^V Produced by UV Irradiation of Polymolybdate(VI) Hydrates

The compounds containing only small amounts of Mo^V produced by UV irradiation of polymolybdates(VI) are described as photochromic or photosensitive salts in Section 2.8.2, see also p. 39.

The following compounds have been shown to be photochromic:

stated formula	polymolybdate type
$[(CH_3)_2NH_2]_2O \cdot 3MoO_3 \cdot H_2O$	"trimolybdate"
$[(CH_3)_4N]_2O \cdot 10MoO_3 \cdot 7H_2O$	"decamolybdate"
$[(C_2H_5)_2NH_2]_2O \cdot 3MoO_3 \cdot H_2O$	"trimolybdate"
$(C_3H_7NH_3)_6[Mo_7O_{24}] \cdot 3H_2O$	heptamolybdate
$(i\text{-}C_3H_7NH_3)_6[Mo_7O_{24}] \cdot 3H_2O$	heptamolybdate
$(i\text{-}C_3H_7NH_3)_6[H_2Mo_8O_{28}] \cdot 2H_2O$	dihydrogenoctamolybdate
$(C_5H_{11}NH)_6[Mo_7O_{24}] \cdot 6H_2O$[a]	heptamolybdate
$(C_5H_{11}NH)_2O \cdot 3MoO_3 \cdot 2H_2O$[a]	"trimolybdate"
$(C_6H_{13}NH)_6[Mo_7O_{24}] \cdot 2H_2O$[b]	heptamolybdate
$(C_6H_{13}NH)_2O \cdot 3MoO_3 \cdot H_2O$[b]	?
$(C_7H_{15}NH)_2O \cdot 3MoO_3 \cdot 3H_2O$[c]	?
$(C_4H_9NH)_6[Mo_7O_{24}] \cdot H_2O$[d]	heptamolybdate
$(C_4H_9NH)_2O \cdot 3MoO_3 \cdot H_2O$[d]	"trimolybdate"
$(C_4H_{10}N_2H_2)O \cdot 2MoO_3 \cdot 5H_2O$[e]	"trimolybdate"
$(C_4H_{10}N_2H_2)O \cdot 2MoO_3 \cdot H_2O$[e]	"trimolybdate"
$(C_4H_{10}N_2H_2)O \cdot 3MoO_3 \cdot H_2O$[e]	"trimolybdate"

[a] $C_5H_{11}N$ = piperidine. – [b] $C_6H_{13}N$ = 4-methylpiperidine. – [c] $C_7H_{15}N$ = 2,6-dimethylpiperidine. – [d] C_4H_9N = pyrrolidine. – [e] $C_4H_{10}N_2$ = piperazine.

The following compounds have been found to be photosensitive:

stated formula	polymolybdate type
$(CH_3NH_3)_2O \cdot 2MoO_3 \cdot 2H_2O$	heptamolybdate
$(CH_3NH_3)_2O \cdot 2MoO_3 \cdot H_2O$	heptamolybdate
$(CH_3NH_3)_2O \cdot 2MoO_3 \cdot 0.5H_2O$	heptamolybdate
$(CH_3NH_3)_2O \cdot 5.87$ to $9.36MoO_3 \cdot xH_2O$	36-molybdate
$(C_2H_5NH_3)_2O \cdot 2MoO_3 \cdot 2H_2O$	"trimolybdate"
$(C_2H_5NH_3)_2O \cdot 8.41MoO_3 \cdot xH_2O$	36-molybdate
$[(C_2H_5)_2NH_2]_2O \cdot 7.30$ to $9.02MoO_3 \cdot xH_2O$	36-molybdate
$(C_7H_{15}NH)_2O \cdot 6.83$ to $9.39MoO_3 \cdot \sim 4H_2O$[a]	36-molybdate?

[a] $C_7H_{15}N = 2,6$-dimethylpiperidine.

For the differences between stated (analytically determined) formula and polymolybdate type (mostly ensured by infrared "finger-prints") see Section 2.8.2.

2.8.2 Molybdate(VI) Hydrates and Solvates

Remark. A number of compounds in this section show, for reasons open to question (presumably errors in the elemental analyses), strong deviations between the acid : base ratio analytically determined and that corresponding to the polymolybdate type (see also p. 39). For this reason in each case both formula (analytically determined) and polymolybdate type (infrared "finger-print" and others) are used to organize the material. The large number of compounds in this section originates mainly from the interest in photochromism that many of the polymolybdates with organic cations show. In this section also the anhydrous compounds are shortly mentioned because they are not described in "Molybdän" Erg.-Bd. B 1, 1975.

2.8.2.1 Monomethylammonium Polymolybdates

$(CH_3NH_3)_2O \cdot 2MoO_3 \cdot 2H_2O$ (Heptamolybdate Type)

An aqueous solution of 350 g of $(NH_4)_6Mo_7O_{24} \cdot 4H_2O$ in 450 mL of methylamine (30%) was stirred for about 6 h. The solution was concentrated under reduced pressure to form a solid, which was then filtered off and dried. Recrystallization of the crude product from water yielded colorless crystals.

The compound was characterized by analysis and infrared spectrum (see the paper). The latter resembles those of the paramolybdates.

The compound is UV sensitive. Under irradiation with UV light ($\geqq$313 nm) the white crystals become reddish brown (or violet?), absorption maximum 480 nm, approximate coloration time (50% reflectance) 1 to 2 min. The colored compound shows an EPR signal due to Mo^V. The compound does not return to the white colored one in the dark and hence is not a photochrome. In solution the blue form shows thermochromism (blue$\rightleftarrows$greenish yellow) with an enthalpy of reaction $\Delta H \approx 15$ kcal/mol. The solubility in water is 330 g/100 mL at 20°C. The compound is stable below 100°C [1].

$(CH_3NH_3)_2O \cdot 2MoO_3 \cdot H_2O$ (Heptamolybdate Type)

The compound was prepared from an aqueous solution containing 2 mol of monomethylamine per 1 mol of MoO_3 by evaporation to dryness and further drying of the residue in a desiccator under vacuum [2].

The compound was characterized by analysis and infrared spectrum (see the paper) [2]. The spectrum strongly resembles those of the paramolybdates.

Thermolysis shows at 78°C the formation of the anhydrous paramolybdate, at 145 to 154°C the formation of anhydrous "tetramolybdate", and at 244 to 348°C the formation of MoO_3. The compound is photosensitive but not photochromic. Irradiation with UV light yields a pink coloration [2].

$(CH_3NH_3)_2O \cdot 2MoO_3 \cdot 0.5H_2O$ (Heptamolybdate Type)

The compound was prepared from a mixture of 12 g of MoO_3 and 10 mL of an aqueous solution of monomethylamine (40%). Crystallization at −6°C is followed by drying under vacuum [2].

The compound was characterized by analysis and infrared spectrum (see the paper) [2]. The spectrum strongly resembles those of the paramolybdates.

The compound is photosensitive but not photochromic. Irradiation with UV light yields a pink coloration [2].

$(CH_3NH_3)_2O \cdot 5.87$ to $9.36MoO_3 \cdot xH_2O$ (36-Molybdate Type)

The compound was prepared by acidification of a solution of 8.1 g of MoO_3 and 10 mL of an aqueous solution of monomethylamine (40%) in 0 to 50 mL of water with 15 to 30 mL of 6N HNO_3 [2].

The salt was characterized by analysis and infrared spectrum (see the paper) [2]. The latter strongly resembles those of the 36-molybdates.

Thermolysis shows at 80°C the formation of the anhydrous salt (36-molybdates, however, give on heating normally the decamolybdates, compare pp. 67/8, 84, and 114) and at 132 to 348°C the formation of MoO_3. The compound is photosensitive but not photochromic. Irradiation with UV light yields a blue coloration [2].

Anhydrous Monomethylammonium Molybdates

In addition to the monomethylammonium molybdate hydrates described above the anhydrous compounds of composition $(CH_3NH_3)_2MoO_4$, $(CH_3NH_3)_2O \cdot 2MoO_3$, $(CH_3NH_3)_6[Mo_7O_{24}]$, and $(CH_3NH_3)_2O \cdot 4MoO_3$ have been prepared and investigated [2].

References:

[1] T. Yamase, T. Ikawa (Bull. Chem. Soc. Japan **50** [1977] 746/9). – [2] F. Arnaud-Neu, M. J. Schwing-Weill (Bull. Soc. Chim. France **1973** 3225/32).

2.8.2.2 Dimethylammonium Polymolybdates

$[(CH_3)_2NH_2]_6[Mo_7O_{24}] \cdot H_2O$ (Heptamolybdate Type)

The compound was prepred by agitation of a cold mixture of 3 g of MoO_3, 10 mL of an alcoholic solution of dimethylamine (33%), and 10 mL of absolute alcohol.

The salt was characterized by analysis and infrared spectrum (see the paper) showing the presence of a heptamolybdate.

Thermolysis shows at 110 to 124°C the formation of the anhydrous "trimolybdate", at 138 to 150°C the formation of the anhydrous "tetramolybdate", and at 250 to 380°C the formation of MoO_3. The compound is neither photochromic nor photosensitive [1].

$\{[(CH_3)_2NH_2]_4Mo_6O_{20}\cdot 2H_2O\}_\infty$ ("Trimolybdate" Type)

A solution of dimethylamine (40%) in 500 mL of water was mixed with 300 g of $(NH_4)_6Mo_7O_{24}\cdot 4H_2O$. After stirring for about 4 h, the solution was concentrated under reduced pressure, and the resultant white solid was allowed to crystallize. The solid was then dissolved in 1500 mL of water, and the solution was warmed to 50 to 60°C while stirred for 2 to 3 h to form a white precipitate. The precipitated product was filtered off, washed with cold water, and dried under vacuum [2]. The compound was also prepared by agitation of a cold mixture of 13.3 g of MoO_3, 10 mL of an aqueous solution of dimethylamine (40%), and 20 mL of water [1]. According to yet another method, a nearly saturated solution of 1 g of $MoO_2[(CH_3)_2NCS_2]_2$ [dioxobis(dimethyldithiocarbamato)molybdenum(VI) complex] in 1000 mL of chloroform was refluxed for 1 to 2 h. After cooling a white product was allowed to crystallize [2]. Instead of chloroform, methylene-dichloride can also be used. The dry powder was extracted with hot water. After repeated extraction, final crystallization in the dark gave fine white crystals [4]. The starting molybdenum complex was prepared using the method described by [3].

The compound was characterized by analysis and infrared spectrum (see the papers). The latter is a typical "trimolybdate" spectrum [1, 2].

The compound crystallizes in the form of a bundle composed of a number of fibrous crystals having their c axes along the needle axis. The constituent crystals are in completely random orientation around the needle axis [9].

The monoclinic compound has the lattice parameters a = 19.800(8), b = 10.294(2), c = 7.605(3) Å, β = 90.68(4)°; Z = 2. Space group $P2_1$-C_2^2 (No. 4). Calculated density 2.39 g/cm³ [9].

The crystal structure has been determined from photographic intensity data collected with a polycrystalline needle by the equi-inclination Weissenberg method around the needle axis. The pattern-fitting technique was applied to obtain structure factors of overlapping reflections. The atomic parameters are listed, R = 0.15. All the Mo atoms are octahedrally surrounded by O atoms, Mo-O distances 1.61 to 2.51 Å. The MoO_6 octahedra are joined together by sharing edges to form $[Mo_6O_{20}^{4-}]_\infty$ chains parallel to the c axis, see **Fig. 55**, p. 134. This is a new type of isopolymolybdate anion differing characteristically from the anion in the Rb compound, see Fig. 7, p. 13. The chains are held together by bridging hydrogen bonds of the type $O\cdots NH_2\cdots O$ and probably $O\cdots H_2O\cdots O$ (the locations of the water molecules were not elucidated) [9]. This structure of the $[Mo_6O_{20}^{4-}]_\infty$ chain (Fig. 55) contains two-bonded O atoms. Such O atoms have been excluded for "trimolybdates" since the typical vibrational spectra of the "trimolybdates" do not show any bands in the range 670 to 875 cm⁻¹ [11] (cf. p. 121).

The compound is stable below 100°C [2]. Thermolysis indicates at 168°C the formation of the anhydrous "tetramolybdate" and at 255 to 380°C the formation of MoO_3 [1].

The compound forms one of the most sensitive and rapid photochromic polymolybdate systems [2, 6, 7]. Irradiation with UV light (≧313 nm [2], 260 to 400 nm [4]) yields a reddish brown coloration [1, 2, 4, 6, 8], absorption maximum 470 nm [2, 4], 480 nm [6, 7], approximate coloration time (50% reflectance) <1 min [2, 6], saturation time approximately 10 min [2, 6, 7]. Approximate return time in the dark under aerobic conditions at room temperature (half-life period) 2 to 3 h [2, 6], at 31°C 67 min, at 40°C 40 min, under oxygen at room temperature 22 min [6], 1 atm of oxygen at 50°C <10 min [2]. Nitrogen blocks decoloration. Hence, decoloration is due to an oxidation reaction. The decomposition of the diffuse reflectance spectra of the irradiated compound shows four Gaussian curves at (1) 752, (2) 568, (3) 463, and (4) 405 nm. The kinetic analysis of the Gaussians (first order reactions) as well as a matrix rank treatment of the optical densities at different wavelengths and different times after exposure to light showed the presence of at least two colored species; the Gaussians (2) and (3) showed a parallel evolution [6, 7].

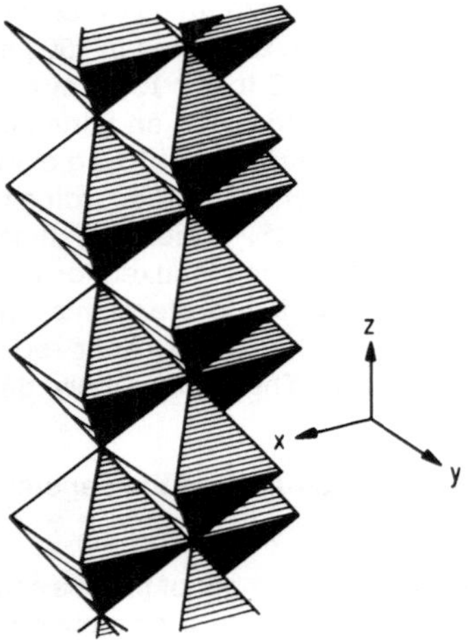

Fig. 55. Structure of the polyhexamolybdate anion $[Mo_6O_{20}^{4-}]_\infty$ in $\{[(CH_3)_2NH_2]_4Mo_6O_{20}\cdot 2H_2O\}_\infty$ [9].

For a free-electron model enabling assignment of the different Gaussian curves which constitute the absorption spectra of the irradiated salt, see the papers. According to this model the four Gaussian constituents are attributed to four different degrees of reduction of the same isopolyanion (number of Mo atoms of oxidation state V in the reduced anion) [6, 7].

An EPR signal of the colored polycrystalline material due to Mo^V was observed [2, 4, 6, 7], $g_1 = 1.885$, $g_2 = 1.930$, $g_3 = 1.935$ [4], $g_1 = 1.873$, $g_2 = 1.926$, $g_3 = 1.937$ [6 to 8]. The same signal was observed at 77 K and for any period of irradiation and during the bleaching. Bleaching kinetics studied by EPR showed again a first order reaction with a half-life period of 2 to 3 h which corresponds to the disappearance of the Gaussians (2) and (3) [and possibly (4)] [6, 7]. Photochromism is restricted to the solid state [4].

The solubility in water is 1.4 g/100 mL at 20°C [2, 5].

$[(CH_3)_2NH_2]_4[Mo_8O_{26}]\cdot 2C_3H_7NO$, C_3H_7NO = N, N-Dimethylformamide (β-Octamolybdate Type)

The compound has been prepared from 7 g MoO_3 heated under reflux for 3 h with 40 mL N, N-dimethylformamide (DMF). The reaction occurs even under rigorously anhydrous conditions and in the absence of atmospheric oxygen. It is accompanied by release of CO, which suggests dissociation of DMF to $NH(CH_3)_2 + CO$. From the yellow-green solution crystallization was induced by addition of ethanol or diethyl ether. Colorless tabular crystals separated after 1 to 2 d in yields up to 25%. The crystals decompose rapidly in moist air and therefore have been coated with Araldite resin.

The compound has been characterized by chemical analysis, IR spectrum, and crystal structure analysis.

The monoclinic crystals have the lattice parameters a = 18.118(4), b = 14.568(3), c = 16.228(3) Å, β = 101.003(3)°; Z = 4. Space group C2/c-C_{2h}^6 (No. 15). $D_m = 2.36$, $D_x = 2.38$ g/cm³. The atomic parameters are listed, R = 0.046.

The octamolybdate anions, $Mo_8O_{26}^{4-}$, are centrosymmetric and possess the β-type structure, see Fig. 10, p. 16. Mean Mo-O bond lengths are: 1.698(8) (terminal O_t), 1.909(8) (single-bridging O_b), 2.091(8) (triple-bridging O), and 2.319(8) Å (five-bridging O). The DMF solvate molecules are H bonded (N···O = 2.73(4) Å) to one of the two crystallographically independent $(CH_3)_2NH_2^+$ cations. Selected bond lengths and angles are listed in the paper. The high apparent thermal parameters of the DMF atoms, combined with large positional uncertainties, strongly suggest that these molecules are disordered.

The IR spectrum shows DMF bands, but rather weakly, and in association with probable N-H bands. There are strong Mo-O_b-Mo bands at 725, 675 cm^{-1}, and the ν(Mo-O_t) bands at 943, 905 cm^{-1} show evidence of splitting [10].

Anhydrous Dimethylammonium Polymolybdates

In addition to the dimethylammonium polymolybdate hydrates described above the anhydrous compounds of composition $[(CH_3)_2NH_2]_2O \cdot 3MoO_3$ [1, 8] and $[(CH_3)_2NH_2]_2O \cdot 4MoO_3$ have been prepared and investigated [1].

References:

[1] F. Arnaud-Neu, M. J. Schwing-Weill (Bull. Soc. Chim. France **1973** 3225/32). – [2] T. Yamase, T. Ikawa (Bull. Chem. Soc. Japan **50** [1977] 746/9). – [3] F. W. Moore, M. L. Larson (Inorg. Chem. **6** [1967] 998/1003). – [4] T. Yamase, T. Ikawa, H. Kokado, E. Inoue (Chem. Letters **1973** 615/6). – [5] T. Yamase, H. Hayashi, T. Ikawa (Chem. Letters **1974** 1055/6).

[6] F. Arnaud-Neu, M. J. Schwing-Weill (Bull. Soc. Chim. France **1973** 3233/9). – [7] F. Arnaud-Neu, M. J. Schwing-Weill (J. Less-Common Metals **36** [1974] 71/8; Chem. Uses Molybdenum Proc. 1st Conf., Reading, Engl., 1973 [1974], pp. 35/8). – [8] F. Arnaud-Neu, M. J. Schwing-Weill (Bull. Soc. Chim. France **1973** 3239/46). – [9] H. Toraya, F. Marumo, T. Yamase (Acta Cryst. B **40** [1984] 145/50). – [10] A. J. Wilson, V. McKee, B. R. Penfold, C. J. Wilkins (Acta Cryst. C **40** [1984] 2027/30).

[11] H. U. Kreusler, A. Förster, J. Fuchs (Z. Naturforsch. **35b** [1980] 242/4).

2.8.2.3 Trimethylammonium Polymolybdates

$[(CH_3)_3NH]_6[Mo_7O_{24}] \cdot 7H_2O$ (Heptamolybdate Type)

The compound was prepared by heterogeneous reaction of 9.16 g of MoO_3 and 20 mL of a methanol solution of trimethylamine (25%); reaction time 2 d.

The salt was characterized by analysis and infrared spectrum (see the paper) showing the presence of a heptamolybdate.

Thermolysis indicates in the range from room temperature to 158°C the formation of the anhydrous tetramolybdate and at 234 to 434°C the formation of MoO_3 *via* reduced molybdenum oxides. The compound is neither photochromic nor photosensitive [1].

$[(CH_3)_3NH]_2O \cdot 3MoO_3 \cdot H_2O$ (Unknown Type)

The compound was prepared by the heterogeneous reaction of 13.7 g of MoO_3, 20 mL of a methanol solution of trimethylamine (25%), and 6 mL of water [1].

The compound was characterized by analysis and infrared spectrum (see the paper) [1]. The spectrum is different from those of the "trimolybdates", and the analytical acid : base ratio of 2.86 also shows strong deviations from the theoretical value of 3.00.

Thermolysis shows at 128 to 154°C the formation of the anhydrous "tetramolybdate" and at 228 to 420°C the formation of MoO_3. The compound is neither photochromic nor photosensitive [1].

Anhydrous Trimethylammonium Polymolybdates

In addition to the trimethylammonium polymolybdate hydrates described above the anhydrous compounds of composition $[(CH_3)_3NH]_2O \cdot 3MoO_3$ and $[(CH_3)_3NH]_2O \cdot 4MoO_3$ have been prepared and investigated [1].

Reference:

[1] F. Arnaud-Neu, M. J. Schwing-Weill (Bull. Soc. Chim. France **1973** 3225/32).

2.8.2.4 Tetramethylammonium Molybdates

$[(CH_3)_4N]_2MoO_4 \cdot 1.5H_2O$

The 1.5-hydrate is formed as the only one in the $(CH_3)_4NOH$-MoO_3-H_2O system at $(CH_3)_4NOH$ to MoO_3 molar ratios varied from 2:1 to 2:10. In specimens with a higher proportion of MoO_3 than the stoichiometric, free molybdic acid is present in addition.

$[(CH_3)_4N]_2MoO_4 \cdot 1.5H_2O$ is prepared by dissolving a stoichiometric amount of MoO_3 in tetramethylammonium hydroxide solution and concentrating on the water bath until the solid phase appears. This, after filtering off, is dried over P_2O_5.

The compound is triclinic with the lattice parameters $a = 13.10 \pm 0.01$, $b = 14.24 \pm 0.02$, $c = 10.23 \pm 0.01$ Å, $\alpha = 95°40' \pm 10'$, $\beta = 109°30' \pm 10'$, $\gamma = 83° \pm 10'$; Z = 6. For a list of d values see the paper. The measured density is 1.87, the calculated density 1.92 g/cm^3.

By heating the hydrate to 140°C the highly hygroscopic anhydrous $[(CH_3)_4N]_2MoO_4$ is obtained. The hydrate readily dissolves in ethanol and butanol and is insoluble in acetone, xylene, and cyclohexane [1].

$[(CH_3)_4N]_2O \cdot 10MoO_3 \cdot 7H_2O$ ("Decamolybdate" Type)

This compound is deposited from a strongly acidified (2 to 3 H^+/MoO_4^{2-}) solution of sodium molybdate upon addition of a solution of tetramethylammonium chloride [2] or bromide [3]. Colorless fine crystals are obtained [2].

The compound was characterized by analysis, infrared and Raman spectra [2].

The crystals become blue-colored in the light; in the dark a return to the original state is observed if the time of exposure to light was not too long. The compound is insoluble in all solvents [2].

$[(CH_3)_4N]_2Na_2[Mo_8O_{26}] \cdot 2H_2O$ (β-Octamolybdate Type)

The compound was prepared by addition of 80 mL of a 0.1M solution of tetramethylammonium bromide (or chloride [2]) to 50 mL of a 0.4M solution of sodium molybdate acidified with 26 mL of 1N H_2SO_4 (degree of acidification 0.8 to 1.5 [3], 1.3 [2]). Crystallization is terminated after about 4 d. The colorless hexagonal crystal plates were filtered off, washed with cold water, ethanol, and ether, and dried under vacuum over P_2O_5. The deuterated compound has been prepared by using anhydrous Na_2MoO_4 in D_2O acidified with DCl [3].

The compound was characterized by analyses (the formula $[(CH_3)_4N]_2Mo_8O_{25} \cdot 5H_2O$ stated in [2, 3] is incorrect [4, 5]), infrared and Raman spectra (see the papers) [2, 3]. For a discussion of the spectra see [2, 3].

The crystals are monoclinic, lattice parameters a = 11.116(7), b = 15.140(10), c = 12.022(7) Å, β = 127.61(6)°, space group C2/m-C_{2h}^3 (No. 12) [5]; Z = 2 [3, 5]. D_m = 2.898, D_x = 2.899 g/cm³ [5]; see also [2, 3]. The atomic coordinates are listed; however, the positions of the H atoms could not be determined [5].

The structure was solved on a twinned crystal (twin plane approximately ($\bar{4}03$) [3]) and refined to R = 0.041. The structure of the polyanion (see Fig. 10, p. 16) consists of $Mo_8O_{26}^{4-}$ blocks composed of eight MoO_6 octahedra sharing 3 to 6 edges with each other. Two groups composed of four MoO_6 octahedra are centrosymmetrically arranged to each other. Mo···Mo distances inside a group of four MoO_6 octahedra are shorter than those between the centrosymmetrical groups. The Mo-O distances depend strongly upon the coordination number of the oxygen atoms: Mo-O(terminal) ≈ 1.70 Å, Mo-O(coordination number 5) ≈ 2.33 Å. Hence, the molybdenum atoms in the octahedral gaps of the packing of the oxygen atoms are displaced in the direction of the terminal oxygen atoms. The positions of the cations and of the water molecules are shown in **Fig. 56**. For the different bond lengths and angles between Mo and O within the polyanion and between the cations, water molecules, and oxygen atoms of the polyanion see the paper [5]. Broad-line NMR measurements and the IR spectrum of the deuterated compound show the absence of OH groups [2, 3].

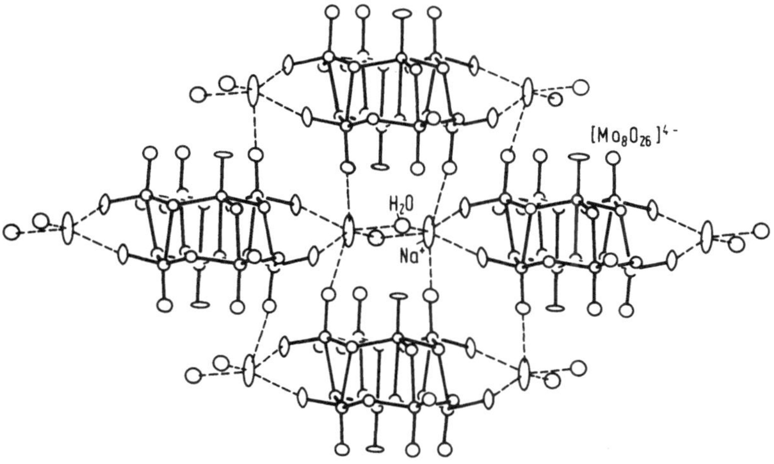

Fig. 56. Distorted octahedral coordination of Na^+ in $[(CH_3)_4N]_2Na_2[Mo_8O_{26}] \cdot 2H_2O$ [5].

Thermogravimetric and DTA investigations show at 309°C the loss of $2H_2O$ and an endothermic peak. At 437°C the compound is completely decomposed to MoO_3, residual H_2O, and presumably $(CH_3)_3N$ and CH_3OH. Two H_2O/formula unit are bound more weakly [3].

References:

[1] V. N. Serezhkin, V. V. Tabachenko, L. B. Serezhkina (Zh. Neorgan. Khim. **21** [1976] 1965/7; Russ. J. Inorg. Chem. **21** [1976] 1080/1). – [2] J. Fuchs (Z. Naturforsch. **28b** [1973] 389/404). – [3] J. Fuchs, I. Knöpnadel, I. Brüdgam (Z. Naturforsch. **29b** [1974] 473/5). – [4] W. G.

Klemperer, W. Shum (J. Am. Chem. Soc. **98** [1976] 8291/3). – [5] J. Fuchs, I. Knöpnadel (Z. Krist. **158** [1982] 165/79).

2.8.2.5 Monoethylammonium Polymolybdates

$(C_2H_5NH_3)_6[Mo_7O_{24}]\cdot 3H_2O$ (Heptamolybdate Type)

The compound was prepared by agitation of a cold mixture of 10 g of MoO_3, 5 mL of an aqueous solution of monoethylamine (70%), and 20 mL of absolute alcohol.

The salt was characterized by analysis and infrared spectrum (see the paper) showing the presence of a heptamolybdate.

Thermolysis shows at 38 to 98°C the loss of the water. At 126°C the anhydrous "tetramolybdate" and at 212 to 360°C MoO_3 is formed. The compound is neither photochromic nor photosensitive [1].

$(C_2H_5NH_3)_2O\cdot 2MoO_3\cdot 2H_2O$ ("Trimolybdate" Type)

The compound was prepared from a mixture of aqueous solutions of ammonium paramolybdate and monoethylammonium chloride by evaporation of a third of the volume and precipitation by cooling [1].

The salt was characterized by analysis and infrared spectrum (see the paper) [1]. The latter, however, is a typical "trimolybdate" spectrum (compare also p. 39).

Thermolysis shows no reaction up to 176°C. At 382°C MoO_3 forms. The compound is photosensitive but not photochromic [1].

$(C_2H_5NH_3)_2O\cdot 4MoO_3\cdot 2H_2O$ (Unknown Type)

The compound was prepared by heating a mixture of 8 g of ammonium "octamolybdate" and 30 mL of a 2 N solution of monoethylammonium chloride under reflux [1]. The "octamolybdate", possibly being ammonium 36-molybdate (compare p. 114), was prepared according to [2].

The salt was characterized by analysis and infrared spectrum (see the paper). The compound is neither photochromic nor photosensitive [1].

$(C_2H_5NH_3)_2O\cdot 8.41MoO_3\cdot xH_2O$ (36-Molybdate Type)

The compound was obtained by strong acidification of a solution of $(C_2H_5NH_3)_2MoO_4$ with 2 N HNO_3 [1].

The salt was characterized by analysis and infrared spectrum (see the paper) [1]. The latter resembles strongly those of the 36-molybdates.

The compound is photosensitive but not photochromic. Irradiation with UV light yields a blue coloration [1].

Anhydrous Monoethylammonium Polymolybdates

In addition to the monoethylammonium polymolybdate hydrates described above the anhydrous compound of composition $(C_2H_5NH_3)_2O\cdot 3MoO_3$ has been prepared and investigated [1].

References:

[1] F. Arnaud-Neu, M. J. Schwing-Weill (Bull. Soc. Chim. France **1973** 3225/32). – [2] M. J. Schwing-Weill (Bull. Soc. Chim. France **1967** 3795/8).

2.8.2.6 Diethylammonium Polymolybdates

$[(C_2H_5)_2NH_2]_2O \cdot 3MoO_3 \cdot H_2O$ ("Trimolybdate" Type)

An aqueous solution of 10 g of $(NH_4)_6Mo_7O_{24} \cdot 4H_2O$ in 20 mL of diethylamine (40%) was stirred for about 4 h. The solution was concentrated under reduced pressure, and the resultant white solid was allowed to crystallize. Recrystallization has been accomplished from warm water as with $[(CH_3)_2NH_2]_2O \cdot 3MoO_3 \cdot H_2O$ (p. 133) [1]. The compound was also obtained by agitation of a cold mixture of 10 g of MoO_3, 5 g of diethylamine, 15 mL of water, and 100 mL of absolute alcohol. According to yet another method the compound was obtained by heating of a mixture of 10.5 g of MoO_3, 7 g of diethylamine, and 40 mL of absolute alcohol under reflux, filtration, and precipitation at 0°C [2].

The compound was characterized by analysis and infrared spectrum (see the papers). The latter is a typical "trimolybdate" spectrum [1, 2].

Thermolysis shows at 50 to 82°C the formation of the anhydrous compound, at 137 to 154°C the formation of the anhydrous "tetramolybdate", and at 212 to 382°C the formation of MoO_3 [2]. According to [1] the compound is stable below 100°C.

The compound forms one of the most sensitive and rapid photochromic polymolybdate systems [1, 3]. Under irradiation with UV light (≧313 nm [1]) the white crystals become reddish brown [1, 2, 4]: absorption maximum 475 nm [1], 480 nm [4], approximate coloration time (50% reflectance) <1 min [1, 4], saturation time approximately 10 min. Approximate return time in the dark under aerobic conditions at room temperature (half-life period) is 2 to 3 h [1], 2 to 3 h at room temperature, 10 min at 40°C, and 15 min under oxygen at room temperature after an irradiation period of 2 min. Nitrogen blocks decoloration, hence, decoloration is due to an oxidation reaction. The decomposition of the diffuse reflectance spectra of the irradiated compound shows five Gaussian curves at (1) 806, (2) 572, (3) 473, (4) 406, and (5) 364 nm. A matrix rank treatment of the optical densities at different wavelengths and different times after exposure to light showed the presence of at least three colored species after an irradiation period of 2 min and of at least five colored species after an irradiation period of 6 min. In the case of the irradiation for 2 min the Gaussians (2) and (3) showed a parallel first order evolution ($t_{1/2} \approx 3$ h); (1) decreased very rapidly during the first hours, while (4) was nearly constant for the first 2 h. The Gaussian (5) could not be evaluated because of an insufficient precision of the measurements. After about 22 h the Gaussian (3) continued to decrease in a first order reaction with $t_{1/2} = 3.3$ d. In the case of the irradiation for 6 min the overall half-life period was much longer, approximately 35 h. The Gaussians (2) and (3) disappeared in first order reactions with $t_{1/2} = 1.5$ and 2.5 d, respectively. The other, less intense Gaussians disappeared still more slowly [4]. For a free-electron model enabling assignment of the different Gaussian curves which constitute the absorption spectra of the irradiated salt, see the papers. According to this model the Gaussians are attributed to differently reduced polyanions of two different types of paramagnetic centers [3, 4].

An EPR signal of the colored polycrystalline material due to Mo^V was observed [1, 3, 4]. The signal and its temporal behavior are complex and due to the superposition of two signals, one ($g_1 = 1.873$, $g_2 = 1.926$, $g_3 = 1.937$) being the same as in case of the dimethylammonium "trimolybdate" monohydrate and the other photochromic polymolybdates. The other signal is characterized by $g_1 = 1.892$, $g_2 = 1.911$, $g_3 = 1.924$ and disappears much more slowly. Bleaching kinetics studied by EPR after an irradiation period of 3 min show two simultaneous first order reactions with $t_{1/2} = 4.6$ h and $t_{1/2} = 2.3$ d, and confirms the spectrophotometric results [3, 4]. See also p. 39.

The solubility in water is 1.0 g/100 mL at 20°C [1].

$[(C_2H_5)_2NH_2]_2O \cdot 7.30$ to $9.02\,MoO_3 \cdot x\,H_2O$ (36-Molybdate Type)

The compound was obtained by strong acidification of a solution of $[(C_2H_5)_2NH_2]_2MoO_4$ with 6 N HCl [2].

The salt was characterized by analysis and infrared spectrum [2]. The latter strongly resembles those of the 36-molybdates.

The compound is photosensitive but not photochromic. Irradiation with UV light yields a blue coloration [2].

Anhydrous Diethylammonium Polymolybdates

In addition to the diethylammonium polymolybdate hydrates described above the anhydrous compounds of composition $[(C_2H_5)_2NH_2]_2O \cdot 3\,MoO_3$ [2, 4] and $[(C_2H_5)_2NH_2]_2O \cdot 4\,MoO_3$ [2] have been prepared and investigated.

References:

[1] T. Yamase, T. Ikawa (Bull. Chem. Soc. Japan **50** [1977] 746/9). – [2] F. Arnaud-Neu, M. J. Schwing-Weill (Bull. Soc. Chim. France **1973** 3225/32). – [3] F. Arnaud-Neu, M. J. Schwing-Weill (J. Less-Common Metals **36** [1974] 71/8; Chem. Uses Molybdenum Proc. 1st Conf., Reading, Engl., 1973 [1974], pp. 35/8). – [4] F. Arnaud-Neu, M. J. Schwing-Weill (Bull. Soc. Chim. France **1973** 3239/46).

2.8.2.7 Triethylammonium Polymolybdates

$[(C_2H_5)_3NH]_3(H_3O)[Mo_8O_{26}] \cdot 2\,H_2O$ (β-Octamolybdate Type)

Colorless prismatic crystals of this compound have been obtained by a similar method to that reported for the diethyl polymolybdate $[(C_2H_5)_2NH_2]_2O \cdot 3\,MoO_3 \cdot H_2O$ (see p. 139, [1]).

The compound has been characterized by chemical analysis and X-ray structure investigation.

The monoclinic crystals have the lattice parameters a = 21.271(9), b = 11.837(1), c = 20.189(9) Å, β = 117.92(5)°; Z = 4. Space group $P2_1/a\text{-}C_{2h}^5$ (No. 14). $D_x = 2.29$ g/cm³. The atomic coordinates are listed in the paper, R = 0.043.

Two crystallographically independent octamolybdate anions are situated at different inversion centers and are stacked alternately along [001]. Their structure is essentially equal to that of β-$[Mo_8O_{26}]^{4-}$, see Fig. 10, p. 16. The terminal Mo-O distances range from 1.686 to 1.717 Å, for other distances see the paper. The H_3O^+ cation connects the two anions with complicated hydrogen bonds. Also the other water molecules and the triethylammonium cations participate in the hydrogen bonding [2].

The compounds of composition **$[(C_2H_5)_3NH]_2O \cdot 4\,MoO_3$** and **$[(C_2H_5)_3NH]_2[Mo_6O_{19}]$** have been obtained only in the anhydrous form [3].

References:

[1] T. Yamase, T. Ikawa (Bull. Chem. Soc. Japan **50** [1977] 746/9). – [2] P. K. Bharadwaj, Y. Ohashi, Y. Sasada, Y. Sasaki, T. Yamase (Acta Cryst. C **40** [1984] 48/50). – [3] J. Fuchs (Z. Naturforsch. **28b** [1973] 389/404, 395).

2.8.2.8 Tetraethylammonium Polymolybdates

The compounds $[(C_2H_5)_4N]_2O \cdot 4MoO_3$ and $[(C_2H_5)_4N]_2[Mo_6O_{19}]$ [1, 2] are known only as anhydrous products. However, later authors were unable to produce $[(C_2H_5)_4N]_2O \cdot 4MoO_3$, instead of which they obtained β-$[(C_2H_5)_4N]_3Na[Mo_8O_{26}]$ [3].

References:

[1] J. Fuchs (Z. Naturforsch. **28b** [1973] 389/404, 395). – [2] J. Fuchs, I. Knöpnadel, I. Brüdgam (Z. Naturforsch. **29b** [1974] 473/5). – [3] W. G. Klemperer, W. Shum (J. Am. Chem. Soc. **98** [1976] 8291/3).

2.8.2.9 Monopropylammonium Polymolybdate $(n\text{-}C_3H_7NH_3)_6[Mo_7O_{24}] \cdot 3H_2O$ (Heptamolybdate Type)

In [1] the compound is erroneously described as $(C_3H_7NH_3)_2O \cdot 2MoO_3 \cdot 2H_2O$ owing to the low accuracy of the elemental analysis [2].

To prepare the compound, an aqueous solution of 50 g of $(NH_4)_6Mo_7O_{24} \cdot 4H_2O$ in 100 mL of propylamine (30%) was stirred for about 6 h. The solution was concentrated under reduced pressure to form a solid, which was then filtered off and dried. Recrystallization of the crude product from water yielded a colorless solid [1]. Single crystals suited for EPR studies were obtained by slow evaporation from aqueous solution [3].

The compound was characterized by analysis and infrared spectrum (see the paper) [1]. The latter resembles those of the paramolybdates.

The triclinic crystals have the lattice parameters a = 14.118(2), b = 15.977(2), c = 11.368(1) Å, α = 90.23(1)°, β = 93.00(1)°, γ = 111.79(1)°; Z = 2. Space group $P\bar{1}\text{-}C_i^1$ (No. 2). D_x = 2.054 g/cm³. The atomic coordinates are listed in the paper, R = 0.035 [2].

The heptamolybdate anion has an approximate point symmetry of 2mm (C_{2v}) and its structure is the same as in the crystals of sodium, potassium, and ammonium salts (see Figs. 1 and 43, pp. 9 and 79, respectively). Two molecules are related by a center of inversion. The Mo-O bonds can be classified into four types by their bond distances: 1.69 to 1.74, 1.87 to 1.99, 2.12 to 2.31, and 2.46 to 2.55 Å. The MoO_6 octahedra are divided into two classes: five octahedra have three types of Mo-O bonds, whereas two octahedra (Mo(5) and Mo(6), the outer ones of the central row of three octahedra) have four types of Mo-O bonds. The hydrophobic groups of the propylammonium cations are brought together around the (½, 0, z) region. The bond distances of the propylammonium cations show rather large deviations from the normal values which is due to their disordered structure. The conformation of the ions may be easily changed in the crystal. All the N-H groups of the propylammonium ion and the O-H groups of the water molecules form hydrogen bonds with the heptamolybdate anions or the water molecules constructing a three-dimensional network. There also exist trifurcated hydrogen bonds. For the individual bond distances and angles see the paper [2]. The hydrogen bonds play an important role in the UV-induced photochromism of the crystals of the compound, see below.

The compound has photochromic properties in the solid state (and in aqueous solution). Under irradiation with UV light (≧313 nm) the white crystals become reddish brown (or violet?) [1, 3] or red [2]: absorption maximum 490 nm, approximate coloration time (50% reflectance) 5 to 7 min, approximate return time in the dark under aerobic conditions at room temperature (half-life period) 4 to 5 h. An EPR signal of the colored polycrystalline material due to Mo^V was observed. The colored sample turned blue when dissolved in deaerated water ($\lambda_{max} = 730$ nm, $\lambda_{sh} = 620$ nm). The solution gave rise to a single intense line (g = 1.926) with six satellite lines, three on each side and equally spaced (a = 52 G), in its EPR spectrum, indicating a hyperfine structure due to isotopes with a nuclear spin of 5/2. The exposure of the blue solution to oxygen brought about the oxidation of the Mo^V to Mo^{VI} with an accompanying bleaching and disappearance of the EPR signal [1]. The direct UV irradiation of the deaerated aqueous solution of the compound (13.6 mmol/L) revealed the presence of two paramagnetic species of $\langle g \rangle =$ 1.921 and 1.910. The signal at $\langle g \rangle = 1.921$ developed weak satellite lines ($\langle A \rangle \approx 50$ G $= 4.5 \times 10^{-3}$ cm^{-1}) due to $^{95,97}Mo$. These values are almost the same as for the photolyte of a solution of the monoisopropylammonium salt, suggesting that the formation of the two paramagnetic species is hardly affected by the variation of the alkylammonium cations [3]. The crystals also turn red on X-ray exposure (as for the data collection for the crystal structure determination, without significant change of the intensities) [2].

The photoreducible Mo atom of the $[Mo_7O_{24}]^{6-}$ anion is considered to be one of the two outer Mo atoms (Mo(5) and Mo(6)) of the central row of three MoO_6 octahedra because each of them has a long Mo-O bond (as is also observed in the photoreducible site of the monoisopropylammonium dihydrogenoctamolybdate crystal, see p. 151) and hence an electronic state different from that of the other Mo atoms. The geometry of the hydrogen bonding scheme at these Mo atoms is shown in **Fig. 57**. There are five hydrogen bonds between the N-H groups of the cations and the bridging oxygen atoms of the Mo(5) and Mo(6) sites. One of the hydrogen bonds is seen to be responsible for the proton transfer from the cation to the anion according to

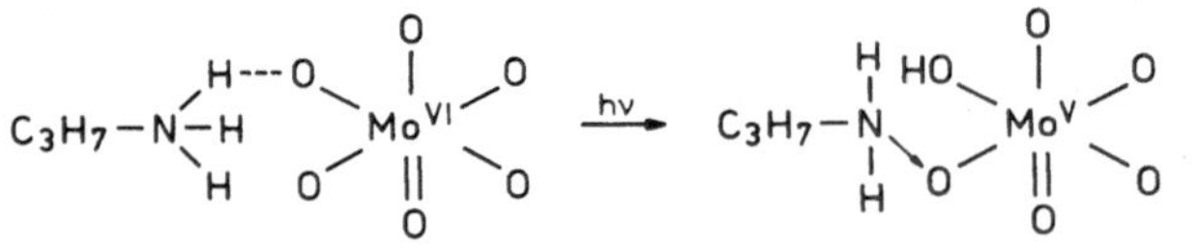

(i) (1)

Since the ammonium heptamolybdate does not exhibit photoreduction on irradiation with UV light, although it has a similar hydrogen bonding scheme, some additional factors have to be taken into account for the photoreduction. It is assumed that a positively charged terminal oxygen atom (?) forms a charge transfer complex with the negatively charged alkylamine nitrogen atom in the complex (i), by which the Mo^V species formed by the UV irradiation would be stabilized. Due to the considerably loose contacts of the propylammonium cations with the neighboring polyanions in the crystal such a charge transfer complex would be easily formed in the solid state. When the crystal is allowed to stand in the dark after irradiation, the charge transfer complex returns to the initial hydrogen bonding complex and its color is bleached. On the other hand, the unsubstituted ammonium cations have more than four hydrogen bonds with neighboring heptamolybdate anions and water molecules. They are closely packed in the crystal. Thus the formation of the charge transfer complex may be inhibited on exposure to UV light [2].

EPR spectra result from a localized paramagnetic site in the molecule. In order to determine the Mo^V center formed by the UV irradiation ($\lambda \geqq 313$ nm) a study has also been made on a reddish brown single crystal of the compound by EPR at room temperature. An orthogonal axis system was chosen in the crystal: the y axis coincident with the c direction, the z axis

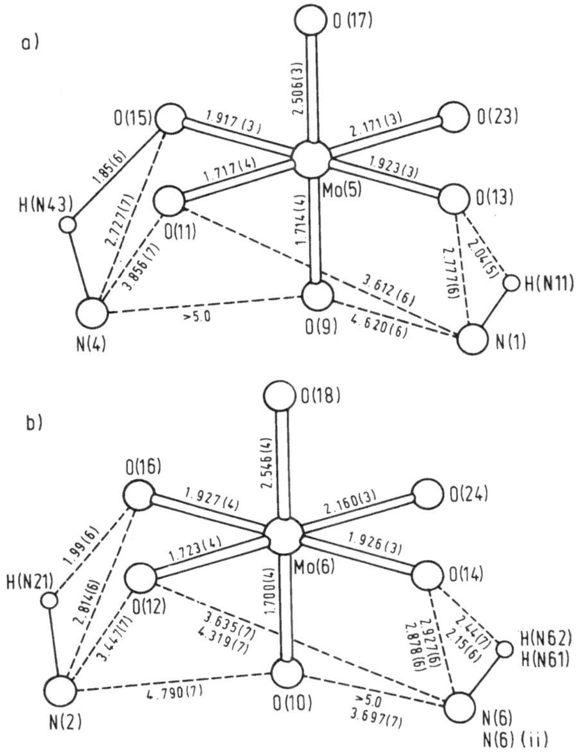

Fig. 57. Bond distances (in Å) and interatomic distances less than 5 Å between nonbonded atoms (in Å) in the environment of the (a) Mo(5) and (b) Mo(6) sites in monopropylammonium heptamolybdate. (In the Mo(6) site, O(14) makes two hydrogen bonds with N(6)-H(N62) (upper values) and N(6)-H(N61) (lower values) in the unit of (ii)) [2].

perpendicular to the flat face of (100), and the x axis perpendicular to y and z (see a figure in the paper). Two intense central lines (1:1) with a line width equal to ca. 5 G result from the superhyperfine coupling with ^1H ($I = 1/2$). On each side of the central spectrum, six lines of low intensity are observed. The splittings between two consecutive lines of the central spectrum are the same as those between two consecutive side bands, and the intensity between a central and a side signal is ca. 5%. These observations indicate that the side-lines result from the hyperfine coupling with the two isotopes $^{95,97}Mo$ in natural abundance (25.15%) which have the same nuclear spin ($I = 5/2$) and nearly the same magnetic moment, so that it is not possible to resolve the separate hyperfine lines, and that each of the hyperfine lines is resolved into two superhyperfine lines due to ^1H. The occurrence of the $^{95,97}Mo$ hyperfine and ^1H superhyperfine structures is characteristic of an unpaired electron localized on one molybdenum atom only. Additionally, the lines are narrow and no exchange broadening is observed. All these observations are in accordance with the paramagnetic site $Mo^VO_5(OH)$ localized in an $[Mo_7O_{24}]^{6-}$ anion resulting from transfer of a hydrogen bonding proton from a nitrogen cation to a bridging oxygen atom. The principal values of the g, hyperfine (A_{Mo}), and superhyperfine (A_H) tensors are $g_1 = 1.895$, $g_2 = 1.925$, $g_3 = 1.937$; $A_{Mo(1)} = 2.34 \times 10^{-3}$, $A_{Mo(2)} = 4.20 \times 10^{-3}$, $A_{Mo(3)} = 6.99 \times 10^{-3}$, $A_{H(1)} = 8.15 \times 10^{-4}$, $A_{H(2)} = 9.14 \times 10^{-4}$, and $A_{H(3)} = 12.5 \times 10^{-4}$ cm^{-1}. Analysis of the EPR parameters indicates the direct spin polarization between the paramagnetic electron orbital of molybdenum and the hydrogen orbital. From the direction of the largest principal value for the A_H tensor, which is approximately parallel to the $H(O) \cdots Mo^V$ direction, the paramagnetic site in the $[Mo_7O_{24}]^{6-}$ anion is determined in correlation with the X-ray crystal structure data in [2]. It is proposed, in accordance with [2], that Mo^V formation occurs at an end of the central row of three MoO_6 octahedra of the $[Mo_7O_{24}]^{6-}$ structure, where the molybdenum atom has a long Mo-O bond (2.5 Å). For the mechanism of the photoreduction see equation (1). It is assumed that the Mo(5) site is less sensitive to photoreduction than the Mo(6) site due to the long distance between O(11) and N(4) or N(1) in the Mo(5) site (see Fig. 57) [3].

The solubility in water is 90 g/100 mL at 20°C. The compound is stable below 100°C [1].

References:

[1] T. Yamase, T. Ikawa (Bull. Chem. Soc. Japan **50** [1977] 746/9). – [2] Y. Ohashi, K. Yanagi, Y. Sasada, T. Yamase (Bull. Chem. Soc. Japan **55** [1982] 1254/60). – [3] T. Yamase (J. Chem. Soc. Dalton Trans. **1982** 1987/91).

2.8.2.10 Tetrapropylammonium Polymolybdates

$[(C_3H_7)_4N]_3Na[Mo_8O_{26}]\cdot 2H_2O$ (β-Octamolybdate Type)

The compound was prepared by addition of a solution of $[(C_3H_7)_4N]Cl$ or $[(C_3H_7)_4N]Br$ to an acidified aqueous solution of sodium molybdate (pH 3 to 4).

The salt was characterized by analysis, Raman and infrared spectroscopy; however, data are not given. The structure of the polymolybdate ion (see Fig. 10, p. 16) has been assigned by "finger-print" Raman and infrared spectroscopy in comparison with other compounds containing the β-$[Mo_8O_{26}]^{4-}$ ion [1].

Anhydrous Propylammonium Polymolybdates

The anhydrous tripropylammonium β-octamolybdate is described in [2].

In addition to the tetrapropylammonium sodium polymolybdate hydrate described above the anhydrous compounds of composition $[(C_3H_7)_4N]_2O\cdot 4MoO_3$ (β-octamolybdate type) and $[(C_3H_7)_4N]_2[Mo_6O_{19}]$ [3, 4] have been prepared and investigated. However, later authors could not produce the compound $[(C_3H_7)_4N]_2O\cdot 4MoO_3$, instead of which they obtained the sodium-containing octamolybdate given above [1].

References:

[1] W. G. Klemperer, W. Shum (J. Am. Chem. Soc. **98** [1976] 8291/3). – [2] J. Fuchs, A. Thiele (Z. Naturforsch. **34b** [1979] 155/9). – [3] J. Fuchs, I. Knöpnadel, I. Brüdgam (Z. Naturforsch. **29b** [1974] 473/5). – [4] J. Fuchs (Z. Naturforsch. **28b** [1973] 389/404, 395, 401/2).

2.8.2.11 Monoisopropylammonium Polymolybdates

2.8.2.11.1 $(i\text{-}C_3H_7NH_3)_6[Mo_7O_{24}]\cdot 3H_2O$ (Heptamolybdate Type)

Preparation

A mixture of 80 g of $(NH_4)_6Mo_7O_{24}\cdot 4H_2O$ in 50 mL of water and 52 mL of isopropylamine (mole ratio 1:1 [2]) was stirred for about 6 h. The solution was concentrated under reduced pressure to form a white solid, which was then filtered off and dried (after washing with ethanol and diethyl ether [2]). Recrystallization of the crude product from water yielded a colorless crystalline material [1, 2].

The compound is in [1, 2] erroneously described as $(i\text{-}C_3H_7NH_3)_2O\cdot 2MoO_3\cdot 2H_2O$ owing to the inaccuracy of the elemental analysis [5].

"Finger-Prints", Spectra

The infrared spectrum of the solid compound [1] (see the paper) resembles those of the paramolybdates.

The Raman spectrum of the aqueous solution [10] is similar to that reported by other authors for the $[Mo_7O_{24}]^{6-}$ ion in aqueous solution (see the description of the ion in "Molybdenum" Suppl. Vol. B 3 [to be published]) with the exception of an additional band at 804 cm^{-1}. The ultraviolet spectrum of the solution shows an absorption maximum at 205 nm ($\varepsilon = 1.56 \times 10^2$ m^2/mol) with shoulders at about 230 nm ($\varepsilon = 86$) and 285 nm ($\varepsilon = 25$) and much weaker absorptions at wavelengths between 310 and 380 nm ($\varepsilon_{313} = 9.7$ and $\varepsilon_{365} = 4.1$) [6, 10].

Crystal Data, Structure

The monoclinic crystals have the lattice parameters a = 23.904(6), b = 10.504(1), c = 20.652(9) Å, $\beta = 115.40(2)°$; Z = 4. Space group P2/n-C_{2h}^4 (No. 13). $D_x = 2.085$ g/cm^3. The atomic coordinates are listed in the paper, R = 0.060 [5].

The heptamolybdate anion has an approximate point symmetry of 2mm (C_{2v}) and its structure ist the same as in the crystals of sodium, potassium, and ammonium salts (see Figs. 1 and 43, pp. 9 and 79, respectively). The corresponding bond distances in heptamolybdates of different cations are in fair agreement. Discrepancies are probably due to different hydrogen bonding schemes. The Mo-O bonds can be classified into four types by their bond distances: 1.69 to 1.74, 1.87 to 1.99, 2.12 to 2.31, and 2.46 to 2.55 Å. The MoO_6 octahedra are divided into two classes: five octahedra have three types of Mo-O bonds, whereas two octahedra (Mo(5) and Mo(6), the outer ones of the central row of three octahedra) have four types of Mo-O bonds. The electronic state of Mo(5) and Mo(6) may be different from that of the other Mo atoms. The bond distances of the isopropylammonium cations show rather large deviations from the normal values, which is due to their disordered structure. The conformation of the ions may be easily changed in the crystal. The cations occupy the space between the heptamolybdate anions and form some hydrogen bonds to the bridging oxygens of the heptamolybdate anions. For the individual bond distances and angles see the paper [5]. The hydrogen bonds play an important role in the UV-induced photochromism of the crystals of the compound, see below.

Photochromism

Behavior of the Polycrystalline Material. The compound has photochromic properties in the solid state and in aqueous solution. Under irradiation with UV light ($\geqq$313 nm [1], 365 nm [2]) the white crystals became reddish violet: absorption maximum 510 nm [1, 2], approximate coloration time (50% reflectance) 5 to 7 min, approximate return time in the dark under aerobic conditions at room temperature (half-life period) 7 to 10 h [1]. Colored samples can be kept indefinitely in the dark in a vacuum at 70°C without changing in any way, while the thermal treatment in oxygen brought about a rapid bleaching. An EPR signal of the colored polycrystalline material due to Mo^V was observed; principal g values are 1.908, 1.930, and 1.944. The photolysis of the isopropylammonium cation in the solid is excluded [2].

The crystals also turn red on X-ray exposure (as for the data collection for the crystal structure determination, without significant change of the intensities) [5].

Aqueous Solution of the Irradiated Material. The colored sample turned blue when dissolved in deaerated water ($\lambda_{max} = 730$ nm, $\lambda_{sh} = 620$ nm) [1, 2]. The solution gave rise to a single intense line (g = 1.926) with six weak satellite lines, three on each side and equally spaced (a = 52 G), in its EPR spectrum, indicating a hyperfine structure due to isotopes with a nuclear spin of 5/2. Assuming that all of the Mo^V in the colored solid was converted into the blue species in water, the extinction coefficient of the Mo^V species, $\varepsilon = 2.3 \times 10^3$ $M^{-1} \cdot cm^{-1}$ at $\lambda_{max} = 730$ nm and 20°C, was determined on the basis of a 230 mg sample showing an EPR signal due to 1.44×10^{15} spin. Beer's law is obeyed below 4×10^{-4} M. The exposure of the blue solution to oxygen brought about the oxidation of Mo^{V} to Mo^{VI} with an accompanying bleaching and disappearance of the EPR signal [1]. For further discussion see the papers [1, 2].

Direct Irradiation of the Aqueous Solution. *EPR Investigations.* The direct UV irradiation of the deaerated solution at pH 5.4 showed the presence of two paramagnetic species. At low concentration ($\leqq$2.7 mM heptamolybdate) two weak signals at g = 1.921 and 1.910 were observed. At high concentrations a well defined signal at g = 1.921 developed with six satellite lines due to ^{95}Mo and ^{97}Mo (nuclear spin I = 5/2, natural abundance 25%). The hyperfine splitting constant is 51 G (4.6×10^{-3} cm^{-1}). The intensity ratio of the signal at g = 1.921 to that at g = 1.910 increased reversibly with increasing temperature, with an accompanying fading of the blue color (blue $\rightleftarrows$ greenish yellow). In combination with the results of the dependence of the initial concentration of the monoisopropylammonium heptamolybdate on the ratio of the concentrations of the blue to the yellow species in the photolyte (see below), these results indicate that the yellow form must consist of at least two different paramagnetic species, one (g = 1.921) of which is involved in the thermochromism with the blue species with $\Delta H \approx$ 15 kcal/mol. Summing up, the dissolution of crystals of the irradiated solid in water leads to the same species as the solution photolysis in its first stage, and hence the photochemical process in the solid state represents the primary step in the solution photolysis [10]. See also [1].

UV Investigations. Steady state photolysis (313 or 365 nm) of a deaerated solution of the compound (pH 5.4) yields a blue (λ_{max} = 730, λ_{sh} = 620 nm) and a yellow species as final photoproducts (see the spectra in the paper; see also [1]). The ratio of the blue to the yellow species increases with the (initial) concentration of the compound. Low concentrated solutions ($\leqq$2.7 mM) show a yellow color (due to an absorption with a long shoulder sloping into the visible region) which turns blue on addition of solid monoisopropylammonium heptamolybdate (or ammonium heptamolybdate), and the blue color deepens further with increasing concentration of additional heptamolybdate (see a figure in the paper) [10]. The blue form shows thermochromism (blue $\rightleftarrows$ greenish yellow, isosbestic point at 480 nm) over the range 0 to 70°C with an enthalpy of reaction $\Delta H \approx 15$ kcal/mol [1]. The quantum yield Φ of the Mo^{V} formation in deaerated solutions ($\Phi \approx 0.3$ to 1, depending on the conditions, in the first 5% of the photolysis; see also [1]) increases slightly with increasing concentration of the heptamolybdate and decreasing light intensity and increases considerably with increasing monoisopropylammonium perchlorate concentration (see a table in the paper). It is little affected by variation of the ionic strength with $NaClO_4$. The photolysis is wavelength independent in the range 313 to 365 nm. During the continued photolysis of the solution acetone (evidenced by ^{1}H NMR investigation) and propylene were formed. Quantum yields of propylene and acetone formations were about 8×10^{-3} and 3×10^{-4}, respectively, for 313 nm photolysis (1.1×10^{-4} $E \cdot dm^{-3} \cdot min^{-1}$) of a deaerated 13.6 mM solution, indicating that there is little destruction of the $i\text{-}C_3H_7NH_3^+$ cations in the photoreaction. Solutions of monoisopropylammonium perchlorate, in the absence of molybdate, did not give any significant yield of propylene and acetone under irradiation [10].

Raman Investigations. The Raman spectra of an UV (313 nm)-irradiated solution containing ca. 5% of the molybdenum as Mo^{V} and excited with 514.5 nm laser light and a 598.8 nm dye-laser, are reported. The comparison of the Raman spectra of the original and the irradiated solution measured with 514.5 nm laser light shows some deviations, in particular in the 310 cm^{-1} region, while the spectrum measured with 598.8 nm dye-laser excitation differs strongly from that using laser light of 514.5 nm, suggesting that the former is the resonance Raman spectrum for the blue species (see above) in the photolyte. For details see the paper [10]. See also [1].

Flash Photolysis. Two intermediates are formed by photoexcitation of the compound in deaerated solution. The first intermediate (I) with a broad absorption at the near-UV-visible range was produced during the lifetime of the flash. Its absorptions disappeared by a transformation to the second intermediate (II), exhibiting $\lambda_{max} = 500$ nm, with a first order rate. The decay rate constant, $(8 \pm 1) \times 10^3\ s^{-1}$, was nearly independent of the concentration (≦30 mM) of the compound, when the decay was followed at 450 nm. The same value was found for the formation of II, followed at 700 nm. At high concentrations (≧10 mM), intermediate II transformed very slowly by means of a pseudo-first-order reaction into the blue species which was the same species as under the steady-state irradiation. A rate constant $(1.5 \pm 0.5) \times 10^{-2}\ s^{-1}$ was obtained for such a decay followed at 500 nm. At low concentrations (≦10 mM), the low absorbance of the spectrum of II caused difficulties in measuring the fate of II. For the transient electronic spectra obtained at the reaction times 0, 0.4 (I), 1 ms (II), (and 10 min) see the paper [10].

Single Crystal EPR Investigations. The EPR spectrum of an irradiated single crystal shows two sets of lines arising from two magnetically inequivalent Mo^V centers A and B. Each of the Mo^V centers exhibits a main line arising from the nonmagnetic ^{96}Mo nucleus, split into two hyperfine lines (1:1) owing to superhyperfine interaction with a hydrogen atom, with six satellite lines due to ^{95}Mo and ^{97}Mo. Each of the hyperfine lines due to ^{95}Mo and ^{97}Mo is resolved into two superhyperfine lines at several orientations of the magnetic field. The occurrence of a six-line hyperfine structure is characteristic of an unpaired electron localized on one molybdenum atom only. Furthermore, the lines are narrow and no exchange broadening is observed. From these results it is inferred that the heptamolybdate ion contains at most one Mo^V center. For the g, $|A_{Mo}|$, and $|A_H|$ tensors for the two Mo^V centers A and B see the paper. These centers differ only in the orientation of the principal molecular g values, with respect to the crystallographic axes. The g_0, $|A_{Mo(0)}|$, and $|A_{H(0)}|$ values of the Mo^V(A) center are fairly close to those of the Mo^V(B) center, which suggests that there is structurally no significant difference between the two magnetically inequivalent Mo^V centers. From the correlation with X-ray crystal structure data, the paramagnetic center can be associated with $Mo^VO_5(OH)$, resulting from the UV-induced proton transfer from a monoisopropylammonium cation to an O atom in an MoO_6 octahedral site (see reaction (1)) [10].

Spin-Trapping Experiments. The spin-trapping technique using 2-methyl-2-nitrosopropane (menp) or N-benzylidene-t-butylamine N-oxide (bbao) as spin traps has been applied to detect short-lived free radicals in the solution photolysis. Photolysis (5 to 15 s, $\lambda = 365$ nm) of the deaerated solution of the compound (68 mM) in the presence of 0.1M menp gave an EPR spectrum (see the paper) which has been assigned to menp spin-adduct radicals of hydroxyl ($^{\bullet}OH$) or perhydroxy ($HO_2^{\bullet}$) radicals. In the absence of the molybdenum compound no such signals were observed. Solutions of the compound in the presence of 0.1M bbao gave after 5 s irradiation a spectrum which was assumed to be due to the $^{\bullet}OH$ adduct. For details see the paper [10].

The Photoreducible Site and the Mechanism of the Photoreduction of Mo^{VI} to Mo^V. The mode of the solid-state photoreaction which corresponds to the primary process for the solution photolysis has been proposed as shown in the equation

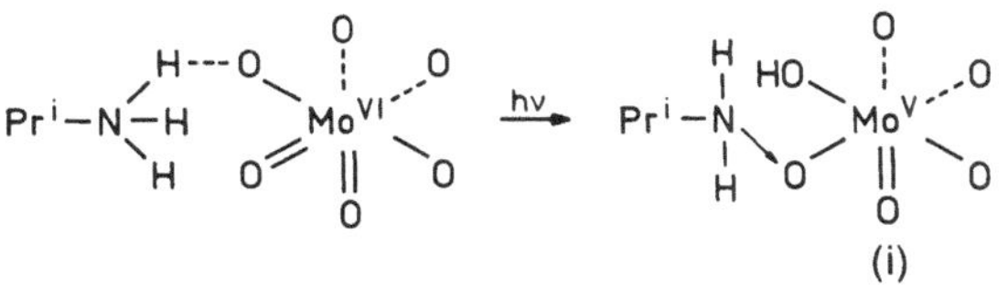

(1)

The reaction proceeds via UV-induced charge transfer in the terminal Mo=O bond with an accompanying transfer of a hydrogen-bonding proton from the isopropylammonium nitrogen atom to a bridging oxygen atom, followed by an interaction of the nonbonding electrons of the amino nitrogen with the terminal oxo group leading to a charge transfer complex [5, 10].

In solution, structure (i) undergoes attack of a solvent water molecule to give rise to the formation of hydroxyl radicals:

$$(i) + H_2O \rightarrow i\text{-}C_3H_7NH_2 +$$ 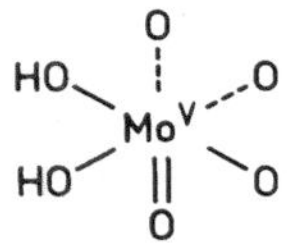 $$+ {}^{\bullet}OH \qquad (2)$$

(ii)

These presumably form H_2O_2. Hence the overall reaction scheme for the initial step in solution is

$$6\,i\text{-}C_3H_7NH_3^+[Mo_7O_{24}]^{6-} + H_2O \xrightarrow{h\nu} 6\,i\text{-}C_3H_7NH_3^+[Mo_7O_{23}(OH)]^{6-} + {}^{\bullet}OH \qquad (3)$$

The evidence of the radicals $^{\bullet}OH$ and $HO_2^{\bullet}$ by the spin-trapping experiments (the radical $HO_2^{\bullet}$ is assumed to be formed by oxidation of H_2O_2 or by a reaction between H_2O_2 and $^{\bullet}OH$) and the increase of Φ with an increase in the $i\text{-}C_3H_7NH_3^+$ concentration (extra $i\text{-}C_3H_7NH_3^+$, see above, which facilitates the formation of the complex (i)) are seen to support the reaction scheme (1) to (3). The quenching process of $^{\bullet}OH$, produced by reactions (2) and (3), is promoted with the readily reducing $i\text{-}C_3H_7NH_2$. Additionally it is assumed that $^{\bullet}OH$ reacts rapidly with $[Mo_7O_{24}]^{6-}$ to yield highly condensed molybdate [10].

The standard redox potential $E^0(Mo^{VI}/Mo^{V})$ for the electrochemically active species exhibits an approximately linear pH dependence with a slope of about −59 mV/pH, which indicates that the anode reaction of the electrochemically active species is given by a protonation/deprotonation process:

$$Mo^{V}\text{–}OH \rightarrow Mo^{VI}\text{=}O + H^+ + e \qquad (4)$$

[8, 10]. In connection with the observation of an $Mo^{V}O_5(OH)$ site in the irradiated single crystal it is assumed that the coordination of the hydroxide to the paramagnetic Mo^{V} atom is also maintained in solution [10].

It is assumed that the species having $g = 1.921$ in the photolyte possesses the same structure as the octahedral Mo^{V} center in $[Mo_7O_{23}(OH)]^{6-}$ for the single crystal. The species having $g = 1.910$ in the photolyte is assumed to be a photoproduct of further reaction of $[Mo_7O_{23}(OH)]^{6-}$, and it is also assumed that the flash photolysis represents the decomposition process for this species. For details see the paper [10].

For the formation mechanism of propylene and acetone as minor photoproducts from $i\text{-}C_3H_7\dot{N}H_2^+$ see the paper [10].

Photogalvanic Effect

The aqueous solution of the compound exhibits photogalvanic behavior based on the photoreduction of Mo^{VI} to Mo^{V} and H_2 formation at the counter electrode for the photogalvanic cell. For the construction of the photogalvanic cell see the paper. **Fig. 58** shows the typical photopotential responses of the compound to illumination cycles. The experimental conditions were: 13.6 mM $(i\text{-}C_3H_7NH_3)_6[Mo_7O_{24}]\cdot 3H_2O$ and 1M $NaClO_4$, pH 5.2, nitrogen bubbling (unless otherwise specified), potentiostatic conditions, illumination wavelength 365 nm (3.6×10^{-7} E/ls). The time course of the photocurrent was similar to that of the photopotential. (The cyclic voltammogram at the platinum electrode did not show any significant current due to $(i\text{-}C_3H_7NH_3)_6[Mo_7O_{24}]\cdot 3H_2O$ at $U_{SCE} = -0.1$ to 1.0 V, SCE = saturated calomel electrode.) The signs of the photostationary potential (E_l) relative to the dark equilibrium potential (E_d) are always negative, indicating that the main electrochemically active species are photoreduced substances. The accumulation of the blue species ($\lambda_{max} = 730$, $\lambda_{sh} = 620$ nm, see above) near the illuminated electrode did not bring about a perfect recovery of the original value of E_d. When oxygen was admitted into the solution E_d was restored to the original value with an accompanying bleaching due to the oxidation of Mo^V to Mo^{VI}. For the solution saturated with oxygen by bubbling it through, $E_l - E_d$ was about half of that for the nitrogen bubbling and the blue coloration was slight. The action spectrum of the photocurrent at $U_{SCE} = 0.7$ V and the absorption spectrum of the aqueous solution of the compound parallel each other (see **Fig. 59**) indicating that the photogalvanic effect is caused by the photoexcitation of $(i\text{-}C_3H_7NH_3)_6[Mo_7O_{24}] \cdot 3H_2O$ (the signals obtained on measuring the action spectrum were corrected to constant photon flux at each wavelength) [6].

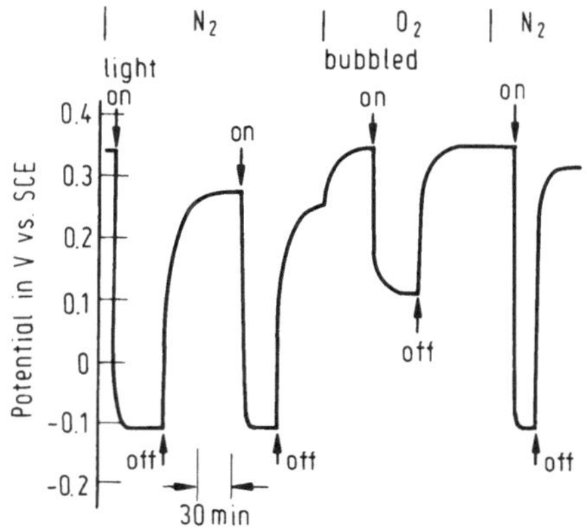

Fig. 58. Photopotential responses of an aqueous solution of $(i\text{-}C_3H_7NH_3)_6[Mo_7O_{24}] \cdot 3H_2O$ to 365 nm light illumination; for the other conditions see the text, SCE = saturated calomel electrode [6].

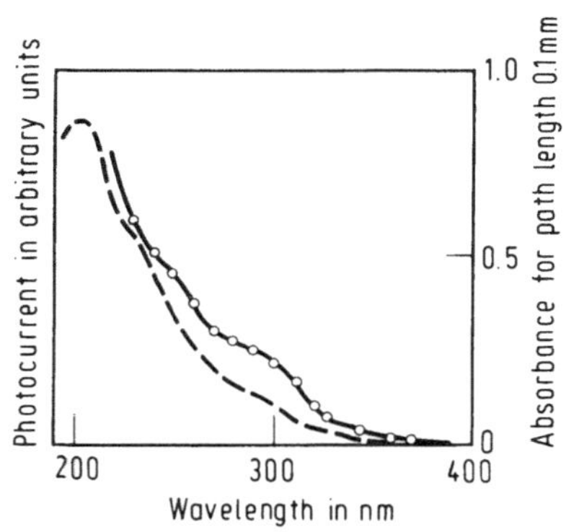

Fig. 59. Action spectrum of the photocurrent (—o—) and absorption spectrum (- - -) of a 1.75 mM aqueous solution of $(i\text{-}C_3H_7NH_3)_6[Mo_7O_{24}] \cdot 3H_2O$ [6].

The photochemical formation of the electrochemically active species leading to the photocurrent can be understood as a quasi first-order process with $\tau \approx 20$ ms. This value has been estimated from the relaxation behavior of the photocurrent using chopped light illumination (chopping frequency $\geqq 8$ Hz) [6].

At $U_{SCE} \geqq 1.1$ V, water is oxidized to oxygen. The standard redox potential $E^0(Mo^{VI}/Mo^V)$ for the electrochemically active species at pH 5.2 is approximately −0.25 V vs. SCE, deduced from the onset of the photocurrent at $U_{SCE} \approx -0.25$ V. Since this value is very close to the standard redox potential of $2H^+/H_2$ (−0.244 V vs. SCE), the photogalvanic cell of this system should be capable of reducing hydronium ions without any applied potential [6]. A photogalvanic cell proposed originally for the hydrogen evolution is described in the papers [6, 8]. The potential of the cathode of this cell at the stationary state was −0.255 V vs. SCE. From the H_2 volume produced (0.04 mL H_2 during an irradiation time of 7 h) and the average photocurrent

(15 μA) the stoichiometric ratio of 2 mol electrons : 1 mol H_2 was calculated [6]. Since the photoreduction to Mo^V in the solution is coupled with a water oxidation to the hydroxyl radical,

$$Mo^{VI}{=}O + H_2O \xrightarrow{h\nu} Mo^V{-}OH + {}^{\bullet}OH \quad (5)$$

the splitting of water by alkylammonium molybdate in photogalvanic application therefore may be given in terms of the half reactions

$$\text{Pt photoanode:} \quad Mo^V{-}OH \rightarrow Mo^{VI}{=}O + H^+ + e \quad (6)$$

$$\text{Pt cathode:} \quad H^+ + e \rightarrow 0.5\,H_2 \quad (7)$$

[6, 8]. (This mechanism represents the C2 category ($H_2O \rightarrow 0.5\,H_2 + {}^{\bullet}OH$) in the classification of methods for photochemical water decomposition [7]) [6]. This is the first example of hydrogen production from water using the photogalvanic effect of molybdenum compounds [8].

In dependence on the pH of the anode solution (monoisopropylammonium heptamolybdate), the photostationary potential (E_l) and the short-circuit photocurrent (i_p) of the photogalvanic cell show a minimum and a maximum, respectively, at pH ≈ 7.4. The quantum yield for H_2 production of the cell at this optimum pH was found to be 0.04 on the basis of the above-mentioned ratio of 2 moles of electrons to 1 mole of H_2. The pH dependence of the operation of the photogalvanic cell is explained by the pH dependence of the redox reaction (6) for the electrochemically active species and the equilibrium reaction between 7- and 1-molybdates (?) (cf. also [10]) in the dark [8]. It must be mentioned that at pH 7.4 molybdate solutions do not contain the heptamolybdate ion (see "Molybdenum" Suppl. Vol. B 3 [to be published], where the ions in equilibria will be described).

To make sunlight available for the photoelectrolysis of water, the weak sensitivity [8, 9] of the alkylammonium heptamolybdate system in the region >400 nm may be compensated by addition of appropriate spectral sensitizers, e.g., riboflavine derivatives that can strongly absorb in the 400 to 500 nm wavelength region. In particular flavine mononucleotide (FMN) was studied. The dye-sensitized photoreduction of Mo^{VI} in deaerated solution containing 13.6 mM monoisopropylammonium heptamolybdate and 8.3×10^{-5} M FMN with 436 nm light was followed by the measurement of the absorbance of the blue species at 730 (λ_{max}) and 620 nm (λ_{sh}). Pertinent quantum yields of Mo^V formation (only the initial values determined at short irradiation times are reported) for solutions of different composition are in the range of ~0.1. FMN degradation ($\Phi_{FMN} \approx 0.01$) competes with the sensitization. The quantum yield of Mo^V formation increased with a decrease in the 436 nm light intensity but was almost unchanged with variations of the alkylammonium heptamolybdate and FMN concentrations and ionic strength. The sensitized photoreduction of $[Mo_7O_{24}]^{6-}$ is not restricted to the alkylammonium cations but occurs also for the NH_4^+ and Na^+ cations. The possibility of the electron transfer from $FMN^{\bullet -}$ or $FMNH^{\bullet}$ to $[Mo_7O_{24}]^{6-}$ in the sensitized photoreduction process is excluded. The result of an EPR spin-trapping investigation suggests a similarity in the reduction process of $[Mo_7O_{24}]^{6-}$ between the sensitized and direct photolyses. In connection with the occurrence of the sensitized photoreduction of $[Mo_7O_{24}]^{6-}$ for NH_4^+ and Na^+ salts it is therefore assumed that the sensitization proceeds via a charge transfer complex between the excited FMN and $[Mo_7O_{24}]^{6-}$ (probably through the isoalloxazine nitrogen atom and the terminal oxo group of the MoO_6 site which is also operative in case of the direct photoreduction of isopropylammonium heptamolybdate). For details see the paper [9]. (For the production of H_2 by a photogalvanic cell with an FMN-EDTA system without participation of a molybdenum compound see [11].)

Other Data

The compound is stable below 100 [1] or 110°C [2]. The solubility in water is 108 g/100 mL at 20°C [1, 2].

2.8.2.11.2 $(i\text{-}C_3H_7NH_3)_6[H_2Mo_8O_{28}]\cdot 2H_2O$ (Dihydrogenoctamolybdate Type)

The compound crystallized from 10 mL of an aqueous solution containing 5 g of isopropylammonium "dimolybdate" dihydrate (later [5] formulated as $(i\text{-}C_3H_7NH_3)_6[Mo_7O_{24}]\cdot 3H_2O$, see the preceding compound) at room temperature when the solution was kept in the dark for two weeks [1].

The compound was characterized by analysis and infrared spectrum [1] (see Fig. 29, p. 27).

The crystals are triclinic; the shape of a single crystal is outlined in [4]. Space group $P\bar{1}\text{-}C_i^1$ (No. 2), lattice parameters a = 10.66(1), b = 12.29(1), c = 9.65(2) Å, α = 104.6(2)°, β = 82.1(5)°, γ = 96.5(2)°; Z = 1. D_x = 2.22 g/cm³. The atomic parameters are listed in the paper [3].

The structure was determined by the heavy-atom method and refined to R = 0.033. The structure of the polyanion $[H_2Mo_8O_{28}]^{6-}$ consists of Mo_8O_{28} blocks composed of eight (distorted) MoO_6 octahedra sharing edges (see Fig. 28, p. 27). It is the same type of arrangement of MoO_6 octahedra as in the octameric units of the ammonium polyoctamolybdate (cf. Fig. 4, p. 12). The bond lengths between Mo and unshared O atoms are 1.70 to 1.73 Å, those between Mo and O atoms shared by two or more neighboring MoO_6 octahedra 1.77 to 2.41 Å. The bond length Mo-OH is 1.97 Å (see Fig. 60, p. 152). The Mo atoms are shifted from the centers of their respective octahedra towards the unshared O atoms owing to the repulsion between neighboring Mo^{VI} atoms with large positive charges (see, however, p. 35). Mo···Mo distances between the edge-sharing octahedra are much longer than the edges of the octahedra, ranging from 3.27 to 3.49 Å. $[H_2Mo_8O_{28}]^{6-}$ anions, $(i\text{-}C_3H_7NH_3)^+$ cations, and the water molecules are held together by a three-dimensional hydrogen bond system (see a figure in the paper). All the NH_3 groups act as donors of hydrogen bonds to the O atoms of the polyanions and/or to water molecules. The water molecules are also bound to the polyanion by weak hydrogen bonds. For the different bond lengths and angles see the paper [3].

The solubility in water is 11 g/100 mL at 20°C. The compound is stable below 100°C [1].

The compound has photochromic properties in the solid state (and in aqueous solution). Under irradiation with UV light (≧313 nm) the white polycrystalline material becomes reddish violet [1, 3], absorption maximum 480 nm, approximate coloration time (50 % reflectance) 3 min, approximate return time in the dark under aerobic conditions at room temperature (half-life period) 15 to 20 h. An EPR signal of the colored polycrystalline material due to Mo^V was observed. The colored sample turned blue when dissolved in deaerated water (λ_{max} = 730 nm, λ_{sh} = 620 nm). The solution gave rise to a single intense line (g = 1.926) with six weak satellite lines, three on each side and equally spaced (a = 52 G), in its EPR spectrum, indicating a hyperfine structure due to isotopes with a nuclear spin of 5/2. The exposure of the blue solution to oxygen brought about the oxidation of Mo^V to Mo^{VI} with an accompanying bleaching and disappearance of the EPR signal [1]. UV irradiation (λ≧313 nm) and EPR investigation of a single crystal showed a 1:2:1 triplet with a line separation of 9.8 to 11 G. The 1:2:1 intensity ratio was retained at all orientations and the triplet splitting showed little anisotropy. Principal values g_1 = 1.955, g_2 = 1.941, g_3 = 1.897. This indicates the existence of two magnetically equivalent hydrogen atoms interacting with Mo^V. In view of the observed g values, the most likely paramagnetic species is

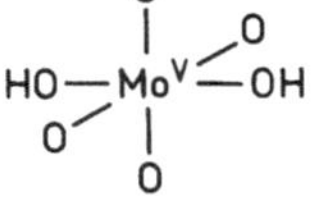

The formation of this species results from internal hydrogen abstraction by oxygen at the Mo(4) [or Mo(4′)] site where the OH group [O(6′)] was originally coordinated (**Fig. 60**, p. 152). Hydrogen transfer from N(3) to O(13) upon UV irradiation is assumed, hence magnetic

equivalence of two H atoms coordinated to two nearly equivalent (*trans*) oxygen atoms associated with an O(13)-Mo(4)-O(6′) angle of 156.8° and with nearly equivalent Mo(4)-O(6′) and Mo(4)-O(13) bond distances results [4].

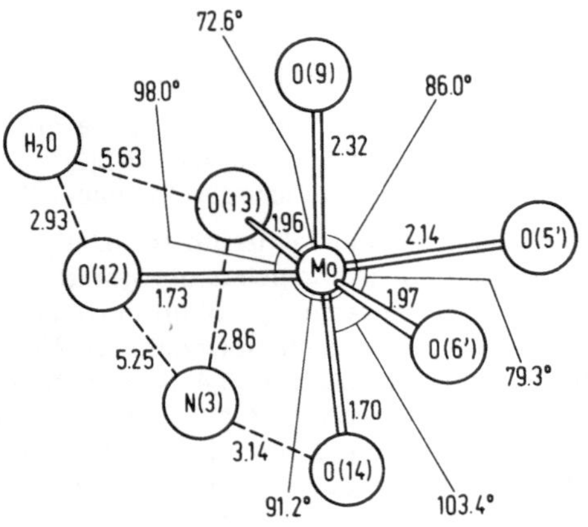

Fig. 60. The environment of the Mo(4) site in $(i\text{-}C_3H_7NH_3)_6[H_2Mo_8O_{28}]\cdot 2H_2O$. Positions of H atoms are omitted for clarity. O(6′) forms the OH group originally present (bond distances in Å) [4].

References:

[1] T. Yamase, T. Ikawa (Bull. Chem. Soc. Japan **50** [1977] 746/9). – [2] T. Yamase, H. Hayashi, T. Ikawa (Chem. Letters **1974** 1055/6). – [3] M. Isobe, F. Marumo, T. Yamase, T. Ikawa (Acta Cryst. B **34** [1978] 2728/31). – [4] T. Yamase (J. Chem. Soc. Dalton Trans. **1978** 283/5). – [5] Y. Ohashi, K. Yanagi, Y. Sasada, T. Yamase (Bull. Chem. Soc. Japan **55** [1982] 1254/60).

[6] T. Yamase, T. Ikawa (Inorg. Chim. Acta **37** [1979] L529/L531). – [7] V. Balzani, L. Moggi, M. F. Manfrin, F. Bolletta, M. Gleria (Science **189** [1975] 852/6). – [8] T. Yamase, T. Ikawa (Inorg. Chim. Acta **45** [1980] L55/L57). – [9] T. Yamase (Inorg. Chim. Acta **54** [1981] L207/L209). – [10] T. Yamase, R. Sasaki, T. Ikawa (J. Chem. Soc. Dalton Trans. **1981** 628/34).

[11] T. Yamase (Photochem. Photobiol. **34** [1981] 111/4).

2.8.2.12 Tetrabutylammonium Polymolybdates

$[(n\text{-}C_4H_9)_4N]_3K[Mo_8O_{26}]\cdot 2H_2O$ (β-Octamolybdate Type)

The compound was obtained by mixing a solution of $\alpha\text{-}[(n\text{-}C_4H_9)_4N]_4[Mo_8O_{26}]$ in acetonitrile with an aqueous solution of KCl. Heating should be avoided since the solution reacts to form $[Mo_6O_{19}]^{2-}$ when heated [1]. To prepare an ^{17}O-enriched compound a saturated solution of 0.1 g of KBr in 0.18 mL of ^{17}O-enriched water was added to a solution of 0.38 g of ^{17}O-enriched $\alpha\text{-}[(n\text{-}C_4H_9)_4N]_4[Mo_8O_{26}]$ in 10 mL of CH_3CN, and the resulting precipitate was immediately filtered off. The clear filtrate yielded 0.05 g of the product as small, colorless needles after 24 h at 0°C [2].

The substance was characterized by analysis, infrared and Raman spectroscopy. For the infrared spectrum (taken as KBr pellet and as solution in CH_3CN) see the paper; no data of the Raman spectrum are given. The structure of the polymolybdate ion (see Fig. 10, p. 16) has been assigned by ("finger-print") Raman and infrared spectroscopy in comparison with other compounds containing the $\beta\text{-}[Mo_8O_{26}]^{4-}$ ion [1].

The ^{17}O Fourier transform NMR spectrum of the ^{17}O-enriched (34 at.%) compound in CH_3CN is reported and assignments of the chemical shifts to the different types of Mo bonded oxygen atoms are made [2, 3].

$[(n\text{-}C_4H_9)_4N]_3NH_4[Mo_8O_{26}]\cdot xH_2O$ (β-Octamolybdate Type)

The compound was obtained by mixing a solution of α-$[(n\text{-}C_4H_9)_4N]_4[Mo_8O_{26}]$ in acetonitrile with an aqueous solution of NH_4Cl. Heating should be avoided since the solution reacts to form $[Mo_6O_{19}]^{2-}$ when heated.

The salt was characterized by analysis, infrared and Raman spectroscopy; however, data and the value of x are not given. The structure of the polymolybdate ion (see Fig. 10, p. 16) has been assigned by "finger-print" Raman and infrared spectroscopy in comparison with other compounds containing the β-$[Mo_8O_{26}]^{4-}$ ion [1].

$[(n\text{-}C_4H_9)_4N]_3Rb[Mo_8O_{26}]\cdot xH_2O$ (β-Octamolybdate Type)

The compound was obtained by mixing a solution of α-$[(n\text{-}C_4H_9)_4N]_4[Mo_8O_{26}]$ in acetonitrile with an aqueous solution of RbCl. Preparation and properties correspond to those of the NH_4 compound above [1].

Anhydrous Butylammonium Molybdates

The anhydrous $(t\text{-}C_4H_9NH_3)_2MoO_4$ can be found in [4, 6] and dibutylammonium heptamolybdate in [5].

In addition to the tetrabutylammonium polymolybdate hydrates described above the anhydrous compounds of composition $[(n\text{-}C_4H_9)_4N]_2[Mo_2O_7]$ [2, 7], α-$[(n\text{-}C_4H_9)_4N]_4[Mo_8O_{26}]$ [1 to 3, 9], and $[(n\text{-}C_4H_9)_4N]_2[Mo_6O_{19}]$ [2, 6, 10] have been prepared and investigated.

References:

[1] W. G. Klemperer, W. Shum (J. Am. Chem. Soc. **98** [1976] 8291/3). – [2] M. Filowitz, R. K. C. Ho, W. G. Klemperer, W. Shum (Inorg. Chem. **18** [1979] 93/103). – [3] V. W. Day, M. F. Fredrich, W. G. Klemperer, W. Shum (J. Am. Chem. Soc. **99** [1977] 952/3). – [4] A. Thiele, J. Fuchs (Z. Naturforsch. **34b** [1979] 145/54). – [5] J. Fuchs, A. Thiele (Z. Naturforsch. **34b** [1979] 155/9).

[6] J. Fuchs (Z. Naturforsch. **28b** [1973] 389/404, 395). – [7] V. W. Day, M. F. Fredrich, W. G. Klemperer, W. Shum (J. Am. Chem. Soc. **99** [1977] 6146/8). – [8] J. Fuchs, I. Knöpnadel, I. Brüdgam (Z. Naturforsch. **29b** [1974] 473/5). – [9] J. Fuchs, H. Hartl (Angew. Chem. **88** [1976] 385/6; Angew. Chem. Intern. Engl. Ed. **15** [1979] 375/6). – [10] J. Fuchs, K. F. Jahr (Z. Naturforsch. **23b** [1968] 1380).

2.8.2.13 Molybdates with Cyclic and Higher Alkyl Ammonium Cations

Monomolybdates

Precipitates of the monomolybdates (or of isopoly molybdates) are formed when organic bases, which are either liquids themselves or, if solid, dissolved in alcohols, chloroform, or benzene + some water, react with added MoO_3. The reaction is carried out by heating to about 40°C and stirring for several hours or up to a few days. What type of molybdate is obtained depends on the organic base. The following monomolybdates and monomolybdate hydrates have been prepared:

(1) tetradecylammonium monomolybdate, $(C_{14}H_{29}NH_3)_2MoO_4$

(2) hexadecylammonium monomolybdate, $(C_{16}H_{33}NH_3)_2MoO_4$

(3) octadecylammonium monomolybdate, $(C_{18}H_{37}NH_3)_2MoO_4$

(4) cyclohexylammonium monomolybdate, $(C_6H_{11}NH_3)_2MoO_4$

(5) N-methylcyclohexylammonium monomolybdate dihydrate, $((C_6H_{11})NH_2(CH_3))_2MoO_4 \cdot 2H_2O$

(6) dicyclohexylammonium monomolybdate dihydrate, $[(C_6H_{11})_2NH_2]_2MoO_4 \cdot 2H_2O$

(7) cycloheptylammonium monomolybdate, $(C_7H_{13}NH_3)_2MoO_4$

(8) cyclooctylammonium monomolybdate, $(C_8H_{15}NH_3)_2MoO_4$

(9) cyclododecylammonium monomolybdate, $(C_{12}H_{23}NH_3)_2MoO_4$

Solvents used for preparation were: ethanol for (9), chloroform for (1) and (3), benzene for (2) [1]. Single crystals are obtained by recrystallization from methanol [1, 2].

Cyclohexylammonium monomolybdate is monoclinic, lattice parameters a = 28.978(9), b = 6.788(2), c = 8.275(3) Å, β = 90.14(2)°; Z = 4. Space group P2/n-C_{2h}^4 (No. 13). D_m = 1.45, D_x = 1.469 g/cm³. Atomic coordinates and figures are given in the paper, R = 0.0715 (anisotropic). In the laminar structure the MoO_4^{2-} anions form distorted tetrahedra in which three O atoms have a mean Mo-O distance of 1.774 Å, the fourth one only 1.729(6) Å. The shortest intermolecular O-O distance is 3.257 Å, the mean O-O distance within one tetrahedron is 2.878 Å. The cyclohexylammonium ions are in the chair form. The two $-NH_3^+$ groups together form six hydrogen bonds to three of the oxygen atoms, the fourth oxygen atom having no hydrogen bridge at all. Each of the cations is connected to three oxygen atoms of three different anions (see figures in the paper). The mean N-O distance is 2.80 Å [2].

Dicyclohexylammonium monomolybdate dihydrate is tetragonal, lattice parameters a = 12.590(4), c = 17.494(3) Å; Z = 4. Space group I$\bar{4}$2d-D_{2d}^{12} (No. 122). D_m = 1.25, D_x = 1.255 g/cm³. For atomic coordinates see a table in the paper, R = 0.0196 (anisotropic). The MoO_4^{2-} anions are nearly ideal tetrahedra. The Mo-O distance is 1.741(3) Å, the O-Mo-O angle has a mean value of 109.7°, and the O-O distance is 2.848 Å. The dicyclohexylammonium cation is present in the chair form. The whole structure is cross-linked in three dimensions by hydrogen bonding, which connects the hydrogen atoms of crystal water and of the ammonium group with the oxygen atoms of the MoO_4^{2-} groups. Each molybdate oxygen atom forms two bridges, one to nitrogen and one to an oxygen atom. The bridges are nearly linear, the N-to-O bridge being shorter than the O-to-O bridge (2.74 and 2.89 Å). The structure is shown in two figures in the paper [2].

IR absorption (1100 to 350 cm^{-1}) and Raman (1100 to 100 cm^{-1}) spectra of the two molybdates are measured as well as the Raman spectra of their concentrated aqueous solutions [2].

Cyclohexylammonium molybdate. The strong line in the Raman spectra at 897 cm^{-1} corresponds to the symmetrical valence mode of the three O atoms at equal distance in the MoO_4^{2-} tetrahedron, whereas the fourth short Mo-O group produces a weak line at a wave number about 20 cm^{-1} higher. In the IR spectrum all these bands are at about 6 cm^{-1} higher wave numbers. The asymmetric stretching mode of the MoO_4^{2-} anion is at the extremely low values of 770 or 771 cm^{-1}. The strong Raman line at 839 cm^{-1} corresponds to the split band in the IR spectrum at 821, 852 cm^{-1}, which possibly stems from factor group splitting. The three deformation modes of the deformed MoO_4^{2-} anion theoretically expected (point group C_{3v}) are at 307, 318, and 329 cm^{-1} [2].

Dicyclohexylammonium molybdate. The symmetrical valence mode (A_1) of the MoO_4^{2-} anion is at 891 cm^{-1}; it is only weakly represented in the IR spectrum. The asymmetric stretching mode is at the low value of 800 cm^{-1}, present only in the IR spectrum. The additional band at 827 cm^{-1} in the Raman spectrum could be explained by dipole-dipole exchange or factor group splitting of A_1. In the latter case also the deformation vibration should be split, but the observed Raman lines (330 and 298 cm^{-1}) may also stem from the deformation vibrations E and F_2 [2].

The compounds listed in the table above and cited here with their numbers are soluble as follows (DMSO = dimethyl sulfoxide):

solvent	solubility	
water	(2), (6) soluble	(4), (5), (7), (8) well soluble
methanol	(1) to (3) soluble	(4) to (9) well soluble
ethanol	(1), (2), (4), (6) soluble	(5), (7) to (9) well soluble
acetonitrile	(2), (4) to (6) soluble	
benzene	(1) to (3) soluble	
chloroform	(1) to (3), (9) weakly soluble	
DMSO	all compounds are soluble	

All the compounds listed are insoluble in acetone, i-butyl methyl ketone, ether, dioxane, ligroin, and cyclohexane [1].

Polymolybdates

The following pentyl- and hexylammonium polymolybdates are known only as anhydrous compounds: $(R_4N)_2O \cdot 4\,MoO_3$ (octamolybdates?) [3] and $(R_4N)_2[Mo_6O_{19}]$ [3, 4] with R = n- and i-C_5H_{11}; hexylammonium heptamolybdate, dihexylammonium heptamolybdate, and N,N-dimethylcyclohexylammonium β-octamolybdate [1].

References:

[1] J. Fuchs, A. Thiele (Z. Naturforsch. **34b** [1979] 155/9). – [2] A. Thiele, J. Fuchs (Z. Naturforsch. **34b** [1979] 145/54). – [3] J. Fuchs, I. Knöpnadel, I. Brüdgam (Z. Naturforsch. **29b** [1974] 473/5). – [4] J. Fuchs (Z. Naturforsch. **28b** [1973] 389/404, 395).

2.8.2.14 Triethanolammonium Polymolybdates

$[(HOC_2H_4)_3NH]_3(NH_4)_3[Mo_7O_{24}] \cdot 4\,H_2O$ (Heptamolybdate Type)

The compound was obtained by treating of a hot suspension of 10.08 g of MoO_3 in 100 mL of water with 4.5 g of triethanolamine and 17 to 18 mL of aqueous ammonia (3%) and concentrating the limpid solution to a syrupy consistency. After cooling and waiting for 24 h the precipitate was filtered off and dried under vacuum. An aqueous suspension of ammonium molybdate and the calculated proportion of triethanolamine boiled and concentrated yields also triethanolammonium triammonium heptamolybdate. For details see the papers.

The compound was characterized by analysis. The analytical results conform to the heptamolybdate "3:7", and not to the "5:12" formulation.

Attempts to prepare mixed salts with a higher proportion of triethanolammonium were unsuccessful, extremely viscous products being obtained in all cases.

$[(HOC_2H_4)_3NH]_2(NH_4)_4[Mo_7O_{24}]\cdot 4H_2O$ (Heptamolybdate Type)

The compound was obtained by salification of MoO_3 suspended in boiling water with triethanolamine and ammonia. To a suspension of 10.08 g of MoO_3 in 60 to 70 mL of boiling water 3 g of triethanolamine and 22 to 23 mL of aqueous ammonia (3%) were added. The suspension was boiled until MoO_3 completely dissolved, and the solution was then concentrated to a small volume. The crystals obtained on cooling were filtered off, washed with a small portion of water, and dried between filter paper. This compound was also obtained by boiling an aqueous solution of triethanolamine saturated with ammonium molybdate and keeping the mixture at 100°C for several hours. The compound was further obtained by letting triethanolamine and a suspension of ammonium molybdate in water stand for 24 h, filtering, and concentrating. For details and a further method of preparation see the papers.

The salt was characterized by analysis. The analytical results conform to the heptamolybdate "3:7", and not to the "5:12" formulation.

$(HOC_2H_4)_3NH(NH_4)_5[Mo_7O_{24}]\cdot 4H_2O$ (Heptamolybdate Type)

The compound was obtained by boiling of a solution of 1.49 g of triethanolamine and 10.57 g of ammonium molybdate in 50 to 70 mL of water, slow concentration to a small volume, and cooling. The white crystals were filtered off, washed with a small portion of water, and air-dried over filter paper. The compound was also formed by treating a suspension of MoO_3 in boiling water with the calculated proportions of triethanolamine and ammonia. For details see the papers.

The salt was characterized by analysis. The analytical results conform to the heptamolybdate "3:7", and not to the "5:12" formulation.

F. Garelli, A. Tettamanzi (Gazz. Chim. Ital. **65** [1935] 1009/15; Atti Accad. Sci. Torino Classe Sci. Fis. Mat. Nat. **70** [1935] 382/90).

2.8.2.15 Arylammonium Polymolybdates

$(XC_6H_4NH_3)_2O\cdot 3MoO_3\cdot 2$ to $3H_2O$, X = H, p-CH_3, p-OCH_3, p-Cl ("Trimolybdate" Type)

To 50 to 100 mL of an aqueous solution of 1.3 to 2.6 g of potassium "trimolybdate" 10 to 20 mL of an aqueous solution of 1 to 2 g of the aromatic amine hydrochloride were added. At pH 4.2 to 4.7 the precipitated salt was stirred for 30 min, filtered off, washed with cold water and alcohol, and dried at 110°C. Before being dried, the salts are fine colorless crystals and contain 2 to 3 moles of water of crystallization per formula weight; when dried, they are faintly yellowish. Yields: 55 to 72%.

The compounds were characterized by analysis. The infrared spectrum of the compound with X = Cl is reported in the paper.

All the arylammonium "trimolybdates" are sparingly soluble in water and readily soluble in dimethyl formamide, dimethyl sulfoxide, and hot alcohol. When the salts are reprecipitated from hot alcohol with acetone, they are isolated in variable composition. Lengthy heating in aqueous alcohol gradually alters their composition until salts with the metamolybdate anion are formed.

All the arylammonium "trimolybdates" are readily diazotized in aqueous HCl solution with sodium nitrite, or in alcohol with isopentyl nitrite. The diazo compounds isolated after this reaction are, however, of variable composition, although the nitrite number after diazotization corresponds exactly to disubstituted "trimolybdates" [1].

$(XC_6H_4NH_3)_2O \cdot 4MoO_3 \cdot 0.33H_2O$, X = H, p-$CH_3$, o-$CH_3$, m-$CH_3$, p-$OCH_3$, o-$OCH_3$, p-Cl, p-Br (β-Octamolybdate Type?)

A solution of 1 to 2 g of the amine in 10 to 20 mL of methanol with 0.9 to 1.8 mL of 35% HCl (pH 3.5 to 3.6) was added gradually with stirring to a solution of 2 to 3 g of sodium (1:4)-molybdate (metamolybdate, $Na_2O \cdot 4MoO_3 \cdot 6H_2O$ (white)) in 20 to 30 mL of water (pH 4.2 to 4.3). The precipitation of the amine metamolybdate occurred at pH 4.0 (method A). A solution of 1 to 4 g of the solid amine hydrochloride in 10 to 40 mL of methanol was added gradually with stirring to a solution of 2 to 8 g of sodium (1:4)-molybdate in 15 to 70 mL of water (pH 4.2 to 4.3). The precipitation of the amine salt occurred at pH 4.0 (method B). A solution of 1 to 2 g of the solid amine hydrochloride in 10 to 20 mL of water at pH 3.0 was added gradually with stirring to a solution of 2 to 4 g of normal sodium molybdate in 5 to 10 mL of glacial acetic acid and 20 to 40 mL of water (pH 4.0). The precipitation of the amine metamolybdate occurred at pH 4.0 (method C). The precipitated salts (obtained by the methods A, B, or C) were stirred for 15 to 30 min, filtered off, washed with methanol, and dried at 110°C. They were recrystallized from methanol. Yields: 72 to 92% according to methods A and B, 54 to 65% according to method C. The weakly basic isomeric nitroanilines do not form metamolybdates.

The salts with X = H, p-OCH_3, o-OCH_3, and p-Br form plates, with o-CH_3 prisms, with p-CH_3 and m-CH_3 microcrystals, and that with p-Cl is an amorphous powder.

The compounds were characterized by analyses. The infrared spectrum of the compound with X = p-OCH_3 is reported. It resembles those of the β-octamolybdates. In the paper the antiquated formula $[H_2Mo_{12}O_{40}]^{6-}$ for the metamolybdate ion is used.

Thermogravimetric studies have been accomplished up to 500°C, thermolysis curves of the aniline and p-anisidine salts are given. The arylammonium "tetramolybdates" are readily diazotized in methanol with isopentyl nitrite, and the diazo compounds could be isolated. For details see the paper (see also p. 158). The arylammonium "tetramolybdates" are sparingly soluble in water, methanol, and ethanol and are insoluble in ether [2].

$[C_6H_5NH(CH_3)_2]_3H_3Mo_7O_{24}$? (Unknown Type)

Addition of hydrochloric acid to a mixture of dimethylanilinium chloride and sodium molybdate in aqueous solution gave a white precipitate when the pH of the solution was nearly 1.5. The precipitate was filtered off, washed with a buffer solution of pH 1.5 and finally with ether, and dried at 105°C. The compound was characterized merely by its Mo content [3]. The formulation of an acid polymolybdate ion as well as the formulation of a heptamolybdate ion is experimentally absolutely unfounded.

References:

[1] V. G. Smirnova, V. V. Kozlov (Zh. Obshch. Khim. **46** [1976] 105/8; J. Gen. Chem. [USSR] **46** [1976] 107/9). – [2] V. V. Kozlov, V. G. Smirnova (Zh. Obshch. Khim. **41** [1971] 901/5; J. Gen. Chem. [USSR] **41** [1971] 908/11). – [3] D. V. Ramana Rao (Anal. Chim. Acta **16** [1957] 1/3).

2.8.2.16 Arenediazonium Polymolybdates

$XC_6H_4N_2[HMo_3O_{10}]$, X = H, p-CH_3, p-OCH_3, p-NO_2, p-Cl ("Trimolybdate" Type)

Diazonium "trimolybdates" were obtained by exchange reaction between diazonium fluoroborates and potassium "trimolybdates". To 50 to 100 mL of an aqueous solution of 1 to 2 g of potassium "trimolybdate" [$(1.9$ to $3.8)\times10^{-3}$ mol of $K_2O \cdot 3MoO_3 \cdot 3H_2O$] 25 to 50 mL of an aqueous solution of 1.2 g of the diazonium fluoroborate were added with stirring. At a pH of 4.3 to 4.5 a crystalline precipitate formed, and after the mixture had been stirred for 30 min the precipitate was filtered off, washed with cold water, alcohol, and ether, and dried in a vacuum-drying oven. For purification the salts were reprecipitated from dimethyl sulfoxide with acetone (or ether) without any change in their composition. The salts are colorless, finely crystalline products. Yields: 51 to 75%.

The compounds are of 1:1 composition (cation: "trimolybdate" anion) as a result of the high basicity of the cation in comparison with the amines themselves. The salts were characterized by analyses and infrared spectra (see the paper) confirming the ionic structure and full analogy with the spectrum of potassium "trimolybdate".

The salts are not soluble in the usual organic solvents, they are slightly soluble in water, butyl alcohol, acetone, and cyclohexane and are soluble in dimethyl formamide and dimethyl sulfoxide.

All the salts couple readily with various azo compounds. For this purpose it is most convenient to use the diazonium salts in 10% aqueous solutions of urea [1].

X	p-CH_3	p-OCH_3	p-NO_2	p-Cl
t_d in °C	99	132	117	130

All the salts couple readily with various azo compounds. For this purpose it is most convenient to use the diazonium salts in 10% aqueous solutions of urea [1].

$(XC_6H_4N_2)_2O \cdot 4MoO_3 \cdot 0.33H_2O$, X = H, p-$CH_3$, o-$CH_3$, m-$CH_3$, p-$OCH_3$, o-$OCH_3$, p-$NO_2$, o-$NO_2$, m-$NO_2$, p-Cl, p-Br ("Tetramolybdate" Type?)

To 25 to 100 mL of an aqueous solution of sodium (1:4)-molybdate (metamolybdate, $Na_2O \cdot 4MoO_3 \cdot 6H_2O$ (white)) [$(3.9$ to $5.1)\times10^{-4}$ mol] were added 25 to 100 mL of an aqueous solution of a diazonium fluoroborate [$(22.5$ to $25.0)\times10^{-4}$ mol] with stirring. The precipitate of the diazonium salt was stirred for 20 to 30 min, filtered off, washed with cold water, alcohol, and ether, and dried in a blackened vacuum desiccator over P_2O_5. For purification the salts were precipitated from dimethyl sulfoxide with acetone. The salts separate in finely crystalline condition and are stable in storage. Yields: 64 to 98%. A solution of 1 to 2 g of the diazonium fluoroborate in 25 to 50 mL of water at pH 3.4 to 3.6 was added gradually with stirring to a solution of 23 g of normal sodium molybdate ($Na_2MoO_4 \cdot 2H_2O$) in 20 to 30 mL of water and 5 to 10 mL of glacial acetic acid (pH 4.0). The precipitation of the diazonium salt occurred at pH 4.0. Stirring was continued for 30 min, and the precipitate was then filtered off, washed with methanol and ether, and vacuum-dried. The salt was reprecipitated from dimethyl sulfoxide with acetone. Yields: 60 to 85%. The compounds have also been prepared by diazotization of the arylammonium metamolybdates at pH 2 to 3. A 0.5 to 1.0 g portion of the arylammonium polymolybdate was dissolved in 40 to 80 mL of methanol, 0.15 to 0.3 mL of 35% HCl was added, and the mixture was cooled to 0 to 5°C and diazotized with 0.3 to 0.6 mL of isopentyl nitrite for 1 h. The diazotization of toluidines and anisidines was conducted in a suspension. In this case the arylammonium metamolybdate that did not react was filtered off, and the solution was added to ether. The colorless precipitate was filtered off, washed with ether, and vacuum-dried. Yields: 42 to 60%.

The compound with X = H forms hexagonal tablets, with p-CH_3 rhombs, with o-CH_3 needles, with o-OCH_3 tetrahedra, and with p-NO_2 light yellow microcrystals.

The compounds were characterized by analyses. The infrared spectrum of the compound with X = o-OCH_3 is reported. In the paper the antiquated formula $[H_2Mo_{12}O_{40}]^{6-}$ for the metamolybdate ion is used. The strong absorption bands corresponding to the vibrations of the diazonium ion $C–\overset{+}{N}≡N$ for the differently substituted compounds are compared.

Decomposition temperatures t_d:

X	H	p-CH_3	o-CH_3	m-CH_3	p-OCH_3	o-OCH_3
t_d in °C	78	98	72	51	132	152

X	p-NO_2	o-NO_2	m-NO_2	p-Cl	p-Br
t_d in °C	93	130	117	115	124

Diazonium metamolybdates are readily soluble in dimethyl formamide, 5 to 35% HCl, 10 to 20% p-toluenesulfonic acid solution, urea solution, and solutions of naphthalenedisulfonic acids. The compounds crystallize from organic solvents without change in composition [2].

References:

[1] V. G. Smirnova, V. V. Kozlov (Zh. Obshch. Khim. **46** [1976] 105/8; J. Gen. Chem. [USSR] **46** [1976] 107/9). – [2] V. V. Kozlov, V. G. Smirnova (Zh. Obshch. Khim. **41** [1971] 901/5; J. Gen. Chem. [USSR] **41** [1971] 908/11).

2.8.2.17 Guanidinium Polymolybdate $[CH_6N_3]_6[Mo_7O_{24}]·H_2O$ (Heptamolybdate Type)

The compound was obtained on cooling a mixture of hot solutions of sodium heptamolybdate (approximately 0.3 M in Mo) and guanidinium chloride in the form of colorless elongated prisms.

The crystals are monoclinic, space group C2/c-C_{2h}^6 (No. 15); lattice parameters a = 11.979(7), b = 15.955(15), c = 19.923(18) Å, β = 92.27(5)°; Z = 4. D_m = 2.46, D_x = 2.50 g/cm^3. The atomic coordinates are listed; R = 0.091.

The heptamolybdate anion has the crystallographic point symmetry of 2, but it approximates the point symmetry mm. The seven MoO_6 octahedra form a compact group by edge sharing (see Fig. 1, p. 9). The octahedra are highly distorted, each Mo atom is displaced from the octahedron center towards the periphery of the anion. Thus, the bonds around each Mo fall into three groups: (1) a short cis-pair (Mo-O 1.70 to 1.75 Å, ∢O-Mo-O 104.1° to 105.1°), terminal except for the central Mo atom which shares all attached O atoms, (2) a pair of medium length (1.89 to 2.03 Å, ∢O-Mo-O 142° to 153°), each cis to both short bonds, (3) a long pair (2.13 to 2.56 Å), each trans to a short bond. The anion may be described in the following alternative way: A boat-like Mo_6O_{18} ring is formed from six MoO_4 tetrahedra sharing corners. Each of these Mo atoms forms an additional long bond to an O atom of a seventh MoO_4 tetrahedron above the center of the boat. Two extra O atoms, each forming four long bonds, complete the sixfold coordination of the Mo atoms.

The lengths of the shorter N···O contacts between cations and anions and their directions close to the cation planes indicate an extensive network of hydrogen bonds linking the cations and anions in three dimensions. Two of the N atoms make only weak contacts with anion O atoms. For a list with other interatomic distances and angles see the paper.

A. Don, T. J. R. Weakley (Acta Cryst. B **37** [1981] 451/3).

2.8.2.18 Piperidinium Polymolybdates

The $C_5H_{11}N$-MoO_3-H_2O System

The system has been investigated at 25°C, see **Fig. 61**. The following solids were found to exist in a stable equilibrium with the solution: $(C_5H_{11}NH)_2O \cdot 4MoO_3$, $(C_5H_{11}NH)_2O \cdot 3MoO_3 \cdot 2H_2O$, $3(C_5H_{11}NH)_2O \cdot 7MoO_3 \cdot 6H_2O$, $(C_5H_{11}NH)_2O \cdot 2MoO_3 \cdot H_2O$, $(C_5H_{11}NH)_2MoO_4$. The "tetramolybdate" and monomolybdate have a congruent solubility, the "trimolybdate", paramolybdate, and "dimolybdate" have no congruent solubility. The monomolybdate is completely insoluble in pure piperidine. The branches of the compounds are characterized as follows [1]:

composition of the solution in wt%		point	solid phases in equilibrium
0.87 $C_5H_{11}N$	1.13 MoO_3	A	"tetramolybdate" + "trimolybdate"
27.0 $C_5H_{11}N$	25.5 MoO_3	B	"trimolybdate" + paramolybdate
38.5 $C_5H_{11}N$	34.0 MoO_3	C	paramolybdate + "dimolybdate"
43.4 $C_5H_{11}N$	37.8 MoO_3	D	"dimolybdate" + monomolybdate

In addition to the compounds found in the system there are the piperidinium polymolybdates $(C_5H_{11}NH)_2O \cdot 2MoO_3$ [2], $(C_5H_{11}NH)_6[Mo_7O_{24}] \cdot 3H_2O$ (p. 161), $(C_5H_{11}NH)_6[Mo_7O_{24}]$ [2 to 5], $(C_5H_{11}NH)_2O \cdot 3MoO_3$ [2 to 4], $(C_5H_{11}NH)_2O \cdot 4MoO_3$ [2 to 4], and $(C_5H_{11}NH)_2O \cdot 4MoO_3 \cdot 2H_2O$ (p. 162) [2].

In the range rich in piperidine (>53.5 wt%, point E) two liquid phases exist in an equilibrium, one containing 31.8 to 10 wt% MoO_3 and the other 0 to 10 wt% [1].

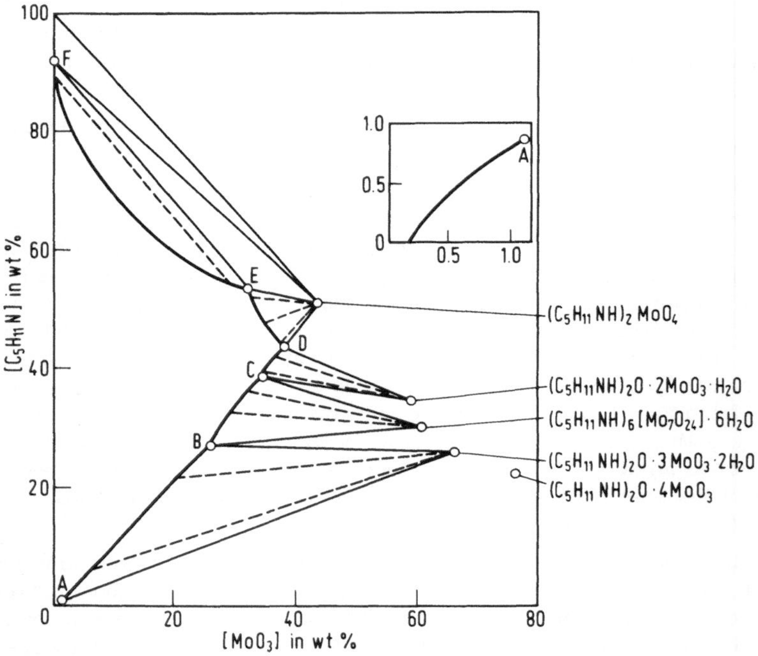

Fig. 61. Phase diagram of the $C_5H_{11}N$-MoO_3-H_2O system at 25°C [1].

$(C_5H_{11}NH)_6[Mo_7O_{24}]\cdot 6\,H_2O$ (Heptamolybdate Type)

The compound was prepared by dissolution of 100 g MoO_3 in a warm mixture of 600 g alcohol of 90 vol% and 120 mL piperidine, filtration of the traces of MoO_3 not dissolved, and distillation of the alcohol until the remaining solution was sufficiently viscous. After 24 h at 40 to 50°C very fine, white crystals deposited which were filtered off and washed with absolute alcohol. The product was twice recrystallized from alcohol of 90 vol%. It forms small parallelepipeds [1, 6].

The salt was characterized by analysis [1, 4], infrared spectrum [3], and X-ray powder pattern [4] (see the papers). The infrared spectrum confirms the presence of a paramolybdate anion [3].

The compound is photochromic. Irradiation with UV light yields a pink [1, 2] or beige [6] coloration ($\lambda_{max}=360$ nm [6]). Coloration requires much more time than in the case of the "trimolybdate". Decoloration is very slow; oxygen accelerates the decoloration. An aqueous solution of the compound shows no coloration on irradiation with UV light [6].

Thermogravimetric investigation shows the formation of the anhydrous "trimolybdate" at 55 to 145°C, the formation of the anhydrous "tetramolybdate" at 155 to 185°C, and the formation of MoO_3 at 225 to 333°C. There are indications of a paramolybdate trihydrate and of the anhydrous paramolybdate as intermediate steps. The trace of a step between the anhydrous "tetramolybdate" and MoO_3 is explained with the temporary occurrence of reduced molybdenum oxides (reduction by the amine) [4].

The compound is soluble in water [6]. It is insoluble in most of the common organic solvents, cold or warm, including absolute alcohol [1, 6]. The solubility in alcohol increases with the concentration of water in the solvent [1, 6] and reaches about 150 g per 100 mL alcohol of 90 vol% at boiling temperature. In cold alcohol of 90 vol% the solubility is very small. This behavior in alcohol of 90 vol% allows recrystallization with yields of 90 to 95% [1].

$(C_5H_{11}NH)_6[Mo_7O_{24}]\cdot 3\,H_2O$ (Heptamolybdate Type)

The compound was precipitated by cooling of a warm aqueous solution containing 45% MoO_3 and 39% piperidine. Very fine, white crystals are obtained [4].

The compound was characterized by analysis [4], infrared spectrum [3], and X-ray powder pattern [4] (see the papers). The infrared spectrum shows the presence of the paramolybdate anion [4]. The compound is neither photochromic nor photosensitive [2].

Thermogravimetric investigations show the formation of the anhydrous paramolybdate (this compound was not observed on the thermal decomposition of the hexahydrate) at 60 to 80°C, the formation of the mixed anhydrous "tri-" and "tetramolybdate" at 135 to 154°C, the formation of the anhydrous "tetramolybdate" at 168 to 190°C, and the formation of MoO_3 at 240 to 365°C [4].

$(C_5H_{11}NH)_2O\cdot 3\,MoO_3\cdot 2\,H_2O$ ("Trimolybdate" Type)

The compound was prepared by heating of ammonium "octamolybdate" 5-hydrate with half of its weight of piperidine and six times its weight of alcohol of 90 vol% for 10 h under reflux [1, 6]. The precipitate was filtered off, washed with absolute alcohol, and dried under vacuum at room temperature [6]. It crystallizes in long, white, opalescent needles [1]. The ammonium "octamolybdate" (see also p. 114) was prepared according to [7].

The salt was characterized by analysis [1, 4], infrared spectrum [3], and X-ray powder pattern [4] (see the papers). The intrared spectrum is a typical "trimolybdate" spectrum. The habit of the crystals also confirms the presence of a "trimolybdate".

The compound is photochromic [1 to 4, 6]. On exposure to light of wavelengths below 400 nm the compound becomes first pink-colored (only one absorption maximum at 480 nm) and later beige-brown (appearance of a second maximum at 380 nm). Other descriptions of the color: reddish brown [2, 8], pink (without restriction of the irradiation period) [4]. The intensities of the two maxima increase with the irradiation time until a saturation value is attained (after about 30 min). At this point the second maximum is almost as high as the first one. The reaction is so sensitive that exposure to sunlight for only a few minutes yields the pink coloration. Decoloration in air is a much slower phenomenon than coloration and requires several days. The decrease is initially much more rapid at 480 than at 380 nm; during the first minutes following the exposition to light the 380 nm maximum is still growing a little. Under nitrogen, the maximum at 380 nm continues to increase, while that at 480 nm decreases until an equilibrium is reached. Decoloration under oxygen or at 55°C is much more rapid than at room temperature in air. Hence, decoloration is due to an oxidation process [6].

Two classes of absorbing centers were postulated: class A ($\lambda_{max}=480$ nm), which develops rapidly, and class B ($\lambda_{max}=480$ and 380 nm), which develops slowly [6]. Later investigations showed that the spectrum is constituted of three Gaussian curves at (2) 546, (3) 467, and (5) 369 nm (the numbers in parentheses refer to the numbering given on p. 40) [8]. The colored sample has the same composition and X-ray powder pattern as the original product, indicating that the colored species are produced only in very small quantities at the surface of the irradiated sample [4]. An EPR signal of the colored polycrystalline material due to Mo^{V} was observed at $g_1=1.880$, $g_2=1.925$, $g_3=1.935$ [8].

Thermogravimetric investigation shows the formation of the anhydrous "tetramolybdate" at 25 to 167°C and the formation of MoO_3 at 249 to 350°C [4].

The compound is slightly soluble in water [6]. Its solubility is not congruent. The compound is insoluble in alcohol and the common organic solvents [1, 6]. It is also insoluble in dimethyl formamide. Hence, purification by recrystallization is impossible [1].

$(C_5H_{11}NH)_2O \cdot 4MoO_3 \cdot 2H_2O$ (β-Octamolybdate Type)

The compound was obtained by hydrolysis of piperidinium "trimolybdate" 2-hydrate for a period of 3 d. The precipitate was washed with water, alcohol, and ether and was finally dried in a desiccator under vacuum [4].

The salt was characterized by analysis [4], infrared spectrum [3], and X-ray powder pattern [4] (see the papers). The infrared spectrum shows unambiguously the presence of a β-octamolybdate. The compound is neither photochromic nor photosensitive [2].

Thermogravimetric investigation shows the formation of the anhydrous salt at 75 to 95°C and the formation of MoO_3 at 248 to 365°C [4].

Anhydrous Piperidinium Polymolybdates

In addition to the polymolybdate hydrates described above the anhydrous compounds of composition $(C_5H_{11}NH)_2O \cdot 2MoO_3$ [2], $(C_5H_{11}NH)_6[Mo_7O_{24}]$ [2 to 5], $(C_5H_{11}NH)_2O \cdot 3MoO_3$ [2 to 4, 8], and $(C_5H_{11}NH)_2O \cdot 4MoO_3$ [2 to 4] have been prepared and investigated.

References:

[1] M. J. Weill (Bull. Soc. Chim. France **1960** 1136/8). – [2] F. Arnaud-Neu, M. J. Schwing-Weill (Bull. Soc. Chim. France **1973** 3225/32; J. Less-Common Metals **36** [1974] 71/8). – [3] M. J. Schwing-Weill, F. Arnaud-Neu (Bull. Soc. Chim. France **1970** 853/60). – [4] M. J. Schwing-Weill

(Bull. Soc. Chim. France **1967** 3801/5). – [5] J. Fuchs, A. Thiele (Z. Naturforsch. **34b** [1979] 155/9).

[6] M. J. Schwing-Weill (Bull. Soc. Chim. France **1965** 2159/63). – [7] J. Byé, M. J. Weill (Bull. Soc. Chim. France **1960** 1130/3). – [8] F. Arnaud-Neu, M. J. Schwing-Weill (Bull. Soc. Chim. France **1973** 3239/46).

2.8.2.19 4-Methylpiperidinium Polymolybdates

$(C_6H_{13}NH)_6[Mo_7O_{24}]\cdot 2H_2O$ (Heptamolybdate Type)

The compound was prepared by adding 12.5 g MoO_3 to a boiling solution of 15 mL 4-methylpiperidine in 75 g alcohol of 90 vol%. The insoluble residue was filtered off and 40 mL alcohol were distilled off. The small residue in the form of crusts was filtered off, and by slow evaporation of the cold solution with continual agitation a white precipitate was formed which was dried under vacuum for 48 h [1].

The compound was characterized by analysis [1], infrared spectrum [2], and X-ray powder pattern [1] (see the papers). The infrared spectrum shows the presence of a paramolybdate anion. The bands at 840 and 630 to 640 cm^{-1} are splitted [1]. The statements on the photochromic behavior are contradictory. According to [1] the compound is neither photochromic nor photosensitive, but according to [3] it is photochromic, yielding a pink coloration.

Thermogravimetric investigations show the formation of the anhydrous paramolybdate at 77 to 92°C, the formation of the anhydrous "trimolybdate" at 129 to 138°C, the formation of the anhydrous "tetramolybdate" at 156 to 172°C, and the formation of MoO_3 at 236 to 348°C [1].

The salt is readily soluble in cold and warm water, and it is soluble in absolute alcohol and in warm dimethyl formamide. It cannot be recrystallized from absolute alcohol due to the formation of the insoluble "trimolybdate" [1].

$(C_6H_{13}NH)_2O\cdot 3MoO_3\cdot H_2O$ (Unknown Type)

The compound was obtained by hydrolysis of anhydrous 4-methyl-piperidinium paramolybdate in seven times its weight of water. The crystals were washed with absolute alcohol, then ether and then dried under vacuum. The anhydrous paramolybdate was prepared by heating a mixture of 8.5 g ammonium "octamolybdate" (for the preparation of this compound see [1, 4], see also p. 114), 5 g 4-methylpiperidine, and 51 g alcohol under reflux until all ammonia was liberated. The small insoluble residue was filtered off. Anhydrous paramolybdate was precipitated by the addition of absolute alcohol, filtered, washed with absolute alcohol, and dried under vacuum [1].

The compound was characterized by analysis [1], infrared spectrum [2], and X-ray powder pattern [1] (see the papers). The infrared spectrum is very different from that of a typical "trimolybdate" [1].

The white compound is photochromic. Irradiation with UV light yields a violet coloration (λ_{max} = 560 nm) [1, 3], which reaches its final value after about 3 h (under the given conditions). Decoloration requires a stream of pure oxygen and is in this case complete after about 3 h [1].

Thermogravimetric investigations show the formation of the anhydrous "trimolybdate" at 42 to 74°C, the formation of the anhydrous "tetramolybdate" at 146 to 158°C, and the formation of MoO_3 at 233 to 348°C [1].

The salt is slightly soluble in cold and warm water. Its solubility at 25°C is not congruent. It is insoluble in most of the usual solvents, but soluble in dimethyl formamide at higher temperatures giving a yellow solution [1].

Anhydrous 4-Methylpiperidinium Polymolybdates

In addition to the hydrates described above there are anhydrous 4-methylpiperidinium polymolybdates of the following compositions: $(C_6H_{13}NH)_6[Mo_7O_{24}]$, $(C_6H_{13}NH)_2O \cdot 3MoO_3$, and $(C_6H_{13}NH)_2O \cdot 4MoO_3$ [1 to 3].

References:

[1] F. Arnaud-Neu, M. J. Schwing-Weill (Bull. Soc. Chim. France **1971** 60/8). – [2] M. J. Schwing-Weill, F. Arnaud-Neu (Bull. Soc. Chim. France **1970** 853/60). – [3] F. Arnaud-Neu, M. J. Schwing-Weill (Bull. Soc. Chim. France **1973** 3225/32, 3225; J. Less-Common Metals **36** [1974] 71/8). – [4] M. J. Schwing-Weill (Bull. Soc. Chim. France **1967** 3795/8).

2.8.2.20 N-Methylpiperidinium Polymolybdates

$(C_6H_{13}NH)_2O \cdot 3MoO_3 \cdot H_2O$ (Unknown Type)

The compound was prepared by adding 12.5 g MoO_3 to a boiling (reflux) solution of 15 mL N-methylpiperidine in 72 g alcohol of 90 vol%. The insoluble residue (MoO_3) was filtered off and 40 mL alcohol were distilled off. The small residue in the form of crusts was filtered off, and by slow evaporation of the cold solution with continual agitation a yellow precipitate was formed which was dried under vacuum for 48 h to give a fine powder [1].

The compound was characterized by analysis [1], infrared spectrum [2], and X-ray powder pattern [1] (see the papers). The infrared spectrum is very different from that of a typical "trimolybdate" [1]. The compound is neither photochromic nor photosensitive [1, 3].

Thermogravimetric investigations show at 50 to 140°C the formation of the anhydrous "tetramolybdate", clearly having the infrared spectrum of this compound but an X-ray powder pattern different from that of the reference compound. The intermediate stages are ill-defined [1].

The solubility in water at 25°C is not congruent. The "trimolybdate" hydrolyzes to give the "tetramolybdate" monohydrate. The compound is insoluble in most of the common organic solvents, particularly in alcohol. It is soluble in warm dimethyl formamide [1].

$(C_6H_{13}NH)_2O \cdot 4MoO_3 \cdot H_2O$ (β-Octamolybdate Type)

The compound was prepared by adding 7.2 g MoO_3 to a warm solution of 12 g N-methylpiperidine in 35 mL water, filtration of the insoluble residue, and addition of 20 mL 2 N HCl. Under continual agitation fine, pale green crystals precipitated which were filtered off, washed with water, alcohol, and ether, and then dried under vacuum. The compound is also formed in a warm aqueous solution of 2M N-methylpiperidinium chloride saturated with solid ammonium paramolybdate. The precipitate formed on cooling was washed with water, alcohol, and ether, and dried under vacuum. The compound was also obtained by hydrolysis of the "trimolybdate" monohydrate [1].

The compound was characterized by analysis [1], infrared spectrum [2], and X-ray powder pattern [1] (see the papers). The infrared spectrum resembles those of the β-octamolybdates. The compound is neither photochromic nor photosensitive [1, 3].

Thermogravimetric investigations show the formation of the anhydrous "tetramolybdate" at 60 to 76°C and the formation of MoO_3 at 214 to 384°C [1]. (For the properties of the anhydrous compound see [1 to 3].)

The solubility in water is 5.5 g/L at 25°C. The solubility is congruent. The compound is insoluble in almost all organic solvents at room temperature. It is soluble in cold and warm dimethyl formamide giving a dark green solution [1].

References:

[1] F. Arnaud-Neu, M. J. Schwing-Weill (Bull. Soc. Chim. France **1971** 60/8). – [2] M. J. Schwing-Weill, F. Arnaud-Neu (Bull. Soc. Chim. France **1970** 853/60). – [3] F. Arnaud-Neu, M. J. Schwing-Weill (Bull. Soc. Chim. France **1973** 3225/32; J. Less-Common Metals **36** [1974] 71/8).

2.8.2.21 2,6-Dimethylpiperidinium Molybdates

$(C_7H_{15}NH)_2MoO_4 \cdot H_2O$

A preparation with a small deficiency of the base has been obtained by dissolving MoO_3 in a mixture of 2,6-dimethylpiperidine and alcohol as described for the "trimolybdate" below. Recrystallization from absolute alcohol further diminishes the part of the base. The stoichiometric composition can be obtained by stirring the base-deficient product for 3 d in an aqueous 30% solution of 2,6-dimethylpiperidine. The shiny white crystals are filtered, washed with ether, and dried in vacuo [1].

The compound was characterized by analysis [1], infrared spectrum [2], and X-ray powder pattern [1] (see the papers). The IR spectrum resembles that of $Na_2MoO_4 \cdot 2H_2O$ [1]. The compound is neither photochromic nor photosensitive [1, 3].

Thermogravimetric investigations show the hydrate to be stable up to 64°C. The anhydrous compound forms between 73 and 104°C, and the "tetramolybdate" between 116 and 217°C. Above 317°C the compound is completely decomposed [1].

The compound is easily soluble in water (~100 g/100 mL H_2O at 25°C) giving a yellow solution. It is soluble in cold and warm dimethyl formamide, but insoluble in all usual organic solvents at room temperature [1].

$(C_7H_{15}NH)_2O \cdot 3MoO_3 \cdot 3H_2O$ (Unknown Type)

The compound was prepared by dissolution of 25 g MoO_3 in a mixture of 35 g 2,6-dimethylpiperidine and 150 mL alcohol of 90 vol%. The small amounts of the insoluble residue were filtered off and 100 mL of alcohol were distilled off. The solid obtained on cooling was filtered off, rapidly washed with ether, and dried. The product was treated with absolute alcohol, and the insoluble fraction was washed with alcohol and ether and then dried under vacuum [1].

The white powder was characterized by analysis [1], infrared spectrum [2], and X-ray powder pattern [1] (see the papers). The infrared spectrum is very different from that of a typical "trimolybdate" [1].

The compound is photochromic. Irradiation with UV light yields a pink coloration [1, 3] (λ_{max} = 380 nm [1]). Decoloration in air is very slow and requires 14 d for "semi-decoloration" [1].

Thermogravimetric investigations show the formation of the monohydrate at 62 to 88°C, the formation of the anhydrous "trimolybdate" at 100 to 108°C, the formation of the anhydrous "tetramolybdate" at 126 to 138°C, and the formation of MoO_3 at 218 to 354°C [1].

The solubility in water is small and incongruent; the compound hydrolyzes to give the "tetramolybdate". The compound is insoluble in cold and warm absolute alcohol, acetone, ether, chloroform, and toluene, and only slightly soluble in warm alcohol of 96 vol%. It is soluble in dimethyl formamide [1].

$[(C_7H_{15}NH)_{1.5}(NH_4)_{0.5}]O \cdot 4MoO_3 \cdot H_2O$ ("Tetramolybdate" Type?)

The compound was obtained by mixing 50 mL of a warm solution of 5 g ammonium paramolybdate and 10 mL 2,6-dimethylpiperidine with continual agitation and neutralization of the solution with 2N HCl. The compound forms very fine, white crystals.

The compound was characterized by analysis, infrared spectrum (not reproduced), and X-ray powder pattern (see the paper). The infrared spectrum is that of a hydrated "tetramolybdate". The compound is not photochromic.

The compound is stable in water. It decomposes above 40°C [1].

$(C_7H_{15}NH)_2O \cdot 6.83$ to $9.39MoO_3 \cdot {\sim}4H_2O$ (36-Molybdate Type?)

The compound was prepared by dissolution of 7.2 g MoO_3 in a hot solution (reflux) of 11.3 g 2,6-dimethylpiperidine in 30 to 130 mL water. The excess of the base was subsequently neutralized with 2N HCl. The ratio $MoO_3:(C_7H_{15}NH)_2O$ depends on the concentration of the solution. Application of 30 and 130 mL water gave values of 6.83 and 8.77, respectively. The compound is a fine, very pale green powder [1].

The compound was characterized by analysis [1], infrared spectrum [2], and X-ray powder pattern [1] (see the papers). The varying composition of the compound does not influence the spectrum and the X-ray powder pattern and is explained by a statistical exchange of the base cations for H_3O^+ ions [1, 2]. This explanation, however, can only be true for the compositions $M^I_2O \cdot {>}9MoO_3$. The infrared spectrum is very similar to that of a 36-molybdate; however, the content of water of crystallization is rather small.

The compound is photosensitive but not photochromic. Irradiation with UV light yields a blue coloration [1, 3].

Thermogravimetric investigation shows the formation of the anhydrous compound at 52 to 164°C and the formation of MoO_3 at 214 to 357°C. However, all the decomposition products occur in a reduced (colored) form [1].

Anhydrous 2,6-Dimethylpiperidinium Molybdates

In addition to the hydrates described above the anhydrous compounds of composition $(C_7H_{15}NH)_2MoO_4$, $(C_7H_{15}NH)_2O \cdot 3MoO_3$, and $(C_7H_{15}NH)_2O \cdot 4MoO_3$ have been prepared and investigated [1 to 3].

References:

[1] F. Arnaud-Neu, M. J. Schwing-Weill (Bull. Soc. Chim. France **1971** 68/78). – [2] M. J. Schwing-Weill, F. Arnaud-Neu (Bull. Soc. Chim. France **1970** 853/60). – [3] F. Arnaud-Neu, M. J. Schwing-Weill (Bull. Soc. Chim. France **1973** 3225/32; J. Less-Common Metals **36** [1974] 71/8).

2.8.2.22 2,2,6,6-Tetramethylpiperidinium Molybdates

$(C_9H_{19}NH)_2O \cdot 4\,MoO_3 \cdot H_2O$ ("Tetramolybdate" Type?)

The compound was prepared by adding 7.2 g MoO_3 to a warm solution of about 12 g 2,2,6,6-tetramethylpiperidine in 35 mL water, filtration of the insoluble residue, and addition of 20 mL 2N HCl. Under continual agitation pale green crystals precipitated which were filtered off, washed with water, alcohol, and ether, and then dried under vacuum [1].

The salt was characterized by analysis [1], infrared spectrum [2], and X-ray powder pattern [1] (see the papers). The infrared spectrum is somewhat different from those of the other "tetramolybdates". The compound is neither photochromic nor photosensitive [1, 3].

Thermographic investigations show the formation of the anhydrous "tetramolybdate" at 72 to 80°C and the formation of MoO_3 at 234 to 366°C [1].

The salt is only slightly soluble in cold but more in warm water. It is insoluble in almost all the common organic solvents, cold or warm, in particular in absolute alcohol and alcohol of 90 vol%. It is soluble in cold and warm dimethyl formamide, giving a yellow solution [1].

Anhydrous 2,2,6,6-Tetramethylpiperidinium Molybdates

In addition to the "tetramolybdate" hydrate described above the anhydrous compounds of composition $(C_9H_{19}NH)_2MoO_4$ [1 to 3], $(C_9H_{19}NH)_2O \cdot 2\,MoO_3$ [1, 3], and $(C_9H_{19}NH)_2O \cdot 4\,MoO_3$ [1 to 3] have been prepared and investigated.

References:

[1] F. Arnaud-Neu, M. J. Schwing-Weill (Bull. Soc. Chim. France **1971** 68/78). – [2] M. J. Schwing-Weill, F. Arnaud-Neu (Bull. Soc. Chim. France **1970** 853/60). – [3] F. Arnaud-Neu, M. J. Schwing-Weill (Bull. Soc. Chim. France **1973** 3225/32; J. Less-Common Metals **36** [1974] 71/8).

2.8.2.23 Pyrrolidinium Polymolybdates

$(C_4H_9NH)_6[Mo_7O_{24}] \cdot H_2O$ (Heptamolybdate Type)

The compound was obtained by heating a mixture of 10 g MoO_3, 10 mL pyrrolidine, and 60 mL alcohol of 90 vol% under reflux for 24 h [1].

The salt was characterized by analysis and infrared spectrum (see the paper) showing the presence of a paramolybdate [1]. The compound is photochromic. Irradiation with UV light yields a pink coloration [1, 2].

Thermolysis shows the formation of the anhydrous paramolybdate at 50 to 64°C, the formation of the anhydrous "trimolybdate" at 138 to 151°C, the formation of the anhydrous "tetramolybdate" at 162 to 176°C, and the formation of MoO_3 at 225 to 370°C [1].

$(C_4H_9NH)_2O \cdot 3\,MoO_3 \cdot H_2O$ ("Trimolybdate" Type)

The compound was prepared by heating a mixture of 8 g ammonium "octamolybdate", 5 mL pyrrolidine, and 100 mL absolute alcohol under reflux [1]. The "octamolybdate", possibly ammonium 36-molybdate (cf. p. 114), was prepared according to [3].

The salt was characterized by analysis and infrared spectrum (see the paper) indicating the presence of a "trimolybdate" [1].

The compound is photochromic. Irradiation with UV light yields a reddish brown coloration [1, 2, 4]. The analysis of the diffuse reflectance spectrum of the irradiated compound shows four Gaussian curves at (1) 800, (2) 574, (3) 459, and (4) 387 nm (the numbers in parentheses refer to the numbering given on p. 40). An EPR signal of the colored polycrystalline material due to Mo^V was observed at $g_1=1.879$, $g_2=1.919$, $g_3=1.934$ [4]. It is one of the most sensitive and rapid photochromic systems [2].

Thermolysis shows the formation of the anhydrous "tetramolybdate" at 142°C and the formation of MoO_3 at 220 to 346°C [1].

Anhydrous Pyrrolidinium Polymolybdates

In addition to the polymolybdate hydrates described above the anhydrous compounds of composition $(C_4H_9NH)_6[Mo_7O_{24}]$, $(C_4H_9NH)_2O \cdot 3MoO_3$, and $(C_4H_9NH)_2O \cdot 4MoO_3$ have been prepared and investigated [1, 2].

References:

[1] F. Arnaud-Neu, M. J. Schwing-Weill (Bull. Soc. Chim. France **1973** 3225/32). – [2] F. Arnaud-Neu, M. J. Schwing-Weill (J. Less-Common Metals **36** [1974] 71/8). – [3] M. J. Schwing-Weill (Bull. Soc. Chim. France **1967** 3795/8). – [4] F. Arnaud-Neu, M. J. Schwing-Weill (Bull. Soc. Chim. France **1973** 3239/46).

2.8.2.24 Piperazinium Molybdates

$(C_4H_{10}N_2H_2)O \cdot 2MoO_3 \cdot 5H_2O$ ("Trimolybdate" Type)

The compound was obtained by heating a mixture of 0.05 mol MoO_3, piperazine, and 50 mL water under reflux [1].

The compound was characterized by analysis and infrared spectrum (see the paper) [1]. The latter, however, is a typical "trimolybdate" spectrum (compare also p. 39).

The compound is photochromic. Irradiation with UV light yields a deep pink coloration [1, 2]. The analysis of the diffuse reflectance spectrum of the irradiated compound shows five Gaussian curves at (1) 769, (2) 571, (3) 463, (4) 415, and (5) 365 nm (the numbers in parentheses refer to the numbering given on p. 40). An EPR signal of the colored polycrystalline material due to Mo^V was observed at $g_1=1.875$, $g_2=1.925$, $g_3=1.937$ [2].

Thermolysis shows the formation of the anhydrous "dimolybdate" at 66°C and the formation of MoO_3 at 230 to 316°C [1].

$(C_4H_{10}N_2H_2)O \cdot 2MoO_3 \cdot H_2O$ ("Trimolybdate" Type)

The compound was obtained by agitation of a cold mixture of 0.1 mol MoO_3, piperazine, 100 mL water, and 600 mL absolute alcohol [1].

The compound was characterized by analysis and infrared spectrum (see the paper) [1]. The latter, however, is a typical "trimolybdate" spectrum as with the pentahydrate above.

The compound is photochromic. Irradiation with UV light yields a deep pink coloration [1, 2]. The analysis of the diffuse reflectance spectrum of the irradiated compound shows the same Gaussian curves as the pentahydrate. An EPR signal of the colored polycrystalline material due to Mo^V was also observed at the same values as for the pentahydrate [2].

Thermolysis shows the formation of the anhydrous "dimolybdate" at 52 to 64°C and the formation of MoO_3 at 232 to 338°C [1].

$(C_4H_{10}N_2H_2)O \cdot 3\,MoO_3 \cdot H_2O$ ("Trimolybdate" Type)

The compound was obtained by hydrolysis of the "dimolybdates" with 5 or 1 equivalents of H_2O (see above) [1].

The compound was characterized by analysis and infrared spectrum (see the paper) showing the presence of a "trimolybdate" [1].

The compound is photochromic. Irradiation with UV light yields a deep pink coloration [1, 2]. The analysis of the diffuse reflectance spectrum of the irradiated compound shows four Gaussian curves at (1) 1000, (2) 568, (3) 485, and (4) 386 nm [2] (the numbers in parentheses refer to the numbering given on p. 40). It is one of the most sensitive and rapid photochromic systems [3].

Thermolysis shows the formation of the anhydrous "trimolybdate" at 74 to 88°C and the formation of MoO_3 at 230 to 308°C [1].

Anhydrous Piperazinium Molybdates

In addition to the polymolybdate hydrates described above the anhydrous compounds of composition $(C_4H_{10}N_2H_2)MoO_4$ [1, 3], $(C_4H_{10}N_2H_2)O \cdot 2\,MoO_3$, and $(C_4H_{10}N_2H_2)O \cdot 3\,MoO_3$ [1 to 3] have been prepared and investigated.

References:

[1] F. Arnaud-Neu, M. J. Schwing-Weill (Bull. Soc. Chim. France **1973** 3225/32). – [2] F. Arnaud-Neu, M. J. Schwing-Weill (Bull. Soc. Chim. France **1973** 3239/46). – [3] F. Arnaud-Neu, M. J. Schwing-Weill (J. Less-Common Metals **36** [1974] 71/8).

2.8.2.25 Pyridinium Polymolybdates

The C_5H_5N-MoO_3-H_2O System

This system has been investigated at 25°C, see **Fig. 62**. No hydrates of pyridinium polymolybdates have been observed, only the anhydrous "trimolybdate" (metastable) and "tetramolybdate" have been found [1]. The preparation and properties of these compounds are described in [1 to 4].

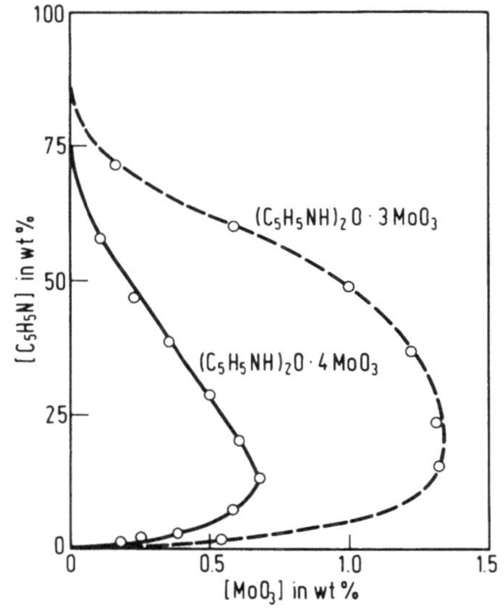

Fig. 62. Solubility isotherms of the C_5H_5N-MoO_3-H_2O system at 25°C (– – – – metastable) [1].

$(C_5H_5NH)_2O \cdot 4MoO_3 \cdot C_5H_5N$ ("Tetramolybdate" Type?)

The compound was prepared by the action of a large excess of pyridine on the dietherate of molybdic hydrochloride, $MoO_3 \cdot 2HCl \cdot 2(C_2H_5)_2O \cdot H_2O$. The reaction mixture must not be shaken continuously [2], to avoid the formation of anhydrous pyridinium "trimolybdate" [1]. (The dietherate of molybdic hydrochloride was prepared according to [5].)

The compound was characterized by analysis [2], infrared spectrum [3], and X-ray powder pattern [2] (see the papers).

Thermogravimetric investigations show the formation of the "tetramolybdate" $(C_5H_5NH)_2O \cdot 4MoO_3$ at 140 to 155°C and the formation of MoO_3 at 218 to 290°C [2].

$(C_5H_5NH)_3(H_3Mo_7O_{24})$? (Unknown Type)

Addition of hydrochloric acid to a mixture of pyridinium chloride and sodium molybdate in aqueous solution gave a white precipitate when the pH of the solution was nearly 1.5. The precipitate was filtered off, washed with a buffer solution of pH 1.5 and finally with ether, and dried at 105°C [6].

The compound was characterized merely by its Mo and pyridinium content. The formulation of an acid heptamolybdate ion is novel.

Anhydrous Pyridinium Polymolybdates

In addition to the anhydrous "tri-" and "tetramolybdate" found in the C_5H_5N-MoO_3-H_2O system (p. 169), there are the 3-methylpyridinium β-octamolybdate, $(CH_3C_5H_4NH)_4[Mo_8O_{26}]$ [7], and the 2- [9], 3- [8], and 4-ethylpyridinium β-octamolybdates [10], $(C_2H_5C_5H_4NH)_4[Mo_8O_{26}]$. Their crystal structures have been investigated.

References:

[1] M. J. Weill (Bull. Soc. Chim. France **1960** 1133/5). – [2] M. J. Schwing-Weill (Bull. Soc. Chim. France **1967** 3799/800). – [3] M. J. Schwing-Weill, F. Arnaud-Neu (Bull. Soc. Chim. France **1970** 853/60, 859). – [4] F. Arnaud-Neu, M. J. Schwing-Weill (Bull. Soc. Chim. France **1973** 3225/32; J. Less-Common Metals **36** [1974] 71/8). – [5] J. Byé, M. J. Weill (Bull. Soc. Chim. France **1960** 1130/3).

[6] D. V. Ramana Rao (Anal. Chim. Acta **16** [1957] 1/3). – [7] P. Román, M. E. Gonzalez-Aguado, C. Esteban-Calderón, M. Martínez-Ripoll, S. García-Blanco (Z. Krist. **165** [1983] 271/6). – [8] P. Román, J. Jaud, J. Galy (Z. Krist. **154** [1981] 59/68). – [9] P. Román, A. Vegas, M. Martínez-Ripoll, S. García-Blanco (Z. Krist. **159** [1982] 291/5). – [10] P. Román, M. Martínez-Ripoll, J. Jaud (Z. Krist. **158** [1982] 141/7).

2.8.2.26 N-Cetylpyridinium Polymolybdate $(C_{16}H_{33}NC_5H_5)_3(H_3Mo_6O_{21})$? (Unknown Type)

The compound was precipitated in the form of white crystals from an ammonium molybdate solution of pH 4 to 5 by addition of a solution of cetylpyridinium chloride [1].

The compound was characterized by analysis. Titration curves of the formation reaction confirm the molar ratio cetylpyridinium:Mo = 1:2 [1]. The formulation as an acid hexamolybdate is based on antiquated ideas of the species existing in aqueous polymolybdate solutions. The analytical results allow other formulations, e.g., $(C_{16}H_{33}NC_5H_5)_2O \cdot 4MoO_3 \cdot H_2O$.

The solubility product is 5.2×10^{-20} at 20°C [1].

Reference:

[1] R. K. Chernova, A. L. Lobachev, Z. R. Muratova (Zh. Neorgan. Khim. **24** [1979] 2415/8; Russ. J. Inorg. Chem. **24** [1979] 1340/2).

2.8.2.27 Bipyridinium Molybdates

Addition of hydrochloric acid to a mixture of 2,2′-bipyridine and sodium molybdate in aqueous solution gave a white precipitate of composition $(C_{10}H_8N_2H_2)_2(H_2Mo_7O_{24})$ when the pH of the solution was 1.92 [1]. Due to a misprint the formula is erroneously given in the paper as $(C_{10}H_8N_2H_2)_2(H_4Mo_7O_{24})$. The compound was characterized merely by its Mo content [1] and hence the formulation is inconclusive. According to [2], the composition of the compound is $(C_{10}H_8N_2H)_2(H_4Mo_7O_{24})$ (another interpretation of the data and correction of the formula given in [1]).

Similarly, a 4,4′-bipyridinium polymolybdate of composition $(C_{10}H_8N_2H)_2(H_4Mo_7O_{24})$ has been obtained from 4,4′-bipyridine and ammonium molybdate in acid medium [2].

The colorless silky 2,2′-bipyridinium molybdate $(C_{10}H_8N_2H_2)MoO_4$ has been prepared in aqueous solution from 2,2′-bipyridine and "H_2MoO_4". It is also formed by the thermal decomposition of the peroxomolybdate $(C_{10}H_8N_2H_2)MoO_6$ (p. 174) [2].

The formulations of the bipyridinium cations in [2] are inconsistent: for the paramolybdate, forming at lower pH values, it has been formulated only as the monoprotonated base $(C_{10}H_8N_2H^+)$, whereas for the monomolybdate, occurring at higher pH values, it has been formulated as the diprotonated base $(C_{10}H_8N_2H_2^{2+})$. Consequently, all formulations based on the ratio base : Mo in the compounds are inconclusive.

References:

[1] D. V. Ramana Rao, S. Pani (Current Sci. [India] **24** [1955] 272). – [2] R. G. Beiles, E. M. Beiles (Zh. Neorgan. Khim. **10** [1965] 1618/23; Russ. J. Inorg. Chem. **10** [1965] 883/6).

2.8.2.28 Quinolinium Polymolybdate $(C_9H_7NH)_3(H_3Mo_7O_{24})$? (Unknown Type)

Addition of hydrochloric acid to a mixture of quinolinium chloride and sodium molybdate in aqueous solution gave a white precipitate when the pH of the solution was nearly 1.5. The precipitate was filtered off, washed with a buffer solution of pH 1.5 and finally with ether, and dried at 105°C. The compound was characterized merely by its Mo content [1]. Later the compound was prepared from quinoline and ammonium paramolybdate in sulfuric acid solution [2]. The formulation of an acid polymolybdate ion as well as the formulation of a heptamolybdate ion is experimentally unfounded.

References:

[1] D. V. Ramana Rao (Anal. Chim. Acta **16** [1957] 1/3). – [2] R. G. Beiles, E. M. Beiles (Zh. Neorgan. Khim. **10** [1965] 1618/23; Russ. J. Inorg. Chem. **10** [1965] 883/6).

2.8.2.29 1,1′-Diethyl-2,2′-quinocarbocyanin (Pinacyanol) Polymolybdate $(C_{25}H_{25}N_2)_3(H_3Mo_6O_{21})$? (Unknown Type)

The intense blue cyanine dye pinacyanol chloride and molybdates at pH 3 to 4 in a mixture of 30% ethanol and 70% water form the pinacyanol molybdate compound in a solid-state reaction as a fine, pink colored suspension which is very stable. Centrifugation yields the solid compound [1].

The suspension was characterized by its visible spectrum, $\lambda_{max} = 520$ nm, $\varepsilon = 4.2 \times 10^4$. The ratio cation:Mo = 1:2 was determined by using molar ratios and isomolar series [1]. The formulation as an acid hexamolybdate is based on antiquated ideas of the species existing in aqueous polymolybdate solutions. The analytical results allow other formulations, e.g., $(C_{25}H_{25}N_2)_2O \cdot 4\,MoO_3 \cdot H_2O$.

The freshly prepared suspensions are quantitatively decomposed by thorium ions with dye recovery. This reaction can be used as indicating the endpoint of a Th^{4+} titration with molybdate (for details see the paper). Organic solvents do not extract pinacyanol molybdate [1].

Reference:

[1] A. K. Babko, D. S. Turova, P. P. Turov (Ukr. Khim. Zh. **35** [1969] 953/7; Soviet Progr. Chem. **35** No. 9 [1969] 42/5).

2.8.3 Peroxomolybdate Hydrates

2.8.3.1 Octylammonium Peroxomolybdate $[(C_8H_{17})_4N]HMoO_6$

In the extraction of Mo from peroxide solutions by tetraoctylammonium sulfate, the extracted species probably is the title compound [1], compare [2].

References:

[1] G. M. Vol'dman, A. N. Zelikman, I. Sh. Khutoretskaya (Izv. Vysshikh Uchebn. Zavedenii Tsvetn. Met. **1978** No. 2, pp. 74/7; C. A. **89** [1978] No. 49561). – [2] A. N. Zelikman, I. G. Kalinina (Zh. Neorgan. Khim. **19** [1974] 1040/5; Russ. J. Inorg. Chem. **19** [1974] 567/70).

2.8.3.2 Pyridinium Peroxomolybdates

There were three peroxomolybdates with the $C_5H_5NH^+$ cation claimed by the first investigation [1]. The formulae given were $(C_5H_5NH)_2MoO_5$, $(C_5H_5NH)_2MoO_6$, and $(C_5H_5NH)_2MoO_7$. Investigations of the vibrational spectra [2, 3] provided little elucidation of the formulae. Two compounds were prepared according to the procedure of [1] but unfortunately no details were given. Their crystal structures revealed the formulae $(C_5H_5NH)_2O[MoO(O_2)_2H_2O]_2$ (≙ "$C_5H_5NHMoO_{6.5}H$") and $(C_5H_5NH)_2[MoO(O_2)_2OOH]_2$ (≙ "$C_5H_5NHMoO_7H$") [4].

$(C_5H_5NH)_2MoO_5$

This pyridinium peroxomolybdate is not prepared directly from pyridine, "molybdic acid", and H_2O_2, but by treating $(C_5H_5NH)_2MoO_6$ with $KMnO_4$ solution. This yields a product from which orange yellow $(C_5H_5NH)_2MoO_5$ is obtained after treatment with H_2O_2. For a practical example of preparation, see the paper [1].

$(C_5H_5NH)_2O[MoO(O_2)_2H_2O]_2$ Pyridinium μ-oxo-tetraperoxo-dioxo-diaquo-dimolybdate(VI)

This compound is prepared by reaction of "molybdic acid" with pyridine in the presence of H_2O_2 at pH < 5. For the superseded formula $(C_5H_5NH)_2MoO_6$, see the paper where experimental proportions of the preparation are also given [1]. Single crystals are obtained by recrystallization in H_2O_2 solution at room temperature [4].

The compound is monoclinic, lattice parameters a = 22.494 ± 0.010, b = 7.192 ± 0.005, c = 17.214 ± 0.012 Å, β = 140.64° ± 0.20°; Z = 4. Space group C2/c–C^6_{2h} (No. 15). For positional and thermal parameters (anisotropic), see the paper, R = 0.034. The anion consists of two bipyramids with a common corner as shown in **Fig. 63**. The anion has C2 symmetry. The Mo-O bond distances (in Å) are: Mo-O_t = 1.674(7), Mo-O_b = 1.917(7), Mo-O_p = 1.955(7), 1.961(8) (2 ×), 1.977(8), Mo-OH_2 = 2.445(7); t = terminal, b = bridging, p = peroxo oxygen atoms. For a detailed discussion of other distances and angles, see the paper [4].

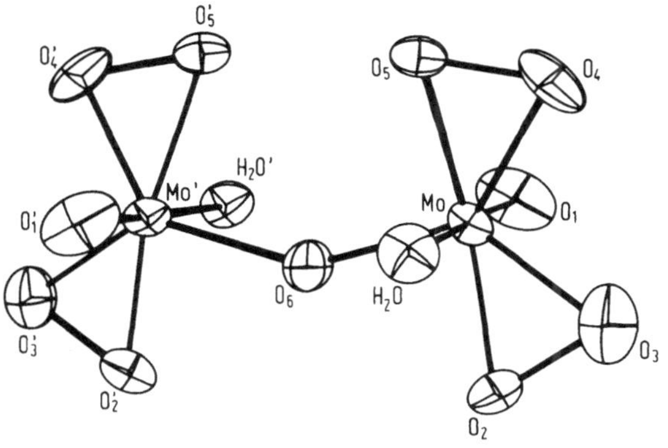

Fig. 63. The structure of the $O[MoO(O_2)_2H_2O]_2^{2-}$ ion [4].

The measured density is 2.14 ± 0.01, the calculated 2.130 g/cm³ [4]. Raman frequencies observed are (in cm⁻¹): 943, ν(Mo=O); 910, ν(O–O); 570, ν(MoO_2, asym); 536, ν(MoO_2, sym); 353, 320 [2].

The compound, assumed to possess the formula "$(C_5H_5NH)_2MoO_6$", loses its active oxygen on heating at 150°C and is converted into $(C_5H_5NH)_2MoO_4$. By prolonged boiling of "$(C_5H_5NH)_2MoO_6$" in water, part of the pyridine is removed and pyridinium paramolybdate is formed. With H_2O_2, "$(C_5H_5NH)_2MoO_6$" reacts to give "$(C_5H_5NH)_2MoO_7$" [1].

$(C_5H_5NH)_2[MoO(O_2)_2OOH]_2$ Pyridinium di-μ-hydroperoxo-tetraperoxo-dioxo-dimolybdate(VI)

This compound is obtained from $(C_5H_5NH)_2O[MoO(O_2)_2H_2O]_2$ with H_2O_2. For a practical example of preparation at pH < 5 and excess H_2O_2, see [1], where the misleading formula $(C_5H_5NH)_2MoO_7$ is used. Single crystals are obtained by crystallizing the $(C_5H_5NH)_2O$-$[MoO(O_2)_2H_2O]_2$ in 30% H_2O_2 solution at −10°C [4].

The compound is triclinic, lattice parameters a = 6.587 ± 0.005, b = 7.620 ± 0.006, c = 10.903 ± 0.009 Å, α = 105.22° ± 0.20°, β = 117.39° ± 0.20°, γ = 100.49° ± 0.20°; Z = 1. Space group P$\bar{1}$–C^1_i (No. 2). For positional and thermal parameters (anisotropic), see the paper, R = 0.029. The structure of the anion is shown in **Fig. 64**, p. 174. The anion has C_i symmetry. The Mo-O bond distances (in Å) are: Mo-O_t = 1.669(6), Mo-O_b = 2.047(6), 2.391(6), Mo-O_p = 1.920(6), 1.925(6), 1.948(6), 1.954(6). For a discussion and other distances and angles, see the paper [4].

The measured density is 2.18 ± 0.01, the calculated 2.192 g/cm³ [4]. Raman frequencies observed are (in cm⁻¹): 943, ν(Mo=O); 910, ν(O–O); 874; 570, ν(MoO_2, asym); 536, ν(MoO_2, sym); 353, 320. The spectrum differs from that of "$(C_5H_5NH)_2MoO_6$" (i.e., possibly $(C_5H_5NH)_2O$-$[MoO(O_2)_2H_2O]_2$) only in the 874 cm⁻¹ frequency which is unassigned [2].

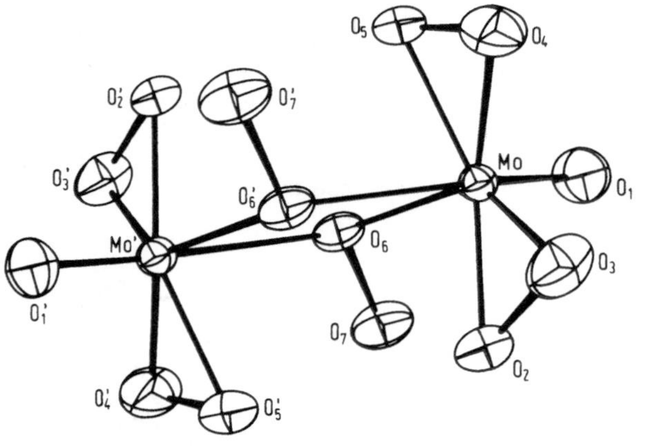

Fig. 64. The structure of the $[MoO(O_2)_2OOH]_2^{2-}$ ion [4].

Experimental results from treatment of "$(C_5H_5NH)_2MoO_7$" over BaO in vacuum by [1] cannot be correct, due to the compositional relationship between $(C_5H_5NH)_2[MoO(O_2)_2OOH]_2$ and $(C_5H_5NH)_2O[MoO(O_2)_2H_2O]_2$.

References:

[1] R. G. Beiles, R. A. Safina, E. M. Beiles (Zh. Neorgan. Khim. **6** [1961] 1612/6; Russ. J. Inorg. Chem. **6** [1961] 825/7). – [2] W. P. Griffith, T. D. Wickins (J. Chem. Soc. A **1968** 397/400). – [3] H. Mimoun, I. Seree de Roch, L. Sajus (Bull. Soc. Chim. France **1969** 1481/92). – [4] J.-M. LeCarpentier, A. Mitschler, R. Weiss (Acta Cryst. B **28** [1972] 1288/98).

2.8.3.3 Bipyridinium Peroxomolybdates

2,2'-Bipyridinium Peroxomolybdate $(C_{10}H_8N_2H_2)MoO_6$

The yellow compound is obtained as a precipitate from the reaction of 2,2'-bipyridine with a solution of "H_2MoO_4" in aqueous H_2O_2.

When heated to 130°C, it loses active oxygen and is converted into the colorless molybdate (p. 171). The compound is sparingly soluble in water; 0.030 g/L at 20°C. The electrical conductivity of the saturated aqueous solution is $2.5 \times 10^{-4}\ \Omega^{-1} \cdot cm^{-1}$. The pH of the solution is in the neutral range and becomes acidic only after half an hour or more, see a figure in the paper. It reacts with NaOH solution only upon heating, e.g., at 80°C.

4,4'-Bipyridinium Peroxomolybdates $(C_{10}H_8N_2H_2)MoO_6$ and $(C_{10}H_8N_2H_2)H_2(MoO_6)_2$

The two compounds have been prepared presumably by a similar method as for the 2,2'-bipyridinium compound, but no details are given. The compounds lose active oxygen on heating to 110°C and are considered to be true peroxides and not peroxohydrates. On treatment with alkali, the bipyridine is recovered unchanged. For potentiometric titration curves of the compounds, see a figure in the paper.

R. G. Beiles, E. M. Beiles (Zh. Neorgan. Khim. **10** [1965] 1618/23; Russ. J. Inorg. Chem. **10** [1965] 883/6).

2.8.3.4 Quinolinium Peroxomolybdates

$(C_9H_7NH)HMoO_6$

For its preparation, peroxomolybdic acid is prepared first by dissolving freshly prepared "H_2MoO_4" (0.5 g) in 30% H_2O_2 (4 mL). Subsequently, alcohol (4 mL) is added, then quinoline (0.6 g) in dilute (1:5) H_2SO_4. The crystals (0.8 g) which gradually separate are filtered off and washed with water and alcohol. The compound is also formed when quinolinium paramolybdate $(C_9H_7NH)_3H_3Mo_7O_{24}$ (p. 171) is reacted with H_2O_2.

The yellowish green compound is a 1:1 salt according to its electrical conductivity in aqueous solutions and, therefore, the formula is given as acid salt. The molar conductivity at dilutions of 500, 1000, 2000 L/mol is 51, 59, 76 $\Omega^{-1}\cdot cm^2$, respectively. The compound contains two active oxygen atoms. It decomposes vigorously when heated [1].

$(C_9H_7NH)HMoO_7$

For its preparation, $(C_9H_7NH)HMoO_6$ is dissolved in a tenfold excess of 30% H_2O_2 and the solvent is evaporated by a stream of air until the greenish yellow crystals separate. They are filtered off and washed with water and alcohol [1].

8-Hydroxyquinolinium Peroxomolybdate $(C_9H_6NO)_2MoO_4$

Contrary to published data (see [2]), 8-hydroxyquinoline in the presence of H_2O_2 reacts with molybdate ions. The 8-hydroxyquinoline (2 g) is dissolved in CH_3COOH solution (1:2, 15 mL) and is added to $(NH_4)_2MoO_4$ (2 g) dissolved in 10% H_2O_2 (30 mL). The precipitate of the compound (3 g) gradually forms. It is filtered off and washed with water and alcohol. Under the microscope, yellow green pyramids and rectangular platelets are distinguished, which are gathered into clusters [1].

It is concluded that the compound contains peroxomolybdenyl cations $MoO_2\cdot 2O^{2+}$ (analogous to the molybdenyl ions MoO_2^{2+}) for which it is assumed that they are in equilibrium with $HMoO_6^-$ in peroxomolybdate solutions. The formula of the compound may, therefore, be written $(C_9H_6NOO)_2MoO_2$ or $(C_9H_6NO)_2MoO_2\cdot 2O$ [1].

The molar electrical conductivity at a dilution of 500 L/mol is 35 $\Omega^{-1}\cdot cm^2$ at 20°C. In aqueous solution, the compound is weakly acidic (pH 5.0) and is gradually hydrolized with the formation of a precipitate [1].

References:

[1] R. G. Beiles, E. M. Beiles (Zh. Neorgan. Khim. **10** [1965] 1618/23; Russ. J. Inorg. Chem. **10** [1965] 883/6). – [2] F. Feigl, I. Raacke (Anal. Chim. Acta **1** [1947] 317/25, 319).

2.8.4 Phosphonium Polymolybdates

Propyltriphenylphosphonium Polymolybdate $[(n\text{-}C_3H_7)(C_6H_5)_3P]_4[Mo_8O_{26}]\cdot H_2O\cdot CH_3CN$
(α-Octamolybdate Type)

The preparation of the compound is not reported.

The compound crystallizes in the monoclinic space group C2/c-C_{2h}^6 (No. 15), lattice parameters a = 28.096(6), b = 14.313(2), c = 27.116(5) Å, β = 121.32(1)°; Z = 4. The crystal structure has been refined to R = 0.031, but the atomic parameters are not given (thermal parameters: anisotropic for all nonhydrogen atoms and isotropic for the hydrogen atoms) [1].

The structure of the polyanion is shown in Fig. 13, p. 17, and **Fig. 65**. Deviations from idealized D_{3d} symmetry are slight. Average Mo-O distances: Mo_I-O_A = 1.696, Mo_I-O_D = 1.904, Mo_I-O_C = 2.425, Mo_{II}-O_C = 1.783, Mo_{II}-O_B = 1.708 Å. Three features of the structure indicate the weakness of the Mo_I-O_C bonds and the potential lability of the MoO_4^{2-} unit within the α-$[Mo_8O_{26}]^{4-}$ ion: (1) The average Mo_I-O_C distance of 2.425 Å implies a bond order of <0.1 according to [2] (according to [3] the bond order would be about 0.3); (2) the significant variation of Mo_I-O_C distances, ranging from 2.369 to 2.444 Å, reflects the ease with which these bonds may be stretched; (3) the average Mo_{II}-O_C distance within the tetrahedral coordination sphere is 1.764 Å, which agrees, within standard deviations, with the average of 1.772 Å found in $Na_2MoO_4 \cdot 2H_2O$ [4]. Thus, the MoO_4^{2-} unit could be reoriented with only a slight deformation of its geometry. The α-$[Mo_8O_{26}]^{4-}$ ion may be accurately represented by the formula $(MoO_4^{2-})_2(Mo_6O_{18})$ which characterizes the weak interactions between the two MoO_4^{2-} ions and a ring of six distorted MoO_4 tetrahedra sharing corners [1].

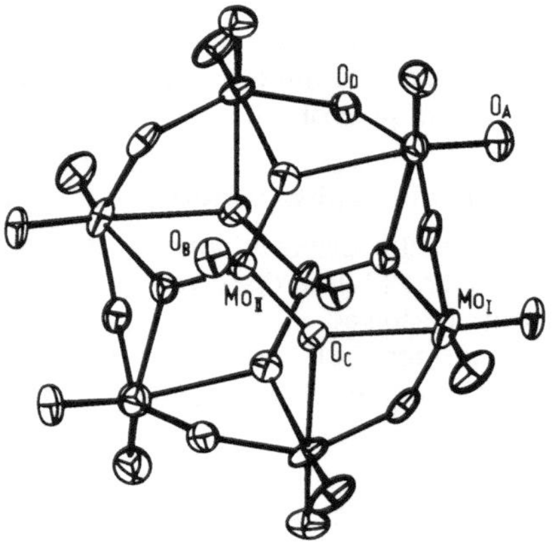

Fig. 65. The structure of the α-octamolybdate ion $[Mo_8O_{26}]^{4-}$ [1].

Anhydrous Phosphonium Polymolybdates

In addition to the phosphonium polymolybdate hydrate described above, the anhydrous tetraphenyl phosphonium hexamolybdate, $[(C_6H_5)_4P]_2[Mo_6O_{19}]$ [5, 7], and the hexakis(dimethylamino)cyclotriphosphazenium hexamolybdate, $\{HN_3P_3[N(CH_3)_2]_6\}_2[Mo_6O_{19}]$ [6], have been prepared and investigated.

References:

[1] V. W. Day, M. F. Fredrich, W. G. Klemperer, W. Shum (J. Am. Chem. Soc. **99** [1977] 952/3). – [2] F. A. Schröder (Acta Cryst. B **31** [1975] 2294/309). – [3] J. Fuchs, I. Knöpnadel (Z. Krist. **158** [1982] 165/79). – [4] K. Matsumoto, A. Kobayashi, Y. Sasaki (Bull. Chem. Soc. Japan **48** [1975] 1009/13). – [5] J. Fuchs (Z. Naturforsch. **28b** [1973] 389/404, 395).

[6] H. R. Allcock, E. C. Bissell, E. T. Shawl (Inorg. Chem. **12** [1973] 2963/8; J. Am. Chem. Soc. **94** [1972] 8603/4). – [7] R. Grase, J. Fuchs (Z. Naturforsch. **33b** [1978] 533/6).

3 Molybdate Hydrates with Alkaline Earth Metals

A review and comparative data for alkali and alkaline earth polymolybdate hydrates are given on pp. 2/47.

3.1 Beryllium Molybdate Hydrates

Older data are given in "Molybdän", 1935, p. 287.

3.2 Magnesium Molybdate Hydrates

Older data are given in "Molybdän", 1935, pp. 288/9.

3.2.1 Magnesium Polymolybdate Hydrates with Mo of Oxidation State <6

Compounds of this type are not reported.

3.2.2 Magnesium Molybdate(VI) Hydrates

3.2.2.1 The $MgO–MoO_3–H_2O$ System

This system has been investigated at 25°C, see **Fig. 66**, p. 178. The following solid compounds have been found to exist in stable equilibria with the solution: $Mg(OH)_2$, $8MgO\cdot MoO_3\cdot 48H_2O$, $4MgO\cdot MoO_3\cdot 12H_2O$, $MgMoO_4\cdot 5H_2O$, $MgO\cdot 3MoO_3\cdot 10H_2O$, MoO_3. The "trimolybdate" 10-hydrate is the only polymolybdate in this sequence [1]. The paramolybdate $3MgO\cdot 7MoO_3\cdot 20H_2O$ was found to exist only in a metastable equilibrium [2]. Because of the slow equilibration when starting from MgO, MoO_3, and H_2O, the initial materials in this investigation were the compounds $MgMoO_4\cdot 5H_2O$, $MgO\cdot 3MoO_3\cdot 10H_2O$, and $Mg_3[Mo_7O_{24}]\cdot 20H_2O$. Selected points on the phase diagram are [1]:

composition of the solution		solid phases in equilibrium
MgO in wt%	MoO_3 in wt%	
0.23	0.64	$Mg(OH)_2 + 8MgO\cdot MoO_3\cdot 48H_2O$
0.55	1.88	$8MgO\cdot MoO_3\cdot 48H_2O + 4MgO\cdot MoO_3\cdot 12H_2O$
3.85	13.15	$4MgO\cdot MoO_3\cdot 12H_2O + MgMoO_4\cdot 5H_2O$
3.50	12.55	$MgMoO_4\cdot 5H_2O$
5.45	40.00	$MgMoO_4\cdot 5H_2O + MgO\cdot 3MoO_3\cdot 10H_2O$
0.145	1.65	$MgO\cdot 3MoO_3\cdot 10H_2O$
0.71	10.12	$MgO\cdot 3MoO_3\cdot 10H_2O + MoO_3$

The **$MgMoO_4$–H_2O subsystem** is investigated from about −2 to 95°C, see **Fig. 67**, p. 178. With the transition points estimated graphically, equilibria are as follows (D = density of the solution, I = $MgMoO_4\cdot 7H_2O$, II = $MgMoO_4\cdot 5H_2O$, III = $MgMoO_4\cdot 2H_2O$):

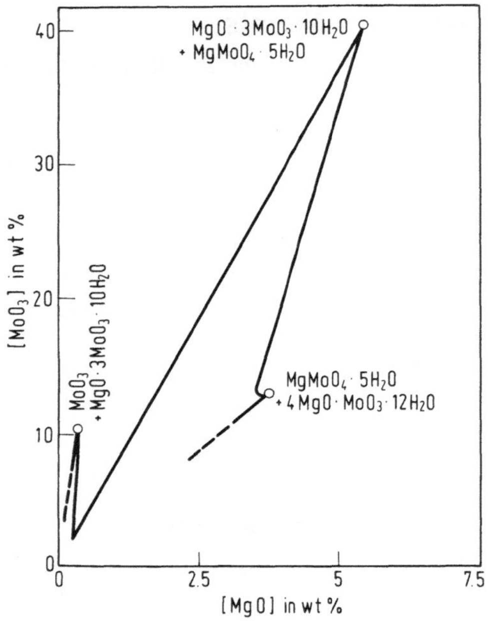

Fig. 66. The $MgO-MoO_3-H_2O$ system at 25°C [1]. (The distinguished points of the solubility curve do not fit well with the coordinates given for these points.)

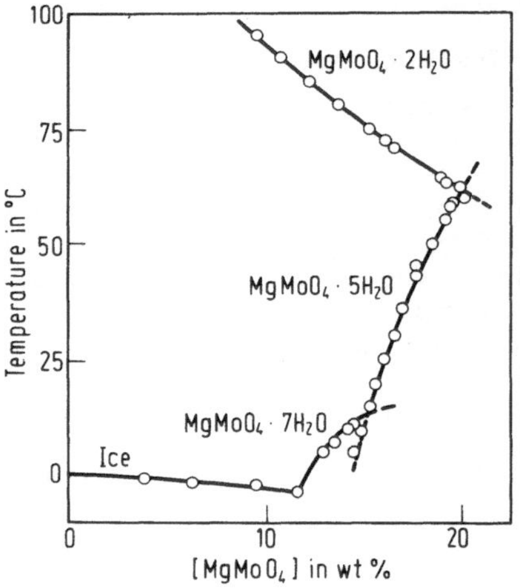

Fig. 67. The $MgMoO_4-H_2O$ subsystem [3].

t in °C	−0.45	−1.25	−1.67	5.0	12.7	15	60.8	63	95
$[MgMoO_4]$ in wt% ...	3.94	9.38	11.55	12.94	15.15	15.26	19.85	19.12	9.38
D in g/cm^3	—	—	1.111	1.127	1.151	1.151	1.192	1.182	1.072
solid compound ...	ice	ice	ice + I	I	I + II	II	II + III	III	III

The solubility of the heptahydrate increases with temperature and the slope for the pentahydrate is also positive but smaller, whereas the dihydrate exhibits retrograde solubility. The heptahydrate reaches equilibrium rapidly from both directions. The pentahydrate dissolves readily but requires more than 1 h with stirring to reach its saturation concentration and its approach from supersaturation is very slow. The dihydrate crystallizes only slowly after a change of temperature. An addition of $MgCl_2$ (26.8 wt%) to the solution lowers the transition temperature between the pentahydrate and dihydrate (by 32.8 K) [3]. The monohydrate $MgMoO_4 \cdot H_2O$ is formed by prolonged (28 to 29 h) heating to 100°C and stirring in a 1 M aqueous solution [4].

Fig. 68 shows the dehydration of the monomolybdates depending on the pressure at a constant temperature of 50°C. At every step, three phases are in equilibrium: water vapor and two solid phases, or water vapor, solution, and a solid phase. Dehydration is also observed at constant pressure on raising the temperature; see the figure for 4 Torr in the paper. In both cases, the trihydrate is observed in addition to the pentahydrate [5]. For the hydrate $MgMoO_4 \cdot 0.5H_2O$, see p. 183.

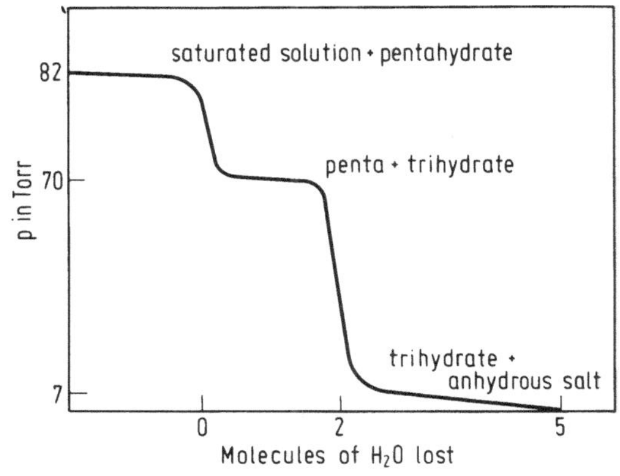

Fig. 68. Dehydration of magnesium monomolybdate hydrates at 50°C and low pressures [5].

References:

[1] G. Teller, J. Byé (Bull. Soc. Chim. France **1959** 1531/2). – [2] G. Teller (Bull. Soc. Chim. France **1959** 1535/6). – [3] J. E. Ricci, W. F. Linke (J. Am. Chem. Soc. **73** [1951] 3603/7). – [4] J. Meullemeestre, É. Pénigault (Bull. Soc. Chim. France **1975** 1925/32). – [5] G. Teller (Bull. Soc. Chim. France **1959** 1533/5).

3.2.2.2 $8MgO \cdot MoO_3 \cdot 48H_2O$

This compound has been found in the MgO-MoO_3-H_2O system (p. 177) and is characterized by its X-ray powder pattern [1] (a line diagram is given in [2]).

On heating at 70°C, a heptahydrate forms which decomposes above 200°C [2].

References:

[1] G. Teller, J. Byé (Bull. Soc. Chim. France **1959** 1531/2). – [2] G. Teller (Bull. Soc. Chim. France **1959** 1533/5).

3.2.2.3 $4MgO \cdot MoO_3 \cdot 12H_2O$

This compound has been found in the MgO-MoO_3-H_2O system (p. 177) and is characterized by its X-ray powder pattern [1] (a line diagram is given in [2]).

On heating at 85°C, a pentahydrate forms which decomposes at 400°C [2].

References:

[1] G. Teller, J. Byé (Bull. Soc. Chim. France **1959** 1531/2). – [2] G. Teller (Bull. Soc. Chim. France **1959** 1533/5).

3.2.2.4 $MgMoO_4 \cdot 7H_2O$

The magnesium molybdate heptahydrate is obtained by evaporation at 0°C of the mother liquor formed in the course of the preparation of the pentahydrate (see below) after filtering off the latter compound. The heptahydrate, in the form of a white material consisting of small crystalline spheres, is dried in the air. The water content (determined thermogravimetrically with several specimens) is between 6.6 and 8.1. It can only be prepared at low temperatures (e.g., at 4°C) and if it is not evaporated to dryness. Otherwise, a mixture with the pentahydrate results, because in the absence of the mother liquor the heptahydrate loses part of its water forming the pentahydrate [1].

For lattice spacings of $MgMoO_4 \cdot 7H_2O$ according to powder diffraction data, see the table in the paper and for the IR absorption spectrum, see the figure. The IR bands in the region 700 to 1100 cm^{-1} in comparison with those of the penta- and trihydrates show that the degree of hydration has little influence on the structure of the molybdate anions [1].

At room temperature, the heptahydrate is unstable [1, 2]. Thermogravimetrically, four steps of dehydration are observed. The first one, starting at room temperature, leads to the pentahydrate; the second (loss of another $2H_2O$) begins at 75°C, the third (loss of $1H_2O$) is at 130°C, and the last step starts at 170°C ($1H_2O$) [1].

References:

[1] J. Meullemeestre, É. Pénigault (Bull. Soc. Chim. France **1975** 1925/32). – [2] G. Teller (Bull. Soc. Chim. France **1959** 1533/5).

3.2.2.5 $MgMoO_4 \cdot 5H_2O$

The compound is prepared in aqueous solution according to $BaMoO_4 + MgSO_4 \cdot 7H_2O \rightarrow BaSO_4 + MgMoO_4 \cdot 5H_2O$. The $BaMoO_4$ is added to the sulfate solution and stirred for 48 h at 60°C. The magnesium molybdate solution is concentrated at 70 to 80°C and finally at room temperature at reduced pressure in the presence of Na_2CO_3 [1].

The well-crystallized compound is prepared in the following manner: $MgO + MoO_3$ (0.1 mol/L each) are added to 50 mL water and the mixture is maintained at 25°C with stirring. After 16 h, practically all solid material is dissolved (save a small amount of MoO_3 which is filtered off). After a time between 24 and 48 h, a precipitate appears, which is separated and dried in the air. The evaporation at room temperature leads to a mixture of the pentahydrate and the heptahydrate. Translucent prisms of the pentahydrate finally often separate, forming a layer, which covers the bottom of the vessel [2].

$MgMoO_4 \cdot 5H_2O$ is triclinic [3], for d values see [2] and for a line diagram see [4]. The lattice parameters are a = 6.529(7), b = 10.706(12), c = 6.341(7) Å, α = 76.44(9)°, β = 109.03(9)°, γ =

90.31(9)°; Z = 2. Space group $P\bar{1}$-C_i^1 (No. 2). The atomic coordinates corresponding to R = 0.065 (anisotropic) are [3]:

atom	position	x	y	z
Mg(1)	1a	0	0	0
Mg(2)	1e	0.5	0.5	0
Mo	2i	0.12850(8)	0.29801(5)	0.67460(9)
O(1)	2i	0.1474(9)	0.1387(5)	0.8375(9)
O(2)	2i	0.3496(8)	0.3750(5)	0.8154(8)
O(3)	2i	0.8824(8)	0.3734(6)	0.6525(9)
O(4)	2i	0.1337(7)	0.2965(5)	0.3980(8)
O(W5)	2i	0.7287(9)	0.1258(5)	0.9116(9)
O(W6)	2i	0.1489(9)	0.0569(5)	0.3057(9)
O(W7)	2i	0.5087(7)	0.3697(5)	0.3069(8)
O(W8)	2i	0.7947(7)	0.4200(5)	0.0219(9)
O(W9)	2i	0.4094(9)	0.8879(6)	0.7235(10)

The two crystallographically different Mg atoms have octahedral coordination by four molecules of water and two oxygen atoms in trans position which belong to molybdate ions, see **Fig. 69** [3]. Each $Mg(H_2O)_4O_2$ octahedron is tied to two MoO_4^{2-} tetrahedra which are bound to other $Mg(H_2O)_4O_2$ octahedra thus forming chains parallel [110]. The chains are held together by the fifth water molecule and by strong hydrogen bridges between oxygen atoms as indicated in Fig. 69, also see a figure in the paper. The MoO_4^{2-} tetrahedra are practically undistorted (T_d symmetry maintained). The mean Mo–O distance, 1.756 Å, corresponds to that determined for a great number of anhydrous molybdates. The $Mg(H_2O)_4O_2$ octahedra are significantly though weakly deformed, cis-angles O–Mg–O range from 87.92° to 92.08° and from 89.00° to 91.00°. The Mg–O distances with O belonging to an MoO_4^{2-} group are shorter than those with O belonging to water molecules (2.024(4) to 2.026(4) Å compared with 2.060(5) to 2.097(5) Å, respectively). In spite of the great similarities between their unit cells, $MgMoO_4 \cdot 5H_2O$ is not isomorphous with $MgSO_4 \cdot 5H_2O$ and $MgCrO_4 \cdot 5H_2O$ [3].

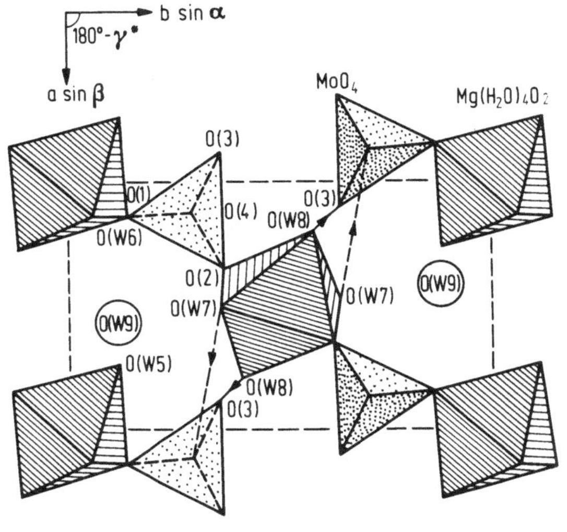

Fig. 69. Projection of the crystal structure of $MgMoO_4 \cdot 5H_2O$ parallel [001] [3].

The measured density is 2.29, the calculated density 2.24 g/cm^3 [3]. The IR absorption spectrum in the range 250 to 1200 cm^{-1} is shown in a figure in [2].

On heating, 3H_2O are lost between 70 and 80°C. The remaining 2H_2O are liberated successively above 120°C [2].

In the $MgMoO_4 \cdot 5H_2O$-$MgSO_4 \cdot 7H_2O$-H_2O system, neither a double salt nor measurable solid solutions are formed at 25°C [5].

References:

[1] M. Audibert, L. Cot, C. Avinens (Compt. Rend. C **273** [1971] 1085/8). – [2] J. Meullemeestre, É. Pénigault (Bull. Soc. Chim. France **1975** 1925/32). – [3] O. Bars, J. Y. Le Marouille, D. Grandjean (Acta Cryst. B **33** [1977] 1155/7). – [4] G. Teller (Bull. Soc. Chim. France **1959** 1533/5). – [5] J. E. Ricci, W. F. Linke (J. Am. Chem. Soc. **73** [1951] 3607/12).

3.2.2.6 $MgMoO_4 \cdot 3H_2O$

In thermogravimetric experiments with $MgMoO_4 \cdot 5H_2O$, the trihydrate is found to form at 77°C and to decompose beginning at 92°C [1]. In the thermogravimetry of the heptahydrate, the trihydrate occurs beginning at 75°C and exists to about 130°C. Contrary to this, it is found that a dry specimen of the pentahydrate heated in this manner is transformed into $MgMoO_4 \cdot 2H_2O$, but if the specimen is wetted, the trihydrate is obtained [2]. For its formation at low pressure, see Fig. 68, p. 179.

The X-ray powder pattern reveals a poor state of crystallization, for lattice spacings, see a table in the paper [2]. A line diagram is given by [1]. The IR spectrum between 250 and 1200 cm^{-1} can be found in [2].

References:

[1] G. Teller (Bull. Soc. Chim. France **1959** 1533/5). – [2] J. Meullemeestre, É. Pénigault (Bull. Soc. Chim. France **1975** 1925/32, 1927).

3.2.2.7 $MgMoO_4 \cdot 2H_2O$

The compound is prepared by refluxing and vigorous stirring of $MoO_3 + MgO$ (1 mol/L each) in water for 24 h. It is obtained as a white precipitate. It is filtered off while hot, washed with ethanol, and dried in the air. Washing with ethanol is important, because a remainder of the mother liquor induces crystallization of the pentahydrate in the time the specimen is dried. The filtration should be carried out at 100°C because the compound is more soluble at lower temperatures [1]. The dihydrate has also been obtained from equimolar amounts of $(NH_4)_6Mo_7O_{24} \cdot 4H_2O$ and MgO made into a paste with H_2O and dried at 110°C for 24 h [2].

The X-ray powder diagram (a list of d values is given in the paper) of the dihydrate resembles that of monoclinic $ZnMoO_4 \cdot 2H_2O$ (p. 216) and the lattice parameters a = 8.579, b = 10.196, c = 5.908 Å, β = 90°23′ are obtained for $MgMoO_4 \cdot 2H_2O$. The IR absorption spectrum (see figure in the paper) also has features similar to that of $ZnMoO_4 \cdot 2H_2O$, especially in the range 700 to 1100 cm^{-1}, i.e., regarding the Mo–O bonds, whereas between 250 and 700 cm^{-1} differences are encountered which stem from the metal-oxygen bonds [1]. For X-ray diffraction and IR absorption of specimens heated at several temperatures up to 900°C, see [2].

On heating to somewhat above 100°C, the dihydrate loses water and the monohydrate is formed [1]. This reaction is accomplished in thermal analysis (DTA and TG) in the interval 110 to 220°C [2].

References:

[1] J. Meullemeestre, É. Pénigault (Bull. Soc. Chim. France **1975** 1925/32). – [2] P. K. Sinhamahapatra, S. K. Bhattacharyya (J. Therm. Anal. **8** [1975] 45/56).

3.2.2.8 $MgMoO_4 \cdot H_2O$ and $MgMoO_4 \cdot 0.5H_2O$

By refluxing and vigorous stirring of an equimolar $MoO_3 + MgO$ suspension in water, the dihydrate forms first and, after 4 d, the monohydrate appears as a new stable phase. It is obtained as a very fine white deposit in 5 d, separated by filtration while hot, washed with ethanol, and dried in the air [1]. $MgMoO_4 \cdot H_2O$ is also prepared by double decomposition of a soluble magnesium salt with an alkali molybdate. It is triclinic and isomorphous with the molybdate monohydrates of Zn (p. 218), Co, Ni, Mn, Fe, for which deficits in H_2O and retention of alkali ions in the lattice as well as a structural disorder are reported [2]. For lattice spacings, see a table in [1], see also [3]. The IR spectrum between 250 and 1200 cm^{-1} is given. It resembles those of the penta- and trihydrate in the anionic part [1].

On heating, the monohydrate begins to lose water at 160°C. It is dehydrated in two steps, the intermediate composition corresponding to $MgMoO_4 \cdot 0.5H_2O$. In thermogravimetric experiments, it is formed at 250°C, and its complete dehydration proceeds gradually up to 400°C. An X-ray powder diagram of the semi-hydrate shows only a few reflections probably due to a poor crystallization, the d values are given. For the IR spectrum, see the paper [1]. According to other authors, the dehydration of the monohydrate proceeds to the anhydrous compound in the interval 250 to 420°C [3].

References:

[1] J. Meullemeestre, É. Pénigault (Bull. Soc. Chim. France **1975** 1925/32). – [2] G. Pezerat (Compt. Rend. C **265** [1967] 368/71). – [3] P. K. Sinhamahapatra, S. K. Bhattacharyya (J. Therm. Anal. **8** [1975] 45/56).

3.2.2.9 $Mg_3[Mo_7O_{24}] \cdot 20H_2O$ (= $3MgO \cdot 7MoO_3 \cdot 20H_2O$), Magnesium Paramolybdate

To prepare the paramolybdate, 37.4 g MgO and 305 g MoO_3 are heated in 1 L H_2O for about 2 h. Dissolution is almost complete. After filtration and rapid cooling, the solution is diluted with 1 L alcohol. A viscous oil separates from the solution which crystallizes after vigorous shaking for 30 min. The crystals are dried at room temperature. The procedure sometimes fails to give the paramolybdate, especially when the solution is shaken for a longer time. Instead, a mixture of the compound $3MgO \cdot 8MoO_3 \cdot 24H_2O$ with a compound assumed to be a "dimolybdate" is formed [1]. The paramolybdate could not be prepared [1] from purely aqueous solutions as described by Ullik [2] (see "Molybdän", 1935, p. 289) although, as can be seen from the correct formula reported, Ullik obviously had prepared this compound.

The compound is characterized by analysis and the X-ray powder diagram (see the paper). The paramolybdate is stable in the dry state [1].

The compound is readily soluble in water (a solution, 40 min vigorously shaken at 25°C, contains 48 wt% MoO_3 and 5.9 wt% MgO ≙ 54 wt% $Mg_3Mo_7O_{24}$). In aqueous solution, the paramolybdate rapidly decomposes, usually to form magnesium "trimolybdate". A second type of decomposition is observed after several hours of vigorous shaking with water in the absence of crystalline nuclei of the "trimolybdate" or in the presence of small amounts of monomolybdate. The product of this decomposition is the (3:8)-molybdate $3MgO \cdot 8MoO_3 \cdot 24H_2O$ [1].

References:

[1] G. Teller (Bull. Soc. Chim. France **1959** 1535/6). – [2] F. Ullik (Sitz.-Ber. Kaiserl. Akad. Wiss. Wien Math. Naturw. Kl. II **55** [1867] 781).

3.2.2.10 $3MgO \cdot 8MoO_3 \cdot 24H_2O$, Magnesium (3:8)-Molybdate

The "trimagnesium octamolybdate" precipitates from highly concentrated aqueous magnesium heptamolybdate solutions after vigorous shaking for several hours. The precipitate is dried on a porous plate. This (3:8)-molybdate occurs sometimes in a mixture with a second compound instead of the paramolybdate (compare the section above).

The compound is characterized by analysis and the X-ray powder diagram (see the paper).

G. Teller (Bull. Soc. Chim. France **1959** 1535/6).

3.2.2.11 $[Mg_2Mo_6O_{20} \cdot 2nH_2O]_\infty$ (= $MgO \cdot 3MoO_3 \cdot nH_2O$), n = 5, 7, 10, 13, or 14; Magnesium "Trimolybdates"

$MgO \cdot 3MoO_3 \cdot 13$ or $14H_2O$. The preparation of this hydrate, described as metastable at ordinary temperatures [1, 2], could not be repeated [3].

$[Mg_2Mo_6O_{20} \cdot 20H_2O]_\infty$ (= $MgO \cdot 3MoO_3 \cdot 10H_2O$). The compound precipitates from a solution of 5 mmol $MgMoO_4 \cdot 5H_2O$ in 50 mL H_2O treated with 10 mmol MoO_3 after about 8 d. The product is separated by filtration and dried in air [3]. A mixture of 0.75 mol (108 g) MoO_3 and 0.25 mol (10 g) MgO in 500 mL H_2O is refluxed for 3 h followed by storage of the mixture at 0°C for several days. The precipitate, obtained in high yield, is filtered and dried in air [3, 4]. The compound could not be obtained by double-exchange reaction between sodium "trimolybdate" and magnesium nitrate [3].

The compound forms fine needles [4] and is characterized by analysis [3, 4], X-ray powder diagram [2] (or the interplanar distances [3]), and infrared spectrum [3] (see the papers). The similarity of the infrared spectrum with those of the "trimolybdates" of sodium, potassium, zinc, etc. shows the same anionic species to be present in all cases [3] (see Fig. 7, p. 13).

TG curves show the 10-hydrate to be stable up to 28°C [2] or 35°C [3] and also show the existence of a 7-hydrate at 56 to 60°C [2] (~87°C [3]) and of a 5-hydrate at 75 to 110°C [2] (100 to 130°C [3]). At a constant temperature of 100°C for 24 h, the 5-hydrate is formed as crystals of good quality. Dehydration is completed at about 240°C. DTA curves confirm the existence of the three steps in the dehydration process [3].

The solubility in water is very small. This "trimolybdate" 10-hydrate is the only polymolybdate found to exist in a stable equilibrium with the solution [4].

$[Mg_2Mo_6O_{20} \cdot 14H_2O]_\infty$ (= $MgO \cdot 3MoO_3 \cdot 7H_2O$). To prepare this compound, the mixture of 1 mol MoO_3 and ⅓ mol MgO in 500 mL H_2O is refluxed with vigorous shaking. After about 1 h, the mixture is an almost limpid solution, strongly supersaturated. The supersaturation can continue several days although, usually, a light white precipitate appears within 24 h. Another week is necessary to get a high yield. The hot solution is filtered, and the solid is washed with alcohol (to avoid admixture of the decahydrate, which can be obtained from the filtrate by cooling) and dried in air. The heptahydrate is the stable phase in equilibrium with the solution at 100°C and, accordingly, is formed only from hot solutions [3].

The compound is characterized by analysis [3], X-ray powder diagram [2] or the interplanar distances, and infrared spectrum [3] (see the papers). The infrared spectrum is similar to that of the 10-hydrate and a typical "trimolybdate" spectrum thus showing the same anionic species to be present as in the other hydrates [3] (see Fig. 7, p. 13).

TG curves show dehydration in two steps, the first beginning at 40°C (although in aqueous suspension the heptahydrate is stable at 100°C). The first step of dehydration is finished at about 130°C (formation of the pentahydrate), the second step at about 250°C (formation of the anhydrous salt). DTA curves confirm the existence of two steps in the dehydration process [3].

$[Mg_2Mo_6O_{20}\cdot10H_2O]_\infty$ (= $MgO\cdot3MoO_3\cdot5H_2O$). This 5-hydrate is formed from the 10-hydrate or 7-hydrate by heating at 100°C for 24 h. It is obtained in the form of crystals of good quality.

The substance is characterized by analysis, the interplanar distances, and infrared spectrum (see the paper). The infrared spectrum is a typical "trimolybdate" spectrum and similar to those of the 10- and 7-hydrates, thus showing the same anionic species to be present as in the other hydrates.

TG curves show a one-step dehydration between 130 and 260°C. DTA curves confirm the one-step dehydration process. On further heating, at 510°C, the compound $MgO\cdot3MoO_3$ decomposes to $2MgO\cdot3MoO_3+MoO_3$ [3].

References:

[1] G. Teller (Diss. Strasbourg Univ. 1959 from [3]). – [2] G. Teller (Bull. Soc. Chim. France **1959** 1533/5). – [3] J. Meullemeestre (Bull. Soc. Chim. France **1978** I-231/5). – [4] G. Teller, J. Byé (Bull. Soc. Chim. France **1959** 1531/2).

3.2.2.12 Magnesium Phase C Polymolybdate[1)]

The reaction of hot, concentrated HNO_3 on solid normal or isopolymolybdates of magnesium or of 1 N HNO_3 with 0.5 M $MgMoO_4$ solution at 100°C for a few minutes yields the magnesium phase C compound in admixture with MoO_3. The magnesium phase C is very unstable and is entirely converted to MoO_3 by boiling in 1 N HCl for 10 min.

M. L. Freedman (J. Chem. Eng. Data **8** [1963] 113/6).

3.2.3 Magnesium Metal Molybdates

3.2.3.1 The $MgMoO_4$-Na_2MoO_4-H_2O System

In this system, in addition to $MgMoO_4\cdot5H_2O$ and $Na_2MoO_4\cdot2H_2O$, the hydrated double salt $Na_2Mg(MoO_4)_2\cdot2H_2O$ is formed at 25°C. It is incongruently soluble with respect to $MgMoO_4\cdot5H_2O$. The curve for saturation with $Na_2MoO_4\cdot2H_2O$ is very short (about 0.1%).

[1)] See the notes added in proof on pp. 7 and 23/4.

Selected values of saturated solution composition (in wt%) and density (in g/cm³) are as follows (m = metastable):

saturated solution $MgMoO_4$	Na_2MoO_4	density	solid phase
13.71	4.84	1.177	$Mg_2MoO_4 \cdot 5H_2O$
12.09	9.14	1.210	$Mg_2MoO_4 \cdot 5H_2O$
10.03	14.86	1.250	$Mg_2MoO_4 \cdot 5H_2O$
9.26	17.18	1.270	$Mg_2MoO_4 \cdot 5H_2O$
8.87	18.45	1.282	$Mg_2MoO_4 \cdot 5H_2O$ (m)
9.28	17.31	1.272	$Mg_2MoO_4 \cdot 5H_2O$ +
9.27	17.27	1.271	$Na_2Mg(MoO_4)_2 \cdot 2H_2O$
8.57	17.91	1.272	$Na_2Mg(MoO_4)_2 \cdot 2H_2O$
6.94	19.85	1.273	$Na_2Mg(MoO_4)_2 \cdot 2H_2O$
1.62	29.30	1.321	$Na_2Mg(MoO_4)_2 \cdot 2H_2O$
0.48	34.74	1.373	$Na_2Mg(MoO_4)_2 \cdot 2H_2O$
0.10	38.26	1.419	$Na_2Mg(MoO_4)_2 \cdot 2H_2O$
0.13	39.25	1.424	$Na_2Mg(MoO_4)_2 \cdot 2H_2O$ +
0.11	39.31	1.429	$Na_2MoO_4 \cdot 2H_2O$

The double salt does not form immediately, requiring, in some cases, several days after seeding. It is suggested that it may become congruently soluble at some temperatures not far from 25°C.

J. E. Ricci, W. F. Linke (J. Am. Chem. Soc. **73** [1951] 3607/12).

3.2.3.2 $Na_2Mg(MoO_4)_2 \cdot 2H_2O$

The compound is formed in the Na_2MoO_4-$MgMoO_4$-H_2O system at 25°C, see above.

$Na_2Mg(MoO_4)_2 \cdot 2H_2O$ is prepared by dissolution of 1 mol $MgMoO_4 \cdot 5H_2O$ per 5 mol $Na_2MoO_4 \cdot 2H_2O$ at 25°C. The double salt is filtered from its mother liquor after 48 h of stirring [1].

The compound is triclinic, lattice parameters a = 6.007, b = 7.382, c = 5.796 Å, α = 100.83°, β = 104.76°, γ = 108.52°, the errors being ±0.005 Å and ±0.05°. For lattice spacings, see a table in the paper. The calculated density is 3.140±0.008 g/cm³. According to the X-ray powder diagrams, $Na_2Mg(MoO_4)_2 \cdot 2H_2O$ is isomorphous with $Na_2Mg(WO_4)_2 \cdot 2H_2O$ and with the compounds $Na_2M^{II}(SeO_4)_2 \cdot 2H_2O$, M^{II} = Mg, Fe, Co, Ni, or Zn [1].

On heating, the hydrated compound is converted into anhydrous $Na_2Mg(MoO_4)_2$ beginning at ~210°C [2]. Solid solution series are formed with $Na_2Mg(WO_4)_2 \cdot 2H_2O$ and with $Na_2Mg(SeO_4)_2 \cdot 2H_2O$ but not with $K_2Mg(MoO_4)_2 \cdot 2H_2O$ [1].

References:

[1] M. Audibert, L. Cot, C. Avinens (Compt. Rend. C **273** [1971] 1085/8). – [2] M. Audibert, S. Peytavin, L. Cot, C. Avinens (Compt. Rend. C **275** [1972] 825/8).

3.2.3.3 $K_2Mg(MoO_4)_2 \cdot 2H_2O$

The compound is prepared by dissolution of the molybdates as with $Na_2Mg(MoO_4)_2 \cdot 2H_2O$ (see above), but stoichiometric amounts of K_2MoO_4 and $MgMoO_4 \cdot 5H_2O$ are dissolved because it is congruently soluble [1].

$K_2Mg(MoO_4)_2 \cdot 2H_2O$ is triclinic, lattice parameters a = 6.477, b = 7.690, c = 5.870 Å, α = 96.66°, β = 108.54°, γ = 110.67°, the errors being ±0.005 Å and ±0.05°. For lattice spacings, see a table in the paper. The calculated density is 3.037 ± 0.008 g/cm³. According to the X-ray powder diagrams, $K_2Mg(MoO_4)_2 \cdot 2H_2O$ is isomorphous with $K_2Mg(WO_4)_2 \cdot 2H_2O$ and with the compounds $K_2M^{II}(CrO_4)_2 \cdot 2H_2O$, M^{II} = Mg, Mn, Co, Ni, Zn, and Cd [1].

On heating, the hydrated compound is converted into anhydrous $K_2Mg(MoO_4)_2$ beginning at ~220°C [2]. Solid solution series are formed with $K_2Mg(WO_4)_2 \cdot 2H_2O$ and $K_2Mg(CrO_4)_2 \cdot 2H_2O$ but not with $Na_2Mg(MoO_4)_2 \cdot 2H_2O$ [1].

References:

[1] M. Audibert, L. Cot, C. Avinens (Compt. Rend. C **273** [1971] 1085/8). – [2] M. Audibert, S. Peytavin, L. Cot, C. Avinens (Compt. Rend. C **275** [1972] 825/8).

3.2.3.4 $(NH_4)_2Mg(MoO_4)_2 \cdot 2H_2O$

The compound is prepared by dissolution of the monomolybdates in the stoichiometric ratio, then by concentration of the resulting solution at 50°C, and crystallization. Single crystals are prepared by slow evaporation at room temperature [1, 2]. They form needles along [001] with hexagonal section [1].

The crystals are monoclinic, lattice parameters a = 6.821(5), b = 14.124(5), c = 5.879(5) Å, β = 110.57(5)°; Z = 2. Space group $P2_1/c\text{-}C_{2h}^5$ (No. 14) [1, 2]. For a list of indexed d values, see [2]. Atomic coordinates according to R = 0.061 (anisotropic) are:

atom	position	x	y	z
N	4e	0.924(3)	0.1340(9)	0.251(3)
Mg	2d	0.5	0	0.5
Mo	4e	0.2820(2)	0.38666(9)	0.4006(3)
O(1)	4e	0.042(2)	0.3239(9)	0.331(3)
O(2)	4e	0.273(3)	0.450(1)	0.138(2)
O(3)	4e	0.305(3)	0.466(1)	0.639(3)
O(4)	4e	0.490(2)	0.307(1)	0.487(3)
O(5)	4e	0.648(3)	0.1297(9)	0.544(3)

The crystal structure of $(NH_4)_2Mg(MoO_4)_2 \cdot 2H_2O$ is of the kröhnkite type, $Na_2Cu(SO_4)_2 \cdot 2H_2O$. Alternating MoO_4 tetrahedra and $MgO_4(OH_2)_2$ octahedra form chains in the c direction, see **Fig. 70**, p. 188. The Mg coordination octahedra are comprised of four oxygen atoms of four MoO_4 tetrahedra and two oxygen atoms of water molecules in a trans position. Each MoO_4 tetrahedron attaches two oxygen atoms of two neighboring octahedra. Thus, a chain results, in which the octahedra are joined to each other by two tetrahedra at each connection. The chains are bound to each other by hydrogen bonds and via the ammonium cations. The Mg coordination octahedron is rather regular; Mg–O distances are between 2.06(1) and 2.12(2) Å. The MoO_4 tetrahedra are slightly deformed with Mo–O between 1.74(1) and 1.78(1) Å. The NH_4^+ cation is surrounded by six oxygen atoms with distances between 2.84(2) and 3.05(2) Å. For other distances and angles, see the paper. The bonding of the water molecules is discussed, together with the IR bands (3220 and 3140 cm^{-1}) [1].

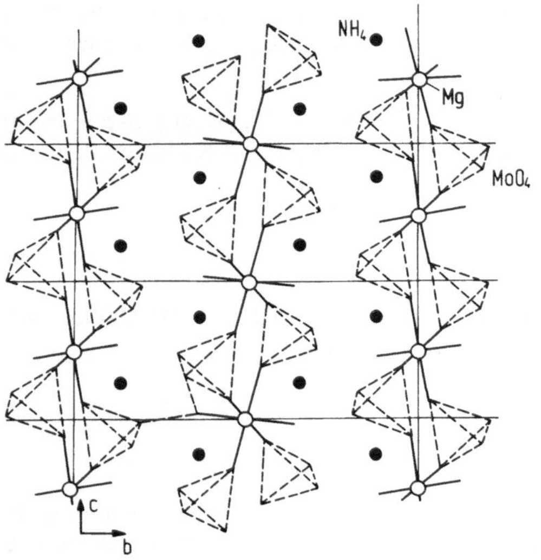

Fig. 70. Projection of the crystal structure of $(NH_4)_2Mg(MoO_4)_2 \cdot 2H_2O$ parallel [100] [1].

The measured density is 2.610 ± 0.008, the calculated 2.608 ± 0.006 g/cm^3 [1, 2]. Decomposition on heating starts at 150°C with loss of H_2O and NH_3 [2].

References:

[1] S. Peytavin, E. Philippot, M. Maurin (J. Solid State Chem. **11** [1974] 71/7). – [2] M. Audibert, S. Peytavin, L. Cot, C. Avinens (Compt. Rend. C **275** [1972] 825/8).

3.2.3.5 $Rb_2Mg(MoO_4)_2 \cdot 2H_2O$

The compound is prepared by dissolving the rubidium and the magnesium monomolybdates in equimolar proportions. The dihydrate precipitates from the concentrated solution on stirring at 50°C.

The X-ray powder pattern of $Rb_2Mg(MoO_4)_2 \cdot 2H_2O$ is analogous to that of the corresponding ammonium salt. It is therefore considered to be also monoclinic, space group $P2_1/c\text{-}C_{2h}^5$ (No. 14). The lattice parameters are (±0.005 Å) a = 6.849, b = 14.263, c = 5.863 Å, β = 111.01° ±0.05°; Z = 2. For a list of indexed d values see the paper. The measured density is 3.420 ±0.008, the calculated 3.424 ± 0.006 g/cm^3. On heating the compound with 10 K/h, its dehydration to anhydrous $Rb_2Mg(MoO_4)_2$ begins at ~150°C.

M. Audibert, S. Peytavin, L. Cot, C. Avinens (Compt. Rend. C **275** [1972] 825/8).

3.2.3.6 $Cs_2Mg(MoO_4)_2 \cdot 4H_2O$

The compound is prepared by dissolution of the monomolybdates in the stoichiometric ratio, concentration of the solution by heating to 50°C, and crystallization. Single crystals are prepared by slow evaporation at room temperature [1, 2].

The crystals are monoclinic, lattice parameters a=7.387(5), b=11.461(5), c=9.409(5) Å, β=122.28(5)°; Z=2. Space group $P2_1/c\text{-}C_{2h}^5$ (No. 14) [1 to 3]. For a list with indexed d values, see [2]. Atomic coordinates corresponding to R=0.068 (anisotropic) are [1, 3]:

atom	position	x	y	z
Cs	4e	0.25670(18)	0.19844(7)	0.07436(7)
Mg	2d	0.5	0	0.5
Mo	4e	0.91807(20)	0.06346(9)	0.24018(14)
O(1)	4e	0.8435(22)	0.0803(18)	0.0326(13)
O(2)	4e	0.1981(17)	0.0748(8)	0.3660(13)
O(3)	4e	0.1712(22)	0.0741(9)	0.7392(17)
O(4)	4e	0.7994(21)	0.1775(9)	0.2893(16)
O(5)	4e	0.5239(22)	0.0538(9)	0.7163(13)
O(6)	4e	0.6330(19)	0.1620(9)	0.4918(14)

The crystal structure of $Cs_2Mg(MoO_4)_2 \cdot 4H_2O$ is similar to those of $Rb_2Mg(CrO_4)_2 \cdot 4H_2O$ and $Cs_2Mg(CrO_4)_2 \cdot 4H_2O$ and shows similarities with $Na_2Zn(SO_4)_2 \cdot 4H_2O$ and $K_2Mn(SO_4)_2 \cdot 4H_2O$ [1 to 3]. The structure contains molybdate tetrahedra and $MgO_2(OH_2)_4$ coordination octahedra having one oxygen in common with each of two MoO_4^{2-} tetrahedra bonded in trans position. These linear arrangements stretch along the b axis; Cs^+ ions are located in gaps of the structure [1, 3], see **Fig. 71** [3]. The $MgO_2(OH_2)_4$ octahedra are rather regular with Mg–O distances between 2.04(1) and 2.12(1) Å. The MoO_4 tetrahedra are slightly deformed, Mo–O

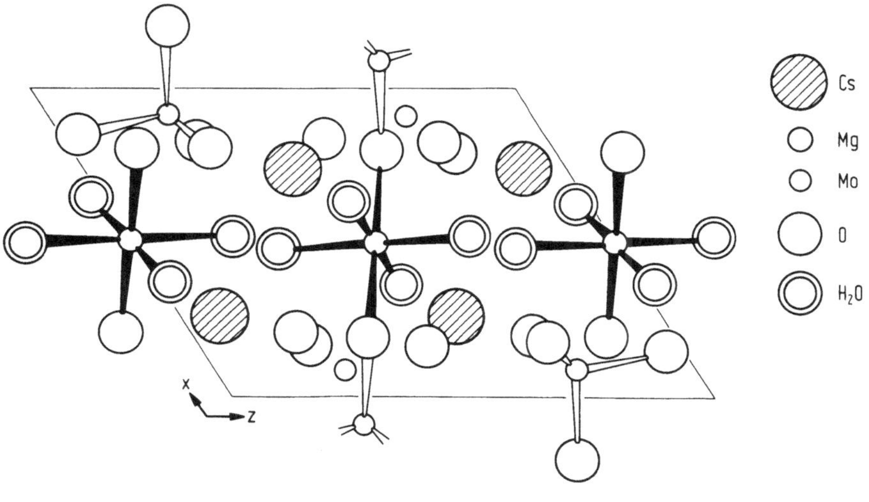

Fig. 71. Projection of the crystal structure of $Cs_2Mg(MoO_4)_2 \cdot 4H_2O$ parallel [010] [3].

being between 1.73(1) and 1.76(1) Å. The Cs^+ cation is surrounded by eleven oxygen atoms with distances between 3.14(1) and 3.41(1) Å. For other distances and bond angles, see the paper. Hydrogen bonding takes place between the water molecules and terminal oxygen atoms of the MoO_4 tetrahedra. From the X-ray structure investigation and the infrared absorption measurements, it is concluded that there are three types of hydrogen bonding [1].

The measured density is 3.366 ± 0.008, the calculated 3.362 ± 0.006 g/cm³ [1 to 3]. On heating the hydrated compound is converted into anhydrous $Cs_2Mg(MoO_4)_2$ beginning at ~105°C [2].

References:

[1] S. Peytavin, E. Philippot, M. Maurin (J. Solid State Chem. **11** [1974] 71/7). – [2] M. Audibert, S. Peytavin, L. Cot, C. Avinens (Compt. Rend. C **275** [1972] 825/8). – [3] S. Peytavin, E. Philippot (Cryst. Struct. Commun. **2** [1973] 355/7).

3.2.3.7 $Na_2Mg_4(OH)_2(MoO_4)_4(H_2O)_2$

A sodium containing sample isomorphous with the corresponding potassium compounds described below has been investigated. Preparation and exact composition are not given. The IR spectrum is discussed together with those of the potassium containing samples.

M. J. Peltre, D. Olivier, H. Pézerat (J. Solid State Chem. **24** [1978] 57/70).

3.2.3.8 $K_2Mg_4(OH)_2(MoO_4)_4(H_2O)_2$

This compound, like the corresponding compounds containing Co, Ni, or Zn (p. 224) instead of Mg, may have a broad range of vacancies up to 10% for Mg, 7% for Mo and K, 1.2% for O; the amount of interstitial H can reach 25%. The cationic vacancies are compensated by anionic vacancies and addition of protons. In spite of this extensive variation, all the compounds crystallize in the same monoclinic type [1]. The K-Mg compounds show the greatest stoichiometric deviations and variations in the lattice parameters of all isomorphous compounds investigated [2].

Samples have been obtained in aqueous solution according to the gross equation $4K_2MoO_4 + 4MgCl_2 + 2KOH + 2H_2O \rightarrow K_2Mg_4(OH)_2(MoO_4)_4(H_2O)_2 + 8KCl$. First, K_2MoO_4 is prepared by slowly dissolving a stoichiometric amount of MoO_3 in 0.02 M KOH at 20°C. Then $MgCl_2$ is added quickly and the mass is heated under reflux at 95°C. Some preparations were carried out at constant pH in 1 M Tris-HCl buffers (Tris = tris-hydroxymethylamino methane). To alter the composition of the compounds, the duration of heating and the concentrations of the MoO_4^{2-}, Mg^{2+}, K^+, and H^+ have been varied. The variation of the composition vs. pH is shown in **Fig. 72** [1, 2].

The compounds are monoclinic. With increasing cell content, the lattice parameters a, b, and β, and the cell volume V increase, the c parameter decreases (see figures in [1]): a = 9.515 to 9.739, b = 6.457 to 6.484, c = 7.882 to 7.831 Å, β = 115.40° to 116.77°, V = 437.5 to 441.5 Å³ (Z = 2), see table in the paper [2].

The crystal structure of the compounds with Mg corresponds to those of the compounds with Co, Ni, or Zn (see p. 224). The atomic coordinates for the Co compound are given in [5]. Space group $C2/m-C_{2h}^3$ (No. 12). The structure is described as a pile of layers parallel (010). The layers are connected by the potassium cations and probably by protons. Each layer is composed of rows of MoO_4^{2-} tetrahedra, and a sheet containing three types of irregular octahedra: The first type contains the Mg atoms. The second type has no cations and shares two faces with two MoO_4^{2-} tetrahedra, which are placed on each side of the layer. The third type also has no cations but is located differently from the second type. The octahedral sheet possesses a structure closely resembling that of the brucite sheet ($Mg(OH)_2$) [1, 2].

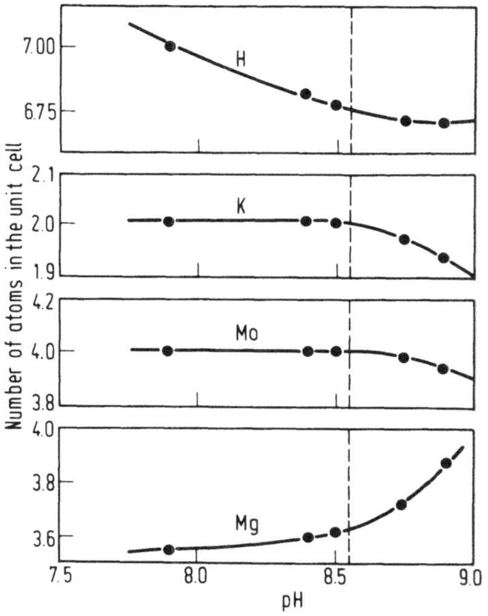

Fig. 72. H, K, Mo, and Mg contents in the unit cell of $K_2Mg_4(OH)_2(MoO_4)_4(H_2O)_2$ vs. pH of the preparation [2].

From the IR absorption spectra, it is concluded that there are two H_2O and two OH groups in the unit cell of the stoichiometric compound [3]. A discussion of the lacunary structure of the nonstoichiometric compounds can be found in [2]; for the role of the compensating OH groups, see [3].

The IR spectra in the range 1000 to 250 cm^{-1} show bands of the vibrations of the MoO_4 tetrahedra, the MgO_6 octahedra, and the H_2O and OH groups as follows:

specimen	ν_1		ν_3			W_α	W_β		ν_2 and ν_1'	
K–Mg	925		870		795	700			440 to 420	
	925	880	865		800	700			440	415
	925		865		780	715	625	(490 sh)	450	
Na–Mg	935		875	840	790		635			425

specimen	ν_3'		ν_4		ν_2'
K–Mg	355 to 345		310	285	
	355 and 335		310	285	
	373		310	290	
Na–Mg	375	345	310	292	282

The frequencies ν_1, ν_2, ν_3, and ν_4 are vibrations of the MoO_4 tetrahedra, ν'_1, ν'_2, ν'_3, are those of the MgO_6 octahedra, and W_α and W_β are bands belonging to OH bonds (α designates groups in stoichiometric compounds and β those for compensation in nonstoichiometric compounds). At higher wave numbers, $(OH)_\alpha$ valence vibrations are found mainly at about 3400 cm^{-1} and about 3280 to 3200 cm^{-1}. In specimens with a high number of vacancies, an $(OH)_\beta$ band appears at 3100 cm^{-1}. In the range of deformation vibrations of H_2O and MgOH, two principal bands are observed at 1600 and 1110 cm^{-1}. At a high number of vacancies, an additional band at 1540 cm^{-1} appears [3].

According to thermogravimetric analysis, dehydration of the nonstoichiometric samples proceeds in two main steps, but in the case of a high number of Mg vacancies in only one step. The DTA curves show endothermal effects at 395°C (one peak) or at lower temperatures, the lowest reported at 310 and 360°C (two peaks). The latter specimen has the most Mg vacancies of all the specimens investigated, while the first is near to stoichiometry in Mg but has vacancies in H, K, and Mo (see table in the paper). For the compounds rich in Mg cation vacancies, the existence of another H_2O type resulting from the presence of pairs of adjacent vacancies is concluded from the DTA and TG results. For compounds poor in Mg vacancies, the presence of new OH groups is inferred, which compensate either the isolated Mg vacancies or the charge deficiency created by K, Mo, and O vacancies. The dehydrated samples have been investigated by X-ray powder diffraction and IR spectroscopy [4].

References:

[1] M. J. Peltre, H. Pézerat (J. Less-Common Metals **36** [1974] 61/9). – [2] M. J. Peltre, H. Pézerat (J. Solid State Chem. **23** [1978] 19/32). – [3] M. J. Peltre, D. Olivier, H. Pézerat (J. Solid State Chem. **24** [1978] 57/70). – [4] M. J. Peltre, H. Pézerat (J. Solid State Chem. **26** [1978] 245/54). – [5] H. Pézerat (Bull. Soc. Franc. Mineral. Crist. **90** [1967] 549/57).

3.3 Calcium Molybdate Hydrates

Older data are given in "Molybdän", 1935, pp. 291/3.

3.3.1 Calcium Polymolybdate Hydrates with Mo of Oxidation State <6

Compounds of this type are not reported.

3.3.2 Calcium Molybdate(VI) Hydrates

Remarks

Most of the calcium polymolybdate hydrates described in this section are very ill-defined and poorly characterized products. In [1], the group of Mokhosoev and coworkers gives, in addition to the polymolybdates described below, the conditions of the preparation and compositions of a large number of other solid phases in the systems $Ca(NO_3)_2$–Na_2MoO_4–H_2O and $Ca(NO_3)_2$–$(NH_4)_6Mo_7O_{24}$–H_2O. In a later paper [2], the solid phases of [1] were investigated by several techniques. As a result of chemical phase analyses, determinations of the conductivity of aqueous extracts, X-ray diffraction patterns of the products in the air-dry state and after heating at 400°C, and the NMR and infrared spectra, the authors conclude that calcium polymolybdate hydrates precipitated from acid solutions are not individual compounds but contain admixtures of $CaMoO_4$ and "H_2MoO_4". It is not clear whether this statement also refers to the compounds of Mokhosoev and coworkers described in Sections 3.3.2.2 to 3.3.2.11. It seems that at least the products $2CaO \cdot 9MoO_3 \cdot 14.8H_2O$, $CaO \cdot 6MoO_3 \cdot 11.7H_2O$, $2CaO \cdot 15MoO_3 \cdot 23H_2O$, and $CaO \cdot 8MoO_3 \cdot 11.9H_2O$ [3] are assumed to be definite compounds.

In the descriptions of the preparation procedures given by Mokhosoev and coworkers, the pH of the molybdate solution and the alkaline earth metal solution sometimes differ considerably. It should be noted that only the pH of the strongly buffered molybdate solution determines the final pH of the solution since the solution of the alkaline earth metal cation is nearly unbuffered.

References:

[1] V. I. Krivobok, M. V. Mokhosoev (Ukr. Khim. Zh. **35** [1969] 1035/8; Soviet Progr. Chem. **35** No. 10 [1969] 22/5). – [2] V. I. Krivobok, M. V. Mokhosoev, A. P. Nakhodnova, T. I. Goloperova (Zh. Prikl. Khim. **45** [1972] 1691/6; J. Appl. Chem. [USSR] **45** [1972] 1766/71). – [3] V. I. Krivobok, M. V. Mokhosoev, A. P. Nakhodnova, R. M. Sedneva, G. Ya. Samsonova (Ukr. Khim. Zh. **38** [1972] 1201/6; Soviet Progr. Chem. **38** No. 12 [1972] 6/9).

3.3.2.1 $\mathbf{CaMoO_4 \cdot nH_2O}$, n = 0.17 to 1

In contrast to the magnesium molybdate, the calcium molybdate (as with the molybdates of Sr and Ba) does not form definite hydrates [1].

A material very similar to the natural mineral powellite (anhydrous $CaMoO_4$) in physical and optical properties, chemical composition, and crystal structure is obtained at 95 to 100°C and pH ≈ 7 from solutions containing Ca^{2+} and MoO_4^{2-}. The products mostly have a substantial content of crystallization water, ranging from 0.17 to 0.63 H_2O, in the structure [2]. On preparation at room temperature, the water contents were, e.g., 4.87 wt% H_2O (4.30 wt% theoretical for the semihydrate) [3] and 4.6 wt% (0.54 H_2O) [1]. Even higher water contents, up to 1 H_2O, are reported depending on the preparation conditions [4].

Water containing products have been obtained when $CaCl_2$ and Na_2MoO_4 in the stoichiometric ratio are dissolved separately, the solutions heated to boiling, and mixed. After cooling, the mixture is maintained for two weeks, then filtered, and the precipitate dried in the air. The parent solution may also be heated in an autoclave for 13 h at 250°C and 40 atm. In all these cases, spherulites are obtained. Small prismatic microcrystals, hundredths of a millimeter in length, with a water content of 5.59 wt% (0.63 H_2O) are formed in a diluted solution in the presence of Fe^{3+}, because the foreign ions prevent instantaneous formation of the calcium molybdate precipitate. $CaMoO_4 \cdot nH_2O$ also crystallizes from weakly acid and weakly alkaline solutions [2]. Precipitation from mixtures of equimolar $Ca(NO_3)_2$ and Na_2MoO_4 solutions in 3 to 4 h gives a yield of 70 to 92% of $CaMoO_4 \cdot nH_2O$. A 15 to 20% excess of $Ca(NO_3)_2$ increases the yield. Ammonium paramolybdate, $(NH_4)_6Mo_7O_{24} \cdot 4H_2O$, can also be used instead of Na_2MoO_4. At 25°C and pH 6 to 9, the normal calcium molybdate is obtained with water contents varying between 0.5 and 0.7 H_2O and at 90°C between 0.5 and 1 H_2O [4]. By another method, CaO or $Ca(OH)_2$ and MoO_3 are stirred together with H_2O in stoichiometric amounts of 0.1 mol/L for 10 d at 25°C. After filtering, a product $CaMoO_4 \cdot 0.54H_2O$ is obtained. With the same starting materials at 2 mol/L by stirring and heating under reflux for 4 d, $CaMoO_4$ with 2.0 wt% H_2O (0.23 H_2O) results [1].

The $CaMoO_4 \cdot nH_2O$ products, like the mineral powellite (isomorphous with scheelite, $CaWO_4$), are tetragonal, space group $I4_1/a\text{-}C_{4h}^6$ (No. 88) [1, 2]. The unit cell dimensions increase with the water content [2]. The following values are reported:

n	0.174	0.23	0.472	0.54	0.631
wt% H_2O	1.55	2.0	4.14	4.6	5.59
a in Å	5.226	5.236 ± 0.002	5.226	5.231 ± 0.002	5.235
c in Å	11.434	11.455 ± 0.005	11.431	11.463 ± 0.006	11.442
Ref.	[2]	[1]	[2, 5]	[1]	[2, 5]

For indexed lattice spacings of the sample with 5.59 wt% H_2O, see [2]. Upon heating at 900°C for 30 min, the parameter a is reduced to 5.225 Å and c to 11.430 Å for the preparations with 4.14 and 5.59 wt% H_2O [2, 5].

From the results of IR absorption measurements, one might infer that some preparations (with 4.14 wt% H_2O) have the scheelite-type structure and others (with 1.55 or 5.59 wt% H_2O) the wolframite-type structure, (Mn, Fe)WO_4. But distortion of part of the scheelite-type structure is more likely [2]. For the IR spectrum of $CaMoO_4 \cdot 0.5H_2O$, see [7]. From ^{1}H NMR measurements, it is concluded that $CaMoO_4 \cdot 0.69H_2O$ contains adsorbed water and OH groups [8].

The density is 6.88 g/cm^3 for the sample with 5.59 wt% H_2O and 7.01 g/cm^3 for that with 4.14 wt% H_2O [2].

The spherulitic products with 1.55 and 4.14 wt% H_2O are colorless or light brown in polarized light. The sample with 4.14 wt% H_2O has refractive indices between 1.904 and 1.911, that with 5.59 wt% H_2O values of 1.883 to 1.904 [2].

Aging of the $CaMoO_4 \cdot nH_2O$ probably would result in partial or complete removal of the water. DTA curves show a sharp endothermic effect at 180°C, indicating loss of water [2]. For DTA and TG curves of $CaMoO_4 \cdot 0.8H_2O$ with one peak at 140°C, see [7]. On thermal analysis, $CaMoO_4 \cdot 0.54H_2O$ (4.6 wt% H_2O, prepared at 25°C) is dehydrated in two steps, first at 40 to 200°C, then at 200 to 700°C. Material with 2 wt% H_2O (prepared at 100°C) shows only the dehydration step at 200 to 700°C [1].

The solubility of $CaMoO_4 \cdot nH_2O$ samples at room temperature in solutions of various pH values was investigated. Stability of the compounds has a maximum at pH 4 to 8, see table and figure in the paper [2]. The heat of solution of $CaMoO_4 \cdot 0.5H_2O$ in H_2O is 9.5 kcal/mol [6]. In uranium containing calcium molybdate hydrate, no isomorphous replacement of Ca by U or of MoO_4^{2-} by UO_4^{2-} is observed [5].

In addition to the stoichiometric $CaO:MoO_3$ (1:1) compound, some products with excess MoO_3 have also been found in the $Na_2MoO_4\text{-}Ca(NO_3)_2\text{-}H_2O$ and $(NH_4)_6Mo_7O_{24}\text{-}Ca(NO_3)_2\text{-}H_2O$ systems: $CaO \cdot 1.07MoO_3 \cdot 0.7H_2O$, $CaO \cdot 1.09MoO_3 \cdot 0.8H_2O$, and $CaO \cdot 1.15MoO_3 \cdot 0.6H_2O$. From ^{1}H NMR, it is concluded that in $CaO \cdot 1.07MoO_3 \cdot 0.7H_2O$ 0.1 H_2O is adsorbed, 0.1 H_2O is water of crystallization, and 0.5H_2O are in the OH form. DTA and TG heating curves are given for $CaO \cdot 1.09MoO_3 \cdot 0.8H_2O$ and $CaO \cdot 1.15MoO_3 \cdot 0.6H_2O$ [7].

References:

[1] J. Meullemeestre (Bull. Soc. Chim. France **1978** I 95/100). – [2] I. G. Zhil'tsova, L. N. Karpova, G. A. Sidorenko, A. A. Valueva, A. D. Dara (Geokhimiya **1969** 147/56; Geochem. Intern. **6** [1969] 104/13). – [3] V. I. Spitsyn, I. A. Savich (Zh. Obshch. Khim. **22** [1952] 1278/81; J. Gen. Chem. [USSR] **22** [1952] 1323/4). – [4] V. I. Krivobok, M. V. Mokhosoev (Ukr. Khim. Zh. **35** [1969] 1035/8; Soviet Progr. Chem. **35** No. 10 [1969] 22/5). – [5] A. D. Dara, G. A. Sidorenko (Geokhimiya **1968** 572/8; Geochem. Intern. **5** [1968] 489).

[6] N. A. A. Mumallah, W. J. Popiel (Anal. Chem. **46** [1974] 2055/6). – [7] V. I. Krivobok, M. V. Mokhosoev, A. P. Nakhodnova, T. I. Goloperova (Zh. Prikl. Khim. **45** [1972] 1691/6; J. Appl. Chem. [USSR] **45** [1972] 1766/71). – [8] V. V. Mank, V. I. Krivobok, I. V. Matyash, M. V. Mokhosoev (Zh. Strukt. Khim. **8** [1967] 227/32; J. Struct. Chem. [USSR] **8** [1967] 196/200).

3.3.2.2 $CaO \cdot 2MoO_3 \cdot 1.6(1.5)H_2O$, Calcium "Dimolybdate"

The compound is precipitated by dropwise addition of ammonium molybdate solution of pH 5.0 (or 6.0) (280 g $(NH_4)_6Mo_7O_{24} \cdot 4H_2O$/L) to calcium nitrate solution of pH 4.3 (280 g $Ca(NO_3)_2 \cdot 4H_2O$/L) at 90°C using 20% excess of the calcium salt calculated to give normal calcium molybdate. The suspension is stirred for a period of 3 to 4 h. The compound is a loose, white powder and is X-ray-amorphous [1].

The shapes of the resonance lines of ^1H NMR spectra obtained at 295 and 77 K from samples dried at characteristic temperatures (obtained from the thermographic curve) indicate that water is present in two states: as adsorbed H_2O ($\Delta H \approx 0.3$ G) and as OH groups ($\Delta H \approx 3$ G), the ratio being 0.6 :1. These conclusions are confirmed by infrared spectra. From these results, the formula of the compound is written $CaMoO_4 \cdot H_2MoO_4 + 0.6H_2O$ (adsorbed) [2, 3].

Thermogravimetric investigation shows that elimination of water is stepwise with an increase in temperature [2, 3], endothermic effects occurring at 128 and 378°C. An exothermic effect between 378 and 586°C is attributed to the formation of anhydrous isopolymolybdate, which melts incongruently at 744 [1] or 764°C [3] forming the normal calcium molybdate and molybdenum trioxide [1].

The compound is insoluble in cold water, sparingly soluble in hot water, and readily soluble in hydrochloric acid (1:3) [1].

References:

[1] M. V. Mokhosoev, Z. E. Batura, V. I. Krivobok (Izv. Akad. Nauk SSSR Neorgan. Materialy **3** [1967] 133/7; Inorg. Materials [USSR] **3** [1967] 108/11). – [2] V. V. Mank, I. V. Matyash, M. V. Mokhosoev, Z. E. Batura (Dokl. Akad. Nauk SSSR **171** [1966] 1113/5; Dokl. Chem. Proc. Acad. Sci. USSR **166/171** [1966] 1181/3). – [3] V. V. Mank, V. I. Krivobok, I. V. Matyash, M. V. Mokhosoev (Zh. Strukt. Khim. **8** [1967] 227/32; J. Struct. Chem. [USSR] **8** [1967] 196/200).

3.3.2.3 $3CaO \cdot 7MoO_3 \cdot 3H_2O$, Calcium Paramolybdate

The compound is precipitated by dropwise addition of ammonium molybdate solution of pH 6.0 (280 g $(NH_4)_6Mo_7O_{24} \cdot 4H_2O$/L) to calcium nitrate solution of pH 4.3 (290 g $Ca(NO_3)_2 \cdot 4H_2O$/L) at 90°C using 20% excess of the calcium salt calculated to give normal calcium molybdate. The suspension is stirred for 3 to 4 h. The compound is an X-ray-amorphous, loose, white powder.

Thermogravimetric investigation shows that elimination of water is terminated at 380°C, endothermic effects occurring at 148 and 368°C. An exothermic effect between 368 and 552°C is attributed to the formation of anhydrous isopolymolybdate, which melts incongruently at 732°C forming the normal calcium molybdate and molybdenum trioxide.

The compound is insoluble in cold water, sparingly soluble in hot water, and readily soluble in hydrochloric acid (1:3).

M. V. Mokhosoev, Z. E. Batura, V. I. Krivobok (Izv. Akad. Nauk SSSR Neorgan. Materialy **3** [1967] 133/7; Inorg. Materials [USSR] **3** [1967] 108/11).

3.3.2.4 $[Ca_2Mo_6O_{20}\cdot 2nH_2O]_\infty$ ($=CaO\cdot 3MoO_3\cdot nH_2O$), n=1, 3, 4, 5, 6; Calcium "Trimolybdates"

$[Ca_2Mo_6O_{20}\cdot 12H_2O]_\infty$ ($=CaO\cdot 3MoO_3\cdot 6H_2O$). This compound is prepared by double-exchange reaction between 5 mmol sodium "trimolybdate" and 50 mL 0.1 M $CaCl_2$ solution. The mixture is vigorously shaken for 3 to 10 d at 25°C and then filtered. The salt crystallizes from the solution after several months storage at room temperature in the form of big conglomerates. The solution shows a propensity for supersaturation. Crystallization is accelerated by the addition of some seed crystals of the salt; however, even in this case, one can obtain more crystal fractions for a long time (up to 12 months). The crystals are dried between filter paper. According to the author's [1] view, this is the only method to get a "trimolybdate" of calcium.

The compound was characterized by analysis, interplanar distances, and infrared spectrum (see the paper). The latter is a typical "trimolybdate" spectrum, hence, the anionic species is assumed to be the same as in the "trimolybdates" of other cations [1].

TG and DTA experiments show that dehydration starts at 35°C. Between 100 and 140°C, the trihydrate is stable and, between 200 and 220°C, the monohydrate is stable. Dehydration is terminated at ~340°C. At this temperature, normal calcium molybdate and molybdenum trioxide are formed [1].

$CaO\cdot 3MoO_3\cdot 5H_2O$. The compound is precipitated by dropwise addition of an ammonium molybdate solution of pH 6.0 (277 g $(NH_4)_6Mo_7O_{24}\cdot 4H_2O$/L) to a calcium nitrate solution of pH 6.0 (443 g $Ca(NO_3)_2\cdot 4H_2O$/L) at 20°C using 20% excess of the calcium salt calculated to give normal calcium molybdate. The suspension is stirred for 3 to 4 h [2]. The compound has also been obtained from the reactants at pH 5 and 25°C [3]. The X-ray-amorphous compound is a loose, white powder [2].

The product was merely characterized by analysis.

$CaO\cdot 3MoO_3\cdot 4(3)H_2O$. The conditions of preparation for these two hydrates are identical. The compounds are precipitated by dropwise addition of ammonium molybdate solution of pH 5.0 (280 g $(NH_4)_6Mo_7O_{24}\cdot 4H_2O$/L) to a calcium nitrate solution of pH 4.3 (280 g $Ca(NO_3)_2\cdot 4H_2O$/L) at 90°C using 20% excess of the calcium salt. The suspension is stirred for 3 to 4 h. The X-ray-amorphous products are loose, white powders [2].

Thermogravimetric investigation of the 4-hydrate shows that elimination of water is terminated at 380°C, endothermic effects occurring at 190, 310, and 378°C. An exothermic effect between 378 and 512°C is attributed to the formation of anhydrous isopolymolybdate, which melts incongruently at 747°C forming the normal calcium molybdate and molybdenum trioxide [2].

The hydrates are insoluble in cold water, sparingly soluble in hot water, and readily soluble in hydrochloric acid (1:3) [2].

$[Ca_2Mo_6O_{20}\cdot 6H_2O]_\infty$ ($=CaO\cdot 3MoO_3\cdot 3H_2O$). This "trimolybdate" trihydrate occurs as a dehydration product when $CaO\cdot 3MoO_3\cdot 6H_2O$ is heated at 100 to 140°C (see above). The compound was characterized by its infrared spectrum (which is a typical "trimolybdate" spectrum) and the interplanar distances (see the paper) [1].

$[Ca_2Mo_6O_{20}\cdot 2H_2O]_\infty$ ($=CaO\cdot 3MoO_3\cdot H_2O$). This compound occurs as a dehydration product when $CaO\cdot 3MoO_3\cdot 6H_2O$ is heated at 200 to 220°C (see above). The compound was characterized by its infrared spectrum (which is a typical "trimolybdate" spectrum) and the interplanar distances (see the paper) [1].

References:

[1] J. Meullemeestre (Bull. Soc. Chim. France **1978** I 236/42). – [2] M. V. Mokhosoev, Z. E. Batura, V. I. Krivobok (Izv. Akad. Nauk SSSR Neorgan. Materialy **3** [1967] 133/7; Inorg. Materials [USSR] **3** [1967] 108/11). – [3] V. I. Krivobok, M. V. Mokhosoev (Ukr. Khim. Zh. **35** [1969] 1035/8; Soviet Progr. Chem. **35** No. 10 [1969] 22/5).

3.3.2.5 $CaO\cdot 4MoO_3\cdot nH_2O$, n = 4, 6, 7, 14, 15; Calcium "Tetramolybdates"

$CaO\cdot 4MoO_3\cdot 14(15)H_2O$. These compounds are precipitated by dropwise addition of ammonium molybdate solution of pH 5.0 (or 4.0) (277 g $(NH_4)_6Mo_7O_{24}\cdot 4H_2O$/L) to a calcium nitrate solution of pH 5.0 (or 4.0) (443 g $Ca(NO_3)_2\cdot 4H_2O$/L) at 20°C using 20% excess of the calcium salt calculated to give normal calcium molybdate. The suspension is stirred for 3 to 4 h. The X-ray-amorphous compounds are loose, white powders [1].

The compounds were merely characterized by analysis.

$CaO\cdot 4MoO_3\cdot 7H_2O$. This compound has been obtained from ammonium paramolybdate and calcium nitrate solutions at pH 4 and 25°C. The precipitate has a mole ratio of $CaO:MoO_3$ = 1:3.8 [2].

$CaO\cdot 4MoO_3\cdot 4(6)H_2O$. The conditions of preparation for these two hydrates are identical. The compounds are precipitated by dropwise addition of ammonium molybdate solution of pH 4.0 (280 g $(NH_4)_6Mo_7O_{24}\cdot 4H_2O$/L) to a calcium nitrate solution of pH 4.3 (280 g $Ca(NO_3)_2\cdot 4H_2O$/L) at 90°C using 20% excess of the calcium salt. The suspension is stirred for 3 to 4 h. The X-ray-amorphous products are loose, white powders [1].

Thermogravimetric investigation of the 6-hydrate shows that elimination of water is terminated at 380°C, endothermic effects occurring at 168 and 376°C. An exothermic effect between 376 and 542°C is attributed to the formation of anhydrous isopolymolybdate, which melts incongruently at 688°C forming the normal calcium molybdate and molybdenum trioxide [1].

References:

[1] M. V. Mokhosoev, Z. E. Batura, V. I. Krivobok (Izv. Akad. Nauk SSSR Neorgan. Materialy **3** [1967] 133/7; Inorg. Materials [USSR] **3** [1967] 108/11). – [2] V. I. Krivobok, M. V. Mokhosoev (Ukr. Khim. Zh. **35** [1969] 1035/8; Soviet Progr. Chem. **35** No. 10 [1969] 22/5).

3.3.2.6 $2CaO\cdot 9MoO_3\cdot 14.8H_2O$ (= $Ca_2Mo_9O_{29}\cdot 14.8H_2O$)

The compound was isolated from the heterogeneous product $CaO\cdot 6MoO_3\cdot 8H_2O$, obtained by precipitation from aqueous $Ca(NO_3)_2$ and $(NH_4)_6Mo_7O_{24}$ solutions. The precipitate was leached by soaking it with a solvent (presumably water) for 6 h with continuous stirring. The solution was separated from the insoluble components and subjected to repeated freezing at −1 to −2°C. The precipitated crystals were separated and dried to constant weight at room temperature. (The exact conditions of preparation of $CaO\cdot 6MoO_3\cdot 8H_2O$ are not reported.)

X-ray diffraction analysis showed the (air-dried) compound to be amorphous. The NMR spectra of the air-dried hydrated compound and of the compound dehydrated at different temperatures (150 and 250°C) indicate the protons to be present as mobile water molecules (ΔH = 0.6 G), hydroxyl groups (ΔH = 2.5 G), and water of crystallization (ΔH = 11 G). According to these results, the air-dried hydrated polymolybdate is assumed to be an "acid" salt which contains hydroxyl groups bound to molybdenum atoms, as well as crystallization and adsorbed water. For a model of the structure assuming hexacoordinated molybdenum, see the paper.

TG and DTA experiments show the elimination of $6.8\,H_2O$ between 25 and 180°C, $7.2\,H_2O$ between 180 and 360°C, and $0.8\,H_2O$ between 310 and 400°C, accompanied by endothermic effects. The anhydrous product is assumed to be a calcium polymolybdate.

V. I. Krivobok, M. V. Mokhosoev, A. P. Nakhodnova, R. M. Sedneva, G. Ya. Samsonova (Ukr. Khim. Zh. **38** [1972] 1201/6; Soviet Progr. Chem. **38** No. 12 [1972] 6/9).

3.3.2.7 $CaO \cdot 5MoO_3 \cdot nH_2O$, n = 4, 7, 18; Calcium "Pentamolybdates"

$CaO \cdot 5MoO_3 \cdot 18H_2O$. The compound is precipitated by dropwise addition of ammonium molybdate solution of pH 3.0 (277 g $(NH_4)_6Mo_7O_{24} \cdot 4H_2O$/L) to calcium nitrate solution of pH 3.0 (443 g $Ca(NO_3)_2 \cdot 4H_2O$/L) at 20°C using 20% excess of the calcium salt calculated to give normal calcium molybdate. The suspension is stirred for 3 to 4 h. The X-ray-amorphous compound is a loose, white powder [1].

Thermogravimetric investigation shows that elimination of the water is terminated at 380°C, endothermic effects occurring at 110, 314, and 380°C. An exothermic effect between 380 and 590°C is due to the formation of anhydrous isopolymolybdate, which melts incongruently at 720°C forming the normal calcium molybdate and molybdenum trioxide [1].

The compound is insoluble in cold water, sparingly soluble in hot water, and readily soluble in hydrochloric acid (1:3) [1].

$CaO \cdot 5MoO_3 \cdot 7H_2O$. Thermogravimetric investigation shows that elimination of water is stepwise with increasing temperature. The shapes of the resonance lines of the ^1H NMR spectra obtained at 295 and 77 K of samples dried at characteristic temperatures (obtained from the thermographic curve) indicate that water is present in two states: as adsorbed H_2O ($\Delta H \approx 0.3$ G) and as OH groups ($\Delta H \approx 3$ G). These conclusions are confirmed by infrared spectra [2]. (There is no reference given to the conditions of preparation of the compound.)

$CaO \cdot 5MoO_3 \cdot 4H_2O$. The compound is prepared by adding ammonium paramolybdate solution of pH 3.0 to calcium nitrate solution of pH 3.0 at 90°C [3].

The shapes of the resonance lines of the ^1H NMR spectra obtained at room temperature of samples dried at characteristic temperatures indicate that water is present in two states: as adsorbed H_2O ($\Delta H \approx 0.24$ G) and as OH groups ($\Delta H \approx 3$ G), the ratio of the integral intensities being 3:1 [3].

Thermogravimetric investigation shows that elimination of water is stepwise with increasing temperature [3].

References:

[1] M. V. Mokhosoev, Z. E. Batura, V. I. Krivobok (Izv. Akad. Nauk SSSR Neorgan. Materialy **3** [1967] 133/7; Inorg. Materials [USSR] **3** [1967] 108/11). – [2] V. V. Mank, I. V. Matyash, M. V. Mokhosoev, Z. E. Batura (Dokl. Akad. Nauk SSSR **171** [1966] 1113/5; Dokl. Chem. Proc. Acad. Sci. USSR **166/171** [1966] 1181/3). – [3] V. V. Mank, V. I. Krivobok, I. V. Matyash, M. V. Mokhosoev (Zh. Strukt. Khim. **8** [1967] 227/32; J. Strukt. Chem. [USSR] **8** [1967] 196/200).

3.3.2.8 $CaO \cdot 6MoO_3 \cdot 11.7H_2O$ (= $CaMo_6O_{19} \cdot 11.7H_2O$), Calcium "Hexamolybdate"

The compound was prepared according to the description given for the preparation of $2CaO \cdot 9MoO_3 \cdot 14.8H_2O$ (p. 197) starting from the heterogeneous product $CaO \cdot 7.2MoO_3 \cdot 12.1H_2O$ obtained by precipitation from aqueous $Ca(NO_3)_2$ and $(NH_4)_6Mo_7O_{24}$ solution. (No reference to the exact conditions of preparation of the starting material $CaO \cdot 7.2MoO_3 \cdot 12.1H_2O$ is given.)

The conclusions drawn from the X-ray diffraction spectrum, infrared, and NMR investigations are the same as those for the structure of $2CaO \cdot 9MoO_3 \cdot 14.8H_2O$.

TG and DTA measurements show the elimination of 5.1 H_2O between 25 and 180°C, 6.3 H_2O between 180 and 360°C, and 0.3 H_2O between 360 and 400°C, accompanied by endothermic effects.

V. I. Krivobok, M. V. Mokhosoev, A. P. Nakhodnova, R. M. Sedneva, G. Ya. Samsonova (Ukr. Khim. Zh. **38** [1972] 1201/6; Soviet Progr. Chem. **38** No. 12 [1972] 6/9).

3.3.2.9 $CaO \cdot 7MoO_3 \cdot 5H_2O$, Calcium "Heptamolybdate"

The compound is precipitated by dropwise addition of ammonium molybdate solution of pH 4.0 (280 g $(NH_4)_6Mo_7O_{24} \cdot 4H_2O$/L) to calcium nitrate solution of pH 4.3 (280 g $Ca(NO_3)_2 \cdot 4H_2O$/L) at 90°C using 20% excess of the calcium salt calculated to give normal calcium molybdate. The suspension is stirred for 3 to 4 h. The X-ray-amorphous compound is a loose, white powder [1].

The shapes of the resonance lines of the ^1H NMR spectra obtained at room temperature and 77 K of samples dried at characteristic temperatures indicate that water is present in two states: as adsorbed H_2O ($\Delta H \approx 0.24$ G) and as OH groups ($\Delta H \approx 3$ G), the ratio of the integral intensities being 2:3. From these results, the formula $CaMo_4O_{13} \cdot 3H_2MoO_4 + 2H_2O$ (adsorbed) is postulated [2].

Thermogravimetric investigation shows that elimination of water is terminated at 380°C, endothermic effects occurring at 150 and 376°C. An exothermic effect between 376 and 550°C is attributed to the formation of anhydrous isopolymolybdate, which melts incongruently at ~748°C forming the normal calcium molybdate and molybdenum trioxide [1].

The compound is insoluble in cold water, sparingly soluble in hot water, and readily soluble in hydrochloric acid (1:3) [1].

References:

[1] M. V. Mokhosoev, Z. E. Batura, V. I. Krivobok (Izv. Akad. Nauk SSSR Neorgan. Materialy **3** [1967] 133/7; Inorg. Materials [USSR] **3** [1967] 108/11). – [2] V. V. Mank, V. I. Krivobok, I. V. Matyash, M. V. Mokhosoev (Zh. Strukt. Khim. **8** [1967] 227/32; J. Struct. Chem. [USSR] **8** [1967] 196/200).

3.3.2.10 $2CaO \cdot 15MoO_3 \cdot 23(21)H_2O$ (= $Ca_2Mo_{15}O_{47} \cdot 23(21)H_2O$)

These compounds were prepared according to the description given for the preparation of $2CaO \cdot 9MoO_3 \cdot 14.8H_2O$ (p. 197) starting from the heterogeneous product $CaO \cdot 9MoO_3 \cdot 15H_2O$ obtained by precipitation from aqueous $Ca(NO_3)_2$ and $(NH_4)_6Mo_7O_{24}$ solution. (No reference to the exact conditions of preparation of the starting material $CaO \cdot 9MoO_3 \cdot 15H_2O$ is given.)

The conclusions drawn from the X-ray diffraction analysis, infrared, and NMR investigations are the same as those for the structure of $2CaO \cdot 9MoO_3 \cdot 14.8H_2O$. The NMR investigations showed that there are $7H_2O$ adsorbed, $12H_2O$ as water of crystallization, and $4H_2O$ in form of OH groups in the 23-hydrate [1].

V. I. Krivobok, M. V. Mokhosoev, A. P. Nakhodnova, R. M. Sedneva, G. Ya. Samsonova (Ukr. Khim. Zh. **38** [1972] 1201/6; Soviet Progr. Chem. **38** No. 12 [1972] 6/9).

3.3.2.11 $CaO \cdot 8MoO_3 \cdot 11.9H_2O$ (= $CaMo_8O_{25} \cdot 11.9H_2O$), Calcium "Octamolybdate"

The compound was prepared according to the description given for the preparation of $2CaO \cdot 9MoO_3 \cdot 14.8H_2O$ (p. 197) starting from the heterogeneous product $CaO \cdot 18MoO_3 \cdot 23H_2O$ obtained by precipitation from aqueous $Ca(NO_3)_2$ and $(NH_4)_6Mo_7O_{24}$ solution. (No reference to the exact conditions of preparation of the starting material $CaO \cdot 18MoO_3 \cdot 23H_2O$ is given.)

The conclusions drawn from the X-ray diffraction spectrum, infrared, and NMR investigations for the structure are the same as those for $2CaO \cdot 9MoO_3 \cdot 14.8H_2O$.

TG and DTA measurements show the elimination of $3.9H_2O$ between 25 and 175°C, $7.2H_2O$ between 175 and 360°C, and $0.8H_2O$ between 360 and 400°C, accompanied by endothermic effects.

V. I. Krivobok, M. V. Mokhosoev, A. P. Nakhodnova, R. M. Sedneva, G. Ya. Samsonova (Ukr. Khim. Zh. **38** [1972] 1201/6; Soviet Progr. Chem. **38** No. 12 [1972] 6/9).

3.3.2.12 Calcium Phase C Polymolybdate[1)]

The reaction of hot, concentrated HNO_3 on solid normal or isopolymolybdates of calcium yields the calcium phase C compound in admixture with MoO_3. The calcium phase C is very unstable and is entirely converted to MoO_3 by boiling in 1 N HCl for 10 min.

M. L. Freedman (J. Chem. Eng. Data **8** [1963] 113/6).

3.3.3 Calcium Peroxomolybdate Hydrates

3.3.3.1 $(CaMoO_6)_2O \cdot 9H_2O$

The compound is prepared by adding $CaMoO_4$ in small portions to a 30% H_2O_2 solution (10 mL) with vigorous stirring. It dissolves forming a dark cherry red peroxomolybdate. When no more $CaMoO_4$ will dissolve, 75 to 100 mL cold acetone is added and the mixture cooled on dry ice. After the red solution has become lemon yellow, the residual $CaMoO_4$ is filtered off rapidly while cooling with dry ice and the filtrate is brought to room temperature. It now becomes dark cherry red again and gradually reverts to yellow with evolution of oxygen and the lemon yellow $(CaMoO_6)_2O \cdot 9H_2O$ precipitates. The crystals are separated, washed with alcohol and ether [1], see also [2].

$(CaMoO_6)_2O \cdot 9H_2O$ forms parallel-sided plates. They retain their crystalline form and the peroxide oxygen for several weeks when they are stored in a desiccator at room temperature. Under high vacuum in a desiccator, the compound loses a large part of its water rapidly and a

1) See the notes added in proof on pp. 7 and 23/4.

little of the peroxide oxygen. After 140 to 160 h, the product is $CaMoO_6 \cdot nH_2O$ with n = 0.25 to 0.33. Therefore, the product $CaMoO_6$ seems to be a true peroxide and the title compound a peroxohydrate of it, best expressed with the formula $(CaMoO_6 \cdot 4H_2O)_2 \cdot H_2O_2$ [1].

The compound dissolves in water, aqueous alcohol, and aqueous acetone; it is almost insoluble in organic solvents. In aqueous solutions, it gradually changes to $CaMoO_4$. With H_2O_2, it immediately forms a cherry red product [1]. For the catalytic decomposition of H_2O_2 in the presence of calcium peroxomolybdate, see [3].

References:

[1] V. A. Shcherbinin, G. A. Bogdanov (Zh. Neorgan. Khim. **4** [1959] 260/71; Russ. J. Inorg. Chem. **4** [1959] 112/8). – [2] G. A. Bogdanov, T. I. Berkengeim, V. A. Shcherbinin (Zh. Fiz. Khim. **30** [1956] 889/95; C.A. **1956** 16315). – [3] V. A. Shcherbinin (Zh. Fiz. Khim. **37** [1963] 1832/40; Russ. J. Phys. Chem. **37** [1963] 986/91).

3.3.3.2 $CaMoO_8 \cdot xH_2O$

The formula is merely assumed from analogy with similar peroxomolybdates (of Na and Sr), because the compound could not be prepared free from impurities, especially H_2O_2, and could not be analyzed for oxygen and water. The red compound is prepared by the action of concentrated H_2O_2 on the yellow $(CaMoO_6)_2O \cdot 9H_2O$ with slight cooling and consists of a fine sparkling crystalline powder. At room temperature, it changes rapidly to the yellow peroxomolybdate.

$CaMoO_8 \cdot xH_2O$ dissolves readily in water and is highly soluble in organic solvents (therefore, it could not be prepared in pure form). The solution in acetone on cooling with dry ice changes from cherry red to the yellow of $(CaMoO_6)_2O \cdot 9H_2O$. At room temperature the cherry red color of this solution is only gradually lost with the evolution of oxygen and crystallization of the yellow peroxomolybdate [1], see also [2, 3].

References:

[1] V. A. Shcherbinin, G. A. Bogdanov (Zh. Neorgan. Khim. **4** [1959] 260/71; Russ. J. Inorg. Chem. **4** [1959] 112/8). – [2] G. A. Bogdanov, T. I. Berkengeim, V. A. Shcherbinin (Zh. Fiz. Khim. **30** [1956] 889/95; C.A. **1956** 16315). – [3] V. A. Shcherbinin (Zh. Fiz. Khim. **37** [1963] 1832/40; Russ. J. Phys. Chem. **37** [1963] 986/91).

3.3.4 Calcium Metal Polymolybdates

3.3.4.1 $K_4Ca[Mo_7O_{24}] \cdot 7H_2O$, Potassium Calcium Paramolybdate

The compound is formed when a potassium paramolybdate solution is treated with calcium ions. To a boiling aqueous solution of 11.5 g (0.0695 mol) $K_2CO_3 \cdot 1.5H_2O$, 10.0 g (0.0695 mol) MoO_3 are added in small portions. After the evolution of CO_2 has ceased, the hot solution is acidified with concentrated HNO_3 to pH 5. The solution is diluted to about 200 mL, cooled to <40°C, and a solution of 8.0 g (0.034 mol) $Ca(NO_3)_2 \cdot 4H_2O$ in 20 mL water is added with stirring; final pH 5. A concentrated solution of KNO_3 is added to increase the rate of crystallization and the yield. The white crystalline salt is filtered onto sintered glass, washed with cold water followed by acetone, and then dried by suction.

The elementary analysis corresponds to the formula given above. Widely varying Ca and K proportions in the course of preparation give the same constant Ca:K proportion, 1:4. Potentiometric titrations, UV spectra, and cryoscopic measurements using fused $Na_2SO_4 \cdot 10H_2O$ show the compound to be a paramolybdate (double salt, completely ionized in water into K^+, Ca^{2+}, and $[Mo_7O_{24}]^{6-}$) and not a heteropolymolybdate.

A single crystal X-ray study shows the compound to be monoclinic, lattice parameters $a=15.35 \pm 0.03$, $b=18.60 \pm 0.04$, $c=10.20 \pm 0.02$ Å, $\beta=96°35' \pm 5'$; Z=4. Space group $P2_1/a\text{-}C^5_{2h}$ (No. 14), $D_m=3.191$ g/cm^3.

Complete dehydration at 500°C gives a white, crystalline product completely soluble in water.

The absorption spectrum between 320 and 500 nm of the $K_4Ca[Mo_7O_{24}]$ in solution is compared with those of other molybdates and is identical with that of $(NH_4)_6[Mo_7O_{24}]$.

O. W. Rollins, L. C. W. Baker (Inorg. Chem. **8** [1969] 397/9).

3.4 Strontium Molybdate Hydrates

Older data are given in "Molybdän", 1935, pp. 294/5.

3.4.1 Strontium Polymolybdate Hydrates with Mo of Oxidation State <6

Compounds of this type are not reported.

3.4.2 Strontium Molybdate(VI) Hydrates

Most of the strontium molybdate hydrates described in this section are very ill-defined and poorly characterized products.

3.4.2.1 $SrMoO_4 \cdot nH_2O$, n=0.15 to 0.6

Strontium molybdate does not form definite hydrates [1].

Precipitation from $Sr(NO_3)_2$ solutions with sodium molybdate or ammonium paramolybdate is carried out as in the case of Ca (see p. 193). At 25°C and pH 8 and 9, hydrated $SrMoO_4$ with 0.6 and 0.5 H_2O, respectively, is obtained, and at 90°C and pH 7.4 to 9, with 0.4 to 0.6 H_2O [2]. Hydrated strontium molybdate is also prepared from a stoichiometric mixture of MoO_3 and SrO or $Sr(OH)_2$ in H_2O (0.1 mol/L each) by stirring for 10 d at 25°C and filtering, and also at 100°C by stirring and heating at reflux for some days. At 25°C, a product with 3.8 wt% H_2O (0.55 H_2O) is obtained and at 100°C, a product with 1.1 wt% H_2O (0.15 H_2O) is obtained [1].

Both the preparations crystallize with the tetragonal scheelite type structure, lattice parameters $a=5.389 \pm 0.003$, $c=12.016 \pm 0.006$ Å, and $a=5.392 \pm 0.002$, $c=12.016 \pm 0.005$ Å, respectively; Z=4. The space group is $I4_1/a\text{-}C^6_{4h}$ (No. 88) [1].

Dehydration of the hydrate with 3.8 wt% H_2O (0.55 H_2O) proceeds in two steps, first between 40 and 200°C, then between 300 and 600°C, whereas the material with 1.1 wt% H_2O (0.15 H_2O) shows only one step at 300 to 600°C [1].

In addition to the stoichiometric ($SrO:MoO_3 = 1:1$) molybdate hydrates, some samples with excess MoO_3 have been obtained; $SrO:MoO_3 = 1:1.1$, 1:1.2, and 1:1.4. They contain more water than the stoichiometric preparations, e.g., $SrO \cdot 1.2MoO_3 \cdot 2H_2O$ [2].

References:

[1] J. Meullemeestre (Bull. Soc. Chim. France **1978** I 95/100). – [2] V. I. Krivobok, M. V. Mokhosoev (Ukr. Khim. Zh. **35** [1969] 1035/8; Soviet Progr. Chem. **35** No. 10 [1969] 22/5).

3.4.2.2 $SrO \cdot 2MoO_3 \cdot nH_2O$, n = 2, 7; Strontium "Dimolybdates"

$SrO \cdot 2MoO_3 \cdot 7H_2O$. The compound is precipitated by adding ammonium paramolybdate solution of pH 6.0 (220 g $(NH_4)_6Mo_7O_{24} \cdot 4H_2O$/L) to strontium nitrate solution of pH 6.0 (330 g $Sr(NO_3)_2$/L) at 20°C using 20% excess of the strontium salt. Mixing time is 7 d. The precipitate is washed with water and then dried in air.

$SrO \cdot 2MoO_3 \cdot 2H_2O$. This compound is prepared by adding ammonium paramolybdate solution of pH 7.0 (300 g $(NH_4)_6Mo_7O_{24} \cdot 4H_2O$/L) to strontium nitrate solution of pH 5.6 (300 g $Sr(NO_3)_2$/L) at 90°C using 20% excess of the strontium salt. Mixing time is 1 h. The precipitate is washed with water and then dried in air.

NMR and IR investigations show the water to be present in three states: adsorbed water, water of crystallization, and OH groups. The loss of water proceeds in steps and is completed upon heating to 370°C, where the sample becomes greenish. Most of the water is lost at 250°C. Strontium isopolymolybdate is formed in an exothermic reaction at 370°C. It melts at 674°C with decomposition to $SrMoO_4$ and MoO_3.

The white friable powder is insoluble in cold water, virtually insoluble in hot water, and slightly soluble in hydrochloric acid (1:3).

M. V. Mokhosoev, V. I. Krivobok, Z. E. Batura (Zh. Neorgan. Khim. **12** [1967] 2677/80; Russ. J. Inorg. Chem. **12** [1967] 1412/4).

3.4.2.3 $3SrO \cdot 8MoO_3 \cdot 4H_2O$, Strontium (3:8)-Molybdate

This compound was precipitated by adding a solution of strontium chloride (using an excess of 10%) to a 1 M solution of sodium molybdate (acidified with 1 N HCl to pH 1.9 to 3.4) and boiling [1]. In this case, the content of water of hydration was not determined. A compound of the same composition was obtained starting from a solution containing MoO_3 and $Sr(OH)_2 \cdot 8H_2O$ in the ratios 3:1 and 4:1 in experiments carried out at 100°C [2].

The X-ray powder pattern is different from that of the "trimolybdate" 4-hydrate (p. 204) [2]. Otherwise, the compound was merely characterized by analysis.

References:

[1] É. Carrière, A. Dautheville (Bull. Soc. Chim. France [5] **10** [1943] 264/5). – [2] J. Meullemeestre (Bull. Soc. Chim. France **1978** I 236/42).

3.4.2.4 $[Sr_2Mo_6O_{20}\cdot 2nH_2O]_\infty$ (= $SrO\cdot 3MoO_3\cdot nH_2O$), n = 1.75, 3, 4, 4.4, 6, 10; Strontium "Trimolybdates"

$SrO\cdot 3MoO_3\cdot 10H_2O$. The compound is precipitated by adding ammonium paramolybdate solution of pH 5.0 (220 g $(NH_4)_6Mo_7O_{24}\cdot 4H_2O$/L) to strontium nitrate solution of pH 5.0 (330 g $Sr(NO_3)_2$/L) at 20°C using 20% excess of the strontium salt. Mixing time is 7 d. The precipitate is washed with water and then dried in air [1]. The product was merely characterized by analysis.

This compound is not incontestably characterized as "trimolybdate" since it has been poorly described and the ratio $SrO:MoO_3$ shows deviations from the correct value [4].

$SrO\cdot 3MoO_3\cdot 6H_2O$. The compound is prepared by adding ammonium paramolybdate solution of pH 6.0 (300 g $(NH_4)_6Mo_7O_{24}\cdot 4H_2O$/L) to strontium nitrate solution of pH 6.0 (300 g $Sr(NO_3)_2$/L) at 90°C using 20% excess of the strontium salt. Mixing time is 1 h. The precipitate is washed with water and then dried in air [1]. The product was merely characterized by analysis.

This compound is not incontestably characterized as "trimolybdate" since it has been poorly described and the ratio $SrO:MoO_3$ shows deviations from the correct value [4].

$SrO\cdot 3MoO_3\cdot 4.4H_2O$. The formation of this compound in the Na_2MoO_4-HNO_3-$Sr(NO_3)_2$ system is described in [6].

NMR studies at 93 to 423 K indicate that the molybdate has protons in various states: H^+, H_2O, and H_3O^+. Parameters for the absorption spectra, internuclear distances, separation of the H nuclei in the H_2O molecules, and the number of the different groups of protons were calculated. The compound can be formulated as dodecamolybdate $H(H_3O)_3Sr_4Mo_{12}O_{42}\cdot 12.6H_2O$ [2]. For the investigation of the water by IR spectroscopy in the 3500 to 3000 cm^{-1} range, see [5].

DTA and TG show the water to be lost in four steps: $0.2H_2O$ at 30 to 70°C, $1.9H_2O$ at 70 to 150°C, $1.3H_2O$ at 150 to 220°C, and $1H_2O$ at 220 to 360°C [5].

$[Sr_2Mo_6O_{20}\cdot 8H_2O]$ (= $SrO\cdot 3MoO_3\cdot 4H_2O$). This "trimolybdate" tetrahydrate is obtained in place of the hexahydrate (see above) when the pH of the two reacting solutions is 5.0 instead of 6.0 [1]. The product was characterized by analysis.

According to [4], the above "trimolybdate" is not incontestably characterized as "trimolybdate" since it has been poorly described and the ratio $SrO:MoO_3$ shows deviations from the correct value. The author [4] prepared $SrO\cdot 3MoO_3\cdot 4H_2O$ by double-exchange reaction between 5 mmol of sodium "trimolybdate" and 50 mL of 0.1 M $Sr(NO_3)_2$ solution. The mixture was vigorously shaken for 5 d at 25°C and the solid collected on a glass filter and dried at room temperature in air [4].

This product was characterized by analysis, interplanar distances, and infrared spectrum (see the paper). The latter is a typical "trimolybdate" spectrum, hence, the anionic species is assumed to be the same as in other "trimolybdates" [4].

The shapes of the resonance lines of ^1H NMR spectra obtained at 295 and 77 K from samples dried at characteristic temperatures (obtained from the thermogravimetric curve) indicate that water is present in three states: adsorbed water ($\Delta H \approx 0.3$ G), water of crystallization ($\Delta H \approx 1$ to 2 G), and OH groups ($\Delta H \approx 3$ G). These conclusions are confirmed by infrared spectra. From these results, the formula of the compound is written $SrMoO_4\cdot H_2MoO_4\cdot MoO_3\cdot 2H_2O + 1H_2O$ (adsorbed) [3].

TG and DTA experiments show the formation of $SrO\cdot 3MoO_3\cdot 3H_2O$ at about 116°C, $SrO\cdot 3MoO_3\cdot 1.75H_2O$ about 177°C, $SrO\cdot 3MoO_3$ (assumed but not established) at about 280°C, and at 345°C, the decomposition of the "trimolybdate" in an endothermic reaction to give $SrMoO_4$ and MoO_3 [4].

$[Sr_2Mo_6O_{20}\cdot 6\,H_2O]_\infty$ (= $SrO\cdot 3\,MoO_3\cdot 3\,H_2O$). The conditions for the preparation of this hydrate are identical with those for the preparation of the hexahydrate (see above) [1]. However, in [7], a pH of 5 is given.

According to [4], the above "trimolybdate" is not incontestably characterized as "trimolybdate" since it has been poorly described and the ratio $SrO:MoO_3$ shows deviations from the correct value. The author [4] prepares $SrO\cdot 3\,MoO_3\cdot 3\,H_2O$ by heating of the 4-hydrate for 2 h at 100°C.

This product was characterized by interplanar distances and infrared spectrum (see the paper). The latter is a typical "trimolybdate" spectrum and hence the anionic species is assumed to be the same as in other "trimolybdates" [4].

NMR and IR investigations show the water to be present in three states: adsorbed water, water of crystallization, and OH groups. The loss of water proceeds in steps and is completed upon heating to 372°C, where the sample becomes greenish. Most of the water is lost at 248°C. Strontium isopolymolybdate is formed in an exothermic reaction at 372°C. It melts at 706°C with decomposition to $SrMoO_4$ and MoO_3 [1].

The white friable $SrO\cdot 3\,MoO_3\cdot 3\,H_2O$ powder is insoluble in cold water, virtually insoluble in hot water, and slightly soluble in hydrochloric acid (1:3) [1].

$[Sr_2Mo_6O_{20}\cdot 3.5\,H_2O]_\infty$ (= $SrO\cdot 3\,MoO_3\cdot 1.75\,H_2O$). This hydrate is formed on heating the "trimolybdate" tetrahydrate (see above) for 15 h at 150°C.

The product was characterized by interplanar distances and infrared spectrum (see the paper). The latter is a typical "trimolybdate" spectrum, hence, the anionic species is assumed to be the same as in other "trimolybdates" [4].

References:

[1] M. V. Mokhosoev, V. I. Krivobok, Z. E. Batura (Zh. Neorgan. Khim. **12** [1967] 2677/80; Russ. J. Inorg. Chem. **12** [1967] 1412/4). – [2] L. A. Pozharskaya, V. G. Pitsyuga, V. I. Krivobok, M. V. Mokhosoev (Izv. Sibirsk. Otd. Akad. Nauk SSSR Ser. Khim. Nauk **1976** No. 4, pp. 96/9; C. A. **85** [1976] No. 153195). – [3] V. V. Mank, I. V. Matyash, M. V. Mokhosoev, Z. E. Batura (Dokl. Akad. Nauk SSSR **171** [1966] 1113/5; Dokl. Chem. Proc. Acad. Sci. USSR **166/171** [1966] 1181/3). – [4] J. Meullemeestre (Bull. Soc. Chim. France **1978** I 236/42). – [5] G. Ya. Samsonova, V. I. Krivobok, M. V. Mokhosoev, T. T. Got'manova, V. D. Kolotyuk (Khim. Tekhnol. Molibdena Vol'frama No. 2 [1974] 144/8 from Ref. Zh. Khim. **1975** No. 6 B 1046).

[6] G. Ya. Samsonova, V. I. Krivobok, M. V. Mokhosoev, R. M. Sedneva (Khim. Tekhnol. Molibdena Vol'frama No. 1 [1971] 197/203 from Ref. Zh. Khim. **1973** No. 19 B 678). – [7] V. I. Krivobok, M. V. Mokhosoev (Ukr. Khim. Zh. **35** [1969] 1035/8; Soviet Progr. Chem. **35** No. 10 [1969] 22/5).

3.4.2.5 $SrO\cdot 4\,MoO_3\cdot nH_2O$, n = 3.8 or 4, 5, 6, 9; Strontium "Tetramolybdates"

$SrO\cdot 4\,MoO_3\cdot 9\,H_2O$. The compound is precipitated by adding ammonium paramolybdate solution of pH 4.0 (220 g $(NH_4)_6Mo_7O_{24}\cdot 4\,H_2O$/L) to strontium nitrate solution of pH 4.0 (330 g $Sr(NO_3)_2$/L) at 20°C using 20% excess of the strontium salt. Mixing time is 7 d. The precipitate is washed with water and then dried in air [1]. The product was merely characterized by analysis.

$SrO\cdot 4\,MoO_3\cdot 6\,H_2O$. The compound is prepared by adding ammonium paramolybdate solution of pH 4.0 (300 g $(NH_4)_6Mo_7O_{24}\cdot 4\,H_2O$/L) to strontium nitrate solution of pH 4.0 (300 g

$Sr(NO_3)_2$/L) at 90°C using 20% excess of the strontium salt. Mixing time is 1 h. The precipitate is washed with water and then dried in air [1]. (It should be mentioned that the conditions for the preparation of this compound are identical with those for the preparation of $SrO \cdot 5MoO_3 \cdot 6H_2O$, compare p. 207.)

NMR and IR investigations show the water to be present in three states: adsorbed water, water of crystallization, and OH groups. The loss of water proceeds in steps and is completed upon heating to 374°C, where the sample becomes greenish. Most of the water is lost at 244°C. Strontium isopolymolybdate is formed in an exothermic reaction at 374°C. It melts at 680°C with decomposition to $SrMoO_4$ and MoO_3 [1].

The white friable $SrO \cdot 4MoO_3 \cdot 6H_2O$ powder is insoluble in cold water, virtually insoluble in hot water, and slightly soluble in hydrochloric acid (1:3) [1].

$SrO \cdot 4MoO_3 \cdot 5H_2O$. This pentahydrate is obtained in place of the hexahydrate (see above) when the pH of the strontium nitrate solution is 5.6 instead of 4.0 [1]. The product was characterized by analysis.

The shapes of the resonance lines of ^1H NMR spectra obtained at 295 and 77 K from samples dried at characteristic temperatures (obtained from the thermogravimetric curve) indicate that water is present in three states: adsorbed water ($\Delta H \approx 0.3$ G), water of crystallization ($\Delta H \approx 1$ to 2 G), and OH groups ($\Delta H \approx 3$ G). These conclusions are confirmed by infrared spectra [2].

$SrO \cdot 4MoO_3 \cdot 3.8(4)H_2O$. The formation of the 3.8-hydrate in the Na_2MoO_4-HNO_3-$Sr(NO_3)_2$-H_2O system is described in [5]. In an earlier publication, the compound was formulated as a 4-hydrate, $SrO \cdot 4MoO_3 \cdot 4H_2O$ [6].

NMR studies of air-dried $SrO \cdot 4MoO_3 \cdot 3.8H_2O$ at 93 to 423 K indicate that the polymolybdate has protons in various states: H^+ and H_2O. Parameters for the absorption spectra, internuclear distances, separation of the H nuclei in the H_2O molecules, and the number of the different groups of protons were calculated. The compound can be formulated as dodecamolybdate $H_2Sr_3Mo_{12}O_{40} \cdot 10.4H_2O$ [3]. For the investigation of the water by IR spectroscopy in the 3500 to 3000 cm^{-1} range, see [4].

DTA and TG show the water to be lost in three steps: $0.4H_2O$ at 80 to 130°C, $1.6H_2O$ at 130 to 230°C, and $1.8H_2O$ at 230 to 400°C [4].

References:

[1] M. V. Mokhosoev, V. I. Krivobok, Z. E. Batura (Zh. Neorgan. Khim. **12** [1967] 2677/80; Russ. J. Inorg. Chem. **12** [1967] 1412/4). – [2] V. V. Mank, I. V. Matyash, M. V. Mokhosoev, Z. E. Batura (Dokl. Akad. Nauk SSSR **171** [1966] 1113/5; Dokl. Chem. Proc. Acad. Sci. USSR **166/171** [1966] 1181/3). – [3] L. A. Pozharskaya, V. G. Pitsyuga, V. I. Krivobok, M. V. Mokhosoev (Izv. Sibirsk. Otd. Akad. Nauk SSSR Ser. Khim. Nauk **1976** No. 4, pp. 96/9; C.A. **85** [1976] No. 153195). – [4] G. Ya. Samsonova, V. I. Krivobok, M. V. Mokhosoev, T. T. Got'manova, V. D. Kolotyuk (Khim. Tekhnol. Molibdena Vol'frama No. 2 [1974] 144/8 from Ref. Zh. Khim. **1975** No. 6 B 1046). – [5] G. Ya. Samsonova, V. I. Krivobok, M. V. Mokhosoev, R. M. Sedneva (Khim. Tekhnol. Molibdena Vol'frama No. 1 [1971] 197/203 from Ref. Zh. Khim. **1973** No. 19 B 678).

[6] V. I. Krivobok, M. V. Mokhosoev (Ukr. Khim. Zh. **35** [1969] 1035/8; Soviet Progr. Chem. **35** No. 10 [1969] 22/5).

3.4.2.6 **$SrO \cdot 5MoO_3 \cdot 6H_2O$**, Strontium "Pentamolybdate"

The compound is precipitated by adding ammonium paramolybdate solution of pH 4.0 (300 g $(NH_4)_6Mo_7O_{24} \cdot 4H_2O$/L) to strontium nitrate solution of pH 4.0 (300 g $Sr(NO_3)_2$/L) at 90°C using 20% excess of the strontium salt. Mixing time is 1 h. The precipitate is washed with water and then dried in air. (It should be mentioned that the conditions for the preparation of this compound are identical with those for the preparation of $SrO \cdot 4MoO_3 \cdot 6H_2O$, compare p. 205.)

NMR and IR investigations show the water to be present in three states: adsorbed water, water of crystallization, and OH groups. The loss of water proceeds in steps and is completed upon heating to 370°C, where the sample becomes greenish. Most of the water is lost at 252°C. Strontium isopolymolybdate is formed in an exothermic reaction at 370°C. It melts at 688°C with decomposition to $SrMoO_4$ and MoO_3.

The white friable $SrO \cdot 5MoO_3 \cdot 6H_2O$ powder is insoluble in cold water, virtually insoluble in hot water, and slightly soluble in hydrochloric acid (1:3).

M. V. Mokhosoev, V. I. Krivobok, Z. E. Batura (Zh. Neorgan. Khim. **12** [1967] 2677/80; Russ. J. Inorg. Chem. **12** [1967] 1412/4).

3.4.3 Strontium Peroxomolybdate Hydrates

3.4.3.1 $SrMoO_5 \cdot nH_2O$

For the preparation of this compound, freshly prepared strontium molybdate is stirred vigorously with concentrated H_2O_2 while cooling with an ice/salt mixture until the product has a creamy color. It is separated, washed, and stored as described for $SrMoO_6 \cdot 3H_2O$ (see below) [1]. In a vacuum desiccator, $SrMoO_5 \cdot nH_2O$ with n = 0.5, 0.33, and 0.25 is formed by decomposition of $SrMoO_6 \cdot 3H_2O$. All the higher peroxomolybdates described below decompose in aqueous solution. $SrMoO_5$ is an intermediate in the decomposition and $SrMoO_4$ is the final product. $SrMoO_5 \cdot nH_2O$ is assumed to be a true peroxide [1, 2].

References:

[1] V. A. Shcherbinin, G. A. Bogdanov (Zh. Neorgan. Khim. **4** [1959] 260/71; Russ. J. Inorg. Chem. **4** [1959] 112/8). – [2] V. A. Shcherbinin, G. A. Bogdanov (Zh. Fiz. Khim. **32** [1958] 1942/50; C.A. **1959** 7739).

3.4.3.2 **$SrMoO_6 \cdot nH_2O$**, n = 0.2, 0.25, 0.33, 0.5, 3

Depending on the method of formation, the yellow peroxomolybdate appears in two forms; either as the true peroxide $SrMoO_6 \cdot nH_2O$ or as the peroxohydrate $SrMoO_5 \cdot H_2O_2 \cdot mH_2O$ [1, 5].

$SrMoO_6 \cdot 3H_2O$ is prepared by mixing equal volumes (10 mL) of cooled, saturated Na_2MoO_4 solution and 30% H_2O_2 solution cooled to −3°C. Mixing is done by adding small portions of the molybdate solution to the hydrogen peroxide with constant stirring. Dark red sodium peroxomolybdate is obtained. Then the same volume (10 mL) of cold saturated $SrCl_2$ solution is added. To avoid decomposition of the peroxides, the solution is cooled by an ice/salt mixture. Red strontium peroxomolybdates are formed by this procedure. They are then converted into the sparingly soluble yellow title compound by cooling slightly and stirring vigorously with a glass rod. By this, the solution loses part of its oxygen and the precipitate absorbs all the water, forming a sticky pale yellow mass. To this is added alcohol or acetone (200 to 250 mL), cooled with dry ice. The mixture is filtered and washed with alcohol and ether while continuously cooling with dry ice. It is best stored on filter paper in a desiccator [1, 2].

Decomposition of $SrMoO_7 \cdot 4H_2O$ in a vacuum desiccator gives $SrMoO_6 \cdot 0.5H_2O$, and decomposition of $SrMoO_8 \cdot 4H_2O$ gives $SrMoO_6 \cdot nH_2O$ with $n = 0.33$, 0.25, and 0.2. Also, in aqueous solutions, the higher peroxomolybdates decompose, forming $CaMoO_6$, which is stable in presence of sufficient H_2O_2 [1, 4].

The measured density of $SrMoO_6 \cdot 3H_2O$ is between 2.922 and 2.938 g/cm^3. $SrMoO_6 \cdot 3H_2O$ powder is canary yellow and appears transparent under the microscope. $SrMoO_6 \cdot 3H_2O$ is stable at room temperature in a desiccator for several weeks. At 60°C, it decomposes at an appreciable rate, and at 78 to 80°C, it decomposes in 3 to 5 min. The decomposition temperature is raised to 90 to 92°C when the water is removed by keeping the compound in a vacuum desiccator. It is readily dehydrated in this way [1, 2].

Dissolution in water at room temperature proceeds slowly. About 1 g per 100 mL is dissolved with slight evolution of oxygen. A colorless solution, probably of $SrMoO_5$, is formed and finally $SrMoO_4$ is precipitated from this solution, after the remaining peroxide oxygen has been lost [1, 2]. The measured heat of decomposition in acidic $KMnO_4$ solution (H_2SO_4 added) is 36.9 kcal/mol. From this, the bond energy of peroxide oxygen is calculated to be 40.05 kcal [3]. In hydrogen peroxide, $SrMoO_6 \cdot 3H_2O$ is reversibly converted to $SrMoO_8 \cdot 4H_2O$ (see below) [1, 2]. For a discussion of the catalytic decomposition of hydrogen peroxide in solutions with the title compound, see [5]. $SrMoO_6 \cdot 3H_2O$ is almost insoluble in alcohol, acetone, ether, and carbon tetrachloride [1, 2].

References:

[1] V. A. Shcherbinin, G. A. Bogdanov (Zh. Neorgan. Khim. **4** [1959] 260/71; Russ. J. Inorg. Chem. **4** [1959] 112/8). – [2] V. A. Shcherbinin, G. A. Bogdanov (Zh. Fiz. Khim. **32** [1958] 1466/71; C. A. **1959** 834). – [3] G. K. Yurchenko, G. A. Bogdanov (Zh. Fiz. Khim. **34** [1960] 2199/201; Russ. J. Phys. Chem. **34** [1960] 1045/6). – [4] V. A. Shcherbinin, G. A. Bogdanov (Zh. Fiz. Khim. **32** [1958] 1942/50; C. A. **1959** 7739). – [5] V. A. Shcherbinin (Zh. Fiz. Khim. **37** [1963] 1832/40; Russ. J. Phys. Chem. **37** [1963] 986/91).

3.4.3.3 $SrMoO_7 \cdot 4H_2O$

This compound is considered to be a peroxohydrate $SrMoO_6 \cdot H_2O_2 \cdot 3H_2O$ of the yellow strontium peroxomolybdate $SrMoO_6 \cdot 3H_2O$ [1, 3].

To prepare this compound, saturated sodium molybdate solution (5 mL) is added to a 20 to 22% H_2O_2 solution (20 mL) at a temperature of −3°C. After 15 to 20 min, saturated $SrCl_2$ solution is added. The mixture is thoroughly stirred and cold alcohol (100 mL) is added while cooling with an ice/salt mixture until the compound is precipitated. It is separated as described for $SrMoO_6 \cdot 3H_2O$ (see above) and stored on dry ice. The compound may also be obtained from $SrMoO_6 \cdot 3H_2O$ and H_2O_2 [1, 2].

The $SrMoO_7 \cdot 4H_2O$ powder is brick red and under the microscope appears as red, glassy particles. The compound is less stable than $SrMoO_8 \cdot 4H_2O$ and decomposes rapidly at 45 to 50°C. Dehydration over H_2SO_4 or P_2O_5 under high vacuum first leads to a loss of oxygen and water until the oxygen content remains at one mole, while the water content continues to fall and becomes less than 0.5 mole after 25 h. Thus, the final composition is about $SrMoO_6 \cdot 0.5H_2O$ [1, 3]. Solubilities and behavior in solution correspond to those of $SrMoO_8 \cdot 4H_2O$ [1, 2].

References:

[1] V. A. Shcherbinin, G. A. Bogdanov (Zh. Neorgan. Khim. **4** [1959] 260/71; Russ. J. Inorg. Chem. **4** [1959] 112/8). – [2] V. A. Shcherbinin, G. A. Bogdanov (Zh. Fiz. Khim. **32** [1958] 1466/71; C.A. **1959** 834). – [3] V. A. Shcherbinin, G. A. Bogdanov (Zh. Fiz. Khim. **32** [1958] 1942/50; C.A. **1959** 7739).

3.4.3.4 $SrMoO_8 \cdot 4H_2O$

This compound, containing the most oxygen of any of the strontium peroxomolybdates, is formed by reaction between $SrMoO_6 \cdot 3H_2O$ and hydrogen peroxide. It is considered to be a peroxohydrate $SrMoO_6 \cdot 2H_2O_2 \cdot 2H_2O$ of the yellow strontium peroxomolybdate $SrMoO_6 \cdot 3H_2O$ [1, 4].

To prepare this compound, freshly prepared $SrMoO_6 \cdot 3H_2O$ is filtered and added in small portions to an at least 30% H_2O_2 solution. The mixture rapidly becomes deep red and precipitation occurs on saturation at room temperature. The mixture occasionally is slightly cooled with ice and kept at 0°C for 10 to 15 min before the deep red crystals of the compound are filtered off [1, 2].

$SrMoO_8 \cdot 4H_2O$ powder under the microscope appears as ruby red crystals in elongated hexagonal parallelepipeds. The color is a dark, deep red with a velvety luster. The measured density is 3.052 to 3.062 g/cm^3 [1, 2].

At room temperature and in the absence of moisture, the powder retains its color and most of its peroxide oxygen for 3 to 4 d. After this time, it is slowly converted into yellow $SrMoO_6$. This process is accelerated by reduction of pressure, increase in temperature, and by moisture. At 50 to 55°C, decomposition is explosive and strongly exothermic. Aqueous $SrMoO_8$ solutions remain dark cherry red for only a short time, changing to yellow red with simultaneous precipitation of yellow $SrMoO_6$ and evolution of oxygen. Solutions in aqueous alcohol or acetone become pale yellow but the original cherry red color is gradually restored on warming to room temperature. These observations are explained by the equilibrium $SrMoO_8 \cdot 4H_2O \rightleftharpoons SrMoO_6 + 2H_2O_2 + 2H_2O$ at low temperature. Also, at room temperature $SrMoO_8 \cdot 4H_2O$ is stabilized by the presence of H_2O_2. Therefore, it is assumed that the compound is a true peroxide only in part and contains some hydrogen peroxide of crystallization [1, 2, 4]. The measured heat of decomposition in acidic $KMnO_4$ solution (H_2SO_4 added) is 84.6 kcal/mol. From this, the bond energy of peroxide oxygen is calculated as 37.3 kcal [3]. For a discussion of the catalytic decomposition of hydrogen peroxide in solutions containing $SrMoO_8 \cdot 4H_2O$, see [5]. The compound is practically insoluble in alcohol, acetone, ether, and carbon tetrachloride, but dissolves readily in aqueous alcohol or acetone, and very readily in water [1, 2, 4].

References:

[1] V. A. Shcherbinin, G. A. Bogdanov (Zh. Neorgan. Khim. **4** [1959] 260/71; Russ. J. Inorg. Chem. **4** [1959] 112/8). – [2] V. A. Shcherbinin, G. A. Bogdanov (Zh. Fiz. Khim. **32** [1958] 1466/71; C.A. **1959** 834). – [3] G. K. Yurchenko, G. A. Bogdanov (Zh. Fiz. Khim. **34** [1960] 2199/201; Russ. J. Phys. Chem. **34** [1960] 1045/6). – [4] V. A. Shcherbinin, G. A. Bogdanov (Zh. Fiz. Khim. **32** [1958] 1942/50; C.A. **1959** 7739). – [5] V. A. Shcherbinin (Zh. Fiz. Khim. **37** [1963] 1832/40; Russ. J. Phys. Chem. **37** [1963] 986/91).

3.5 Barium Molybdate Hydrates

Older data are given in "Molybdän", 1935, pp. 296/8.

3.5.1 Barium Polymolybdate Hydrates with Mo of Oxidation State <6

Compounds of this type are not reported.

3.5.2 Barium Molybdate(VI) Hydrates

3.5.2.1 The BaO-MoO_3-H_2O System

Upon increasing acidification, solutions of sodium molybdate spontaneously deposit $BaMoO_4 \cdot nH_2O$, $Ba_3Mo_7O_{24} \cdot xH_2O$, $BaO \cdot 2.8MoO_3 \cdot nH_2O$ (assumed to be $BaO \cdot 3MoO_3 \cdot n'H_2O$), and $BaO \cdot 4MoO_3 \cdot xH_2O$ on addition of a solution of $BaCl_2$ (at room temperature), see **Fig. 73a** [1].

In the presence of 3M NaCl as supporting salt, such solutions spontaneously deposit $BaMoO_4 \cdot nH_2O$ and $BaO \cdot 3MoO_3 \cdot n'H_2O$ only (**Fig. 73b**) [1].

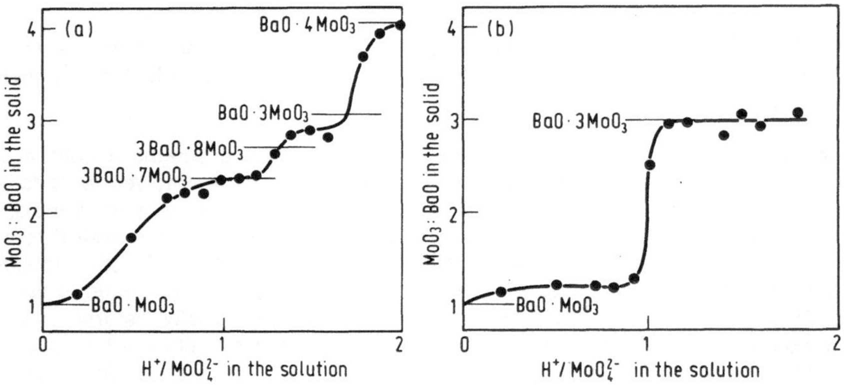

Fig. 73. Composition of spontaneously precipitated barium molybdates dependent on the acidity of the solution. (a) In the absence of a supporting salt, (b) in the presence of 3M NaCl as supporting salt [1].

There are some indications that the compound $BaO \cdot 2.8MoO_3 \cdot nH_2O$ should be interpreted as (3:8)-molybdate $3BaO \cdot 8MoO_3 \cdot n''H_2O$ not as "trimolybdate" $BaO \cdot 3MoO_3 \cdot n'H_2O$ [2] (compare p. 211).

In another investigation of the Ba^{2+}-Na_2MoO_4-H_2O system, the following compounds were precipitated at 90°C and pH between 9 and 4: $BaMoO_4 \cdot 0.2H_2O$, $BaMoO_4 \cdot 0.3H_2O$, $BaO \cdot 1.2MoO_3 \cdot 0.8H_2O$, $BaO \cdot 1.5MoO_3 \cdot 2H_2O$, $BaO \cdot 2.3MoO_3 \cdot 2H_2O$, and $BaO \cdot 2.5MoO_3 \cdot 3H_2O$ [3].

References:

[1] M. Haeringer, G. Goldstein, P. Lagrange, J. P. Schwing (Bull. Soc. Chim. France **1967** 723/8). – [2] J. Meullemeestre (Bull. Soc. Chim. France **1978** I 236/42). – [3] V. I. Krivobok, M. V. Mokhosoev (Ukr. Khim. Zh. **35** [1969] 1035/8; Soviet Progr. Chem. **35** No. 10 [1969] 22/5).

3.5.2.2 $BaMoO_4 \cdot nH_2O$, n = 0.2 to 0.33

Barium molybdate does not form definite hydrates [1].

Precipitation from $Ba(NO_3)_2$ solutions with sodium molybdate or ammonium paramolybdate is carried out as in the case of Ca (see p. 193). At 90°C and pH 7.4 and 7.5, hydrated $BaMoO_4$ is obtained with 0.3 and 0.2 H_2O, respectively [2]. From MoO_3 and BaO or $Ba(OH)_2$ in stoichiometric mixtures in H_2O, hydrated barium molybdate is prepared in the same way as the strontium (see p. 202) or calcium compounds. At 25°C, a product with 2.0 wt% H_2O (0.33 H_2O) is obtained and at 100°C, products with 1.2 to 1.8 wt% H_2O (0.2 to 0.3 H_2O) are obtained [1]. For preparations with excess MoO_3, $BaO \cdot 1.2MoO_3 \cdot 0.8H_2O$ and $BaO \cdot 1.5MoO_3 \cdot 2H_2O$, precipitated at pH 6.5 and 6.8, respectively, see [2].

The sample with 2.0 wt% H_2O has the tetragonal lattice parameters a = 5.577 ± 0.002, c = 12.804 ± 0.006 Å; that with 1.2 to 1.8 wt% H_2O, a = 5.577 ± 0.003, c = 12.816 ± 0.008 Å; Z = 4. Space group, as with the corresponding calcium and strontium compounds, is $I4_1/a\text{-}C_{4h}^6$ (No. 88); scheelite type structure [1].

Dehydration on thermal analysis of the sample with 2.0 wt% H_2O begins at 160°C, while that of the sample with 1.2 to 1.8 wt% H_2O begins at 170 to 180°C. In both cases, dehydration continues up to 400°C [1].

References:

[1] J. Meullemeestre (Bull. Soc. Chim. France **1978** I 95/100). – [2] V. I. Krivobok, M. V. Mokhosoev (Ukr. Khim. Zh. **35** [1969] 1035/8; Soviet Progr. Chem. **35** No. 10 [1969] 22/5).

3.5.2.3 $Ba_3[Mo_7O_{24}] \cdot xH_2O$ (= $3BaO \cdot 7MoO_3 \cdot xH_2O$), Barium Heptamolybdate

To prepare this compound, a mixture of 100 mL of 0.1 M solution of sodium molybdate acidified with HCl to 1.0 to 1.2 H^+/MoO_4^{2-} and 20 mL of 0.2 M solution of barium chloride is vigorously shaken for 5 (to 30) min. The precipitate is filtered, washed, and dried, first in air and subsequently for 24 h under vacuum in the presence of silica gel. The compound was characterized by its X-ray powder diagram [1].

A sample of composition $BaO \cdot 2.3MoO_3 \cdot 2H_2O$ has been obtained by precipitation of $Ba(NO_3)_2$ with sodium molybdate or ammonium paramolybdate of pH 5 (properties of the compound are not given) [2].

References:

[1] M. Haeringer, G. Goldstein, P. Lagrange, J. P. Schwing (Bull. Soc. Chim. France **1967** 723/8). – [2] V. I. Krivobok, M. V. Mokhosoev (Ukr. Khim. Zh. **35** [1969] 1035/8; Soviet Progr. Chem. **35** No. 10 [1969] 22/5).

3.5.2.4 $3BaO \cdot 8MoO_3 \cdot 9H_2O$, ~$BaO \cdot 3MoO_3 \cdot 3H_2O$?, Barium (3:8)-Molybdate, Barium "Trimolybdate"

The compound is prepared by vigorous shaking of a mixture of 15 mmol MoO_3 and 5 mmol $Ba(OH)_2 \cdot 8H_2O$ in 50 mL H_2O at 25°C for 15 d. The solid is a viscid mud, difficult to dry in air. The composition of the product was $BaO \cdot 2.8MoO_3 \cdot 3.2H_2O$ [1]. A mixture of 100 mL of 0.1 M Na_2MoO_4 solution acidified with HCl to 1.4 to 1.6 H^+/MoO_4^{2-} and 20 mL of 0.2 M $BaCl_2$ solution is vigorously shaken for 5 (to 30) min. The precipitate is filtered, washed, and dried, first in air and subsequently for 24 h under vacuum in the presence of silica gel. This product is reported by the authors to have the formula $BaO \cdot 3MoO_3 \cdot xH_2O$, however, according to diagrams given in the paper (see Fig. 73a, p. 210) its composition is $BaO \cdot {\sim}2.8MoO_3 \cdot xH_2O$. Preparation in the presence of 3M NaCl as supporting salt shows the composition to be $BaO \cdot 3MoO_3 \cdot xH_2O$ (see Fig. 73b, p. 210) [2]. Byé [3] has noted that in the acidity range 0.63 to 1.1 the composition of the deposited solid is $3BaO \cdot 8MoO_3 \cdot xH_2O$ rather than $BaO \cdot 3MoO_3 \cdot xH_2O$; the latter occurs in the acidity range 1.3 to 2.0 [3]. A precipitate having the composition $BaO \cdot 2.66MoO_3 \cdot 3H_2O$ ($= 3BaO \cdot 8MoO_3 \cdot 9H_2O$) was obtained in experiments to prepare the barium phase C [4]. The formula $3BaO \cdot 8MoO_3 \cdot xH_2O$ was also stated by [5].

The compound was characterized by X-ray powder diagrams [1, 2] (which are identical for both preparations [1]) and infrared spectrum. The latter (see Fig. 6, p. 12) is rather different from a typical "trimolybdate" spectrum [1].

The analytical deviations from the ratio $BaO:3MoO_3$ and the differing infrared spectra were interpreted to show that the product might possibly be a (3:8)-molybdate $3BaO \cdot 8MoO_3 \cdot 9H_2O$ [1]. The only interpretation given by [2] is a "trimolybdate". However, the infrared spectrum shows indeed some resemblance to that of the ammonium (3:8)-molybdate (Fig. 6, p. 12).

TG and DTA investigations show the behavior on dehydration (see the paper) to be different from that of the calcium "trimolybdate" $CaO \cdot 3MoO_3 \cdot 6H_2O$ and strontium "trimolybdate" $SrO \cdot 3MoO_3 \cdot 4H_2O$. The structure of the ${\sim}BaO \cdot 3MoO_3 \cdot 3H_2O$ is preserved up to 200°C. At higher temperatures (~330°C), the compound $BaMo_2O_7$ forms which decomposes at temperatures above 650°C to give $BaMoO_4$ and MoO_3 [1].

References:

[1] J. Meullemeestre (Bull. Soc. Chim. France **1978** I 236/42). – [2] M. Haeringer, G. Goldstein, P. Lagrange, J. P. Schwing (Bull. Soc. Chim. France **1967** 723/8). – [3] J. Byé (Ann. Chim. [Paris] [11] **20** [1945] 463/550, 539/42). – [4] M. L. Freedman (J. Chem. Eng. Data **8** [1963] 113/6). – [5] É. Carrière, R. Lasri (Compt. Rend. **207** [1938] 1048/9).

3.5.2.5 $BaO \cdot 4MoO_3 \cdot xH_2O$, Barium "Tetramolybdate"

A mixture of 100 mL of 0.1 M Na_2MoO_4 solution acidified with HCl to $2.0H^+/MoO_4^{2-}$ and 20 mL of 0.2 M $BaCl_2$ solution is vigorously shaken for 5 (to 30) min. The precipitate is filtered, washed, and dried, first in air and subsequently for 24 h under vacuum in the presence of silica gel [1]. The compound was characterized by its X-ray powder diagram [1]. The compound is briefly mentioned in [2] to be metastable.

References:

[1] M. Haeringer, G. Goldstein, P. Lagrange, J. P. Schwing (Bull. Soc. Chim. France **1967** 723/8). – [2] J. Byé (Ann. Chim. [Paris] [11] **20** [1945] 463/550, 539/42).

3.5.2.6 $Ba_4[Mo_{36}O_{112}(H_2O)_{16}] \cdot {\sim}64H_2O$ ($= BaO \cdot 9MoO_3 \cdot {\sim}20H_2O$), Barium 36-Molybdate

Crystals of the compound are obtained from a suspension of $BaMoO_4$ acidified to $Z \approx 1.8$ (~$1.9H^+$ added/MoO_4^{2-}; pH ≈ 1) [1].

The compound was characterized by analysis and Raman spectrum [2].

The X-ray investigation gave a triclinic unit cell. Based on Raman spectroscopic "fingerprints", the formula of the polymolybdate ion was proposed to be $[Mo_{36}O_{112}]^{8-}$ corresponding to the sodium compound [2]. In light of the crystallographic determination of the structure of the potassium compound (see p. 84) and considering the identity of the Raman spectra of these compounds, the polymolybdate ion must now be formulated as $[Mo_{36}O_{112}(H_2O)_{16}]^{8-}$

The crystals readily effloresce with conversion into the "decamolybdate" [1].

References:

[1] K. H. Tytko, B. Schönfeld (Z. Naturforsch **30b** [1975] 471/84). – [2] K. H. Tytko, B. Schönfeld, B. Buss, O. Glemser (Angew. Chem. **85** [1973] 305/7; Angew. Chem. Intern. Ed. Engl. **12** [1973] 330/2).

3.5.2.7 $BaO \cdot 22MoO_3 \cdot 14H_2O$, Barium Phase C Polymolybdate[1)]

A slurry of $MoO_3 \cdot 2H_2O$ in 1 M $BaCl_2$ solution was heated for several hours at 50°C. The precipitate produced had the composition $BaO \cdot 2.66MoO_3 \cdot 3H_2O$. This polymolybdate was converted to the compound $BaO \cdot 22MoO_3 \cdot 14H_2O$ by boiling with concentrated HNO_3. The compound was characterized by its X-ray powder diagram as a phase C compound.

M. L. Freedman (J. Chem. Eng. Data **8** [1963] 113/6).

1) See the notes added in proof on pp. 7 and 23/4.

4 Molybdate Hydrates with Subgroup 2 Metals

4.1 Zinc Molybdate Hydrates

Older data are given in "Molybdän", 1935, pp. 298/300.

4.1.1 The ZnO-MoO_3-H_2O System

This system has been studied by mixing at 25°C ZnO, MoO_3, and H_2O in various ratios and with equilibration times up to years. The solubility curves of both the stable and metastable compounds are shown in **Figs. 74** and **75**; the concentrations at the special points are given in the following table:

point	[ZnO] in g/L	[MoO_3] in g/L	compounds in equilibrium
A	—	1.31	MoO_3
B	4.59	31.1	MoO_3, $ZnMo_3O_{10}\cdot5H_2O$
C	2.51	14.1	$ZnMo_3O_{10}\cdot5H_2O$, $ZnMo_2O_7\cdot5H_2O$
D	0.89	1.59	$ZnMo_2O_7\cdot5H_2O$, $Zn_3(OH)_2(MoO_4)_2$
E	0.610	1.00	$Zn_3(OH)_2(MoO_4)_2$, $Zn_2(OH)_2MoO_4$
F	0.016	0.013	$Zn_2(OH)_2MoO_4$, ZnO
G	0.007	—	ZnO
H	1.68	2.87	$Zn_{5.2}Mo_2O_{11.2}\cdot5.4H_2O$, $ZnMoO_4\cdot0.5H_2O$
I	1.28	2.28	$ZnMo_2O_7\cdot5H_2O$, $ZnMoO_4\cdot0.5H_2O$
J	1.34	2.51	$ZnMoO_4\cdot0.5H_2O$, $Zn_3(OH)_2(MoO_4)_2$
K	0.425	0.739	$Zn_2(OH)_2MoO_4$, $Zn_{5.2}Mo_2O_{11.2}\cdot5.4H_2O$
L	0.100	0.131	ZnO, $Zn_{5.2}Mo_2O_{11.2}\cdot5.4H_2O$
M	0.74	1.27	$Zn_3(OH)_2(MoO_4)_2$, $Zn_{5.2}Mo_2O_{11.2}\cdot5.4H_2O$

The stable compounds are MoO_3 (in equilibrium with the solution along curve AB of Fig. 74), $ZnMo_3O_{10}\cdot5H_2O$ (BC), $ZnMo_2O_7\cdot5H_2O$ (CD), $Zn_3(OH)_2(MoO_4)_2$ (DE in Fig. 75), $Zn_2(OH)_2MoO_4$ (EF), and ZnO (FG). Some of the curves were traceable into regions where the compounds are metastable. Thus CD is continued to I, ED to J, and GF to L. Solubility curves of metastable compounds are JIH for $ZnMoO_4\cdot0.5H_2O$ and LKMH for the nonstoichiometric $Zn_{5.2}Mo_2O_{11.2}$ $\cdot5.4H_2O$. In addition $ZnMoO_4\cdot2H_2O$ and $ZnMo_2O_7\cdot2H_2O$ (not shown in the diagrams) are also metastable in this system [1, 2]. Earlier, precipitates of molar ratios $ZnO:MoO_3=2:1$ to $1:4$ were reported from a study of the system as a function of the pH of the mother liquor at room temperature, but no mention was made about their water content [3]. Other zinc molybdate hydrates not found in these investigations of the system are also described in the following sections. Solubility data for $ZnMoO_4$ are reported in "Molybdän" Erg.-Bd. B 2, 1976, pp. 64/5.

References:

[1] J. Meullemeestre, É. Pénigault (Bull. Soc. Chim. France **1974** 9/12). – [2] J. Meullemeestre, É. Pénigault (Bull. Soc. Chim. France **1974** 13/6). – [3] É. Carrière, H. Guiter, M. Annouar (Bull. Soc. Chim. France **1948** 261/2).

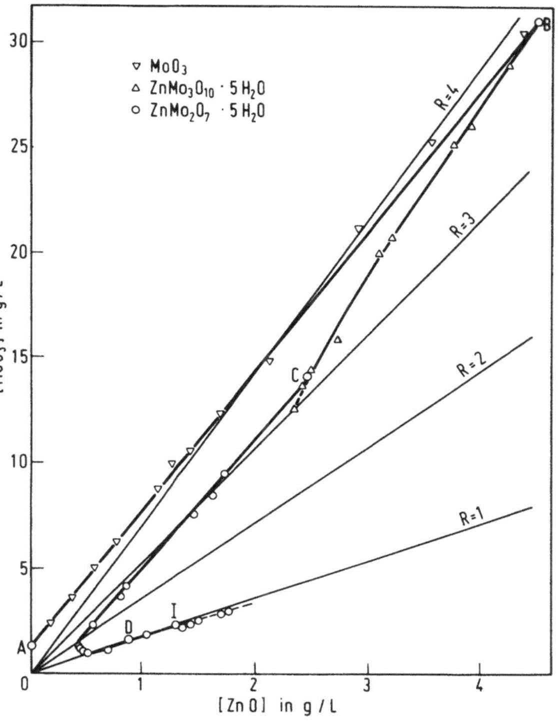

Fig. 74. Solubility curves in the ZnO-MoO$_3$-H$_2$O system with R = MoO$_3$/ZnO ≧ 1 (for the points A to I see the table) [1].

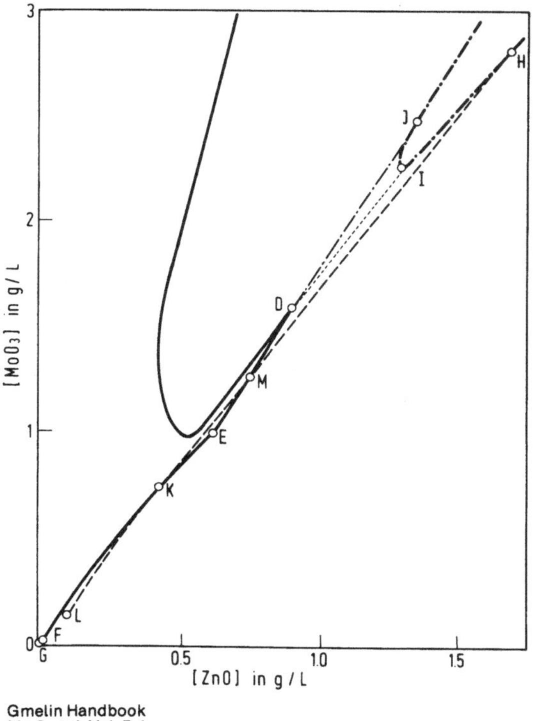

Fig. 75. Solubility curves in the ZnO-MoO$_3$-H$_2$O system with R = MoO$_3$/ZnO ≦ 1 (for the points D to M see the table) [2].

4.1.2 $Zn_{5.2}Mo_2O_{11.2}\cdot5.4H_2O$ (=$5.2ZnO\cdot2MoO_3\cdot5.4H_2O$)

This compound occurs in the ZnO-MoO_3-H_2O system at 25°C in a metastable state, see above. The white compound with the $ZnO:MoO_3$ molar ratio (~2.57 to 2.6):1 is formed when 1 to 2 mmol ZnO are stirred with MoO_3 in the molar ratio $ZnO:MoO_3$ of about 2:1 in 50 mL of H_2O at 25°C for some 7 to 21 d [1]. It crystallizes easily and has been observed at least intermediately under similar conditions at $ZnO:MoO_3$ molar ratios down to 1.02:1 [2]. According to [1, 2] this compound has been obtained by double decomposition of Na_2MoO_4 and a soluble zinc salt in alkaline media by [3].

The powder pattern of $Zn_{5.2}Mo_2O_{11.2}\cdot5.4H_2O$ has been indexed on the basis of a rhombohedral lattice with the parameters a=8.021 Å and α=44°59′, which corresponds to a=6.137 and c=21.59 Å for the hexagonal cell [1]. Earlier a=6.16, c=21.55 Å were obtained [4]. According to [1] $Zn_{5.2}Mo_2O_{11.2}\cdot5.4H_2O$ seems to be isomorphous with $M_{2-x}(H_3O)_xZn_2Mo_2O_9$, M=K, NH_4 (see p. 222) [4]. For the indexed powder pattern see the paper [1].

The IR spectrum shows a broad absorption maximum at 3300 cm^{-1} and two small sharp peaks at 3592 and 3565 cm^{-1} in the range of the OH stretching frequency. These two peaks correspond to two similar absorption maxima in the spectrum of $Zn_2(OH)_2MoO_4$ (p.220). Thus some of the water is thought to be present as hydroxyl groups. Below 1000 cm^{-1} there are absorption bands at 932, 862, 755, 450, and 340 cm^{-1}. For a drawing of the spectrum see the paper [1].

The dry compound is stable under ordinary conditions. On heating it gradually loses water, starting at 90°C. Between 250 and 265°C a sharp endothermic decrease in weight takes place, corresponding to ⅕ of the water initially present. At higher temperatures up to about 390°C there is again a gradual decrease in weight. DTA curves run at higher heating rates confirm the TG results. An exothermic effect between 350 and 450°C is attributed to the crystallization of ZnO, formed by decomposition. X-ray powder studies at continuously rising temperatures show the initial structure to be preserved up to 270°C. After that $Zn_3Mo_2O_9$ is formed [1]. $Zn_{5.2}Mo_2O_{11.2}\cdot5.4H_2O$ reacts with water at 25°C forming $Zn_2(OH)_2MoO_4$ and ZnO. Complete reaction requires several months [2].

References:

[1] J. Meullemeestre, É. Pénigault (Bull. Soc. Chim. France **1975** 1920/4). – [2] J. Meullemeestre, É. Pénigault (Bull. Soc. Chim. France **1974** 13/6). – [3] H. Pezerat (Compt. Rend. **261** [1965] 5490/4). – [4] H. Pezerat, I. Mantin, S. Kovacevic (Compt. Rend. C **263** [1966] 60/3).

4.1.3 $ZnMoO_4\cdot2H_2O$ (=$ZnO\cdot MoO_3\cdot2H_2O$)

This compound occurs in a metastable state in the ZnO–MoO_3–H_2O system at 25°C [1]. Rectangular prisms of $ZnMoO_4\cdot2H_2O$ crystallize in several days from a solution obtained by vigorously stirring 3 g of an equimolar mixture of ZnO and MoO_3 in 50 mL H_2O for 4 h at 25°C with subsequent filtering. The yield is about 0.3 g [2]. $ZnMoO_4\cdot2H_2O$ has also been obtained on mixing 0.5 M aqueous solutions of Na_2MoO_4 and $Zn(NO_3)_2$ at 22°C and pH 5 to 6. At higher pH values coprecipitation of $Zn(OH)_2$ takes place, whereas at lower pH polymolybdates form. Samples precipitated at higher temperatures show a reduced water content (1.62 H_2O at 100°C) [3].

The powder pattern of $ZnMoO_4\cdot2H_2O$ has been indexed using a monoclinic unit cell with a=8.546(15), b=10.263(21), c=5.887(7) Å, β=90.57(14)° [2, 4]; Z=4 [2] (Z=2 is erroneously given by [4]). A single crystal structure determination (R=0.041) gave the space group $P2_1-C_2^2$ (No. 4). Groups of two edge-sharing $ZnO_4(H_2O)_2$ octahedra at z≈0 in the structure are linked by slightly deformed MoO_4 tetrahedra at $z \approx \pm 0.2$ to form layers parallel to (001), see **Fig. 76**. The oxygen atom sharing has been represented by $[ZnO_{2/3}O_{2/2}(H_2O)_2MoO_{1/3}O_{2/2}O]^{2\infty}$. The O atoms coordinated to just one Zn are from the water of crystallization. Formation of hydrogen bonds to unshared O atoms of the MoO_4 tetrahedra and H_2O molecules of neighboring layers provides the bonding in the third dimension. Mean distances are Zn-O = 2.081, Zn-OH_2 = 2.125, Mo-O = 1.766 Å. For tables of positional parameters and interatomic distances see the paper. $ZnMoO_4 \cdot 2\,H_2O$ is isomorphous with $MgMoO_4 \cdot 2\,H_2O$ (see p. 182) [4]. The pycnometric density is 3.38 ± 0.05, the calculated density 3.36 g/cm^3 [2, 4].

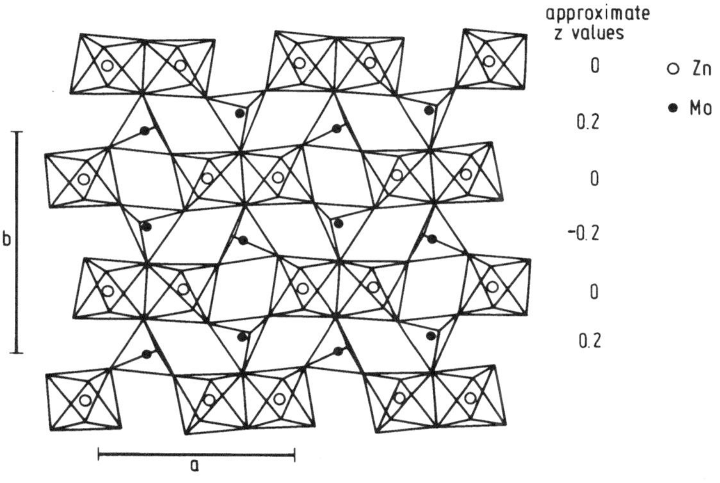

Fig. 76. The crystal structure of $ZnMoO_4 \cdot 2\,H_2O$ projected on (001) [4].

The IR spectrum has at least four poorly resolved absorption bands between 942 and 760 cm^{-1} and three from 428 to 297 cm^{-1}. For the spectrum see the paper [2].

The dry compound is stable under ordinary conditions. According to thermogravimetric analyses with a heating rate of 15 K/h, H_2O starts to split off rather rapidly at 90°C and from 120°C on with a decreasing rate. The weight becomes nearly constant at about 250°C. A DTA run at 500 K/h shows a large endothermic effect at 150°C and smaller ones at 180 and 265°C. X-ray powder studies at continuously rising temperatures show the room temperature crystal structure to be destroyed at 90°C. At least two intermediate compounds occur during dehydration but not the hemihydrate $ZnMoO_4 \cdot 0.5\,H_2O$. At 350°C the anhydrous α-$ZnMoO_4$ is obtained [2]. From other TG and DTA experiments with a heating rate of 400 K/h the decrease in weight and endothermal effect start at 163°C, the maximum Δt being at 278 ± 10°C. The sample weight becomes constant above this temperature [3]. $ZnMoO_4 \cdot 2\,H_2O$ is metastable in water at 25°C and transforms, for example, during five weeks of stirring to $ZnMoO_4 \cdot 0.5\,H_2O$ and finally, after several months, to $ZnMo_2O_7 \cdot 5\,H_2O$ and $Zn_3(OH)_2(MoO_4)_2$ [2].

References:

[1] J. Meullemeestre, É. Pénigault (Bull. Soc. Chim. France **1974** 9/12). – [2] J. Meullemeestre, É. Pénigault (Bull. Soc. Chim. France **1975** 1920/4). – [3] A. N. Zobnina, I. P. Kislyakov, N. V. Strelina (Izv. Akad. Nauk SSSR Neorgan. Materialy **2** [1966] 507/10; Inorg. Materials [USSR] **2** [1966] 437/9). – [4] J. Y. Le Marouille, O. Bars, D. Grandjean (Acta Cryst. B **36** [1980] 2558/60).

4.1.4 $ZnMoO_4 \cdot nH_2O$ (= $ZnO \cdot MoO_3 \cdot nH_2O$), n≈0.5 to 1.15

The hemihydrate $ZnMoO_4 \cdot 0.5H_2O$ occurs in a metastable state in the ZnO-MoO_3-H_2O system at 25°C, see p. 214.

A white compound with the ideal composition $ZnMoO_4 \cdot H_2O$, yet usually with an H_2O deficit, has been obtained by double decomposition of solutions of Na_2MoO_4 and a soluble zinc salt mixed in $Na_2O:MoO_3$ molar ratios between 0.6 and 1 [1, 2]. The same $ZnMoO_4 \cdot nH_2O$ has also been prepared by refluxing for 15 to 20 h mixtures of 0.1 mol ZnO and 0.103 to 0.3 mol MoO_3 in 0.5 L H_2O. The water content of different samples varies between 0.85 and 1.15 [3]. The rate of $ZnMoO_4 \cdot H_2O$ formation has been determined between 303 and 363 K in ZnO-MoO_3 pellets, pressed at 750 kg/cm², with a 3:1 molar ratio and 9 wt% H_2O. The activation energy is 5.7 kcal/mol and the reaction is diffusion controlled [4].

The needle-like crystals, obtained by double decomposition, are twinned with at least three twin domains having c as the common needle axis and $(1\bar{1}0)$ as twin planes. The lattice parameters of the triclinic unit cell are a=10.29, b=6.91, c=6.93 Å, α=105°14′, β=117°44′, γ=84°45′ [2]. For an indexed powder pattern and calculated d values see [3]. The IR spectrum of $ZnMoO_4 \cdot 0.8H_2O$ between 1100 and 250 cm^{-1} is depicted in the paper [3].

On storing in air under normal conditions, the water content of $ZnMoO_4 \cdot nH_2O$ tends to n≈0.8. Heating the 0.8 hydrate at 160°C for 5 h gives $ZnMoO_4 \cdot 0.5H_2O$ which rehydrates under ambient conditions. According to heating experiments (TG, DTA, X-ray diffraction at increasing temperature) the compound contains both adsorbed water and water of crystallization. The first is split off on heating up to 160°C without altering of the powder pattern, giving $ZnMoO_4 \cdot 0.5H_2O$. Above 220°C the remaining H_2O is lost and α-$ZnMoO_4$ forms [3]. $ZnMoO_4 \cdot 0.5H_2O$ is thermodynamically metastable in the presence of water. It decomposes gradually into $Zn_3(OH)_2(MoO_4)_2$ and $ZnMo_2O_7 \cdot 5H_2O$ when stirred with water for 75d [5].

References:

[1] H. Pezerat (Compt. Rend. **261** [1965] 5490/3). – [2] H. Pezerat (Compt. Rend. C **265** [1967] 368/71). – [3] J. Meullemeestre, É. Pénigault (Bull. Soc. Chim. France **1972** 3669/74). – [4] T. F. Romanyuk, D. I. Chemodanov (Izv. Vysshikh Uchebn. Zavedenii Khim. Khim. Tekhnol. **21** [1978] 1307/9; C.A. **90** [1979] No. 157628). – [5] J. Meullemeestre, É. Pénigault (Bull. Soc. Chim. France **1974** 9/12).

4.1.5 $ZnMo_2O_7 \cdot 5H_2O$ (= $ZnO \cdot 2MoO_3 \cdot 5H_2O$)

The pentahydrate is a stable compound in the ZnO-MoO_3-H_2O system at 25°C, see p. 214. White $ZnMo_2O_7 \cdot 5H_2O$ is obtained by stirring 0.032 mol ZnO and 0.064 mol MoO_3 in 400 mL H_2O for 21d at 25°C [1]. With shorter reaction times there might still be present some of the intermediately formed $ZnMo_2O_7 \cdot 2H_2O$ [2]. The pentahydrate forms very small, elongated crystals, which have been characterized by their X-ray powder pattern. For a table of the d values see the paper [1].

The IR spectrum shows absorption bands over the total region between 950 and 400 cm^{-1} (for a plot of the spectrum see the paper), and it has not been possible to identify the type of the anion present by comparison with IR spectra of other dimolybdates [1].

According to studies of the thermal degradation $ZnMo_2O_7 \cdot 5H_2O$ loses water starting at about 50°C, with a maximum rate at about 80°C, and is completely transformed into the dihydrate after several hours at 105°C [1].

References:

[1] J. Meullemeestre, É. Pénigault (Bull. Soc. Chim. France **1972** 868/72). – [2] J. Meullemeestre, É. Pénigault (Bull. Soc. Chim. France **1974** 9/12).

4.1.6 $ZnMo_2O_7 \cdot 2H_2O$ (= $ZnO \cdot 2MoO_3 \cdot 2H_2O$)

The dihydrate occurs in a metastable state in the ZnO-MoO_3-H_2O system at 25°C [1]. $ZnMo_2O_7 \cdot 2H_2O$ is formed intermediately on reacting ZnO and MoO_3 in H_2O at 25°C at molar ratios between 1:3.18 and 1:0.96 [1]. It can be prepared by heating the pentahydrate for several hours at 105°C [2].

The X-ray powder pattern is different from that of the pentahydrate (for the d values see the paper), while the IR spectra between 1000 and 250 cm^{-1} are identical. Thermal degradation of the dihydrate into the anhydrous $ZnMo_2O_7$ starts at about 200°C and is complete between 250 and 300°C. The anhydrous $ZnMo_2O_7$ rehydrates easily in moist air [2].

References:

[1] J. Meullemeestre, É. Pénigault (Bull. Soc. Chim. France **1974** 9/12). – [2] J. Meullemeestre, É. Pénigault (Bull. Soc. Chim. France **1972** 868/72).

4.1.7 $ZnMo_3O_{10} \cdot 5H_2O$ (= $ZnO \cdot 3MoO_3 \cdot 5H_2O$)

The pentahydrate occurs in the ZnO-MoO_3-H_2O system at 25°C; it dissolves incongruently, see p. 214. $ZnMo_3O_{10} \cdot 5H_2O$ is obtained when 0.2 mol ZnO and 0.6 mol MoO_3 in 0.5 L H_2O are refluxed and stirred vigorously for 2 d. After filtering the hot mixture, a white, very finely crystallized solid, a neutral zinc molybdate, deposits rather rapidly from the light green liquid. The solution is then held at room temperature in a closed flask. After one or two weeks silky, white crystals of the pentahydrate appear, very thin and up to 2 cm long. Cooling the mother liquor of the pentahydrate to 0°C precipitates crystals with another habit and higher water content (5.4 to 9.2 H_2O). Their X-ray powder pattern and IR absorption spectra are identical with those of the pentahydrate. If the initial molar ratio $ZnO:MoO_3$ is greater than 1:3 or less than 1:4, there will be contamination by $ZnMoO_4 \cdot nH_2O$ or MoO_3, respectively.

The d values are given in the paper. The IR spectrum in the range 1100 to 250 cm^{-1} shows three major absorption bands at about 920, 650, and 520 cm^{-1} and agrees with the IR spectra of $Na_2Mo_3O_{10} \cdot 4H_2O$ and $K_2Mo_3O_{10} \cdot 3H_2O$ (see pp. 64 and 80, respectively); for a plot see the paper. According to TG and DTA $ZnMo_3O_{10} \cdot 5H_2O$ is thermally dehydrated in two steps. The intermediate phase, the 3.75 hydrate, can be obtained from the pentahydrate after 17 h at 97°C.

J. Meullemeestre, É. Pénigault (Bull. Soc. Chim. France **1972** 868/72).

4.1.8 $ZnMo_3O_{10} \cdot 3.75H_2O$ (= $ZnO \cdot 3MoO_3 \cdot 3.75H_2O$)

This hydrate is the intermediate product (at 97°C) of the thermal dehydration of the pentahydrate. It may also be obtained by drying the pentahydrate for two weeks over silica gel.

The X-ray powder pattern (see the paper for a table of d values) is different from that of the pentahydrate, while the IR spectra of both of the compounds in the range 1100 to 250 cm^{-1} are identical. The 3.75 hydrate is not very stable at elevated temperatures. It degrades slowly even at about 100°C, but loss of water is nearly complete only at 200°C. Powder patterns of samples heated up to 145°C are still characteristic of $ZnMo_3O_{10}\cdot3.75H_2O$. Between 175 and 330°C the samples show a rather diffuse and poorly resolved pattern, which has been attributed to anhydrous $ZnMo_3O_{10}$. Higher temperatures form $ZnMoO_4$ plus $2MoO_3$.

J. Meullemeestre, É. Pénigault (Bull. Soc. Chim. France **1972** 868/72).

4.1.9 $Zn_2(OH)_2MoO_4$ ($=2ZnO\cdot MoO_3\cdot H_2O$)

This compound occurs in the ZnO-MoO_3-H_2O system at 25°C, see p. 214. White $Zn_2(OH)_2MoO_4$ is formed as prismatic needles when 0.1 mol ZnO and 0.05 mol MoO_3 in 0.5 L H_2O are refluxed and vigorously stirred for 4 h [1].

An orthorhombic unit cell with a = 17.62, b = 14.89, c = 12.71 Å has been obtained from the X-ray powder pattern. For a table with intensities, observed and calculated line positions see the paper. The IR spectrum shows two sharp absorption bands at 3590 and 3565 cm^{-1}, which have been attributed to stretching vibrations of hydroxyl groups, because no absorption could be observed in the range of the H_2O bending vibration at ~1600 cm^{-1}. This leads to the formulation of this compound with hydroxyl groups. There are five absorption bands in the range from 930 to 661 cm^{-1} and three at 457, 335, and 320 cm^{-1}, respectively; for a plot of the spectrum see the paper [1].

TG and DTA curves show the compound to be dehydrated between 320 and 355°C. After heating at 376°C $Zn_3Mo_2O_9$ and ZnO have been detected by X-ray diffraction [1]. A saturated solution contains only ~32 mg $Zn_2(OH)_2MoO_4$/L [2].

References:

[1] J. Meullemeestre, É. Pénigault (Bull. Soc. Chim. France **1972** 1702/7). – [2] J. Meullemeestre, É. Pénigault (Bull. Soc. Chim. France **1974** 13/6).

4.1.10 $Zn_3(OH)_2(MoO_4)_2$ ($=3ZnO\cdot2MoO_3\cdot H_2O$)

This compound occurs in the ZnO-MoO_3-H_2O system at 25°C, see p. 214. $Zn_3(OH)_2(MoO_4)_2$ is obtained as twinned needles, when 0.075 mol ZnO and 0.05 mol MoO_3 in 0.5 L H_2O are refluxed and vigorously stirred for 4 h [1]. A compound of the same composition also forms on double decomposition of an Na_2MoO_4 solution plus a soluble zinc salt with an initial mole ratio $Na_2O:MoO_3$ between 0.6 and 1. The reaction conditions were modified "to favor hydrolysis" [2].

A table with d values is given in [1]. The IR spectrum shows no absorption bands in the range of the H_2O bending vibration, thus the sharp absorption band at 3545 cm^{-1} has been attributed to the stretching vibration of hydroxyl groups. There are also absorption bands in the ranges from 950 to 690 cm^{-1} and from 458 to 292 cm^{-1}, but no conclusion has been drawn concerning the structure of the anion. For a plot of the spectrum see the paper [1]. The thermal analyses show that all H_2O is split off between 305 and 345°C, forming $Zn_3Mo_2O_9$ [1].

References:

[1] J. Meullemeestre, É. Pénigault (Bull. Soc. Chim. France **1972** 1702/7). – [2] H. Pezerat (Compt. Rend. **261** [1965] 5490/3).

4.1.11 Zinc Peroxomolybdates $ZnMoO_6 \cdot 9H_2O$ and $ZnMoO_5 \cdot nH_2O$

The orange, unstable $ZnMoO_6 \cdot 9H_2O$ can be obtained from 10 mL of an 80% H_2O_2 solution at −10°C to which 5 mL of a saturated aqueous Na_2MoO_4 and then 7.5 mL of a saturated $ZnSO_4$ solution are added. After cooling the solution for 2 h with solid CO_2 the peroxomolybdate is precipitated by adding cold (−20°C) ethanol, followed by washing with cold alcohol and ether. At very high H_2O_2 concentrations two highly unstable, red peroxomolybdates of probable compositions $ZnMoO_7$ and $ZnMoO_8$ are formed. Their water contents are not reported.

On standing in the air or on heating $ZnMoO_6 \cdot 9H_2O$ decomposes to cream colored $ZnMoO_5 \cdot nH_2O$ and later to normal white zinc molybdate. Zinc peroxomolybdates are thought to be the rate-determining species involved in the catalytic decomposition of H_2O_2 in the presence of Zn^{2+} and MoO_4^{2-}, which is a first-order reaction and has an activation energy of 12.8 kcal/mol.

M. V. Savina, G. A. Bogdanov, G. L. Petrova, G. K. Yurchenko (Zh. Fiz. Khim. **37** [1963] 746/52; Russ. J. Phys. Chem. **37** [1963] 391/4).

4.1.12 $[Zn(NH_3)_4]MoO_4 \cdot 2H_2O$ (= $ZnO \cdot MoO_3 \cdot 2H_2O \cdot 4NH_3$)

This compound has been obtained by passing gaseous NH_3 through an aqueous solution of $ZnMoO_4$ at 5 to 10°C.

The colorless crystals occur in two habits: short hexagonal columns with $\{10\bar{1}0\}$ faces or $\{11\bar{2}1\}$ trigonal pyramids. Usually the faces appear to be striated horizontally. The d values are given in the paper. The Mohs hardness is 2.5. The crystals are optically uniaxial with an average refractive index of 1.718 and show a weak pleochroism in slightly yellow colors. Comparison of the IR spectrum between 3600 and 250 cm^{-1} with the spectra of other molybdates indicates that there are MoO_4^{2-} ions in the structure of $[Zn(NH_3)_4]MoO_4 \cdot 2H_2O$.

Thermal degradation takes place in four steps: at 50°C about 8.6 wt% (≙1 NH_3 and 0.6 H_2O) has been lost and on further heating at 120°C 17.5 wt% (≙2 NH_3 and 1.3 H_2O), at 240°C 0.4 NH_3, and at 290°C 0.5 NH_3. Above 290°C anhydrous $ZnMoO_4$ is obtained. Solutions contain the ion pair $[Zn(NH_3)_4]^{2+} \cdot MoO_4^{2-}$ which has the instability constant $(1.65 \pm 0.85) \times 10^{-4}$.

A. N. Zelikman, I. G. Kalinina, L. V. Myakisheva, B. L. Egorov (Izv. Vysshikh Uchebn. Zavedenii Tsvetn. Met. **1978** No. 1, pp. 71/5; C.A. **88** [1978] No. 177938).

4.2 Metal Zinc Molybdates

4.2.1 $NaZn(OH)MoO_4$ (= $Na_2O \cdot 2ZnO \cdot 2MoO_3 \cdot H_2O$)

This compound has been obtained in very small yields by hydrothermal synthesis from solid $NaZn_2(OH)(MoO_4)_2 \cdot H_2O$ (see p. 224) and aqueous 4 M Na_2MoO_4. A mixture in the ratio 1 g solid : 10 mL solution was placed inside a Teflon-lined stainless steel bomb and held at 220 to 230°C for 3 weeks. Crystals of the product had to be sorted out by handpicking from unreacted $NaZn_2(OH)(MoO_4)_2 \cdot H_2O$ [1].

The crystals are orthorhombic, the lattice parameters are a = 7.850(1), b = 9.2922(8), c = 6.148(1) Å; Z = 4 [1]. A reinvestigation showed Pnam-D_{2h}^{16} (No. 62) [2] to be the right space group and not the lower symmetry Pna2_1-C_{2v}^{9} (No. 33) [1]. Atomic coordinates (R = 0.025) [2]:

atom	position	x	y	z
Na	4c	0.36433(5)	0.3104(3)	0.75
Zn	4b	0.5	0	0.5
Mo	4c	0.12379(5)	0.17058(5)	0.25
O(1,2)	8d	0.1140(4)	0.2785(3)	0.0153(4)
O(3)	4c	0.4529(7)	0.4480(6)	0.25
O(4)	4c	0.8691(5)	0.4357(4)	0.75
O(5)	4c	0.3125(5)	0.0617(5)	0.25
H	4c	0.783(13)	0.457(10)	0.75

In the structure ZnO_6 octahedra share trans edges and form linear chains parallel to [001]. Except for one per octahedron all oxygen atoms of these chains are shared with three of the apices of the MoO_4 tetrahedra, thus linking the chains perpendicular to [001]. The oxygen atom O(4), which is coordinated to Zn but not to Mo, has been identified as a hydroxyl O. Sodium occupies seven-coordinated voids. Drawings of the structure (in Pna2_1) are given in the paper [1]. The interatomic distances are more regular in Pnam than those obtained in Pna2_1: Mo-O = 1.736 to 1.794, Zn-O = 1.943 to 2.246, Na-O = 2.287 to 2.716, O(4)-H = 0.71, and H···O(3) = 2.05 Å [2].

The density determined by a flotation method is ~4 g/cm³, in good agreement with the calculated density 3.965 g/cm³ [1].

References:

[1] A. Clearfield, R. Gopal, C. H. Saldarriaga-Molina (Inorg. Chem. **16** [1977] 628/31). – [2] R. E. Marsh, V. Schomaker (Inorg. Chem. **18** [1979] 2331/6).

4.2.2 $M_{2-x}(H_3O)_xZn_2Mo_2O_9$ (= (2−x)$M_2O \cdot 4ZnO \cdot 4MoO_3 \cdot 3xH_2O$), M = K, NH_4; x = 0.6 to 0.9 for NH_4

These compounds with M = K and NH_4 have been designated as phases φ_y [1]. $H_2M_2Zn_2Mo_2O_{10}$ is given as the ideal formula of these nonstoichiometric compounds [2, 3]. The compound NaZn(OH)MoO_4 (see above) has also been assumed to be a member of the φ_y phases; however, it has a completely different crystal structure and is stoichiometric [4]. The white compounds have been obtained by double decomposition of 1:1 molar mixtures of semiconcentrated solutions of potassium or ammonium molybdates, M_2MoO_4, and a soluble zinc salt. The potassium salt precipitates for initial M_2O : MoO_3 (M = K) molar ratios between 1.4 and 1.6, the ammonium salt (M = NH_4) between 0.6 and 4.2 [1]. Chemical analysis of four different samples of the NH_4 compound shows that the sum of NH_4 and O never exceeds 11 and that NH_3 contents range between 1.1 and 1.4 per rhombohedral unit cell, which leads to the formulation as $(NH_4)_{2-x}(H_3O)_xZn_2Mo_2O_9$ with x = 0.6 to 0.9 [2, 3].

Single crystals frequently are bullet-shaped (in French "obus" [3]) with diameters up to 0.1 mm and the longest dimension along the threefold axis. X-ray studies of powders and single crystals yielded rhombohedral symmetry and the space group $R\bar{3}m$-D_{3d}^5 (No. 166). The (hexagonal) lattice parameters depend on the precipitation conditions; average values of four samples are a = 6.100, c = 21.51 Å for K and a = 6.113, c = 21.80 Å for NH_4 [3]. For the NH_4 compound the lattice parameters vary within the following limits: $6.103 \leqq a \leqq 6.121$ and $21.67 \leqq c \leqq 21.89$ Å [2, 3]. The crystal structure has been determined on a sample with a = 6.104, c = 21.67 Å; Z = 3 (D_x = 3.59 g/cm³). Atomic positions are (R = 0.09) [2]:

atom	position	occupancy	x	y	z
N	3b	3	0	0	0.5
Zn	9e	6	0.5	0	0
Mo	6c	6	0	0	0.917
O(1)	6c	6	0	0	0.173
O(2)	6c	6	0	0	0.298
O(3)	18h	18	0.1587	−0.1587	0.0624

The proposed crystal structure is built up of sheets, which may be deduced from brucite-type layers, consisting of very distorted ZnO_6 octahedra sandwiched between two sheets of MoO_4 tetrahedra, see **Fig. 77**. These layers are stacked rhombohedrally and linked by the NH_4^+ ions. As the proposed structure affords only one NH_4^+ ion per asymmetric unit (instead of 1.1 to 1.4 [2, 3]) additional NH_4^+ ions are considered to replace H_2O molecules statistically. There are 33.3% vacancies on the Zn position. The role of the hydrogen ions is less clear [2].

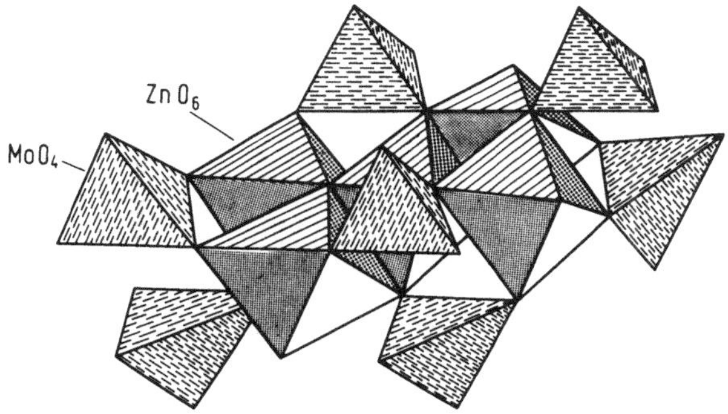

Fig. 77. Projection of the characteristic sheet of the $M_{2-x}(H_3O)_xZn_2Mo_2O_9$ structure [2].

The compounds $M_{2-x}(H_3O)_xZn_2Mo_2O_9$ are slightly soluble in cold water and do not decompose on 1h boiling in H_2O. On heating they lose H_2O and NH_3 between 200 and 300°C [3].

References:

[1] H. Pezerat (Compt. Rend. **261** [1965] 5490/3). – [2] H. Pezerat (Bull. Soc. Franc. Mineral. Crist. **90** [1967] 549/57; Structure Reports, Vol. A 32, 1967, pp. 301/2). – [3] H. Pezerat, I. Mantin, S. Kovacevic (Compt. Rend. C **263** [1966] 60/3). – [4] A. Clearfield, R. Gopal, C. H. Saldarriaga-Molina (Inorg. Chem. **16** [1977] 628/31).

4.2.3 $MZn_2(OH)(MoO_4)_2 \cdot H_2O$ ($= M_2O \cdot 4ZnO \cdot 4MoO_3 \cdot 3H_2O$), M = Na, K

White $NaZn_2(OH)(MoO_4)_2 \cdot H_2O$ has been prepared from 1M aqueous solutions of $ZnSO_4$ or $ZnCl_2$ and Na_2MoO_4 by dropwise addition of one of the solutions to the other, with heating and stirring. The mixture was refluxed for 24 h. Recrystallization of the polycrystalline product under hydrothermal conditions (160°C for 5 weeks) using 1M Na_2MoO_4 solution as the liquid phase yielded colorless single crystals [1]. The compounds with M = Na and K, generally with some deviations from stoichiometry, have been obtained from equimolar mixtures of moderately concentrated solutions of Na_2MoO_4 and a soluble Zn salt at various pH values. The Na compound precipitates for initial $M_2O:MoO_3$ molar ratios between 1.0 and 1.6 and the K compound between 0.7 and 1.4 [2]. The intermediately formed $Zn(OH)_2$ reacts gradually over 5 to 15 d to form the product phase. The lattice parameters of samples taken at the beginning of the precipitation differed from those at the end [3].

X-ray investigations [1, 4] show the crystal structure of $NaZn_2(OH)(MoO_4)_2 \cdot H_2O$ to be monoclinic, space group C2/m-C_{2h}^3 (No. 12); lattice parameters are a = 9.436(2), b = 6.338(1), c = 7.679(1) Å, β = 115.8(1)° at 24 ± 2°C; Z = 2 [1]. The nonstoichiometric samples have lattice parameters a = 9.45, b = 6.34, c = 7.66 Å, β = 115°56′ for M = Na and a = 9.61, b = 6.50, c = 7.88 Å, β = 116°2′ for M = K [3]. The crystal structure was first determined using the isomorphous K-Co compound [4]. The results were confirmed with more precision and reliability (R = 0.043 for 789 reflections) for a hydrothermally grown crystal, which in the course of the refinement was found to be stoichiometric. The atomic parameters are [1]:

atom	position	x	y	z
Na	2d	0	0.5	0.5
Zn	4e	0.25	0.25	0
Mo	4i	0.0812(1)	0	0.2879(1)
O(1)	4i	0.2335(9)	0	0.5198(11)
O(2)	8j	0.4647(6)	0.2744(9)	0.2602(6)
O(3)	4i	0.1695(9)	0	0.1164(11)
O(4)	4i	0.6480(8)	0	0.0820(9)

As for the structure proposed for $(NH_4)_{2-x}(H_3O)_xZn_2Mo_2O_9$, the crystal structure of $NaZn_2(OH)(MoO_4)_2 \cdot H_2O$ can be deduced from brucite-type layers [4]. The octahedrally coordinated Zn atoms lie in the (001) plane and are linked to each other by common edges, thus forming chains parallel to [010]. The chains are connected to each other by tetrahedral molybdate groups, forming zinc-molybdate groups parallel to the (001) plane. The layers are linked together by Na^+ ions lying halfway between them. The placement of the H atoms was based on indirect considerations and resulted in the formula $NaZn_2(OH)(H_2O)(MoO_4)_2$. Interatomic distances in Å: Zn-O = 2.088 to 2.143, average 2.116(28); Mo-O = 1.737 to 1.839, average 1.773(45); Na-O = 2.448 and 2.583. For tables with other interatomic distances and angles see the paper [1]. Portions of the structure are shown in [1], see also [4], for models of the hydrogen bonds see [5, 6]. Other aspects of this structure type with respect to the observed nonstoichiometry have been studied extensively in isomorphous compounds with M^{II} = Mg, Co, Ni in place of Zn [6 to 8]; for compounds with Mg see p. 190.

The density measured by flotation is ~3.9 ± 0.1 g/cm³, the calculated density is 4.089 g/cm³, for M = Na [1].

IR spectra of nonstoichiometric $MZn_2(OH)(MoO_4)_2 \cdot H_2O$ and of samples in which H is replaced by D [5, 6] yield the following assignments for the range below 1000 cm^{-1} (absorption maxima in cm^{-1}; ν refers to MoO_4 tetrahedra, ν' to ZnO_6 octahedra) [6]:

M	ν_1	ν_3	OH vibration	ν_2 and ν'_1	ν'_3	ν_4	ν'_2
Na	935	875, 835, 795		405	335	300	
K	925	885, 845, 775	715	420	360	325	300

Absorption frequencies due to OH and H_2O vibrations are summarized as follows [6]:

M	valence	deformation	rocking-wagging of H_2O or bending of OH
Na	3360, 3140	1595, 1135	
K	3440, 3190	1570, 1010	715
K(deuterized)	2550, 2370	1160, 745	530

For plots of the spectra of isomorphous compounds see the papers [5, 6]. The ratio $\nu_{OH}:\nu_{OD} = 1.35$ for all vibrations. The frequency of the deformation vibration of H_2O is regarded as particularly low, which is the more striking due to the fact that all H^+ (or D^+) are involved in hydrogen bonding forming micro chains O···H-O-H···O-H···O [5, 6].

From thermal analyses $NaZn_2(OH)(MoO_4)_2 \cdot H_2O$ loses 0.4wt% at 170°C, 1wt% at 200°C, and 6.4wt% at 405°C [1]. The K compound does not lose water below 150°C. At 280°C it is transformed into $K_{2-x}(H_3O)_xZn_2Mo_2O_9$ (see above). Both the Na and K compounds are insoluble in water and resist hydrolysis even in boiling water [3].

References:

[1] A. Clearfield, M. J. Sims, R. Gopal (Inorg. Chem. **15** [1976] 335/8). – [2] H. Pezerat (Compt. Rend. **261** [1965] 5490/3). – [3] H. Pezerat, I. Mantin, S. Kovacevic (Compt. Rend. C **262** [1966] 95/8). – [4] H. Pezerat (Bull. Soc. Franc. Mineral. Crist. **90** [1967] 549/57; Structure Reports, Vol. A 32, 1967, pp. 301/2). – [5] M. J. Peltre, J. Guignard, H. Pezerat (J. Chim. Phys. **76** [1979] 104/6).

[6] M. J. Peltre, D. Olivier, H. Pezerat (J. Solid State Chem. **24** [1978] 57/70). – [7] M. J. Peltre, H. Pezerat (J. Solid State Chem. **23** [1978] 19/32). – [8] M. J. Peltre, H. Pezerat (J. Solid State Chem. **26** [1978] 245/54).

4.2.4 Ammonium Hexamolybdozincates

$(NH_4)_4[Zn(OH)_6Mo_6O_{18}] \cdot nH_2O$ ($= 2(NH_4)_2O \cdot ZnO \cdot 6MoO_3 \cdot (n+3)H_2O$), n = 4 or 5; **$(t\text{-}C_4H_9NH_3)_4[Zn(OH)_6Mo_6O_{18}]$**

For the preparation of $(NH_4)_4[Zn(OH)_6Mo_6O_{18}] \cdot 5H_2O$ a solution of 0.034 mol ammonium paramolybdate $(NH_4)_6Mo_7O_{24} \cdot 4H_2O$ in 250 mL H_2O is boiled down to the point of incipient precipitation. After filtration and subsequent cooling to room temperature, 0.034 mol $ZnSO_4 \cdot 7H_2O$ in 150 mL H_2O and 0.034 mol NH_4Cl are added. The mixture is held overnight in a refrigerator. The precipitated polymolybdates are then filtered off and washed with water. When the washings are added to the mother liquor, it becomes turbid and $(NH_4)_4[Zn(OH)_6Mo_6O_{18}] \cdot 5H_2O$ precipitates [1]. A tetrahydrate has been prepared by addition of 10 g $ZnCl_2$ in 50 mL H_2O to a solution of 50 g ammonium paramolybdate in 200 mL H_2O, both at 60 to 70°C. $(NH_4)_4[Zn(OH)_6Mo_6O_{18}] \cdot 4H_2O$ crystallized after standing for some time [2].

$(t\text{-}C_4H_9NH_3)_4[Zn(OH)_6Mo_6O_{18}]$ has been prepared in organic solvents. To 25 mL of an 0.06 M solution of t-butyl ammonium molybdate, $(t\text{-}C_4H_9NH_3)_2MoO_4$, in CH_3OH were added 0.5 mL glacial acetic acid and 8.5 mL of 0.03 M $Zn(NO_3)_2 \cdot 6H_2O$ in $(CH_3)_2CO$. The product precipitated as fine crystals after prolonged standing. The crystals can be recrystallized from methanol, but with a very bad yield [3].

X-ray powder photographs show the crystal structure of $(NH_4)_4[Zn(OH)_6Mo_6O_{18}] \cdot 5H_2O$ to be the same as that of other heteropoly salts of the same formula type, where Zn is replaced by Cu, Ni, or Co, and also very similar to the crystal structure of, for example, potassium hexamolybdochromate(III) (cf. p. 353). For d values see the papers [1, 4]. X-ray photoelectron spectra of $(NH_4)_4[Zn(OH)_6Mo_6O_{18}] \cdot 4H_2O$ samples at least partially dehydrated at 100°C in an ultrahigh vacuum show binding energies as follows (in eV, relative to C 1s = 285 eV): Mo $3d_{5/2}$ = 233.1, O 1s = 531.9, Zn $3s_{1/2}$ = 139.8, N 1s = 402.2. From the agreement between these values and those for other salts $M^{I}_{(6-n)}[M^{n+}(OH)_6Mo_6O_{18}] \cdot mH_2O$ with $M^{n+} = Al^{3+}$, Ga^{3+}, Cr^{3+}, Fe^{3+}, Co^{3+}, Rh^{3+}, Co^{2+}, Ni^{2+}, and Cu^{2+}, it was concluded that these heteropoly ions are isostructural. The NMR spectrum at 80 K shows a signal consisting of two superimposed lines. The line with a width of 4 G confirms the model that the protons form hydroxyl groups at the oxygen atoms coordinating the Zn atoms [2]. The IR and Raman spectral wave numbers (in cm^{-1}) are:

$C_4H_9NH_3$ [3] IR	Raman	NH_4, $5H_2O$ [1,4] IR	NH_4, $4H_2O$ [2] IR	assignment [2]
		1675	1665	$\delta(H_2O)$
		1625	1630	
		1412	1440	$\delta(NH_4)$
			1405	
1035	1035		1030	$\delta(OH)$
		967		
935	944	933	930	$\nu(MoO_2)$
	923	925		
915		908		
	897	895	900	
885		875	875	$\nu(MoOMo)$ $\nu(MoOZn)$
		810	800	
	740			
640	657		630	
590			570	
536	545		550	
			470	

Three bands at 3500, 3400, and 3170 cm^{-1} have been found for the NH_4 compound [1] and some other bands at low wave numbers down to 106 cm^{-1} for the $C_4H_9NH_3$ compound [3]. For plots of the spectra see the papers [1, 4]. The ammonium salts are diamagnetic [2, 4].

The ammonium pentahydrate salt loses water in three steps [1, 4]: $3H_2O$ are split off between 35 and 60°C, $2H_2O$ between 100 and 110°C, and the constitutional water ($3H_2O$) at 130 to 150°C. The enthalpy changes are 16.3 for the first step, 5.25 for the second, and 71 kcal/mol for the third. The activation energy of the last step is 32.2 kcal/mol. At 180°C ammonium trimolybdate is obtained, which decomposes on further heating [4]. Above 400°C the residual compounds are $ZnMoO_4$ and MoO_3 [1].

The $C_4H_9NH_3$ compound loses H_2O and $C_4H_9NH_2$ completely up to 220°C. The remaining material contains a bluish black molybdenum oxide. The $C_4H_9NH_3$ compound is soluble in H_2O, methanol, and dimethylsulfoxide [3].

References:

[1] A. La Ginestra, A. Delli Quadri (Atti Accad. Nazl. Lincei Classe Sci. Fis. Mat. Nat. Rend. [8] **41** [1966] 521/6). – [2] L. P. Kazanskii, S. Holguin Quinones, B. N. Ivanov-Emin, L. A. Filatenko (Koord. Khim. **4** [1978] 1676/83; Soviet J. Coord. Chem. **4** [1978] 1279/86). – [3] J. Fuchs, I. Brüdgam (Z. Naturforsch. **32b** [1977] 403/7). – [4] A. La Ginestra, F. Giannetta, P. Fiorucci (Gazz. Chim. Ital. **98** [1968] 1197/212).

4.3 Cadmium Molybdate Hydrates

4.3.1 $CdMoO_4 \cdot nH_2O$ (= $CdO \cdot MoO_3 \cdot nH_2O$), $n \approx 1.3$ to 1.6?

Samples with a water content after air drying of 1.3 to 1.6 H_2O per formula unit have been obtained by mixing approximately equimolar amounts of aqueous 0.1 or 0.5 M $CdCl_2$ and Na_2MoO_4 solutions at pH 5.5 and 25°C, with equilibration for 48 h [1]. A similar method, mixing of equal amounts of 0.1 M cadmium salt and molybdate solutions and stirring for several days, yields nearly anhydrous $CdMoO_4$ (~0.15 wt% H_2O). The identical result was achieved by stirring a slurry of equimolar amounts of CdO and MoO_3 for several days [2]; cf. also "Molybdän" Erg.-Bd. B 2, 1976, p. 68. On heating $CdMoO_4 \cdot nH_2O$ loses water at 100°C [1].

References:

[1] A. N. Zobnina, I. P. Kislyakov (Izv. Vysshikh Uchebn. Zavedenii Khim. Khim. Tekhnol. **13** [1970] 143/7; C.A. **73** [1970] No. 41306). – [2] J. Meullemeestre (Bull. Soc. Chim. France **1978** I 95/100).

4.3.2 $[Cd(MoO_4)_3]^{4-}$

Acid solutions (pH 3 to 4) of $Cd(NO_3)_2$ and $(NH_4)_2MoO_4$ form a complex of composition $[Cd(MoO_4)_3]^{4-}$. The dissociation constant has been determined by solubility, conductivity, and ion exchange methods with values ranging from 1.3×10^{-5} to 6.7×10^{-5}.

L. I. Lebedeva, G. V. Kuznetsova (Vestn. Leningr. Univ. Fiz. Khim. **22** No. 4 [1967] 106/12; C.A. **67** [1967] No. 60465).

4.3.3 $Cd_5(OH)_2(MoO_4)_4 \cdot 0.8H_2O$ (= $5CdO \cdot 4MoO_3 \cdot 1.8H_2O$)

Mixtures of 0.1 M aqueous solutions of Na_2MoO_4 and excess $CdCl_2$ yield at pH 6 to 8.5 a solid of composition $Cd_5(OH)_2(MoO_4)_4 \cdot nH_2O$ with $n \approx 0.8$, after equilibration for 48 h. Water is lost on heating at 110°C. The anhydrous compound decomposes at 355°C into CdO, $CdMoO_4$, and H_2O.

A. N. Zobnina, I. P. Kislyakov (Izv. Vysshikh Uchebn. Zavedenii Khim. Khim. Tekhnol. **13** [1970] 143/7; CA. **73** [1970] No. 41306).

4.3.4 Cadmium Peroxomolybdates $CdMoO_8 \cdot nH_2O$, $CdMoO_7 \cdot nH_2O$, $CdMoO_6 \cdot nH_2O$, $CdMoO_5 \cdot nH_2O$

Brick red $CdMoO_6 \cdot nH_2O$ has been obtained by mixing 5 mL saturated aqueous Na_2MoO_4 with 10 mL 80% H_2O_2 at −10°C and adding a saturated $CdSO_4$ solution with cooling. The compound was precipitated by ethanol cooled to −20°C, filtered off, and washed with cold ethanol and ether. The freshly prepared product contains $9H_2O$, which do not, of course, all represent water of crystallization. At room temperature or in air the very unstable compound decomposes to pale green $CdMoO_5 \cdot nH_2O$ and then to $CdMoO_4$. When the concentration of H_2O_2 in the solution is very high, hydrates of extremely unstable peroxomolybdates, $CdMoO_8 \cdot nH_2O$ and $CdMoO_7 \cdot nH_2O$, appear in solution.

G. A. Bogdanov, M. V. Savina (Zh. Fiz. Khim. **38** [1964] 1539/44; Russ. J. Phys. Chem. **38** [1964] 834/7).

4.4 Metal Cadmium Molybdates

4.4.1 $LiCd(OH)MoO_4$ ($= Li_2O \cdot 2CdO \cdot 2MoO_3 \cdot H_2O$)

Transparent yellowish crystals of this compound have been obtained as an intermediate product of the exchange reaction, $CdMoO_4 + 2LiCl \rightarrow CdCl_2 + Li_2MoO_4$, under hydrothermal conditions at 400 to 450°C, a temperature gradient of 15 K, 20 to 25 wt% LiCl solution, and a filling factor of 65 to 75% [1]. Chemical analysis reveals a considerable deviation from stoichiometry: $Li_{0.8}Cd(OH)_{0.98}MoO_4 \cdot 0.16H_2O$ [2]. The compound has also been formulated as $Li_5Cd_6(OH)_5(MoO_4)_6 \cdot H_2O$ [1], a description which is not in agreement with the structural data.

The crystals have rhombohedral shapes with well developed faces [1]. X-ray powder diffraction studies [1] (for the powder pattern and d values see the papers [1, 2]) and a single crystal structure determination [2] show $LiCd(OH)MoO_4$ to be monoclinic, the space group is $P2_1/a\text{-}C_{2h}^5$ (No. 14); lattice parameters a = 7.91(10), b = 5.95(20), c = 9.77(10) Å, β = 107°30(30)′; Z = 4. All atoms occupy the general position 4e; R = 0.16. However, according to Structure Reports recalculations of some Cd-O, Mo-O distances seem to indicate that the published coordinates refer to the $P2_1/c$ setting:

atom	x	y	z
Li	0.422	0.602	0.173
Cd	0.061	0.735	0.122
Mo	0.315	0.078	0.963
O(1)	0.435	0.925	0.120
O(2)	0.355	0.378	0.971
O(3)	0.285	0.5[?] 71	0.235
O(4)	0.146	0.410	0.515
O(5)(OH)	0.107	0.930	0.374

The oxygen sublattice is in good approximation a hexagonal close packed arrangement, the hexagonal nets of which are parallel to ($20\bar{1}$). Some of the octahedral interstices are occupied by Cd atoms, thus forming zig-zag chains of edge-sharing CdO_6 octahedra along [010]. The chains are connected by common corners into sheets parallel to (100). The Mo and probably also the Li atoms are in tetrahedral interstices between the sheets. Their coordination tetrahedra share corners among themselves and with the CdO_6 octahedra, see **Fig. 78**. The interatomic distances Mo-O range from 1.69 to 1.83, Cd-O from 2.23 to 2.45, Li-O from 1.92 to 2.07 Å. The measured density is 4.26, the calculated density 4.49 g/cm³ for the idealized formula [2].

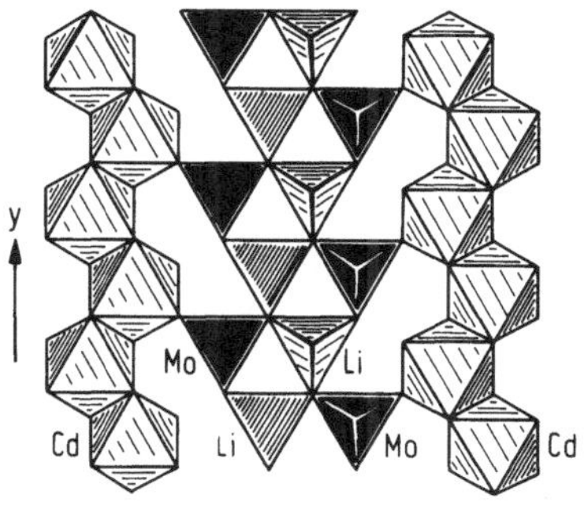

Fig. 78. The crystal structure of $LiCd(OH)MoO_4$ [2].

According to IR spectra protons are bound in hydroxyl groups; the peak at 3430 cm^{-1} is assigned to the stretching mode of OH. Absorption bands at 840 and 786 cm^{-1} have been attributed to the MoO_4 tetrahedra while the bands at 1160 and 1097 cm^{-1} have been assigned to LiO_4 tetrahedra [2].

Thermal analyses confirm that there are no H_2O molecules in the crystal structure [2]; plots up to ~700°C are given in [1]. The crystals are insoluble in water but readily soluble in dilute HCl [1].

References:

[1] L. N. Dem'yanets (Izv. Akad. Nauk SSSR Neorgan. Materialy **5** [1969] 2164/7; Inorg. Materials [USSR] **5** [1969] 1847/9). – [2] B. M. Kobtsev, Yu. A. Kharitonov, E. A. Pobedimskaya, N. V. Belov (Dokl. Akad. Nauk SSSR **179** [1968] 84/7; Soviet Phys.-Dokl. **13** [1968] 193/5; Structure Reports Vol. A 33, 1968, pp. 337/8).

4.5 Mercury Molybdate Hydrates

4.5.1 Mercury Peroxomolybdates $HgMoO_8 \cdot nH_2O$, $HgMoO_6 \cdot nH_2O$, $HgMoO_5 \cdot nH_2O$

The brown $HgMoO_8 \cdot nH_2O$ can be obtained at −10°C by addition of 1.5 to 2 mL of a saturated solution of $Hg(NO_3)_2$ in 6 N HNO_3 to a mixture of 5 mL saturated Na_2MoO_4 and 10 mL 80% H_2O_2. The precipitate was washed several times with cold ethanol and ether. The freshly prepared product has a large water content, n = 50 to 60. It can be stored only at the dry ice temperature. During drying over P_2O_5 in a refrigerator it is transformed into beige $HgMoO_6 \cdot nH_2O$, n ≈ 1.5 to 2, which can be stored at −10°C. In air at room temperature it decomposes to pale beige $HgMoO_5 \cdot nH_2O$, n = 1 to 2, and then to yellow $HgMoO_4$. In solution $HgMoO_5$ decomposes between 55 and 75°C with an activation energy of 25 kcal/mol. $HgMoO_6 \cdot nH_2O$ and $HgMoO_5 \cdot nH_2O$ can be obtained directly by using 30 to 80% H_2O_2.

G. A. Bogdanov, M. V. Savina, G. L. Petrova (Zh. Fiz. Khim. **37** [1963] 1258/63; Russ. J. Phys. Chem. **37** [1963] 671/4).

5 Molybdate Hydrates with Main Group 3 Metals

5.1 Aluminium Molybdate Hydrates

Older data are given in "Molybdän", 1935, p. 303.

5.1.1 The 2 Al^{III}-Mo^{V} Associate in Aqueous Solution

Addition of $Al_2(SO_4)_3$ to an aqueous 4.92 M HCl solution of 0.144 M Mo^V (chloride or sulphate) at 25°C increases the ^{1}HNMR relaxation rate. Also, the absorption of light at 445 nm is influenced, and the polarogram shows a new reduction wave at the more positive potential $E_{1/2} = -0.605$ V, compared to -0.745 V (relative to the saturated calomel electrode) for Mo^V in 4 to 6 M HCl. The formation of a paramagnetic ion associate with the molar ratio Al : Mo = 2 : 1 is deduced from these data. The association constant is given as $K_{ass} = 39.6$ in 5.3 M HCl and 57.3 in 10 M H_2SO_4.

The UV absorption spectrum of Mo^V shows that the Al^{3+} ions have no influence on the Mo←O charge-transfer band. Thus, the reaction of Mo^V with Al^{3+} takes place through the bridging ligands OH^- and H_2O: Mo^V-OH^-(H_2O)-Al^{3+}.

Z. A. Saprykova, N. D. Chichirova, N. G. Kulikova (Zh. Neorgan. Khim. **27** [1982] 100/3; Russ. J. Inorg. Chem. **27** [1982] 57/9).

5.1.2 $2Al_2O_3 \cdot 2MoO_3 \cdot 13H_2O$, $2Al_2O_3 \cdot 3MoO_3 \cdot 14H_2O$

Compositions $\mathbf{2Al_2O_3 \cdot 2MoO_3 \cdot 13H_2O}$ (or $2Al_2O_3 \cdot 2.2MoO_3 \cdot 13H_2O$ [2]) precipitate from mixtures of aqueous solutions of $Al(NO_3)_3$ and Na_2MoO_4 at pH 4.3 to 6.3 for initial Al : Mo molar ratios between 3 : 7 and 1 : 9. The precipitate is X-ray amorphous [1]. ^{1}HNMR measurements at 22°C show the protons of adsorbed water to have interproton distances of 1.54 ± 0.1 Å and those of the OH groups 2.5 Å. From the temperature dependence of the NMR signal between 93 and 373 K, the potential barriers between equilibrium positions of water were calculated to be 7.4 ± 0.2 kcal/mol [2]. The average refractive index n = 1.598 [1]. Heating the compound up to 550°C decomposes it to $Al_2(MoO_4)_3$ and Al_2O_3 [1].

At pH values between 3.8 and 4.3 and the initial Al : Mo molar ratio between 9 : 1 and 3.6 : 6.4, an X-ray amorphous precipitate of composition $\mathbf{2Al_2O_3 \cdot 3MoO_3 \cdot 14H_2O}$ is obtained [1]. Between pH 3.3 and 3.9, the precipitate composition was reported to be independent of the initial $Al_2O_3 : MoO_3$ ratio between 1 : 1 and 1 : 12. Drying at 105 to 110°C yields a microcrystalline hygroscopic powder [3]. The compound has been characterized by the solid state ^{1}HNMR spectrum analogous to the $2Al_2O_3 \cdot 2MoO_3 \cdot 13H_2O$ [2]. The refractive index n = 1.642 [1].

Heating up to 550°C decomposes the compound to $Al_2(MoO_4)_3$ and Al_2O_3. The solubility in water is 0.05 g/L at 25°C [1]. For solubilities at 18, 50, and 75°C, see [3]. In 2 N to 8 N HCl, the solubility increases with the HCl concentration (0.17 to 0.55 g/L, respectively) forming the heteropoly anion $[Al(OH)_6Mo_6O_{18}]^{3-}$ (see p. 232). In 2 N to 8 N NaOH, the solubility is similar to that in the HCl solutions, but forming MoO_4^{2-} [1]. For solubilities in 1% solutions of NH_4Cl, NH_4NO_3, and $(NH_4)_2SO_4$ at 25°C, see [3].

References:

[1] L. F. Klyuevskii, M. V. Mokhosoev (Khim. Tekhnol. Molibdena Vol'frama **1974** 116/22; Ref. Zh. Khim. **1975** No. 4V18). – [2] L. A. Pozharskaya, V. G. Pitsyuga, M. V. Mokhosoev, L. F. Klyuevskii (Tr. Buryat. Inst. Estestv. Nauk Buryat. Fil. Sib. Otd. Akad. Nauk SSSR No. 14 [1977] 394/9; C.A. **94** [1981] No. 53356). – [3] Yu. V. Morachevskii, L. G. Shipunova (Vopr. Anal. Khim. Mineral. Veshchestv **1966** 159/64; C.A. **66** [1967] No. 32367).

5.2 Hexamolybdoaluminates

Older data are given in "Molybdän", 1935, pp. 373/4.

5.2.1 General Data

While in the older literature the compounds were formulated as $M_3H_6[Al(MoO_4)_6] \cdot n\,H_2O$ (see, e.g., "Molybdän", 1935, pp. 373/4), later there has been some controversy about the composition and the structure of the free acid and the salts (as well as of related compounds where Al^{3+} is replaced by trivalent transition metal ions or Zn^{2+}). The anion has been formulated as $[AlO_6Mo_6O_{18}H_6]^{3-}$ [1], $[AlO_6Mo_6O_{15}]_n^{3n-}$ with $n \geqq 2$ [2], $[AlMo_6O_{21}]^{3-}$ [3], $[Al(HMoO_4)_6]^{3-}$ [4], and $[Al(OH)_6Mo_6O_{18}]^{3-}$ [5, 6]. The last formulation is now generally accepted. The structure is of the Anderson-Evans type, which has been determined by single crystal techniques, for example, on a sample with Cr^{3+} as hetero atom (cf. p. 352). A ring of six edge-sharing MoO_6 octahedra provides at its center a site of octahedral coordination at which the hetero atom (Al) is located. For a general consideration of the Anderson-Evans type structure, see, for example, [7, 8]. The location of the protons was first determined by ^{1}H NMR [5]. Titration with NaOH shows these protons not to be acidic (the heteropoly anion decomposes) [2].

X-ray powder and single crystal investigations suggest an orthorhombic unit cell of crystal class mmm with the lattice parameters $a = 11.00 \pm 0.05$ and $b = 12.17 \pm 0.05$ Å for the Rb and Cs compounds. Because of polytypism, the third parameter could not be determined, but preliminary values of c = 6 and 13 Å are reported. From crystal optical observations, it is concluded that the symmetry becomes triclinic when the compounds lose water in air [9]. The known IR spectra are similar for all compounds and confirm the assumed Anderson-Evans structure. However, the assignments given are only approximations: Absorption bands at ~580 and 630 to 650 cm^{-1} have been attributed to AlO_6 stretching vibrations [9, 10] or vibrations of the Mo-O-Mo bridging bonds [6], absorption bands at ~950 and ~900 cm^{-1} are assigned to symmetric and antisymmetric vibrations of the terminal Mo=O bonds [6, 9, 10], and absorption bands at ~1100, ~1600, and 3400 cm^{-1} to OH and H_2O bending and stretching modes [4, 6, 9 to 11].

On heating, the compounds lose the water of crystallization at various temperatures. The anhydrous compounds decompose on further heating according to $M_3[Al(OH)_6Mo_6O_{18}] \rightarrow MAl(MoO_4)_2 + M_2Mo_2O_7 + 2\,MoO_3 + 3\,H_2O$ [10, 12]. The Li and Na compounds are said to decompose at 316 to 322°C [10]. For the K, Rb, and Cs compounds, the temperature of the endothermic effect in the thermograms increases from ~180°C for K to ~215°C for Rb and Cs. The amorphous decomposition products crystallize between 280 and 300°C [12]. For the K compound, see also the thermogram in [3].

On heating the compounds in D_2O on a water bath or in an autoclave at 175°C for 8 h, only the water of crystallization undergoes deuterations as shown by the IR absorption spectra [9].

EPR measurements show that there are two Mo^V in the Mo_6 species in γ-irradiated samples [13].

$[Al(OH)_6Mo_6O_{18}]^{3-}$ in Solutions. The formation of the heteropoly anion from MoO_4^{2-} and Al^{3+} has been investigated in aqueous solutions of various acidities. The formation also depends on the rate at which the Al^{3+} solution is added [14].

Derivative polarograms at various pH values establish a stability range of the $[Al(OH)_6Mo_6O_{18}]^{3-}$ anion from pH 2 to 6. A chromatographic method with a saturated NaCl solution as liquid phase gives the instability constant $K = [Al^{3+}][Mo_6O_{21}^{6-}]/[AlMo_6O_{21}^{3-}]$ as $K = 6.3 \times 10^{-6}$ at pH 4 [3]. Other estimates are $K = 10^{-20}$ from studying the reaction of Al^{3+} with $[PMo_{12}O_{40}]^{3-}$ in weakly acidic solution by measurement of the optical absorption [15] and

$K \approx 10^{-19}$ from measurements of the electrical conductivity of solutions with 0.003 M Al^{3+} and 0.003 to 0.007 M MoO_4^{2-} [16].

In polarographic experiments, the hexamolybdoaluminate anion shows no reduction wave at positive potentials relative to the saturated calomel electrode. Voltammograms show the reduced species produced to be unstable in solution [17].

References:

[1] Y. Shimura, H. Ito, R. Tsuchida (J. Chem. Soc. Japan **75** [1954] 560/2; C. A. **1954** 13422). – [2] L. C. W. Baker, G. Foster, W. Tan, F. Scholnick, T. P. McCutcheon (J. Am. Chem. Soc. **77** [1955] 2136/42). – [3] A. Duca, T. Budiu (Rev. Roumaine Chim. **11** [1966] 817/22). – [4] B. N. Ivanov-Emin, L. A. Filatenko, M. F. Yushchenko, B. E. Zaitsev, et al. (Koord. Khim. **1** [1975] 1332/4; Soviet J. Coord. Chem. **1** [1975] 1112/4). – [5] T. Wada (Compt. Rend. B **263** [1966] 51/4).

[6] L. P. Kazanskii, S. Holguin Quinones, B. N. Ivanov-Emin, L. A. Filatenko (Koord. Khim. **4** [1978] 1676/83; Soviet J. Coord. Chem. **4** [1978] 1279/86). – [7] T. J. R. Weakley (Struct. Bonding [Berlin] **18** [1974] 131/76, 146/7). – [8] M. T. Pope (Heteropoly and Isopoly Oxometalates, Springer, Berlin 1983, pp. 21/3). – [9] L. A. Filatenko, B. N. Ivanov-Emin, S. Holguin Quinones, B. E. Zaitsev, et al. (Zh. Neorgan. Khim. **18** [1973] 799/803; Russ. J. Inorg. Chem. **18** [1973] 419/21). – [10] B. N. Ivanov-Emin, L. A. Filatenko, M. F. Yushchenko, B. E. Zaitsev, A. I. Ezhov (Koord. Khim. **3** [1977] 1382/5; Soviet J. Coord. Chem. **3** [1977] 1079/81).

[11] S. Holguin Q., C. A. Cruz R., A. Campero C. (Rev. Inst. Mexicano Petrol. **5** [1973] 57/62; C. A. **80** [1977] No. 43531). – [12] B. N. Ivanov-Emin, L. A. Filatenko, S. Holguin Quinones, G. Z. Kaziev (Zh. Neorgan. Khim. **23** [1978] 2378/82; Russ. J. Inorg. Chem. **23** [1978] 1311/4). – [13] I. V. Potapova, L. P. Kazanskii, V. I. Spitsyn (Issled. Svoistv Primen. Geteropolikislot Katal. **1978** 135/9; Ref. Zh. Khim. **1979** No. 1V86). – [14] G. Wiese, J. Fuchs (Z. Naturforsch. **22b** [1967] 469/73). – [15] L. I. Lebedeva, Kam Kyong (Vestn. Leningr. Univ. Fiz. Khim. **23** [1968] 127/31; C. A. **69** [1968] No. 70696).

[16] Z. G. Golubtsova, L. I. Lebedeva (Izv. Vysshikh Uchebn. Zavedenii Khim. Khim. Tekhnol. **14** [1971] 947/8; C. A. **76** [1972] No. 7277). – [17] G. A. Tsigdinos, C. J. Hallada (J. Less-Common Metals **36** [1974] 79/93, 87).

5.2.2 $H_3Al(OH)_6Mo_6O_{18} \cdot 3.5H_2O$ ($= Al_2O_3 \cdot 12MoO_3 \cdot 16H_2O$)

The solid acid has been obtained by evaporating over concentrated H_2SO_4 at room temperature an aqueous solution [1], prepared by cation exchange from solutions of the NH_4 salt [1, 2]. A solution can also be obtained by dissolving $2Al_2O_3 \cdot 3MoO_3 \cdot 14H_2O$ (p. 230) in acids [3]. The crystals of the acid are transparent and irregularly shaped with a greenish blue tint due to a certain amount ($\leqq 0.1\%$) of reduced Mo [1]. In another work, the acid has been obtained as a white solid [2]. The content of the water of crystallization varies from 2 to $10H_2O$ and is readily lost when the crystals are stored, without disruption of the crystal form [1].

The X-ray powder pattern has been indexed on the basis of a face-centered cubic cell with the lattice parameter $a = 4.68 \pm 0.015$ Å. Some unindexed lines, among which is the strongest one, have been assumed to be superstructure lines; for d values see the paper [1]. The optically isotropic crystals have refractive indices between 1.6 and 1.7. For an illustration of the IR spectrum between 4000 and 400 cm^{-1}, see the paper [1], assignments of the bands are given on p. 231.

After heating to 140°C, the water of crystallization is given off, and the water of constitution is lost at 275 to 300°C, with decomposition to $Al_2(MoO_4)_3$ and MoO_3 [1].

In aqueous solution, the compound is a tribasic acid with pK≈2 to 3 for all protons. With excess bases, the acid decomposes to $Al(OH)_3$ and MoO_4^{2-}. It is insoluble in organic solvents [2].

References:

[1] B. N. Ivanov-Emin, L. A. Filatenko, M. F. Yushchenko, B. E. Zaitsev, et al. (Koord. Khim. **1** [1975] 1332/4; Soviet J. Coord. Chem. **1** [1975] 1112/4). – [2] L. C. W. Baker, G. Foster, W. Tan, F. Scholnick, T. P. McCutcheon (J. Am. Chem. Soc. **77** [1955] 2136/42). – [3] L. F. Klyuevskii, M. V. Mokhosoev (Khim. Tekhnol. Molibdena Vol'frama **1974** 116/22; Ref. Zh. Khim. **1975** No. 4V18).

5.2.3 $Li[H_2Al(OH)_6Mo_6O_{18}] \cdot 11\,H_2O$ (= $Li_2O \cdot Al_2O_3 \cdot 12\,MoO_3 \cdot 30\,H_2O$)

This compound forms from stoichiometric amounts of aqueous solutions of Li_2MoO_4 and $Al(NO_3)_3$ at pH≈4.5 at room temperature. After mixing, the solution was evaporated at 70 to 90°C to an ~1:1 ratio of the volumes of precipitate and solution. The crystals obtained are colorless quadratic platelets or rhombs.

According to EPR studies at 18°C, the compound is diamagnetic. The magnetic susceptibility has been determined by the Faraday method to be $\chi = -4.918 \times 10^{-6}\,cm^3/g$. The IR spectrum between 4000 and 250 cm^{-1} (KBr technique) is plotted in the paper. The bands are compared with those of other molybdoaluminates, for an assignment, see p. 231.

The solubility of $Li[H_2Al(OH)_6Mo_6O_{18}] \cdot 11\,H_2O$ in water at 25±1°C is 0.134 mol/L, giving a solution density of 1.09 g/cm^3. The molar electrical conductances increase from 242 to 340 $\Omega^{-1} \cdot cm^2 \cdot mol^{-1}$ for decreasing concentrations of 0.005 to 0.0005 mol/L.

S. Holguin Q., C. A. Cruz R., A. Campero C. (Rev. Inst. Mexicano Petrol. **5** No. 3 [1973] 57/62; C.A. **80** [1974] No. 43531).

5.2.4 $Li_3[Al(OH)_6Mo_6O_{18}] \cdot 8\,H_2O$ (= $3\,Li_2O \cdot Al_2O_3 \cdot 12\,MoO_3 \cdot 22\,H_2O$)

This compound forms on neutralizing an aqueous solution of $H_3Al(OH)_6Mo_6O_{18}$ with Li_2CO_3 and evaporation to crystallization on a water bath. The transparent colorless crystals show well-developed tetrahedral forms. They are optically anisotropic with high interference colors and are probably of low symmetry. For the assignment of the IR bands between 4000 and 500 cm^{-1}, see p. 231.

The compound is said to form several hydrates with closely spaced decomposition temperatures between 130 and 140°C.

B. N. Ivanov-Emin, L. A. Filatenko, M. F. Yushchenko, B. E. Zaitsev, A. I. Ezhov (Koord. Khim. **3** [1977] 1382/5; Soviet J. Coord. Chem. **3** [1977] 1079/81).

5.2.5 $Na_3[Al(OH)_6Mo_6O_{18}] \cdot n\,H_2O$ (= $3\,Na_2O \cdot Al_2O_3 \cdot 12\,MoO_3 \cdot (2n+6)H_2O$), n = 4, 5, 8

$Na_3[Al(OH)_6Mo_6O_{18}] \cdot 8\,H_2O$ has been prepared by mixing solutions of sodium molybdate acidified to pH 4.5 and of $Al(NO_3)_3$ [1]. The pentahydrate has been obtained by neutralizing an aqueous $H_3Al(OH)_6Mo_6O_{18}$ solution with Na_2CO_3 solution followed by evaporation on a water bath [2]. A tetrahydrate has been prepared by Tsigdinos' [3] procedure from sodium paramolybdate and a soluble Al salt. Crystalline products were collected after two weeks and, after washing, dried in vacuo over P_2O_5 for 3d [4]. Interaction in aqueous solution between Na_2MoO_4 and $Al_2(SO_4)_3$ has been studied at different pH values and mixing rates [5].

$Na_3[Al(OH)_6Mo_6O_{18}]\cdot 5H_2O$ has been obtained as acicular crystals which form spherulites. From preliminary optical studies, the symmetry is probably monoclinic [2]. Plots of IR spectra of the octa- and the pentahydrate between 4000 and 250 cm^{-1} are given in the papers [1,2], for assignments, see p. 231. On heating, dehydration of the pentahydrate proceeds like that of the Li compound [2], see above.

The solubility of the octahydrate in water at 25°C is 0.172 mol/L, the density of the saturated solution is 1.133 g/cm^3 [1]. ^{17}O NMR studies in aqueous solution at pH 4.1, 25°C, and 1.9 mol Mo/L showed two signals at −831 and −376 ppm with line widths of 20 ppm (a positive chemical shift is upfield from pure H_2O) [6]. In a later paper, line widths of 100 and 50 ppm, respectively, are given [4]. The first signal has been attributed to the 12 terminal O atoms (O-Mo), while the second is thought to arise from the 6 doubly bridging O atoms (Mo-O-Mo) [6].

References:

[1] S. Holguin Q., C. A. Cruz R., A. Campero C. (Rev. Inst. Mexicano Petrol. **5** No. 3 [1973] 57/62; C.A. **80** [1977] No. 43531). – [2] B. N. Ivanov-Emin, L. A. Filatenko, M. F. Yushchenko, B. E. Zaitsev, A. I. Ezhov (Koord. Khim. **3** [1977] 1382/5; Soviet J. Coord. Chem. **3** [1977] 1079/81). – [3] G. A. Tsigdinos (Method. Chim. **8** [1976] 552/65). – [4] M. Filowitz, R. K. C. Ho, W. G. Klemperer, W. Shum (Inorg. Chem. **18** [1979] 93/103). – [5] G. Wiese, J. Fuchs (Z. Naturforsch. **22b** [1967] 469/73).

[6] M. Filowitz, W. G. Klemperer, L. Messerle, W. Shum (J. Am. Chem. Soc. **98** [1976] 2345/6).

5.2.6 $K_3[Al(OH)_6Mo_6O_{18}]\cdot 7H_2O$ ($=3K_2O\cdot Al_2O_3\cdot 12MoO_3\cdot 20H_2O$)

This compound forms on dropwise addition of aqueous 0.2 M $KAl(SO_4)_2\cdot 12H_2O$ to a stirred boiling solution of 30 g potassium paramolybdate in 170 mL H_2O. The addition is stopped when a permanent cloudiness appears. On cooling, white crystals separate. They can be recrystallized from hot water [1]. The compound is also obtained from other Al salts ($Al_2(SO_4)_3$ [2], $Al(NO_3)_3$ [3]) and molybdates (e.g., K_2MoO_4) in solutions at pH 4 to 5 [2, 3]. The crystals effloresce and reversibly lose up to 2 H_2O when stored in air [4].

An X-ray line diagram of a sample prepared by the method of [1] is given in [5]. The solid state ^1H NMR spectrum at −160°C shows two superimposed lines with $\Delta H = 12.0 \pm 0.4$ and 4.5 ± 0.3 G, which were attributed to H_2O and OH protons, respectively. The secondary moment $\Delta H^2 = 19.5 \pm 0.8\,G^2$ of the H_2O signal corresponds to an interproton distance of 1.57 Å. At temperatures above −60°C, the H_2O line narrows and becomes indistinguishable from the OH line. This is explained by the rotation of the H_2O molecules. From the temperature dependence of the spectrum, an activation energy of 5.5 kcal/mol has been calculated for the breakage of the hydrogen bonds [6]. According to XPS studies on samples that had been heated previously to 100°C in the spectrometer, the binding energies are: Al 2p = 74.8, Mo $3d_{5/2}$ = 232.9, O 1s = 531.1, K $2p_{3/2}$ = 293.0 (all values in eV relative to C 1s = 285 eV) [2]. The compound has been characterized by its IR spectrum in the range 4000 to 300 cm^{-1} [2, 3, 5, 7] which is plotted in [3, 5, 7], for the assignments, see p. 231.

According to thermal analyses, the water of crystallization is lost in two well-resolved steps of three and four molecules of H_2O at 80 to 144 and 144 to 180°C. In other work, the anhydrous compound has been obtained at 200°C after 36 h [1]. A thermogram is given in [5]. The solubility in water at 25 ± 0.5°C is 0.0131 mol/L [3, 7].

References:

[1] L. C. W. Baker, G. Foster, W. Tan, F. Scholnick, T. P. McCutcheon (J. Am. Chem. Soc. **77** [1955] 2136/42). – [2] L. P. Kazanskii, S. Holguin Quinones, B. N. Ivanov-Emin, L. A. Filatenko (Koord. Khim. **4** [1978] 1676/82; Soviet J. Coord. Chem. **4** [1978] 1279/86). – [3] S. Holguin Q., C. A. Cruz R., A. Campero C. (Rev. Inst. Mexicano Petrol. **5** No. 3 [1973] 57/62; C.A. **80** [1977] No. 43531). – [4] B. N. Ivanov-Emin, L. A. Filatenko, S. Holguin Quinones, G. Z. Kaziev (Zh. Neorgan. Khim. **23** [1978] 2378/82; Russ. J. Inorg. Chem. **23** [1978] 1311/4). – [5] A. Duca, T. Budiu (Rev. Roumaine Chim. **11** [1966] 817/22).

[6] T. Wada (Compt. Rend. B **263** [1966] 51/4). – [7] L. A. Filatenko, B. N. Ivanov-Emin, S. Holguin Quinones, B. E. Zaitsev, et al. (Zh. Neorgan. Khim. **18** [1973] 799/803; Russ. J. Inorg. Chem. **18** [1973] 419/21).

5.2.7 $(NH_4)_3[Al(OH)_6Mo_6O_{18}]\cdot 7H_2O$ ($=3(NH_4)_2O\cdot Al_2O_3\cdot 12MoO_3\cdot 20H_2O$)

This compound has been prepared from aqueous solutions of $NH_4Al(SO_4)_2\cdot 12H_2O$ [1] or $Al(NO_3)_3$ [2] and ammonium paramolybdate by the same method as described for the corresponding K compound (see above) and recrystallized from water [1].

Binding energies for $Mo3d_{5/2}=233.0$ eV and $O1s=530.8$ eV relative to $C1s=285.0$ eV were determined by X-ray photoelectron spectroscopy on an anhydrous sample [3]. ^1H NMR spectra at 20°C showed one resonance signal with a line width of $\Delta H=1.8\pm 0.2$ G [4]. The compound is diamagnetic at 22°C [1]. The IR spectrum has been measured between 4000 and 300 cm^{-1} [2, 5] and plotted in [2], for the assignments, see p. 231. The UV spectrum between 200 and 400 nm of the aqueous solution shows a single absorption band at 230 nm with $\log\varepsilon=4.5$ [6].

According to thermal analysis, water of crystallization is lost in two steps. Complete decomposition with loss of nitrogen takes place in the range 265 to 295°C [7].

In water at 25 ± 0.5°C, the solubility is 0.0218 mol/L [2, 5]. The activity coefficients in aqueous solutions have been calculated from the osmotic coefficients measured at 37°C by vapor pressure osmometry, with results similar to those for $LaCl_3$ solutions. The activity coefficients decrease from 0.798 to 0.607 as the solution molalities increase from 0.003 to 0.02. Calculations using the Debye-Hückel theory give the interionic distance as 4.3 Å [8].

References:

[1] L. C. W. Baker, G. Foster, W. Tan, F. Scholnick, T. P. McCutcheon (J. Am. Chem. Soc. **77** [1955] 2136/42). – [2] S. Holguin Q., C. A. Cruz R., A. Campero C. (Rev. Inst. Mexicano Petrol. **5** No. 3 [1973] 57/62; C.A. **80** [1977] No. 43531). – [3] V. N. Molchanov, L. P. Kazanskii, E. A. Torchenkova, V. I. Spitsyn (Izv. Akad. Nauk SSSR Ser. Khim. **1978** 1248/51; Bull. Acad. Sci. USSR Div. Chem. Sci. **27** [1978] 1085/7). – [4] T. Wada (Compt. Rend. B **263** [1966] 51/4). – [5] L. A. Filatenko, B. N. Ivanov-Emin, S. Holguin Quinones, B. E. Zaitsev, et al. (Zh. Neorgan. Khim. **18** [1973] 799/803; Russ. J. Inorg. Chem. **18** [1973] 419/21).

[6] Y. Shimura, H. Ito, R. Tsuchida (J. Chem. Soc. Japan **75** [1954] 560/2; C.A. **1954** 13422). – [7] B. N. Ivanov-Emin, L. A. Filatenko, S. Holguin Quinones, G. Z. Kaziev (Zh. Neorgan. Khim. **23** [1978] 2378/82; Russ. J. Inorg. Chem. **23** [1978] 1311/4). – [8] E. Meyer Jr., R. Huckfeldt (J. Phys. Chem. **74** [1970] 164/7).

5.2.8 $Rb_3[Al(OH)_6Mo_6O_{18}]\cdot 7H_2O$ ($=3Rb_2O\cdot Al_2O_3\cdot 12MoO_3\cdot 20H_2O$)

This compound has been prepared by double decomposition from mixtures of saturated aqueous solutions of $(NH_4)_3[Al(OH)_6Mo_6O_{18}]\cdot 7H_2O$ and RbCl in excess. The precipitated crystals contained no ammonium ions [1, 2].

The colorless transparent crystals are well-formed prisms with rhombic sections. They are stable only in contact with the mother liquor and effloresce in air. The refractive indices are greater than 1.80. The rhombic sections have bright interference colors, which indicate considerable differences in the refractive indices. Preliminary data on the crystal class and unit cell have been determined [1], see the general data on p. 231. The IR spectrum between 4000 and 300 cm^{-1} has been determined [1, 2], for a plot see [2]; the assignments are given on p. 231.

Thermal analysis shows the compound to lose its water of crystallization in two steps at about 125 and 170°C, the second endothermic effect being a poorly resolved double peak [3]. The solubility in water is 0.0103 mol/L at 25 ± 0.5°C [1, 2]. The molar electrical conductivity of aqueous solutions at 25°C increases from 327.5 to 423.3 $\Omega^{-1}\cdot cm^2\cdot mol^{-1}$ on diluting from 0.005 to 0.0005 mol/L [1, 2].

References:

[1] L. A. Filatenko, B. N. Ivanov-Emin, S. Holguin Quinones, B. E. Zaitsev, et al. (Zh. Neorgan. Khim. **18** [1973] 799/803; Russ. J. Inorg. Chem. **18** [1973] 419/21). – [2] S. Holguin Q., C. A. Cruz R., A. Campero C. (Rev. Inst. Mexicano Petrol. **5** No. 3 [1973] 57/62; C.A. **80** [1977] No. 43531). – [3] B. N. Ivanov-Emin, L. A. Filatenko, S. Holguin Quinones, G. Z. Kaziev (Zh. Neorgan. Khim. **23** [1978] 2378/82; Russ. J. Inorg. Chem. **23** [1978] 1311/4).

5.2.9 $Cs_3[Al(OH)_6Mo_6O_{18}]\cdot 7H_2O$ ($=3Cs_2O\cdot Al_2O_3\cdot 12MoO_3\cdot 20H_2O$)

This compound precipitates free of NH_4^+ ions on double decomposition of mixtures of saturated aqueous solutions of $(NH_4)_3[Al(OH)_6Mo_6O_{18}]\cdot 7H_2O$ and excess CsCl [1, 2].

The crystals are colorless and transparent prisms with rhombic sections. They effloresce in air but are stable in contact with the mother liquor. The refractive indices are greater than 1.80. The rhombic sections have bright interference colors, which indicate considerable differences in the refractive indices, $n_\gamma - n_\alpha \approx 0.006$. For preliminary results on crystal class and lattice parameters [1], see the general data on p. 231. The compound has been characterized by the IR spectrum between 4000 and 300 cm^{-1} [1, 2], for a plot, see [2], assignments are given on p. 231.

Thermograms show the water of crystallization to be lost at ~118°C [3]. The solubility in water is 0.0059 mol/L at 25 ± 0.5°C [1, 2]. The molar electrical conductivity in aqueous solutions increases from 321.5 to 418.6 $\Omega^{-1}\cdot cm^2\cdot mol^{-1}$ on diluting from 0.005 to 0.0005 mol/L [1].

References:

[1] L. A. Filatenko, B. N. Ivanov-Emin, S. Holguin Quinones, B. E. Zaitsev, et al. (Zh. Neorgan. Khim. **18** [1973] 799/803; Russ. J. Inorg. Chem. **18** [1973] 419/21). – [2] S. Holguin Q., C. A. Cruz R., A. Campero C. (Rev. Inst. Mexicano Petrol. **5** No. 3 [1973] 57/62; C.A. **80** [1977] No. 43531). – [3] B. N. Ivanov-Emin, L. A. Filatenko, S. Holguin Quinones, G. Z. Kaziev (Zh. Neorgan. Khim. **23** [1978] 2378/82; Russ. J. Inorg. Chem. **23** [1978] 1311/4).

5.3 Gallium Molybdate Hydrate

5.3.1 $Ga_2O_3 \cdot 3MoO_3 \cdot 14H_2O$

A solid of this composition (probably prepared by methods similar to those used for the Al compounds, p. 230) has been investigated by ^1H NMR at room temperature. The interproton distance for the adsorbed water is H-H = 1.54 ± 0.1 Å. H-H distances of 2.07 and 2.60 Å have been assigned to OH groups around Ga and Mo, respectively. The temperature dependence yields an estimate of the potential barrier between different equilibrium positions of the water molecules of 8.2 ± 0.2 kcal/mol.

L. A. Pozharskaya, V. G. Pitsyuga, M. V. Mokhosoev, L. F. Klyuevskii (Tr. Buryat. Inst. Estestv. Nauk Buryat. Fil. Sib. Otd. Akad. Nauk SSSR No. 14 [1977] 394/9; C.A. **94** [1981] No. 53356).

5.4 Hexamolybdogallates

5.4.1 General Data

The hexamolybdogallic acid and its alkali salts are isomorphous with the corresponding hexamolybdoaluminic acid and its salts. Also, many other properties of the gallium heteropoly compounds are very similar to those of the aluminium heteropoly compounds, and therefore are reported together in many publications. Salts with alkaline earth metals are known only for the hexamolybdogallic acid.

X-ray powder patterns and rotation photographs give the same orthorhombic crystal class and the same lattice parameters for $M_3[Ga(OH)_6Mo_6O_{18}] \cdot 7H_2O$ with M = NH_4, Rb, Cs as for the molybdoaluminates, see p. 231 [1]. IR spectra are known for all the compounds [1 to 7]. Generally, absorption bands (at least two) at 950 to 880 cm^{-1} are assigned to Mo=O stretching vibrations [1 to 7], while absorption bands at about 650 cm^{-1} have been attributed to either Mo-O-Mo [4 to 7] or Ga-O [1, 3] stretching vibrations. A band at about 580 cm^{-1} is also assigned to Ga-O [1, 4, 6, 7] or Mo-O [5] stretching vibrations. An absorption band at about 450 cm^{-1} was attributed to Mo-O vibrations [6]. H_2O and OH bending modes are observed at 1615 to 1650 cm^{-1} and at ~1100 cm^{-1} [2, 3, 5 to 7], while stretching modes of water appear at 3000 to 3500 cm^{-1} [1, 3, 6, 7].

The anhydrous alkali salts decompose on heating at temperatures similar to those for the corresponding molybdoaluminates [3, 8], see the thermograms in [9]. For the deuteration with D_2O, see [1].

Reactions in Solutions. Optical density measurements show consecutive formation of heteropoly species in aqueous solutions of $Ga_2(SO_4)_3$ and Na_2MoO_4 at room temperature. The optimum pH value is 4 (acetate/HCl buffer). The reaction is complete after 15 to 20 min. Depending on the concentrations, the species with atomic ratios Ga : Mo = 1:3, 1:6, 1:9, and 1:12 form [10, 11]. The 1:12 species has also been obtained by passing a solution of $Ga(NO_3)_3$ and $(NH_4)_2MoO_4$ through a cation exchange resin in the H^+ form. The eluate had a pH of 1.5 [7]. The instability constants are 1×10^{-10} for the $GaMo_6$, 5×10^{-16} for the $GaMo_9$, and 2×10^{-21} for the $GaMo_{12}$ species. Salts of the $GaMo_3$, $GaMo_9$, and $GaMo_{12}$ species have not been isolated [10, 11]. On addition of pyridine, the solution with the 1:12 species gives precipitates with a Ga : Mo molar ratio >1:24, which, however, were not studied [7].

Cryoscopic measurements in $Na_2SO_4 \cdot 10H_2O$ melts at 33°C showed no dissociation or hydrolysis of the $[Ga(OH)_6Mo_6O_{18}]^{3-}$ anion at concentrations as low as 0.005 molal. In 0.0001 M solutions in aqueous 1 and 3 M $NaClO_4$ at pH 4.6 and 25°C, a pK = 48 ± 1 was determined for the reaction $[Ga(OH)_6Mo_6O_{18}]^{3-} \rightarrow Ga^{3+} + 6MoO_4^{2-} + 6H^+$ [12]. The reaction of $[PMo_{12}O_{40}]^{3-}$ with $Ga(NO_3)_3$ in aqueous solutions at pH 3.3 to 4 is reported to form the hexamolybdogallate ion. An instability constant pK≈18.67 was calculated [13]. The maximum absorption of the $GaMo_6$ species in solution is found at 240 to 250 [10, 11] and 250 to 260 nm [13], see also [7].

The heteropoly anion is almost completely decomposed by 1 N HCl or 1 N H_2SO_4 [11]. Organic acids capable of forming complexes with the central metal atom (oxalic, citric, tartaric) have little disruptive influence [10, 11]. The anion is decomposed by 9 equivalents of NaOH forming MoO_4^{2-} and $Ga(OH)_3$ [14].

References:

[1] L. A. Filatenko, B. N. Ivanov-Emin, S. Holguin Quinones, B. E. Zaitsev, et al. (Zh. Neorgan. Khim. **18** [1973] 799/803; Russ. J. Inorg. Chem. **18** [1973] 419/21). – [2] B. N. Ivanov-Emin, L. A. Filatenko, M. F. Yushchenko, B. E. Zaitsev, et al. (Koord. Khim. **1** [1975] 1332/4; Soviet J. Coord. Chem. **1** [1975] 1112/4). – [3] B. N. Ivanov-Emin, L. A. Filatenko, M. F. Yushchenko, B. E. Zaitsev, A. I. Ezhov (Koord. Khim. **3** [1977] 1382/5; Soviet J. Coord. Chem. **3** [1977] 1079/81). – [4] L. P. Tsyganok, T. V. Kleinerman (Zh. Neorgan. Khim. **21** [1976] 3196/8; Russ. J. Inorg. Chem. **21** [1976] 1762/3). – [5] L. P. Kazanskii, S. Holguin Quinones, B. N. Ivanov-Emin, L. A. Filatenko (Koord. Khim. **4** [1978] 1676/83; Soviet J. Coord. Chem. **4** [1978] 1279/86).

[6] M. V. Mokhosoev, L. V. Tumurova, L. G. Maksimova (Zh. Neorgan. Khim. **27** [1982] 109/12; Russ. J. Inorg. Chem. **27** [1982] 62/4). – [7] L. G. Maksimova, L. V. Tumurova, M. V. Mokhosoev, S. R. Sambueva (Zh. Neorgan. Khim. **25** [1980] 2416/20; Russ. J. Inorg. Chem. **25** [1980] 1336/8). – [8] B. N. Ivanov-Emin, Ya. I. Rabovik (Zh. Neorgan. Khim. **3** [1958] 2429/32; Russ. J. Inorg. Chem. **3** No. 10 [1958] 272/5). – [9] B. N. Ivanov-Emin, L. A. Filatenko, S. Holguin Quinones (Zh. Neorgan. Khim. **23** [1978] 2378/82; Russ. J. Inorg. Chem. **23** [1978] 1311/4). – [10] L. P. Tsyganok, B. E. Reznik, T. V. Kleinerman (Zh. Neorgan. Khim. **19** [1974] 2233/8; Russ. J. Inorg. Chem. **19** [1974] 1222/5).

[11] L. P. Tsyganok, T. V. Kleinerman (Zh. Neorgan. Khim. **19** [1974] 2239/43; Russ. J. Inorg. Chem. **19** [1974] 1225/8). – [12] O. W. Rollins (J. Inorg. Nucl. Chem. **33** [1971] 75/80). – [13] M. V. Mokhosoev, L. V. Tumurova, L. V. Mitypova (Zh. Neorgan. Khim. **22** [1977] 2734/7; Russ. J. Inorg. Chem. **22** [1977] 1484/6). – [14] O. W. Rollins, J. E. Earley (J. Am. Chem. Soc. **81** [1959] 5571/2).

5.4.2 $H_3Ga(OH)_6Mo_6O_{18}\cdot 8H_2O$ ($=Ga_2O_3\cdot 12MoO_3\cdot 25H_2O$)

Solid $H_3Ga(OH)_6Mo_6O_{18}\cdot 8H_2O$ precipitates on evaporation in a desiccator over concentrated H_2SO_4 at room temperature of an aqueous solution of the acid [1] prepared by ion exchange from the ammonium salt [1, 2]. The shape and color of the crystals obtained are like those of the corresponding Al compound (p. 232). The water content varies between 2 and 10 H_2O [1].

The X-ray powder pattern shows the compound to be isomorphous with $H_3Al(OH)_6Mo_6O_{18}\cdot 3.5H_2O$. The IR spectrum between 4000 and 400 cm^{-1} is plotted in the paper [1], for the assignments, see p. 237.

When stored at room temperature, the crystals effloresce without disruption of their shape. On heating, the water of crystallization is lost below 140°C, the water of constitution, 4.5 H_2O, at 275 to 300°C with decomposition into $Ga_2(MoO_4)_3$ and MoO_3 [1].

Aqueous solutions are stable under normal conditions. Titration curves show the acid to be moderately strong and the three dissociation constants to differ very little in magnitude [2].

References:

[1] B. N. Ivanov-Emin, L. A. Filatenko, M. F. Yushchenko, B. E. Zaitsev, et al. (Koord. Khim. **1** [1975] 1332/4; Soviet J. Coord. Chem. **1** [1975] 1112/4). – [2] O. W. Rollins, J. E. Earley (J. Am. Chem. Soc. **81** [1959] 5571/2).

5.4.3 $Li_3[Ga(OH)_6Mo_6O_{18}] \cdot 6H_2O$ ($=3Li_2O \cdot Ga_2O_3 \cdot 12MoO_3 \cdot 18H_2O$)

This compound is prepared by neutralizing a solution of $H_3Ga(OH)_6Mo_6O_{18}$ with an Li_2CO_3 solution, with subsequent evaporation on a water bath to crystallization.

$Li_3[Ga(OH)_6Mo_6O_{18}] \cdot 6H_2O$ forms hygroscopic, transparent, in part rectangular crystals, which show blue green and violet red extinction colors. Transverse extinction indicates triclinic symmetry. The IR spectrum has been recorded between 4000 and 400 cm^{-1}, for the assignments, see p. 237. The compound is said to form several discrete hydrates with closely spaced decomposition temperatures at 130 to 140°C.

B. N. Ivanov-Emin, L. A. Filatenko, M. F. Yushchenko, B. E. Zaitsev, A. I. Ezhov (Koord. Khim. **3** [1977] 1382/5; Soviet J. Coord. Chem. **3** [1977] 1079/81).

5.4.4 $Na_3[Ga(OH)_6Mo_6O_{18}] \cdot nH_2O$ ($=3Na_2O \cdot Ga_2O_3 \cdot 12MoO_3 \cdot (2n+6)H_2O$), n = 2.5, 4, and 5

White $Na_3[Ga(OH)_6Mo_6O_{18}] \cdot nH_2O$ has been obtained in 77% yield by the reaction of stoichiometric amounts of sodium molybdate and gallium nitrate: An aqueous Na_2MoO_4 solution was acidified with HNO_3 to pH ~5 and an acidic $Ga(NO_3)_3$ solution was added dropwise with stirring. The mixture was brought to pH ~4.4 by addition of NaOH and kept on a steam bath overnight. On addition of acetone, the compound separated as a heavy, viscous oil. Sodium nitrate was separated by repeated extraction by acetone with intermediate addition of water. The nitrate free solution was then evaporated and dried under vacuum at 105°C [1]. The compound also forms from solutions of the acid and Na_2CO_3 as described for the Li compound [2]. It has been recrystallized from water [3]. The content of water of crystallization has been given as 4 [2], 5 [3], or, after vacuum drying at 105°C, as 2.5 [1].

The tetrahydrate has been characterized by the IR spectrum between 4000 and 400 cm^{-1} which is plotted in [2], see [3] for some listed bands; the assignments are given on p. 237. On heating, $Na_3[Ga(OH)_6Mo_6O_{18}] \cdot 4H_2O$ loses its water of crystallization at 140°C [2]. The compound is readily soluble in H_2O. A basicity of 3 was confirmed by potentiometric measurements [3].

References:

[1] O. W. Rollins (J. Inorg. Nucl. Chem. **33** [1971] 75/80). – [2] B. N. Ivanov-Emin, L. A. Filatenko, M. F. Yushchenko, B. E. Zaitsev, A. I. Ezhov (Koord. Khim. **3** [1977] 1382/5; Soviet J. Coord. Chem. **3** [1977] 1079/81). – [3] L. P. Tsyganok, T. V. Kleinerman (Zh. Neorgan. Khim. **21** [1976] 3196/8; Russ. J. Inorg. Chem. **21** [1976] 1762/3).

5.4.5 $K_3[Ga(OH)_6Mo_6O_{18}] \cdot 7H_2O$ ($=3K_2O \cdot Ga_2O_3 \cdot 12MoO_3 \cdot 20H_2O$)

This compound has been prepared from aqueous solutions of $Ga_2(SO_4)_3$ and potassium paramolybdate. Adding the former solution gradually to the latter formed a white flocculent precipitate which would dissolve in excess $Ga_2(SO_4)_3$ solution. Then hydrochloric acid was added until the initial precipitate dissolved, and the mixture was heated to boiling. Cooling yielded square colorless crystals [1]. They effloresce in air, reversibly losing up to $2H_2O$ [2]. They are stable in contact with the mother liquor [2, 3].

Samples have been dehydrated at 100°C under ultrahigh vacuum conditions in the spectrometer and investigated by XPS. The binding energies (in eV, relative to C1s = 285 eV) are $Mo3d_{5/2}$ = 233.1, $Ga3p_{3/2}$ = 106.51, O1s = 533.1, $K2p_{3/2}$ = 293.0. The ^1H NMR spectrum at 80 K shows two superimposed lines, the narrower of which has a width of 4 G. This indicates protons free of interaction with water molecules and thus confirms the structural model of the anion [4]. The IR spectral bands between 3550 and 530 cm^{-1} are listed [3], for the assignments, see p. 237. The compound is reported to lose $7H_2O$ at 110 to 120°C [1], or about $3H_2O$ at 115°C and 3 to 4 more at 167°C [2].

The solubility in H_2O at 25°C is 0.0185 mol/L [3]. The compound behaves as a four-ion electrolyte according to conductivity measurements. At 25°C, the molar electric conductivity increases from 352.4 to 400.2 $\Omega^{-1} \cdot cm^2 \cdot mol^{-1}$ as the concentration decreases from 0.002 to 0.0005 mol/L [1].

References:

[1] B. N. Ivanov-Emin, Ya. I. Rabovik (Zh. Neorgan. Khim. **3** [1958] 2429/32; Russ. J. Inorg. Chem. **3** No. 10 [1958] 272/5). – [2] B. N. Ivanov-Emin, L. A. Filatenko, S. Holguin Quinones (Zh. Neorgan. Khim. **23** [1978] 2378/82; Russ. J. Inorg. Chem. **23** [1978] 1311/4). – [3] L. A. Filatenko, B. N. Ivanov-Emin, S. Holguin Quinones, B. E. Zaitsev, et al. (Zh. Neorgan. Khim. **18** [1973] 799/803; Russ. J. Inorg. Chem. **18** [1973] 419/21). – [4] L. P. Kazanskii, S. Holguin Quinones, B. N. Ivanov-Emin, L. A. Filatenko (Koord. Khim. **4** [1978] 1676/83; Soviet J. Coord. Chem. **4** [1978] 1279/86).

5.4.6 $(NH_4)_3[Ga(OH)_6Mo_6O_{18}] \cdot nH_2O$ (= $3(NH_4)_2O \cdot Ga_2O_3 \cdot 12MoO_3 \cdot (2n+6)H_2O$), n = 5 and 7

The heptahydrate $(NH_4)_3[Ga(OH)_6Mo_6O_{18}] \cdot 7H_2O$ precipitates from aqueous solutions of $Ga_2(SO_4)_3$ and ammonium paramolybdate as square colorless crystals as described for the corresponding K compound [1]. The pentahydrate forms on dropwise addition of a $Ga(NO_3)_3$ solution to a nearly boiling solution of ammonium paramolybdate. The pH is kept at 4 to 6 by addition of NH_3 solution. On cooling, the white crystalline compound separates. Seeding the hot solution markedly aids crystallization. The compound can be recrystallized from hot water [2].

X-ray powder investigations show the pentahydrate to be isomorphous with the corresponding molybdochromate (p. 354) [2]. The crystals of the heptahydrate effloresce in air, but are stable in contact with the mother liquor. They are well-formed prisms with rhombic sections and are optically positive. Refractive indices are >1.80. The birefringence is $n_\gamma - n_\alpha = 0.030$. IR spectral bands between 3600 and 536 cm^{-1} are listed [3], for the assignments, see p. 237. Heating liberates water of crystallization between 110 and 120°C [1]. Thermal dehydration is also described as proceeding in two steps, the final decomposition with loss of NH_3 and constitutional water taking place between 265 and 295°C [4].

The solubility in water at 24 ± 0.5°C is 0.0207 mol/L [3]. In solution, the salt behaves as a four-ion electrolyte according to conductivity measurements [1].

References:

[1] B. N. Ivanov-Emin, Ya. I. Rabovik (Zh. Neorgan. Khim. **3** [1958] 2429/32; Russ. J. Inorg. Chem. **3** No. 10 [1958] 272/5). – [2] O. W. Rollins, J. E. Earley (J. Am. Chem. Soc. **81** [1959] 5571/2). – [3] L. A. Filatenko, B. N. Ivanov-Emin, S. Holguin Quinones, B. E. Zaitsev, et al. (Zh. Neorgan. Khim. **18** [1973] 799/803; Russ. J. Inorg. Chem. **18** [1973] 419/21). – [4] B. N. Ivanov-Emin, L. A. Filatenko, S. Holguin Quinones (Zh. Neorgan. Khim. **23** [1978] 2378/82; Russ. J. Inorg. Chem. **23** [1978] 1311/4).

5.4.7 $C_5H_5NH(NH_4)_2[Ga(OH)_6Mo_6O_{18}] \cdot 3H_2O$ (=$2C_5H_5N \cdot 2(NH_4)_2O \cdot Ga_2O_3 \cdot 12MoO_3 \cdot 13H_2O$)

This compound has been obtained by adding a 10% aqueous solution of pyridine to aqueous $(NH_4)_3[Ga(OH)_6Mo_6O_{18}] \cdot nH_2O$. Mixing of the two solutions gave an immediate precipitate. The crystalline product was obtained by evaporation of the liquid phase.

The powder pattern has been indexed using an orthorhombic unit cell with the lattice parameters a=11.82, b=13.80, c=16.68 Å; Z=4. A line diagram is given in the paper. The signals of ^1H NMR performed at liquid nitrogen temperature could be split into several components. The resonance line with $\Delta H \approx 1.6$ G was assigned to OH protons, $\Delta H \approx 5.4$ to pyridinium protons, $\Delta H \approx 12$ G to H_2O, and the narrowest component to NH_4^+ protons. The density determined pycnometrically at 25°C is 2.87 and from the X-ray data 2.94 g/cm³. The IR spectrum between 4000 and 400 cm^{-1} plotted in the paper corresponds to those of other hexamolybdogallates, for the assignments, see p. 237.

When heated, there are endothermic effects at 160, 210, and 310°C attributed to the loss of water of crystallization, pyridine, and ammonia, respectively. The degradation of constitutional water is accompanied by exothermal crystallization processes at 430°C.

L. G. Maksimova, L. V. Tumurova, M. V. Mokhosoev, S. R. Sambueva (Zh. Neorgan. Khim. **25** [1980] 2416/20; Russ. J. Inorg. Chem. **25** [1980] 1336/8).

5.4.8 $(C_5H_5NH)_3[Ga(OH)_6Mo_6O_{18}] \cdot 3H_2O$ (= $6C_5H_5N \cdot Ga_2O_3 \cdot 12MoO_3 \cdot 15H_2O$)

This pyridinium salt has been prepared from aqueous solutions of pyridine (10%) and $H_3Ga(OH)_6Mo_6O_{18}$. Mixing of the two solutions gave an immediate precipitate. Evaporation of the liquid phase resulted in the crystalline product.

The X-ray powder pattern has been indexed on the basis of an orthorhombic unit cell with the lattice parameters a=12.79, b=13.54, c=18.27 Å; Z=4. A line diagram is given in the paper. ^1H NMR spectra at liquid nitrogen temperature gave signals that could be separated into several components of different line widths ΔH. The following assignments have been made: $\Delta H \approx 1.6$ G to OH protons, $\Delta H \approx 7.2$ G to pyridinium protons, $\Delta H \approx 12$ G to H_2O protons. The density determined pycnometrically at 25°C is 2.71 and from X-ray data 2.79 g/cm³. The IR spectrum between 4000 and 400 cm^{-1} plotted in the paper agrees with those of other hexamolybdogallates, for the assignments, see p. 237.

Upon heating, the compound loses $3H_2O$ and 2 pyridine molecules at about 210°C and 1 pyridine molecule between 280 and 330°C. The endothermic loss of the hydroxyl groups between 370 and 400°C is followed by exothermic crystallization processes between 430 and 450°C.

L. G. Maksimova, L. V. Tumurova, M. V. Mokhosoev, S. R. Sambueva (Zh. Neorgan. Khim. **25** [1980] 2416/20; Russ. J. Inorg. Chem. **25** [1980] 1336/8).

5.4.9 $Rb_3[Ga(OH)_6Mo_6O_{18}]\cdot 7H_2O$ ($=3Rb_2O\cdot Ga_2O_3\cdot 12MoO_3\cdot 20H_2O$)

This compound precipitates from mixtures of saturated aqueous solutions of $(NH_4)_3[Ga(OH)_6Mo_6O_{18}]$ and excess RbCl. The precipitate was washed with ice water and dried in the air [1].

The compound forms well-developed prismatic crystals with rhombic sections. Orthorhombic symmetry has been proposed from X-ray rotation and Laue photographs (see p. 237). The refractive indices are larger than 1.80. The birefringence $n_\gamma - n_\alpha \approx 0.030$ to 0.050. The IR bands between 3500 and 535 cm^{-1} are listed [1], for the assignments, see p. 237.

When stored in air, the crystals reversibly lose up to $2H_2O$ [1]. Upon heating, water of crystallization is lost in two steps of about 3 and $4H_2O$, respectively, at 115 and 167°C [2]. The solubility in water at 24 ± 0.5°C is 0.0089 mol/L. The molar electrical conductivity at 25°C increases from 333.7 $\Omega^{-1}\cdot cm^2\cdot mol^{-1}$ at 0.005 mol/L to 428.9 $\Omega^{-1}\cdot cm^2\cdot mol^{-1}$ at 0.0005 mol/L [1].

References:

[1] L. A. Filatenko, B. N. Ivanov-Emin, S. Holguin Quinones, B. E. Zaitsev, et al. (Zh. Neorgan. Khim. **18** [1973] 799/803; Russ. J. Inorg. Chem. **18** [1973] 419/21). – [2] B. N. Ivanov-Emin, L. A. Filatenko, S. Holguin Quinones (Zh. Neorgan. Khim. **23** [1978] 2378/82; Russ. J. Inorg. Chem. **23** [1978] 1311/4).

5.4.10 $Cs_3[Ga(OH)_6Mo_6O_{18}]\cdot 7H_2O$ ($=3Cs_2O\cdot Ga_2O_3\cdot 12MoO_3\cdot 20H_2O$)

This compound forms on double decomposition of saturated aqueous solutions of $(NH_4)_3[Ga(OH)_6Mo_6O_{18}]$ and CsCl as described for the corresponding Rb compound [1].

It forms well-developed prismatic crystals with rhombic sections. Orthorhombic symmetry has been proposed from X-ray rotation and Laue photographs (see p. 237). The crystals are optically negative and their refractive indices are larger than 1.80; the birefringence $n_\gamma - n_\alpha$ is between 0.025 and 0.030. IR bands between 3560 and 550 cm^{-1} are listed [1], for the assignments, see p. 237.

When stored in air, the crystals reversibly lose up to two molecules of water [1]. On heating, water of crystallization is lost in three steps: between room temperature and 115°C, between 115 and 140°C, and between 140 and 180°C [2]. The solubility in water at 24 ± 0.5°C is 0.0047 mol/L. The molar electrical conductivity at 25°C increases from 320.7 $\Omega^{-1}\cdot cm^2\cdot mol^{-1}$ at 0.005 mol/L to 419.2 $\Omega^{-1}\cdot cm^2\cdot mol^{-1}$ for concentrations of 0.0005 mol/L [1].

References:

[1] L. A. Filatenko, B. N. Ivanov-Emin, S. Holguin Quinones, B. E. Zaitsev, et al. (Zh. Neorgan. Khim. **18** [1973] 799/803; Russ. J. Inorg. Chem. **18** [1973] 419/21). – [2] B. N. Ivanov-Emin, L. A. Filatenko, S. Holguin Quinones (Zh. Neorgan. Khim. **23** [1978] 2378/82; Russ. J. Inorg. Chem. **23** [1978] 1311/4).

5.4.11 $Ba[HGa(OH)_6Mo_6O_{18}]\cdot 3H_2O$ ($=2BaO\cdot Ga_2O_3\cdot 12MoO_3\cdot 13H_2O$)

This compound precipitates from an aqueous solution of $H_3Ga(OH)_6Mo_6O_{18}$ at pH 2.5 on addition of a $Ba(NO_3)_2$ solution. (The corresponding Ca and Sr compounds could not be prepared by this method.)

The crystals are isometric or elongated prisms. Twins along the long side of the prism are obtained from the melt. Both X-ray data and IR data between 4000 and 400 cm^{-1} (for a plot see the paper) show the hexamolybdogallate ion in this compound to be isostructural with the other hexamolybdogallates. For assignments of the IR bands, see p. 237.

M. V. Mokhosoev, L. V. Tumurova, L. G. Maksimova (Zh. Neorgan. Khim. **27** [1982] 109/12; Russ. J. Inorg. Chem. **27** [1982] 62/4).

5.4.12 $NH_4M^{II}[Ga(OH)_6Mo_6O_{18}]\cdot nH_2O$ (= $(NH_4)_2O\cdot 2M^{II}O\cdot Ga_2O_3\cdot 12MoO_3\cdot(2n+6)H_2O$, M^{II} = Ca, Sr, Ba; n = 4, 7, 10

These compounds with M^{II} = Ca (n = 10), Sr (n = 7), and Ba (n = 4) were obtained from an aqueous solution of $(NH_4)_3[Ga(OH)_6Mo_6O_{18}]\cdot 7H_2O$, to which an aqueous solution of $M^{II}(NO_3)_2$ was added slowly. To prevent precipitation of normal molybdates, the mixtures were acidified with 0.1 N HNO_3 to pH 2.7 to 2.8. After stirring for 30 to 40 min, the precipitation was complete after further 5, 1.5, or 0.5 h on a boiling water bath, for M = Ca, Sr, and Ba, respectively. The Ca and Sr compounds may be recrystallized from hot water. This is not possible for the Ba compound because of its limited solubility.

The crystals (no cation specified) are described as transparent, colorless, isometric or slightly elongated prisms. Perfect cleavage with respect to the prism is observed. X-ray powder patterns have been indexed on the basis of an orthorhombic unit cell with the lattice parameters a = 18.15, b = 15.16, c = 9.63 Å for $NH_4Ca[Ga(OH)_6Mo_6O_{18}]\cdot 10H_2O$ and a = 17.44, b = 14.94, c = 10.83 Å for $NH_4Ba[Ga(OH)_6Mo_6O_{18}]\cdot 4H_2O$; Z = 4. The densities have been determined pycnometrically at 25°C in CCl_4 and from the X-ray data, D_m = 3.19, D_x = 3.24 g/cm^3 for Ca and D_m = 3.05, D_x = 3.16 g/cm^3 for Ba. Crystal optical studies showed the crystals to be optically biaxial negative at room temperature, $n_\alpha \approx 1.688(2)$ and $n_\gamma \approx 1.758(2)$. The angle 2V is approximately 15° to 20°. Pleochroism from colorless to blue is observed. The IR spectra between 4000 and 500 cm^{-1} (for figures, see the paper) are nearly identical for the three compounds and correspond to those of other hexamolybdogallates, for the assignments, see p. 237. The observed band at 1400 cm^{-1} corresponds to the NH_4^+ bending modes.

M. V. Mokhosoev, L. V. Tumurova, L. G. Maksimova (Zh. Neorgan. Khim. **27** [1982] 109/12; Russ. J. Inorg. Chem. **27** [1982] 62/4).

5.5 Indium Molybdate Hydrates

5.5.1 $In_2(MoO_4)_3\cdot nH_2O$ (= $In_2O_3\cdot 3MoO_3\cdot nH_2O$), n = 2, 7, and 9

$In_2(MoO_4)_3\cdot 9H_2O$ has been prepared in a yield of 55% of the theoretical value by mixing aqueous solutions of $InCl_3$ (pH = 3.0 to 3.3) and Na_2MoO_4 (pH = 6.0 to 6.2). The precipitate is washed with a dilute acid (pH = 3.35) and dried to constant weight at 40°C [1]. A heptahydrate has been obtained by reaction of $InCl_3$ and Na_2MoO_4 in a solution adjusted to pH = 4 by ammonia [2]. In the older literature, a dihydrate is described, precipitated from an In salt solution by ammonium molybdate [3].

According to ^1H NMR spectra recorded at 22°C of a sample of composition $In_2O_3\cdot 2.8MoO_3\cdot 9.6H_2O$, the compound contains adsorbed water with the interproton distance H-H = 1.60 Å. After dehydration at 215°C, an H-H distance of 2.11 Å between OH groups is found. From the temperature dependence of the line shapes, the height of the potential barrier between equilibrium positions of H_2O was estimated to be 8.4 kcal/mol [4].

The IR spectrum of $In_2(MoO_4)_3 \cdot 7H_2O$ has been recorded between 4000 and 400 cm^{-1}. Absorption bands at 3000 to 3600 and 1610 cm^{-1} are attributed to stretching and bending vibrations of H_2O, while absorption bands at 980 and 950 to 850 cm^{-1} are assigned to ν(Mo=O). Also, a band at 550 cm^{-1} is attributed to H_2O vibrations [2].

References:

[1] L. F. Klyuevskii, M. V. Mokhosoev (U.S.S.R. 397476 [1973]; C.A. **81** [1974] No. 39559). – [2] M. V. Mokhosoev, L. V. Mitypova, L. V. Tumurova (Izv. Sibirsk. Otd. Akad. Nauk SSSR Ser. Khim. Nauk **1980** No. 5, pp. 46/50; C.A. **94** [1981] No. 24179). – [3] C. Renz (Ber. Deut. Chem. Ges. **34** [1901] 2763/5). – [4] L. A. Pozharskaya, V. G. Pitsyuga, M. V. Mokhosoev, L. F. Klyuevskii (Tr. Buryat. Inst. Estestv. Nauk Buryat. Fil. Sib. Otd. Akad. Nauk SSSR No. 14 [1977] 394/9; C.A. **94** [1981] No. 53356).

5.6 Molybdoindates

5.6.1 Species in Aqueous Solutions

The reaction of 1.0×10^{-3} M $In_2(SO_4)_3$ and 5.0×10^{-3} M Na_2MoO_4 solutions in the pH range 2.5 to 4.5 (adjusted with 0.1 N H_2SO_4) has been investigated spectrophotometrically in the range of the two absorption bands at 250 and 310 nm. Equilibrium is reached after 2.5 to 3 h. The complete formation of the heteropoly species needs up to a 40-fold MoO_4^{2-} excess. Optimum reaction is found at pH 3.5. The heteropoly species found have In:Mo ratios of 1:6, 1:9, and 1:18, the dominant one is the 1:9 species. The average stability constant of the latter is 2×10^9.

L. V. Tumurova, L. V. Mitypova, M. V. Mokhosoev (Koord. Khim. **5** [1979] 824/7; Soviet J. Coord. Chem. **5** [1979] 646/9).

5.6.2 $(NH_4)_3[In(HMoO_4)_4MoO_4] \cdot 3H_2O$ (= $3(NH_4)_2O \cdot In_2O_3 \cdot 10MoO_3 \cdot 10H_2O$)

This compound has been prepared from an aqueous solution of $In(NO_3)_3$ which was added with vigorous stirring to a hot 10% aqueous solution of ammonium paramolybdate. The solution was heated on a water bath for 3 h and then slowly filtered. The precipitate was washed with cold (5°C) water and dried in air.

A microscopic observation at a magnification of 320 indicated the product to be single phase. The fine crystals have the form of tablets or well bounded rhombic grains and are of trigonal symmetry. The refractive indices are greater than 1.78, evidently between 1.80 and 1.85; birefringence $n_\gamma - n_\alpha = 0.007$. The IR spectrum has been recorded and an assignment proposed; a plot between 1400 and 400 cm^{-1} is given in the paper. Octahedral coordination by O has been proposed for In and two types of tetrahedrally coordinated Mo.

Drying over sulfuric acid causes a weight loss of $2H_2O$, while by drying at 110°C to constant weight all of the H_2O is removed. On further heating, there is at first only a slight weight loss, but at 300 to 360°C the nitrogen is lost completely. The final product of the thermal decomposition is $In_2(MoO_4)_3$.

The compound is incongruently soluble in water.

B. N. Ivanov-Emin, L. A. Filatenko, B. E. Zaitsev, A. I. Ezhov (Zh. Neorgan. Khim. **18** [1973] 974/6; Russ. J. Inorg. Chem. **18** [1973] 512/3).

5.6.3 $(NH_4)_5[InMo_{12}O_{40}]\cdot15H_2O$ (= $5(NH_4)_2O\cdot In_2O_3\cdot 24MoO_3\cdot 30H_2O$)

This compound has been obtained by the method of [1]. An acidic $InCl_3$ solution is added dropwise to a nearly boiling solution of ammonium paramolybdate. After heating for 2 to 3h on a water bath and changing the pH to about 4 to 6 by addition of ammonia, a white crystalline powder precipitated on cooling, which was washed with ice water and dried in the air [2].

The IR spectrum has been recorded between 4000 and 400 cm^{-1}. Absorption bands at 3600 to 3000, at 1620 to 1600, and at 1410 to 1400 cm^{-1} are attributed to vibrations $\nu(H_2O)$, $\delta(H_2O)$, and $\delta(NH_4^+)$, respectively. Bands at 950, 910 to 880, 820, and 640 cm^{-1} are assigned to Mo-O stretching vibrations, absorption bands at 550 and 485 cm^{-1} to H_2O and In-O vibrations, respectively. Solutions of pH 4 show a broad absorption band around 250 nm [2].

Weight loss upon heating starts at 60°C. DTA curves show four endothermic effects at 140, 245, 305, and 350°C, which are followed by an exothermal effect at 380°C. The first two steps correspond to the loss of 8 and $2H_2O$, then the loss of NH_3 follows. The last effect has been attributed to the crystallization of the decomposition products [2].

Spectroscopic investigations between 230 and 290 nm of acidic solutions show the heteropoly anion to be decomposed at pH 1.1 [2].

References:

[1] O. W. Rollins, J. E. Earley (J. Am. Chem. Soc. **81** [1959] 5571/2). – [2] M. V. Mokhosoev, L. V. Mitypova, L. V. Tumurova (Izv. Sibirsk. Otd. Akad. Nauk Ser. Khim. Nauk **1980** No. 5, pp. 46/50; C.A. **94** [1981] No. 24179).

5.6.4 $(NH_4)_{5-x}Na_x[InMo_{12}O_{40}]\cdot10H_2O$ (= $(5-x)(NH_4)_2O\cdot xNa_2O\cdot In_2O_3\cdot 24MoO_3\cdot 20H_2O$), $x=1$ to 4

When the filtrate of the $In_2(MoO_4)_3\cdot7H_2O$ (see p. 243) is mixed with NH_3 solution, a white crystalline precipitate is obtained. Depending on the NH_3 concentration, the composition ranges from $(NH_4)_4Na[InMo_{12}O_{40}]$ to $NH_4Na_4[InMo_{12}O_{40}]$. A pure Na molybdoindate could not be obtained by the method described for the NH_4 compound.

Bands in the IR spectrum of $(NH_4)_3Na_2[InMo_{12}O_{40}]\cdot10H_2O$ between 4000 and 400 cm^{-1} are assigned as for the NH_4 compound, see above. DTA and TG curves show the water and NH_3 to be lost in four steps at 140, 210, 280, and 350°C, as with the NH_4 compound. At 380°C, the decomposition products crystallize exothermally.

M. V. Mokhosoev, L. V. Mitypova, L. V. Tumurova (Izv. Sibirsk. Otd. Akad. Nauk Ser. Khim. Nauk **1980** No. 5, pp. 46/50; C.A. **94** [1981] No. 24179).

5.7 Thallium Molybdate Hydrates

Older data are given in "Molybdän", 1935, p. 304.

5.7.1 $TlIn(MoO_4)_2 \cdot 4\,H_2O$ ($= Tl_2O \cdot In_2O_3 \cdot 4\,MoO_3 \cdot 8\,H_2O$)

This compound precipitates on addition of an aqueous sodium molybdate solution with pH 7.7 to a stirred aqueous solution, which is acidified to pH 1.8 to 2.0 and equimolar in $TlNO_3$ and $In(NO_3)_3$, at 20°C.

Air dried samples are X-ray amorphous. On heating, loss of water starts at 100°C and ends at 350 to 450°C, which indicates that some of the water is bound as OH groups. An exothermic effect at 290°C, which is also accompanied by a weight loss, has been attributed to a crystallization process. The anhydrous compound melts at 860°C.

A. P. Perepelitsa, M. V. Mokhosoev (Zh. Neorgan. Khim. **25** [1980] 2848/50; Russ. J. Inorg. Chem. **25** [1980] 1572/3).

6 Molybdate Hydrates with Subgroup 3 Metals

6.1 Rare Earth Molybdate Hydrates

6.1.1 General Data

6.1.1.1 $[MMoO_4]^+$ Complexes in Solution, M = La to Er and Yb

During studies of the solubility of rare earth molybdates $M_2(MoO_4)_3 \cdot nH_2O$ with M = La to Er and Yb in weakly acidic aqueous MCl_3 solutions the formation of soluble complexes $[MMoO_4]^+$ has been observed [1]. These complexes form in an initial reaction when an aqueous solution of ammonium paramolybdate is added to an aqueous MCl_3 solution in the concentration range Mo:M<1.5 for M = La, Sm, Er, and Yb. With increasing paramolybdate concentrations the $M_2(MoO_4)_3 \cdot nH_2O$ molybdates (see below) precipitate [2]. In basic medium the compounds $M(OH)MoO_4 \cdot nH_2O$ are formed (p. 249) [3, 4]. The stability constants K, determined by conductivity measurements at pH 5.5 to 6.5, 25°C, and an ionic strength of 0.0012, are given in the following table (Std. Dev. = standard deviation) [1]:

M	La	Ce	Pr	Nd	Sm	Eu	Gd	Tb	Dy	Ho	Er	Yb
K in 10^{-4}	1.67	2.63	2.63	5.55	7.15	5.55	2.50	3.23	3.13	2.08	1.82	1.70
Std. Dev.	0.27	0.35	0.45	1.30	1.15	1.30	0.40	0.23	0.42	0.25	0.25	0.20

Solid compounds of the type $[MMoO_4]A$ with A = anion have not been isolated.

References:

[1] N. K. Davidenko, G. A. Komashko, K. B. Yatsimirskii (Zh. Neorgan. Khim. **13** [1968] 117/22; Russ. J. Inorg. Chem. **13** [1968] 58/61). – [2] B. N. Ivanov-Emin, E. I. Zakharikova (Zh. Neorgan. Khim. **14** [1969] 145/9; Russ. J. Inorg. Chem. **14** [1969] 75/8). – [3] A. M. Golub, K. S. Aganiyazov (Ukr. Khim. Zh. **35** [1969] 1239/43; Soviet Progr. Chem. **35** No. 12 [1969] 1/4). – [4] A. M. Golub, A. P. Perepelitsa, A. A. Govorov (Zh. Neorgan. Khim. **16** [1971] 660/5; Russ. J. Inorg. Chem. **16** [1971] 352/5).

6.1.1.2 $M_2(MoO_4)_3 \cdot nH_2O$ (= $M_2O_3 \cdot 3MoO_3 \cdot nH_2O$), M = Sc, Y, La to Sm, and Gd to Lu; n = 0.5 to 6

The rare earth molybdate hydrates $M_2(MoO_4)_3 \cdot nH_2O$ (except M = Pm and Eu) are usually prepared by mixing aqueous solutions of alkali molybdates with neutral or weakly acidic aqueous solutions of $M(NO_3)_3$ or MCl_3, preferably but not necessarily in the stoichiometric ratio [1, 2]. A procedure proposed for obtaining well-defined crystallized samples is the precipitation at low supersaturation by dropping simultaneously and slowly both solutions into a stirred 5% aqueous solution of NH_4Cl at 80°C [3]. At low molar ratios of M_2O_3:MoO_3 or low pH values higher molybdates (see p. 248), hydroxide compounds (see p. 249), or double molybdates with alkali metals (see p. 274) are likely to be coprecipitated. Trihydrates can also be obtained by hydration in a moist atmosphere of the hygroscopic high-temperature modifications of the compounds $M_2(MoO_4)_3$ with M = Y, Dy to Lu [4, 5].

From the X-ray powder diagrams the compounds $M_2(MoO_4)_3 \cdot 4H_2O$ with M = Y, Gd to Lu are isomorphous [2]. Two different structure types [5] are reported for $M_2(MoO_4)_3 \cdot 3H_2O$ from hydration in moist atmosphere, called K (M = Y, Ho to Lu) [4] and K′ (M = Dy to Lu) [5]. No crystal structures of these compounds are known.

Solid state ^{1}H NMR spectra of $M_2(MoO_4)_3 \cdot 6H_2O$ with M = La, Ce, Pr, and Nd show three different types of protons. The data have been interpreted by assuming OH^- groups in these compounds according to the formula $M_4(OH)_4Mo_6O_{22} \cdot 10H_2O$, cf. also p. 253 [6]. This assumption has been questioned because it is said to be inconsistent with chemical and IR-spectroscopic observations made on $M_2(MoO_4)_3 \cdot 4H_2O$ with M = Y, Gd to Lu. Densities have been determined for the compounds $M_2(MoO_4)_3 \cdot 4H_2O$ with M = Gd to Lu. They increase linearly with the atomic number from 4.13 g/cm³ for M = Gd to 4.42 g/cm³ for M = Lu [2].

IR spectra show five to six absorption bands between 1000 and 600 cm^{-1}, which are attributed to vibrations of distorted MoO_4^{2-} tetrahedra supposedly of C_{2v} symmetry (cf. also p. 268) [7]. While for some hydrates (with M = Ce and Er) only one absorption band at ~1630 cm^{-1} is mentioned between 2000 and 1100 cm^{-1} [7, 8], there are always three, at 1630, 1490, and ~1400 cm^{-1}, in the spectra of $M_2(MoO_4)_3 \cdot 4H_2O$ with M = Y or Gd to Lu [2].

The tetrahydrates with Y and Gd to Lu can be dehydrated over P_2O_5 [2]. Thermal dehydration takes place stepwise. It starts at ~80°C and is reported to be complete at temperatures ranging from 200 to 450°C. The amorphous dehydration products crystallize between 500 and 600°C, see for example [2].

The compounds dissolve in MCl_3 solutions forming the $[MMoO_4]^+$ complexes for M = La to Er, Yb (see p. 247) [9].

References:

[1] B. N. Ivanov-Emin, E. I. Zakharikova (Zh. Neorgan. Khim. **14** [1969] 145/9; Russ. J. Inorg. Chem. **14** [1969] 75/8). – [2] E. M. Avzhieva, I. V. Shakhno, V. E. Plyushchev, G. N. Voronskaya, K. I. Petrov (Zh. Neorgan. Khim. **20** [1975] 1561/5; Russ. J. Inorg. Chem. **20** [1975] 874/6). – [3] S. Pajakoff (Monatsh. Chem. **100** [1969] 1350/61, 1357). – [4] K. Nassau, H. J. Levinstein, G. M. Loiacono (J. Phys. Chem. Solids **26** [1965] 1805/16, 1809/11). – [5] E. Ya. Rode, G. V. Lysanova, V. G. Kuznetsov, L. Z. Gokhman (Zh. Neorgan. Khim. **13** [1968] 1295/302; Russ. J. Inorg. Chem. **13** [1968] 678/82).

[6] A. M. Golub, K. S. Aganiyazov, V. V. Mank, M. V. Mokhosoev (Teor. Eksperim. Khim. **4** [1968] 705/8; Theor. Exptl. Chem. [USSR] **4** [1968] 453/5). – [7] B. E. Zaitsev, E. I. Zakharikova, B. N. Ivanov-Emin, G. I. Cherenkova (Zh. Neorgan. Khim. **14** [1969] 1493/6; Russ. J. Inorg. Chem. **14** [1969] 781/3). – [8] M. J. Schwing-Weill (Bull. Soc. Chim. France **1972** 1754/61). – [9] N. K. Davidenko, G. A. Komashko, K. B. Yatsimirskii (Zh. Neorgan. Khim. **13** [1968] 117/22; Russ. J. Inorg. Chem. **13** [1968] 58/61).

6.1.1.3 $\mathbf{M_2Mo_7O_{24} \cdot nH_2O}$ (= $M_2O_3 \cdot 7MoO_3 \cdot nH_2O$), M = Sc, Y, La, Pr to Lu; n = 4, 6 to 10

Mixing aqueous solutions of $M(NO_3)_3$ and ammonium or sodium paramolybdate with 5 h stirring at room temperature yields the compounds $M_2Mo_7O_{24} \cdot nH_2O$ with M = Sc [1], Y, La, Pr to Lu [2] and n = 4 [1] and from 6 to 10 [2]. Generally the molar ratio of $M_2O_3 : MoO_3$ in the precipitates is 1:7 when the initial ratio of $MoO_3 : M_2O_3 \leqq 7$ and the pH is ~3 to ~6 [2]. The water content of the samples dried over $CaCl_2$ is higher (n = 7 to 10) than of those dried over H_2SO_4 or P_2O_5 (n = 6 to 7) [2].

X-ray powder diagrams of $CaCl_2$-dried samples show rather poor contrast for M = La, Pr to Dy, while for M = Y, Ho to Lu the diagrams show strong lines and prove the heptahydrates to be isomorphous [2].

According to DTA the $M_2Mo_7O_{24} \cdot 4H_2O$ with M = Sc, Y, Er are dehydrated at temperatures from 140 to 170°C [1]. The $CaCl_2$-dried samples with n = 7 to 10 lose water in one step at temperatures between 183 and 228°C for M = La, Pr to Gd, and Dy, and in two steps from 210 to 230 and from 234 to 264°C for M = Y, Tb, and Ho to Lu [2].

References:

[1] A. M. Golub, A. P. Perepelitsa, V. I. Maksin (Izv. Vysshikh Uchebn. Zavedenii Khim. Khim. Tekhnol. **14** [1971] 815/7; C.A. **76** [1972] No. 9894). – [2] S. S. Antonova, I. V. Shakhno, V. E. Plyushchev (Izv. Vysshikh Uchebn. Zavedenii Khim. Khim. Tekhnol. **15** [1972] 1289/92; C.A. **78** [1973] No. 10950).

6.1.1.4 $M(OH)MoO_4 \cdot nH_2O$ (= $M_2O_3 \cdot 2MoO_3 \cdot (2n+1)H_2O$), M = Sc, Y, La to Lu; n = 0, 1, 1.5, and 2

The **hydrates** are prepared by adding an aqueous solution of $M(NO_3)_3$ to an aqueous solution of $Na_2MoO_4 + NaOH$ or $K_2MoO_4 + KOH$ to a final overall molar ratio of $M^{3+} : MoO_4^{2-} : OH^- = 1:1:1$ [1 to 3].

^1H NMR spectra of samples dried at 120°C with M = La to Nd [1] and at 150°C with M = Sc, Y, Er [2] are interpreted to show two types of protons, belonging to H_2O and OH^-. Gravimetric and differential thermal analyses show H_2O to split off between 140 and 160°C for M = Sc, Y, Er [2], below 200°C for M = Sm to Lu [3], or in two steps between 210 and 240°C (1 H_2O) and between 340 and 360°C (0.5 H_2O) for M = La to Nd [1].

The **anhydrous** $M(OH)MoO_4$ compounds with M = La, Pr to Er are formed under hydrothermal conditions, in stainless steel autoclaves equipped with titanium inserts, between 450 and 550°C at pressures between 1200 and 1400 atm (corresponding to 55 to 70% filling), with reaction times from 160 to 200 h. Starting materials are M_2O_3-MoO_3 mixtures with molar ratios $M_2O_3 : MoO_3 = 2:1$ to $5:1$ in 20 to 25% LiCl or KCl, 5% NH_4Cl, or 15 to 20% lithium molybdate solutions. At temperatures below 525°C reaction products consist mainly of $M(OH)MoO_4$ with minor amounts of molybdenum oxides (black or brown, platy or acicular crystals), which become dominant at reaction temperatures above 550°C. In lithium molybdate solutions at 525 to 550°C the compounds $LiM(MoO_4)_2$ crystallize simultaneously for M = La, Pr, Nd [4].

The $M(OH)MoO_4$ crystals are up to 1.5 mm in size and show monoclinic symmetry with a somewhat flattened habit with the faces {010} (largest), {110}, {120}, {111}, {131}, etc. The compounds $M(OH)MoO_4$ are isomorphous with $Nd(OH)WO_4$ (see [5]). The lattice parameters are in the ranges a = 5.2 to 5.4, b = 12.3 to 12.8, c = 6.6 to 6.9 Å, β = 113° [4].

^1H NMR investigations of the compounds with M = La to Lu confirmed the presence of OH groups. The line widths ΔH of the La to Eu group molybdates are very narrow and increase linearly with the atomic weight. The ΔH of the Gd to Yb group molybdates are broader and vary periodically. Maxima are observed for Gd, Dy, Er, and Yb, which contain odd numbers of 4f electrons [6]. IR spectra of crystalline specimens show absorption bands at 3570 to 3560 cm^{-1} due to stretching vibrations of the OH group and absorption bands at ~950 cm^{-1} [4].

Temperatures of decomposition, presumably according to $2M(OH)MoO_4 \rightarrow M_2Mo_2O_9 + H_2O$, have been measured by DTA. They range from 520 to 700°C [1 to 4].

References:

[1] A. M. Golub, K. S. Aganiyazov (Ukr. Khim. Zh. **35** [1969] 1239/43; Soviet Progr. Chem. **35** No. 12 [1969] 1/4). – [2] A. M. Golub, A. P. Perepelitsa, A. A. Govorov (Zh. Neorgan. Khim. **16** [1971] 660/5; Russ. J. Inorg. Chem. **16** [1971] 352/5). – [3] A. P. Perepelitsa, V. N. Solomakha (Zh. Neorgan. Khim. **18** [1973] 28/30; Russ. J. Inorg. Chem. **18** [1973] 14/5). – [4] V. I. Protasova,

L. Yu. Kharchenko, P. V. Klevtsov (Izv. Akad. Nauk SSSR Neorgan. Materialy **9** [1973] 421/3; Inorg. Materials [USSR] **9** [1973] 374/6). – [5] R. F. Klevtsova, S. V. Borisov (Kristallografiya **14** [1969] 904/7; Soviet Phys. Cryst. **14** [1969] 776/8).

[6] A. P. Perepelitsa, A. M. Kalinichenko (Ukr. Khim. Zh. **43** [1977] 543/5; Soviet Progr. Chem. **43** No. 5 [1977] 98/100).

6.1.2 Scandium Molybdates

6.1.2.1 $\mathbf{Sc_2(MoO_4)_3 \cdot 0.5\,H_2O}$ ($= Sc_2O_3 \cdot 3\,MoO_3 \cdot 0.5\,H_2O$)

General data for $M_2(MoO_4)_3 \cdot n\,H_2O$ molybdates are given on p. 247.

$Sc_2(MoO_4)_3 \cdot 0.5\,H_2O$ precipitates upon mixing equimolar amounts of aqueous solutions of $Sc(NO_3)_3$ and Na_2MoO_4 in the pH range 5 to 6.5. The precipitate is washed with water and dried in air for 24 h [1]. A sample of unspecified water content was formed on adding the stoichiometric amount of ammonium molybdate to an aqueous solution of $ScCl_3$ at pH 3.75 and keeping the mixture for 3 h at 80 to 90°C for equilibration [2]. Precipitation at 80°C from a 2.5% aqueous NH_4NO_3 solution at pH 4 and drying at 120°C gave the anhydrous compound [3].

The compound is X-ray amorphous [1, 2]. The thermogram up to 700°C differs slightly from those of the trihydrates $M_2(MoO_4)_3 \cdot 3\,H_2O$ with M = Y, Tb, Er, Yb [1].

References:

[1] A. M. Golub, A. P. Perepelitsa, A. A. Govorov (Zh. Neorgan. Khim. **16** [1971] 660/5; Russ. J. Inorg. Chem. **16** [1971] 352/5). – [2] L. N. Komissarova, V. M. Shatskii, N. P. Anoshina, I. I. Volodin (Vestn. Mosk. Univ. Khim. **26** [1971] 562/7; Moscow Univ. Chem. Bull. **26** No. 5 [1971] 35/8). – [3] S. Pajakoff (Monatsh. Chem. **100** [1969] 1350/61, 1360/1).

6.1.2.2 $\mathbf{Sc_2Mo_7O_{24} \cdot 4\,H_2O}$ ($= Sc_2O_3 \cdot 7\,MoO_3 \cdot 4\,H_2O$)

General data for $M_2Mo_7O_{24} \cdot n\,H_2O$ molybdates are given on p. 248.

The white $Sc_2Mo_7O_{24} \cdot 4\,H_2O$ has been obtained upon mixing stoichiometric amounts of aqueous HNO_3 solutions of 0.1 M $Sc(NO_3)_3$ and 0.1 M Na_2MoO_4 at pH 3.6. ^1H NMR shows the compound to contain adsorbed water and water of crystallization. The water is completely lost at 140°C. The anhydrous compound is decomposed into $Sc_2(MoO_4)_3$ and MoO_3 at 438°C.

A. M. Golub, A. P. Perepelitsa, V. I. Maksin (Izv. Vysshikh Uchebn. Zavedenii Khim. Khim. Tekhnol. **14** [1971] 815/7; C.A. **76** [1972] No. 9894).

6.1.2.3 $\mathbf{Sc(OH)MoO_4 \cdot n\,H_2O}$ ($= Sc_2O_3 \cdot 2\,MoO_3 \cdot (2n+1)H_2O$), n = 0 and 1

General data for $M(OH)MoO_4 \cdot n\,H_2O$ molybdates are given on p. 249.

$Sc(OH)MoO_4 \cdot H_2O$ was prepared by mixing equimolar amounts of aqueous solutions of $Sc(NO_3)_3$, Na_2MoO_4, and NaOH at pH > 10.5 and room temperature. The white precipitate was washed with water, alcohol, and ether and dried for 24 h. The monohydrate is dehydrated between 140 and 160°C to form $Sc(OH)MoO_4$ [1]. A sample dehydrated at 200°C in vacuo gave a quite broad ^1H NMR line compared with $Y(OH)MoO_4$. This is explained by structures with different inter-proton distances [2]. Between 580 and 650°C the anhydrous $Sc(OH)MoO_4$ is decomposed forming $Sc_2Mo_2O_9 + H_2O$ [1].

References:

[1] A. M. Golub, A. P. Perepelitsa, A. A. Govorov (Zh. Neorgan. Khim. **16** [1971] 660/5; Russ. J. Inorg. Chem. **16** [1971] 352/5). – [2] A. P. Perepelitsa, A. M. Kalinichenko (Ukr. Khim. Zh. **43** [1977] 543/5; Soviet Progr. Chem. **43** No. 5 [1977] 98/100).

6.1.2.4 $Sc(OH)Mo_4O_{13} \cdot 6H_2O$ ($= Sc_2O_3 \cdot 8MoO_3 \cdot 13H_2O$)

A compound of this composition precipitates from mixtures of 0.1 M aqueous solutions of $Sc(NO_3)_3$ and Na_2MoO_4, both brought to pH 2.5 by HNO_3. The presence of OH is inferred from the ^1H NMR spectrum. According to TG and DTA curves adsorbed water is lost at 80°C and the water of crystallization at 178°C, where the compound is already beginning to decompose. The OH is lost at 526°C with decomposition into $Sc_2(MoO_4)_3 + MoO_3$.

A. M. Golub, A. P. Perepelitsa, V. I. Maksin (Izv. Vysshikh Uchebn. Zavedenii Khim. Khim. Tekhnol. **14** [1971] 815/7; C.A. **76** [1972] No. 9894).

6.1.3 Yttrium Molybdates

6.1.3.1 $Y_2(MoO_4)_3 \cdot nH_2O$ ($= Y_2O_3 \cdot 3MoO_3 \cdot nH_2O$), n = 1 to 4

General data for $M_2(MoO_4)_3 \cdot nH_2O$ molybdates are given on p. 247.

Hydrated samples of $Y_2(MoO_4)_3 \cdot nH_2O$ precipitate, when concentrated (2.1 M [1] or nearly saturated [2]) aqueous solutions of M_2MoO_4 with M = Li [1, 3], Na [1, 4, 5], or K [6] are slowly added to aqueous 0.1 to 0.4 M $Y(NO_3)_3$ solutions at 25°C, pH values 3.86 to 5.87 [3], 5.45 to 5.72 [1], or 5 to 6.5 [4], and mole ratios $Y_2O_3 : MoO_3 = 1:3$ [2, 4] or between 2:1 and 1:3 [1, 3, 6]. Precipitation with Li_2MoO_4 is said to avoid formation of $LiY(MoO_4)_2 \cdot nH_2O$ under these conditions [2]. The water content has been determined after drying over H_2SO_4 [2], $CaCl_2$ [5], or at 30°C [1] as n = 4 or air dried for 24 h as n = 3 [4]. A dihydrate precipitates upon simultaneous and dropwise addition of stoichiometric amounts of an aqueous yttrium salt and molybdate solutions to a stirred 5% aqueous NH_4Cl solution of pH 4 to 4.5 at 80°C and drying the precipitate at 100°C. This procedure, precipitation at low supersaturation, is said to yield well-defined crystallized samples [7].

$Y_2(MoO_4)_3 \cdot nH_2O$ is also formed within several weeks from anhydrous $Y_2(MoO_4)_3$ by absorption of moisture from air of 40% relative humidity, giving n = 2.39 [8] or n = 2.8 [9]. Formation of a di- and a monohydrate has been reported during heating of the tetrahydrate up to 170 and from 170 to 220°C, respectively [1, 2].

$Y_2(MoO_4)_3 \cdot 4H_2O$ is reported to be crystalline, for a line diagram see the paper [2]. The trihydrate has been found to be X-ray amorphous [4] and the 2.39 hydrate crystalline [8]. The X-ray powder pattern of the 2.8 hydrate [9] is similar to that of the K structure type (2.39 hydrate) of [8] (for d values see the paper [9]) and resembles that reported for the anhydrous $Y_2(MoO_4)_3$ [1]. The density of the isometric crystals of $Y_2(MoO_4)_3 \cdot 2.8H_2O$ is 3.49 g/cm^3. The crystals have been observed in two different habits: the isometric crystals (size 15 to 20 μm) are optically uniaxial positive, whereas the platelets (> 0.1 mm) are biaxial negative. The refractive indices have been found as $n_\alpha = 1.799$ and $n_\gamma = 1.805$ for both modifications [9]. The IR spectrum between 3700 and 400 cm^{-1} of the tetrahydrate has been recorded, all absorption bands below 1000 cm^{-1} are attributed to vibrations of the MoO_4 tetrahedra (for a tracing of the spectrum see the paper) [2].

According to TG and DTA $Y_2(MoO_4)_3 \cdot 4H_2O$ loses 2 mol H_2O below 170, one further mole below 220, and becomes anhydrous at 295°C [1], while according to [2] dehydration takes place in two steps from 80 to 245 and from 270 to 435°C. $Y_2(MoO_4)_3 \cdot 3H_2O$ is reported to be dehydrated between 150 and 200 [4] or between 60 and 120°C [8]. Starting at 70°C, dehydration of $Y_2(MoO_4)_3 \cdot 2.8H_2O$ is complete at 220°C according to high-temperature X-ray investigations, yet minor weight losses have been observed up to 300°C [9].

References:

[1] E. M. Avzhieva, I. V. Shakhno, V. E. Plyushchev (Izv. Vysshikh Uchebn. Zavedenii Khim. Khim. Tekhnol. **12** [1969] 1307/10; C.A. **72** [1970] No. 85711). – [2] E. M. Avzhieva, I. V. Shakhno, V. E. Plyushchev, G. N. Voronskaya, K. I. Petrov (Zh. Neorgan. Khim. **20** [1975] 1561/5; Russ. J. Inorg. Chem. **20** [1975] 874/6). – [3] E. M. Avzhieva, I. V. Shakhno, V. E. Plyushchev (Izv. Vysshikh Uchebn. Zavedenii Khim. Khim. Tekhnol. **15** [1972] 1131/5; C.A. **77** [1972] No. 144456). – [4] A. M. Golub, A. P. Perepelitsa, A. A. Govorov (Zh. Neorgan. Khim. **16** [1971] 660/5; Russ. J. Inorg. Chem. **16** [1971] 352/5). – [5] F. Zambonini (Bull. Soc. Franc. Mineral. **38** [1915] 206/64, 228).

[6] E. M. Avzhieva, I. V. Shakhno, V. E. Plyushchev (Izv. Vysshikh Uchebn. Zavedenii Khim. Khim. Tekhnol. **13** [1970] 611/4; C.A. **73** [1970] No. 62111). – [7] S. Pajakoff (Monatsh. Chem. **100** [1969] 1350/61, 1357). – [8] K. Nassau, H. J. Levinstein, G. M. Loiacono (J. Phys. Chem. Solids **26** [1965] 1805/16, 1810/1). – [9] N. V. Gul'ko, A. G. Karaulov, N. M. Taranukha, A. M. Gavrish (Izv. Akad. Nauk SSSR Neorgan. Materialy **9** [1973] 1766/9; Inorg. Materials [USSR] **9** [1973] 1572/5).

6.1.3.2 $Y_2Mo_7O_{24} \cdot nH_2O$ ($=Y_2O_3 \cdot 7MoO_3 \cdot nH_2O$), n = 4, 6, and 7

General data for $M_2Mo_7O_{24} \cdot nH_2O$ molybdates are given on p. 248.

$Y_2Mo_7O_{24} \cdot nH_2O$ has been prepared by mixing aqueous 0.1 M $Y(NO_3)_3$ and 0.1 M Na or NH_4 paramolybdate solutions in stoichiometric amounts for 5 h [1] at pH 3.6 adjusted with HNO_3 [2, 3]. After washing with acetone and ether the compound is dried over $CaCl_2$ to give the heptahydrate and over H_2SO_4 or P_2O_5 to give the hexahydrate [1]. A tetrahydrate is obtained by [3].

The X-ray powder line diagrams given [1, 2] show the compound to be isomorphous with the corresponding compounds of Ho to Lu. The pycnometric density of the heptahydrate is 3.810 ± 0.015 g/cm³ [1].

The tetrahydrate has been found to dehydrate at ~140°C [3]. The water-rich samples dehydrate in two endothermic steps at 230 and 261°C [1]. The amorphous dehydration products crystallize at 420 to 430°C giving MoO_3 and Y molybdates [1, 3].

References:

[1] S. S. Antonova, I. V. Shakhno, V. E. Plyushchev (Izv. Vysshikh Uchebn. Zavedenii Khim. Khim. Tekhnol. **15** [1972] 1289/92; C.A. **78** [1973] No. 10950). – [2] S. S. Antonova, I. V. Shakhno, V. E. Plyushchev (Izv. Vysshikh Uchebn. Zavedenii Khim. Khim. Tekhnol. **15** [1972] 12/5; C.A. **76** [1972] No. 132181). – [3] A. M. Golub, A. P. Perepelitsa, V. I. Maksin (Izv. Vysshikh Uchebn. Zavedenii Khim. Khim. Tekhnol. **14** [1971] 815/7; C.A. **76** [1972] No. 9894).

6.1.3.3 $Y(OH)MoO_4 \cdot nH_2O$ ($= Y_2O_3 \cdot 2MoO_3 \cdot (2n+1)H_2O$), n = 0 and 1

General data for $M(OH)MoO_4 \cdot nH_2O$ molybdates are given on p. 249.

$Y(OH)MoO_4 \cdot H_2O$ was obtained and investigated as described for the corresponding Sc compound (p. 250) [1]. The compound has also been obtained from aqueous $K_2MoO_4 + KOH$ solution with varying $Y(NO_3)_3$ amounts [2]. ^1H NMR spectra have been measured with increasing temperature. Air-dried specimens show water of crystallization and OH groups. The water of crystallization is lost between 140 and 160°C [1]. Narrow line widths are characteristic for samples dehydrated at 200°C in vacuo [3]. The OH groups remain even after heating for 30 min at 600°C and disappear only on heating to 650 to 700°C, where $Y_2Mo_2O_9$ and H_2O form [1], see also [2].

References:

[1] A. M. Golub, A. P. Perepelitsa, A. A. Govorov (Zh. Neorgan. Khim. **16** [1971] 660/5; Russ. J. Inorg. Chem. **16** [1971] 352/5). – [2] A. P. Perepelitsa, V. N. Solomakha (Zh. Neorgan. Khim. **18** [1973] 28/30; Russ. J. Inorg. Chem. **18** [1973] 14/5). – [3] A. P. Perepelitsa, A. M. Kalinichenko (Ukr. Khim. Zh. **43** [1977] 543/5; Soviet Progr. Chem. **43** No. 5 [1977] 98/100).

6.1.4 Lanthanum Molybdates

6.1.4.1 $La_2(MoO_4)_3 \cdot nH_2O$ ($= La_2O_3 \cdot 3MoO_3 \cdot nH_2O$), n = 2 and 6

General data for $M_2(MoO_4)_3 \cdot nH_2O$ molybdates are given on p. 247.

$La_2(MoO_4)_3 \cdot 6H_2O$ has been prepared by mixing moderately concentrated solutions of $La(NO_3)_3$ and Na_2MoO_4 in the stoichiometric ratio. The compound is dried at room temperature in the air [1]. A dihydrate is obtained by mixing stoichiometric amounts of 0.05 M $La(NO_3)_3$ and 0.075 M Na_2MoO_4 solutions and drying the precipitate at 105 to 110°C in air [2] or by simultaneous dropwise addition of stoichiometric amounts of lanthanum salt and molybdate solutions to a stirred 5% aqueous NH_4Cl solution of pH 4 to 4.5 at 80°C and subsequently drying the solid at 100°C. The latter procedure is said to yield well-defined crystallized samples [3]. Amperometric, conductometric, and pH titrations show the formation of $La_2(MoO_4)_3 \cdot nH_2O$ (n not given) from $La(NO_3)_3$ and Na_2MoO_4 solutions to take place between pH 5.0 and 6.0 [4].

$La_2(MoO_4)_3 \cdot nH_2O$ is said to precipitate also from mixtures of aqueous $LaCl_3$ and $(NH_4)_6Mo_7O_{24}$ solutions at pH ≈ 5 and a starting concentration of 1.82×10^{-3} M La^{3+}, if the mole ratio La : Mo is between 1:1.5 and 1:5 [5]. However, amperometric, conductometric, and pH titrations of $La(NO_3)_3$ and sodium paramolybdate solutions show the formation and precipitation of $La_2Mo_7O_{24} \cdot nH_2O$ (see below) in the pH range 3.5 to 4.5 [4].

From ^1H NMR spectra taken at 77 K and room temperature of samples dried at room temperature, 120, and 195°C, it was concluded that there are three different types of protons in the structure of $La_2(MoO_4)_3 \cdot 6H_2O$: OH^- groups, $2H_2O$ in the vicinity of La^{3+}, and $3H_2O$ near MoO_4^{2-}. To account for this the formula $La_4(OH)_4Mo_6O_{22} \cdot 10H_2O$ has been proposed [1, 6]. IR spectra of the hexahydrate show absorption bands typical for H_2O and also bands attributed to OH^- groups [1].

On heating, $La_2(MoO_4)_3 \cdot 6H_2O$ is dehydrated gradually between 20 and 520°C. DTA and TG curves together with the ^1H NMR results indicate that H_2O is lost in three steps: $3H_2O$ from 20 to 120, $2H_2O$ between 120 and 200, and $1H_2O$ from 200 to 550°C [1].

The solubility product of $La_2(MoO_4)_3 \cdot 2H_2O$ in water at 25°C is 2.2×10^{-21}. $La_2(MoO_4)_3 \cdot 2H_2O$ dissolves in aqueous $LaCl_3$ solutions with formation of the soluble $[LaMoO_4]^+$ complex, see

p. 247. In aqueous 4.2×10^{-5} to 5.0×10^{-1} M Na_2MoO_4 solutions the solubility of the dihydrate decreases to 3.6×10^{-6} M (at 3.20×10^{-4} M Na_2MoO_4) and then remains almost constant. Soluble molybdo complexes could not be detected. In the solid phase the La:Mo ratio increases to 2:5 at higher Na_2MoO_4 concentrations [2].

References:

[1] A. M. Golub, K. S. Aganiyazov, V. V. Mank, Ya. Ya. Shcherbak (Teor. Eksperim. Khim. **3** [1967] 494/7; Theor. Exptl. Chem. [USSR] **3** [1967] 286/8). – [2] N. K. Davidenko, G. A. Komashko, K. B. Yatsimirskii (Zh. Neorgan. Khim. **13** [1968] 117/22; Russ. J. Inorg. Chem. **13** [1968] 58/61). – [3] S. Pajakoff (Monatsh. Chem. **100** [1969] 1350/61, 1357). – [4] R. S. Saxena, M. L. Mittal (Z. Anorg. Allgem. Chem. **324** [1963] 208/13; Acta Chim. [Budapest] **34** [1962] 193/201). – [5] B. N. Ivanov-Emin, E. I. Zakharikova (Zh. Neorgan. Khim. **14** [1969] 145/9; Russ. J. Inorg. Chem. **14** [1969] 75/8).

[6] A. M. Golub, K. S. Aganiyazov, V. V. Mank, M. V. Mokhosoev (Teor. Eksperim. Khim. **4** [1968] 705/8; Theor. Exptl. Chem. [USSR] **4** [1968] 453/5).

6.1.4.2 $La_2Mo_7O_{24}\cdot nH_2O$ (= $La_2O_3\cdot 7MoO_3\cdot nH_2O$), n = 7 and 10

General data for $M_2Mo_7O_{24}\cdot nH_2O$ molybdates are given on p. 248.

$La_2Mo_7O_{24}\cdot nH_2O$ precipitates from mixtures of aqueous 0.1 M solutions of $La(NO_3)_3$ and $Na_6Mo_7O_{24}$ or $(NH_4)_6Mo_7O_{24}$ at pH ≈ 3, adjusted by HNO_3 [1, 2]. Amperometric, conductometric [3], and pH titrations [4] of $La(NO_3)_3$ and sodium paramolybdate in aqueous and ethanolic (up to 50%) solutions show the lanthanum paramolybdate to be formed in the pH range 3.5 to 4.5. Also with sodium dimolybdate only $La_2Mo_7O_{24}\cdot nH_2O$ is formed in this pH range [3, 4]. After washing with acetone and ether the decahydrate is obtained over $CaCl_2$ and the heptahydrate over H_2SO_4 or P_2O_5 [1].

An X-ray powder line diagram is given in the paper. The pycnometric density of the decahydrate is 3.344 ± 0.015 g/cm³. The water is lost in one step at 183°C. The amorphous dehydration product crystallizes at 412°C [1].

References:

[1] S. S. Antonova, I. V. Shakhno, V. E. Plyushchev (Izv. Vysshikh Uchebn. Zavedenii Khim. Khim. Tekhnol. **15** [1972] 1289/92; C.A. **78** [1973] No. 10950). – [2] S. S. Antonova, I. V. Shakhno, V. E. Plyushchev, G. A. Zenkova (Uch. Zap. Mosk. Inst. Tonkoi Khim. Tekhnol. **1** [1971] 99/103; C.A. **77** [1972] No. 131389). – [3] R. S. Saxena, M. L. Mittal (Z. Anorg. Allgem. Chem. **324** [1963] 208/13). – [4] R. S. Saxena, M. L. Mittal (Acta Chim. [Budapest] **34** [1962] 193/201).

6.1.4.3 $La(OH)MoO_4\cdot nH_2O$ (= $La_2O_3\cdot 2MoO_3\cdot(2n+1)H_2O$), n = 0 and 1.5

General data for $M(OH)MoO_4\cdot nH_2O$ molybdates are given on p. 249.

$La(OH)MoO_4\cdot 1.5H_2O$ has been prepared by addition of the stoichiometric amount of aqueous 0.5 M $La(NO_3)_3$ solution to a boiling solution equimolar in Na_2MoO_4 and NaOH. The amorphous precipitate was washed with water and dried in air [1]. The formation of $La(OH)MoO_4\cdot nH_2O$ has been observed during the reaction of $LaCl_3$ with ammonium paramolybdate solutions at Mo:La = 1 [2]. The 1.5-hydrate loses its water of crystallization in two steps, $1H_2O$ at 240°C and $0.5H_2O$ at 350°C [1]. Anhydrous $La(OH)MoO_4$ is also obtained by hydrothermal synthesis from La_2O_3/MoO_3 mixtures as described for the corresponding Nd compound (p. 259) [3]. The standard enthalpy of formation of $La(OH)MoO_4$ was determined as $\Delta H^\circ_{f,298} =$ 33.76 kcal/mol by measuring the heats of reaction of dissolving $La(OH)MoO_4$ and Cs_2MoO_4 in aqueous HCl [4].

The flattened crystals of anhydrous $La(OH)MoO_4$ are isomorphous with the monoclinic Nd compound, lattice parameters a = 5.39, b = 12.78, c = 6.90 Å (all ± 0.03 Å), β = 113°. The ^1H NMR spectrum of the partly dehydrated compound (dried at 120°C) shows both a narrow and a broad component, which are assigned to hydroxyl and water protons, respectively [3]. The spectrum of the anhydrous compound (from dehydration of the hydrate at 200°C in vacuo) contains only the narrow line [5]. The IR spectrum, which is similar to that of the Nd compound, shows also the existence of the OH groups [3].

On heating, anhydrous $La(OH)MoO_4$ decomposes at 540 [1] to 545°C [3], possibly into $La_2Mo_2O_9$ and H_2O.

References:

[1] A. M. Golub, K. S. Aganiyazov (Ukr. Khim. Zh. **35** [1969] 1239/43; Soviet Progr. Chem. **35** No. 12 [1969] 1/4). – [2] B. N. Ivanov-Emin, E. I. Zakharikova (Zh. Neorgan. Khim. **14** [1969] 145/9; Russ. J. Inorg. Chem. **14** [1969] 75/8). – [3] V. I. Protasova, L. Yu. Kharchenko, P. V. Klevtsov (Izv. Akad. Nauk SSSR Neorgan. Materialy **9** [1973] 421/3; Inorg. Materials [USSR] **9** [1973] 374/6). – [4] Yu. L. Suponitskii, A. G. Dyunin, V. I. Protasova (Deposited Doc. VINITI-467-81 [1980] 1/8; C.A. **96** [1982] No. 169775). – [5] A. P. Perepelitsa, A. M. Kalinichenko (Ukr. Khim. Zh. **43** [1977] 543/5; Soviet Progr. Chem. **43** No. 5 [1977] 98/100).

6.1.5 Cerium Molybdates

6.1.5.1 $Ce_2^{III}(MoO_4)_3 \cdot nH_2O$ (= $Ce_2O_3 \cdot 3MoO_3 \cdot nH_2O$), n = 3, 4.5, and 6

General data for $M_2(MoO_4)_3 \cdot nH_2O$ molybdates are given on p. 247.

$Ce_2(MoO_4)_3 \cdot nH_2O$ precipitates upon mixing aqueous solutions of $Ce(NO_3)_3$ and Na_2MoO_4 in the stoichiometric ratio [1 to 6]. Fresh precipitates are white and gelatinous, but later become yellow and crystalline. The state of hydration has been determined to be n = 6 (air dried) [2], 4.5 (vacuum dried at room temperature) [3, 4], or 3 (air dried) [1]; it has not been specified for the precipitates obtained from $CeCl_3$ and Na_2MoO_4 solutions [5, 6].

For tables of d values see the papers [3, 4]. ^1H NMR spectra of air-dried specimens showed asymmetric lines which have been resolved into three components of different line widths. From this it was concluded that protons are present as OH^- and two types of differently bound H_2O. Samples dried at 230°C show a new broad component which was also observed in samples that had been vacuum dried at 150°C and rehydrated in room air. The formula $Ce_4(OH)_4Mo_6O_{22} \cdot 10H_2O$ has been proposed, as for the corresponding La compound (p. 253). The IR spectrum of an air-dried sample of the hexahydrate is characterized by the absorption at 1640 cm^{-1} due to the H_2O bending vibration [2]. The IR absorption spectrum between 1700 and 300 cm^{-1} of $Ce_2(MoO_4)_3 \cdot 4.5H_2O$ is plotted, but assignments are not given [4]. The reflection spectrum in the UV and visible range between 210 and 700 nm shows an absorption maximum at about 400 nm and a shoulder at 270 nm, for a plot see the paper [3].

DTA curves of the hexahydrate show two endothermic effects at 140 and at 245°C. The sample weight becomes constant at about 300°C according to thermogravimetry. In connection with the ^1H NMR data the DTA curves are interpreted by assuming a stepwise dehydration of at first 2, then 3, and finally $1H_2O$ [2]. Crystals of $Ce_2(MoO_4)_3 \cdot 4.5H_2O$ are reported to be stable up to

60°C [3], whereas according to [4] dehydration starts at 40°C. Up to 155°C, 1.5 H_2O are lost, and most of the remaining water content is gone at 200°C [4]. Dehydration is reported to be complete at 285°C [4] or after 72 h at 200°C [3].

$Ce_2(MoO_4)_3 \cdot nH_2O$ dissolves in an aqueous $CeCl_3$ solution and forms the soluble $[CeMoO_4]^+$ complex (see p. 247) [7].

References:

[1] F. Zambonini (Bull. Soc. Franc. Mineral. **38** [1915] 206/64, 210). – [2] A. M. Golub, K. S. Aganiyazov, V. V. Mank, M. V. Mokhosoev (Teor. Eksperim. Khim. **4** [1968] 705/8; Theor. Exptl. Chem. [USSR] **4** [1968] 453/5). – [3] A. Castellan, J. C. J. Bart, A. Bossi, P. Perissinoto, N. Giordano (Z. Anorg. Allgem. Chem. **422** [1976] 155/72). – [4] M. J. Schwing-Weill (Bull. Soc. Chim. France **1972** 1754/61). – [5] M. C. Saxena, A. K. Bhattacharya (Proc. Natl. Acad. Sci. India A **30** [1961] 44/50; C.A. **56** [1962] 2946).

[6] M. C. Saxena, A. K. Bhattacharya (Proc. Natl. Acad. Sci. India A **30** [1961] 200/5; C.A. **56** [1962] 9496). – [7] N. K. Davidenko, G. A. Komashko, K. B. Yatsimirskii (Zh. Neorgan. Khim. **13** [1968] 117/22; Russ. J. Inorg. Chem. **13** [1968] 58/61).

6.1.5.2 $\mathbf{Ce^{III}(OH)MoO_4 \cdot nH_2O}$ ($=Ce_2O_3 \cdot 2MoO_3 \cdot (2n+1)H_2O$), n = 0 and 1.5

General data for $M(OH)MoO_4 \cdot nH_2O$ molybdates are given on p. 249.

$Ce(OH)MoO_4 \cdot 1.5H_2O$ is precipitated from an equimolar $Na_2MoO_4 + NaOH$ solution by $Ce(NO_3)_3$ solution as described for the corresponding La compound (p. 254) [1].

The 1.5-hydrate dehydrates in two steps, at 220°C (1 H_2O) and 340°C (0.5 H_2O) [1]. The ^1H NMR spectrum of a sample dehydrated in vacuo at 200°C shows the narrow line characteristic for protons of the OH group. This kind of line has also been found for the La to Eu compounds [2]. Anhydrous $Ce(OH)MoO_4$ decomposes at 520 [1] or 600°C [3], presumably into $Ce_2Mo_2O_9$ and H_2O.

References:

[1] A. M. Golub, K. S. Aganiyazov (Ukr. Khim. Zh. **35** [1969] 1239/43; Soviet Progr. Chem. **35** No. 12 [1969] 1/4). – [2] A. P. Perepelitsa, A. M. Kalinichenko (Ukr. Khim. Zh. **43** [1977] 543/5; Soviet Progr. Chem. **43** No. 5 [1977] 98/100). – [3] V. I. Protasova, L. Yu. Kharchenko, P. V. Klevtsov (Izv. Akad. Nauk SSSR Neorgan. Materialy **9** [1973] 421/3; Inorg. Materials [USSR] **9** [1973] 374/6).

6.1.5.3 $\mathbf{Ce^{IV}(MoO_4)_2 \cdot nH_2O}$ ($=CeO_2 \cdot 2MoO_3 \cdot nH_2O$), n = 1, 3, 7

X-ray amorphous $Ce(MoO_4)_2 \cdot 7H_2O$ has been prepared by grinding a slurry of 4.5 g $Ce(SO_4)_2 \cdot 4H_2O$ and 6.3 g $Na_2MoO_4 \cdot 2H_2O$ in 40 mL of water in a mortar [1]. The amorphous heptahydrate has also been obtained from acidic (pH = 0) ceric ammonium nitrate and ammonium molybdate solutions at 30°C. The precipitate was left overnight in the mother liquor, then filtered, washed, and dried in the air [2]. Yellow $Ce(MoO_4)_2 \cdot 3H_2O$ precipitated upon addition of a solution of 3.2 g ammonium molybdate in 250 mL of water to a solution of 10 g ceric nitrate in 1 L of water at pH 1.45 and 45°C [3]. A monohydrate can be prepared from a melt of $Ce(SO_4)_2 \cdot 4H_2O$, $Na_2MoO_4 \cdot 2H_2O$, $NaNO_3$, and KNO_3. After grinding and washing in H_2O, $Ce(MoO_4)_2 \cdot H_2O$ remained as an insoluble residue [1]. Cerium(IV) molybdate preparations of various compositions and water contents were prepared by mixing ceric sulfate and ammonium or sodium molybdate solutions on a steam bath. The precipitates were filtered, washed with water, and dried over P_2O_5 [4].

The IR spectrum of the amorphous heptahydrate shows sharp bands at 3340 (OH stretching), 1610 (OH bending), and 1100 cm^{-1} (Ce-O-H bending mode). Broad bands in the 800 to 500 cm^{-1} region are due to overlapping of molybdate and Ce-O-H bands [2]. The IR spectrum of the trihydrate measured between 4000 and 625 cm^{-1} shows corresponding bands at 3400, 1620, and 1160 cm^{-1}. The broad band at 1000 to 750 cm^{-1} is attributed to Mo-O stretching vibrations [3]. Samples heated to 600°C still show bands at 3340, 1610, and 1100 cm^{-1} and in the 800 to 500 cm^{-1} region [2].

Thermogravimetric and differential thermal analyses show the hydrates to lose all water between 50 and 250°C [1], with the maximum endothermic effect at 150 [2] or 170°C [3]. $Ce(MoO_4)_2 \cdot H_2O$ is scarcely soluble in H_2O. It decomposes in acidic or basic media [1]. The amorphous Ce^{IV} molybdate precipitates have been used successfully as ion exchangers [2, 4].

References:

[1] A. M. Golub, V. I. Maksin, A. P. Perepelitsa (Zh. Neorgan. Khim. **20** [1975] 867/70; Russ. J. Inorg. Chem. **20** [1975] 486/8). – [2] S. K. Srivastava, R. P. Singh, S. Agrawal, S. Kumar (J. Radioanal. Chem. **40** [1977] 7/15). – [3] P. K. Bhattacharya, S. K. Bhattacharyya (J. Indian Chem. Soc. **53** [1976] 561/3). – [4] A. K. De, S. K. Das (Separ. Sci. **11** [1976] 183/91; C.A. **84** [1976] No. 112153).

6.1.6 Praseodymium Molybdates

6.1.6.1 $Pr_2(MoO_4)_3 \cdot nH_2O$ (= $Pr_2O_3 \cdot 3MoO_3 \cdot nH_2O$), n = 2, 3.5, and 6

General data for $M_2(MoO_4)_3 \cdot nH_2O$ molybdates are given on p. 247.

$Pr_2(MoO_4)_3 \cdot 6H_2O$ has been prepared by double decomposition of $Pr(NO_3)_3$ and Na_2MoO_4 in aqueous solution with air drying [1]. The 3.5-hydrate has also been obtained from $Pr(NO_3)_3$ and Na_2MoO_4 solutions but dried over concentrated H_2SO_4 [2]. $Pr_2(MoO_4)_3 \cdot 2H_2O$ has been obtained by dropwise addition of Pr^{3+} and molybdate solutions to a stirred hot 5% aqueous solution of NH_4Cl and subsequent drying at 100°C [3].

To explain the ^1H NMR spectra of the hexahydrate, which indicate the presence of three types of protons, the formulation $Pr_4(OH)_4Mo_6O_{22} \cdot 10H_2O$ has been proposed, as for the corresponding La compound (p. 253) [1]. IR spectra of $Pr_2(MoO_4)_3 \cdot 6H_2O$ were too complex for complete assignment, yet the deformation vibration of H_2O at 1640 cm^{-1} has been recognized [1].

In accordance with the ^1H NMR investigations the thermogravimetric and differential thermal analyses have been interpreted by assuming a dehydration in three steps of at first 2, then 3, and finally $1H_2O$ at somewhat higher temperatures than those of the corresponding La compound; for a plot of the curves see the paper [1]. In aqueous $PrCl_3$ solution the soluble $[PrMoO_4]^+$ complex formed (see p. 247) [4].

References:

[1] A. M. Golub, K. S. Aganiyazov, V. V. Mank, M. V. Mokhosoev (Teor. Eksperim. Khim. **4** [1968] 705/8; Theor. Exptl. Chem. [USSR] **4** [1968] 453/5). – [2] F. Zambonini (Bull. Soc. Franc. Mineral. **38** [1915] 206/64, 220). – [3] S. Pajakoff (Monatsh. Chem. **100** [1969] 1350/61, 1357). – [4] N. K. Davidenko, G. A. Komashko, K. B. Yatsimirskii (Zh. Neorgan. Khim. **13** [1968] 117/22; Russ. J. Inorg. Chem. **13** [1968] 58/61).

6.1.6.2 **$Pr_2Mo_7O_{24} \cdot nH_2O$** ($=Pr_2O_3 \cdot 7MoO_3 \cdot nH_2O$), n = 7 and 9

General data for $M_2Mo_7O_{24} \cdot nH_2O$ molybdates are given on p. 248.

$Pr_2Mo_7O_{24} \cdot nH_2O$ has been prepared and investigated together with the corresponding compounds of Y, La, and Nd to Lu by the same methods. It precipitates from stoichiometric mixtures of 0.1 M aqueous solutions of $Pr(NO_3)_3$ and ammonium or sodium paramolybdate after 5 h stirring at room temperature. After washing with acetone and ether the nonahydrate is obtained over $CaCl_2$ and the heptahydrate over H_2SO_4 or P_2O_5.

An X-ray powder line diagram is given in the paper. The compound is isomorphous with the corresponding Nd to Dy compounds. The pycnometric density of the nonahydrate is 3.483 ± 0.015 g/cm³ at 25°C. The water is lost in one step at 216°C and the amorphous dehydration product crystallizes at 426°C. A DTA curve up to 750°C is given in the paper.

S. S. Antonova, I. V. Shakhno, V. E. Plyushchev (Izv. Vysshikh Uchebn. Zavedenii Khim. Khim. Tekhnol. **15** [1972] 1289/92; C.A. **78** [1973] No. 10950).

6.1.6.3 **$Pr(OH)MoO_4 \cdot nH_2O$** ($=Pr_2O_3 \cdot 2MoO_3 \cdot (2n+1)H_2O$), n = 0 and 1.5

General data for $M(OH)MoO_4 \cdot nH_2O$ molybdates are given on p. 249.

$Pr(OH)MoO_4 \cdot 1.5H_2O$ is prepared from aqueous $Pr(NO_3)_3$ solution and an equimolar $Na_2MoO_4 + NaOH$ solution as described for the corresponding La compound (p. 254). The 1.5-hydrate dehydrates in two steps at 210°C ($1H_2O$) and 360°C ($0.5H_2O$) [1]. Anhydrous $Pr(OH)MoO_4$ has also been obtained by hydrothermal synthesis from Pr_2O_3 and MoO_3 as described for the corresponding Nd compound (p. 259) [2].

The flattened crystals of anhydrous $Pr(OH)MoO_4$ are isomorphous with the monoclinic Nd compound, lattice parameters a = 5.37, b = 12.57, c = 6.85 Å (all ± 0.03 Å), β = 113° [2]. The ^1H NMR spectrum of a sample obtained from the hydrate by dehydration in vacuo at 200°C shows the narrow line characteristic of OH protons which has been observed for the whole La to Eu group [3]. The presence of the OH group has also been established by the IR spectrum [2].

Anhydrous $Pr(OH)MoO_4$ decomposes at 560 [1] or 620°C [2], forming presumably $Pr_2Mo_2O_9$ and H_2O.

References:

[1] A. M. Golub, K. S. Aganiyazov (Ukr. Khim. Zh. **35** [1969] 1239/43; Soviet Progr. Chem. **35** No. 12 [1969] 1/4). – [2] V. I. Protasova, L. Yu. Kharchenko, P. V. Klevtsov (Izv. Akad. Nauk SSSR Neorgan. Materialy **9** [1973] 421/3; Inorg. Materials [USSR] **9** [1973] 374/6). – [3] A. P. Perepelitsa, A. M. Kalinichenko (Ukr. Khim. Zh. **43** [1977] 543/5; Soviet Progr. Chem. **43** No. 5 [1977] 98/100).

6.1.7 Neodymium Molybdates

6.1.7.1 **$Nd_2(MoO_4)_3 \cdot nH_2O$** ($=Nd_2O_3 \cdot 3MoO_3 \cdot nH_2O$), n = 2, 4, and 6

General data for $M_2(MoO_4)_3 \cdot nH_2O$ molybdates are given on p. 247.

$Nd_2(MoO_4)_3 \cdot 6H_2O$ has been prepared by mixing aqueous solutions of $Nd(NO_3)_3$ and Na_2MoO_4 in the stoichiometric ratio. The precipitate is washed and air dried [1]. The tetrahydrate has also been obtained from $Nd(NO_3)_3$ and Na_2MoO_4 solutions but dried over concentrated H_2SO_4 [2]. $Nd_2(MoO_4)_3 \cdot 2H_2O$ is formed by precipitation at 80°C in a 5% aqueous NH_4Cl solution and subsequently drying the precipitate at 100°C [3].

Showing three different types of protons, the ^1H NMR spectrum of the hexahydrate has been interpreted by assuming the presence of hydroxyl protons as well as two types of crystal water, suggesting the formula $Nd_4(OH)_4Mo_6O_{22} \cdot 10H_2O$ as for the corresponding La compound (p. 253) [1]. Aside from the characteristic H_2O deformation vibration at 1640 cm^{-1}, the IR spectrum of $Nd_2(MoO_4)_3 \cdot 6H_2O$ proved too complex for assignment [1].

According to TG and DTA, thermal dehydration of the hexahydrate takes place in three steps of 2, 3, and 1 H_2O between ~40 and ~250°C [1].

In aqueous $NdCl_3$ solutions the soluble $[NdMoO_4]^+$ complex forms, see p. 247 [4].

References:

[1] A. M. Golub, K. S. Aganiyazov, V. V. Mank, M. V. Mokhosoev (Teor. Eksperim. Khim. **4** [1968] 705/8; Theor. Exptl. Chem. [USSR] **4** [1968] 453/5). – [2] F. Zambonini (Bull. Soc. Franc. Mineral. **38** [1915] 206/64, 218). – [3] S. Pajakoff (Monatsh. Chem. **100** [1969] 1350/61, 1357). – [4] N. K. Davidenko, G. A. Komashko, K. B. Yatsimirskii (Zh. Neorgan. Khim. **13** [1968] 117/22; Russ. J. Inorg. Chem. **13** [1968] 58/61).

6.1.7.2 $\mathbf{Nd_2Mo_7O_{24} \cdot nH_2O}$ (= $Nd_2O_3 \cdot 7MoO_3 \cdot nH_2O$), n = 6 and 8

General data for $M_2Mo_7O_{24} \cdot nH_2O$ molybdates are given on p. 248.

$Nd_2Mo_7O_{24} \cdot nH_2O$ has been prepared and investigated as described for the corresponding Pr compound (p. 258). The pycnometric density of the octahydrate is 3.548 ± 0.015 g/cm³. The compound dehydrates in one step at 214°C and the amorphous products crystallize at 432°C.

S. S. Antonova, I. V. Shakhno, V. E. Plyushchev (Izv. Vysshikh Uchebn. Zavedenii Khim. Khim. Tekhnol. **15** [1972] 1289/92; C.A. **78** [1973] No. 10950).

6.1.7.3 $\mathbf{Nd(OH)MoO_4 \cdot nH_2O}$ (= $Nd_2O_3 \cdot 2MoO_3 \cdot (2n+1)H_2O$), n = 0 and 1.5

General data for $M(OH)MoO_4 \cdot nH_2O$ molybdates are given on p. 249.

$Nd(OH)MoO_4 \cdot 1.5H_2O$ is prepared from aqueous $Nd(NO_3)_3$ solution plus an equimolar $Na_2MoO_4 + NaOH$ solution as described for the corresponding La compound (p. 254). The 1.5-hydrate dehydrates in two steps at 230°C (1 H_2O) and 340°C (0.5 H_2O) [1].

Anhydrous $Nd(OH)MoO_4$ has also been prepared by hydrothermal synthesis from mixtures of Nd_2O_3 and MoO_3 in a Ti-lined autoclave at 450 to 525°C and 1200 to 1400 atm in 160 to 200 h. The solvents were chloride solutions of Li^+, K^+ (20 to 25%) or NH_4^+ (5%), or also their mixtures. Crystallization was also performed in a 15 to 20% lithium molybdate solution. The largest crystals (up to 1.5 mm) were obtained in mixed solutions of LiCl or KCl (20%) + NH_4Cl (5%). At higher temperatures (550°C) molybdenum oxides were present as admixtures and in lithium molybdate solutions (525 to 550°C) $LiNd(MoO_4)_2$ also forms [2].

Intergrowths of the crystals are formed as druses, spherical bodies, and brushes. In lithium molybdate solution the crystals are very thin and foliated. The most developed simple form is the {010} pinacoid; other growth forms are the prisms {110}, {120}, etc. A stereographic projection is given in the paper. $Nd(OH)MoO_4$ is monoclinic, lattice parameters a = 5.37, b = 12.56, c = 6.83 Å (all ± 0.03 Å), β = 113°. The molybdate is isomorphous with the correspond-

ing tungstate $Nd(OH)WO_4$ [3], space group $P2_1/c\text{-}C_{2h}^5$ (No. 14). X-ray powder data show th
Er molybdates to be isomorphous with the Pr to Gd tungstates [2].

The narrow line in the ^1H NMR spectrum of the dehydrated compound (200°C, vacu
characteristic for OH protons and has been observed in the whole La to Eu group moly
[4]. The IR spectrum has been recorded between 3700 and 800 cm^{-1} and is plotted in the
The band corresponding to the stretching vibration of the OH group is displaced from 3
the 3570 to 3560 cm^{-1} region [2].

Anhydrous $Nd(OH)MoO_4$ decomposes at 590 [1] or 655 ± 10°C [2], forming presu
$Nd_2Mo_2O_9$ and H_2O. A thermogram is given in [2].

References:

[1] A. M. Golub, K. S. Aganiyazov (Ukr. Khim. Zh. **35** [1969] 1239/43; Soviet Progr. Ch
No. 12 [1969] 1/4). – [2] V. I. Protasova, L. Yu. Kharchenko, P. V. Klevtsov (Izv. Akad. Nauk
Neorgan. Materialy **9** [1973] 421/3; Inorg. Materials [USSR] **9** [1973] 374/6). – [3] R. F. Klev
S. V. Borisov (Kristallografiya **14** [1969] 904/7; Soviet Phys.-Cryst. **14** [1969] 776/8). – [4
Perepelitsa, A. M. Kalinichenko (Ukr. Khim. Zh. **43** [1977] 543/5; Soviet Progr. Chem. **43**
[1977] 98/100).

6.1.8 Samarium Molybdates

6.1.8.1 $Sm_2(MoO_4)_3 \cdot 2H_2O$ (= $Sm_2O_3 \cdot 3MoO_3 \cdot 2H_2O$)

General data for $M_2(MoO_4)_3 \cdot nH_2O$ molybdates are given on p. 247.

$Sm_2(MoO_4)_3 \cdot 2H_2O$ is obtained when aqueous solutions of $Sm(NO_3)_3$ and alkali mol
are added simultaneously dropwise to a 5% aqueous solution of NH_4Cl at 80°C. The prec
is dried at 100°C [1]. Amperometric and conductometric titrations of aqueous soluti
$Sm(NO_3)_3$ and Na_2MoO_4 of various concentrations show the compound to be formed at
to 6.0. The enthalpy change of the reaction is $\Delta H = -188.8$ cal/mol $Sm_2(MoO_4)_3 \cdot nH_2O$

An $Sm_2(MoO_4)_3$ hydrate of unspecified water content has been recovered from mixt
aqueous $SmCl_3$ and $(NH_4)_6Mo_7O_{24}$ solutions with molar ratios Sm : Mo = 2 : 3 at 25°C
contrast amperometric and conductometric titrations of aqueous $Sm(NO_3)_3$ and Na_6
solutions of various concentrations show that only $Sm_2Mo_7O_{24} \cdot nH_2O$ forms at pH 3.5 to 4

$Sm_2(MoO_4)_3 \cdot nH_2O$ reacts with aqueous $SmCl_3$ solutions to form a soluble [Sm
complex (cf. p. 247) [4].

References:

[1] S. Pajakoff (Monatsh. Chem. **100** [1969] 1350/61, 1357). – [2] C. M. Gupta, M. P
(Bull. Chem. Soc. Japan **41** [1968] 1268/70). – [3] B. N. Ivanov-Emin, E. I. Zakharikov
Neorgan. Khim. **14** [1969] 145/9; Russ. J. Inorg. Chem. **14** [1969] 75/8). – [4] N. K. Davi
G. A. Komashko, K. B. Yatsimirskii (Zh. Neorgan. Khim. **13** [1968] 117/22; Russ. J. Inorg.
13 [1968] 58/61).

6.1.8.2 $Sm_2Mo_7O_{24} \cdot nH_2O$ ($= Sm_2O_3 \cdot 7MoO_3 \cdot nH_2O$), n = 6 and 8

General data for $M_2Mo_7O_{24} \cdot nH_2O$ molybdates are given on p. 248.

$Sm_2Mo_7O_{24} \cdot nH_2O$ forms upon mixing 0.1 M [1, 2] or 0.01 M [3] aqueous solutions of $Sm(NO_3)_3$ and ammonium or sodium paramolybdate in the stoichiometric ratio at pH 3.55 (adjusted by HNO_3) [3] as described for the corresponding Pr compound (p. 258) [1]. Amperometric and conductometric titrations of $Sm(NO_3)_3$ and $Na_6Mo_7O_{24}$ solutions of various concentrations show the compound to be formed at pH 3.5 to 4.5. The enthalpy change of the formation reaction is $\Delta H = -183.0$ cal/mol $Sm_2Mo_7O_{24}$ [4].

The pycnometric density of the octahydrate is 3.553 ± 0.015 g/cm³. The compound dehydrates in one step at 214°C and the amorphous dehydration product crystallizes at 432°C [1].

References:

[1] S. S. Antonova, I. V. Shakhno, V. E. Plyushchev (Izv. Vysshikh Uchebn. Zavedenii Khim. Khim. Tekhnol. **15** [1972] 1289/92; C.A. **78** [1973] No. 10950). – [2] S. S. Antonova, I. V. Shakhno, V. E. Plyushchev (Izv. Vysshikh Uchebn. Zavedenii Khim. Khim. Tekhnol. **15** [1972] 12/5; C.A. **76** [1972] No. 132181). – [3] S. S. Antonova, I. V. Shakhno, V. E. Plyushchev, G. A. Zenkova (Uch. Zap. Mosk. Inst. Tonkoi Khim. Tekhnol. **1** No. 3 [1971] 99/103; C.A. **77** [1972] No. 131389). – [4] C. M. Gupta, M. P. Joshi (Bull. Chem. Soc. Japan **41** [1968] 1268/70).

6.1.8.3 $Sm(OH)MoO_4 \cdot nH_2O$ ($= Sm_2O_3 \cdot 2MoO_3 \cdot (2n+1)H_2O$), n = 0 and 2

General data for $M(OH)MoO_4 \cdot nH_2O$ molybdates are given on p. 249.

$Sm(OH)MoO_4 \cdot 2H_2O$ has been prepared from aqueous KOH-K_2MoO_4 solutions by reaction with aqueous $Sm(NO_3)_3$ solution [1]. The formation of $Sm(OH)MoO_4 \cdot nH_2O$ has been observed during the reaction of $SmCl_3$ with ammonium paramolybdate solutions at Mo : Sm = 1 [2]. On heating the dihydrate up to 200°C all the water of crystallization is lost [1]. Anhydrous $Sm(OH)MoO_4$ is also formed under hydrothermal conditions from Sm_2O_3/MoO_3 mixtures as described for the corresponding Nd compound (p. 259) [3].

The flattened crystals of anhydrous $Sm(OH)MoO_4$ are isomorphous with the monoclinic Nd compound, lattice parameters a = 5.33, b = 12.54, c = 6.81 Å (all ± 0.03 Å), β = 113° [3]. ^1H NMR [4] and IR spectra [3] show the presence of OH groups in the structure.

$Sm(OH)MoO_4$ decomposes at 675°C [3] in accordance with the results of [1] into a samarium oxomolybdate and H_2O.

References:

[1] A. P. Perepelitsa, V. N. Solomakha (Zh. Neorgan. Khim. **18** [1973] 28/30; Russ. J. Inorg. Chem. **18** [1973] 14/5). – [2] B. N. Ivanov-Emin, E. I. Zakharikova (Zh. Neorgan. Khim. **14** [1969] 145/9; Russ. J. Inorg. Chem. **14** [1969] 75/8). – [3] V. I. Protasova, L. Yu. Kharchenko, P. V. Klevtsov (Izv. Akad. Nauk SSSR Neorgan. Materialy **9** [1973] 421/3; Inorg. Materials [USSR] **9** [1973] 374/6). – [4] A. P. Perepelitsa, A. M. Kalinichenko (Ukr. Khim. Zh. **43** [1977] 543/5; Soviet Progr. Chem. **43** No. 5 [1977] 98/100).

6.1.9 Europium Molybdates

6.1.9.1 $Eu_2Mo_7O_{24}\cdot nH_2O$ (= $Eu_2O_3\cdot 7MoO_3\cdot nH_2O$), n = 6 and 8

General data for $M_2Mo_7O_{24}\cdot nH_2O$ molybdates are given on p. 248.

$Eu_2Mo_7O_{24}\cdot nH_2O$ has been prepared and investigated as described for the corresponding Pr compound (p. 258). The pycnometric density of the octahydrate is 3.631 ± 0.015 g/cm³. The compound dehydrates in one step at 211°C, and the amorphous dehydration product crystallizes at 420°C.

S. S. Antonova, I. V. Shakhno, V. E. Plyushchev (Izv. Vysshikh Uchebn. Zavedenii Khim. Khim. Tekhnol. **15** [1972] 1289/92; C.A. **78** [1973] No. 10950).

6.1.9.2 $Eu(OH)MoO_4\cdot nH_2O$ (= $Eu_2O_3\cdot 2MoO_3\cdot (2n+1)H_2O$), n = 0 and 2

General data for $M(OH)MoO_4\cdot nH_2O$ molybdates are given on p. 249.

$Eu(OH)MoO_4\cdot 2H_2O$ precipitates from stoichiometric mixtures of aqueous solutions of $Eu(NO_3)_3$, K_2MoO_4, and KOH. On heating up to 200°C the water of crystallization is released [1]. Anhydrous $Eu(OH)MoO_4$ has also been obtained by hydrothermal synthesis from Eu_2O_3/MoO_3 mixtures as described for the corresponding Nd compound (p. 259) [2].

The flattened crystals of anhydrous $Eu(OH)MoO_4$ are isomorphous with the monoclinic Nd compound, lattice parameters a = 5.30, b = 12.54, c = 6.73 Å (all ± 0.03 Å), β = 113° [2]. ^{1}H NMR [3] and IR spectra [2] show the presence of OH groups in the structure.

$Eu(OH)MoO_4$ decomposes at 695°C [2] in accordance with the results of [1] into a europium oxomolybdate and H_2O.

References:

[1] A. P. Perepelitsa, V. N. Solomakha (Zh. Neorgan. Khim. **18** [1973] 28/30; Russ. J. Inorg. Chem. **18** [1973] 14/5). – [2] V. I. Protasova, L. Yu. Kharchenko, P. V. Klevtsov (Izv. Akad. Nauk SSSR Neorgan. Materialy **9** [1973] 421/3; Inorg. Materials [USSR] **9** [1973] 374/6). – [3] A. P. Perepelitsa, A. M. Kalinichenko (Ukr. Khim. Zh. **43** [1977] 543/5; Soviet Progr. Chem. **43** No. 5 [1977] 98/100).

6.1.10 Gadolinium Molybdates

6.1.10.1 $Gd_2(MoO_4)_3\cdot 4H_2O$ (= $Gd_2O_3\cdot 3MoO_3\cdot 4H_2O$)

General data for $M_2(MoO_4)_3\cdot nH_2O$ molybdates are given on p. 247.

$Gd_2(MoO_4)_3\cdot 4H_2O$ has been prepared by slowly mixing aqueous solutions of lithium or sodium molybdate (2 M to saturated) with 0.1 M $Gd(NO_3)_3$ solution in the stoichiometric ratio at pH ≦ 4.64, with stirring for ~2 h at room temperature. The content of water has been determined after drying the precipitate over $CaCl_2$ or H_2SO_4 [1, 2].

X-ray powder diffraction studies confirmed $Gd_2(MoO_4)_3\cdot 4H_2O$ to be a single phase and isomorphous with the tetrahydrates of the yttrium subgroup rare earth molybdates (Y, Gd to Lu). An X-ray powder line diagram is given in [2], for the d values see [1]. The density was determined as 4.13 g/cm³ [2]. The IR spectrum between 3700 and 400 cm^{-1} has been recorded and does not show the presence of OH^- groups. All bands in the range 950 to 700 cm^{-1} can be attributed to stretching vibrations of MoO_4^{2-} groups. For a plot of the spectrum see the paper [2].

According to DTA water splits off in at least two steps from 100 to 240 and from 300 to 335°C [2] or in three steps at 170, 220, and 330°C [1]. The anhydrous amorphous product crystallizes between 510 and 625°C [1, 2]. TG and DTA curves are given in [1]. The tetrahydrate is also degraded by prolonged storage over P_2O_5 [2]. In aqueous $GdCl_3$ solution the soluble $[GdMoO_4]^+$ complex forms (see p. 247) [3].

References:

[1] E. M. Avzhieva, I. V. Shakhno, V. E. Plyushchev, V. I. Arzhaeva, V. V. Duplitskaya (Izv. Vysshikh Uchebn. Zavedenii Khim. Khim. Tekhnol. **13** [1970] 1237/41; C.A. **74** [1971] No. 60283). – [2] E. M. Avzhieva, I. V. Shakhno, V. E. Plyushchev, G. N. Voronskaya, K. I. Petrov (Zh. Neorgan. Khim. **20** [1975] 1561/5; Russ. J. Inorg. Chem. **20** [1975] 874/6). – [3] N. K. Davidenko, G. A. Komashko, K. B. Yatsimirskii (Zh. Neorgan. Khim. **13** [1968] 117/22; Russ. J. Inorg. Chem. **13** [1968] 58/61).

6.1.10.2 $\mathbf{Gd_2Mo_7O_{24} \cdot nH_2O}$ ($=Gd_2O_3 \cdot 7MoO_3 \cdot nH_2O$), n = 6 and 8

General data for $M_2Mo_7O_{24} \cdot nH_2O$ molybdates are given on p. 248.

$Gd_2Mo_7O_{24} \cdot nH_2O$ has been prepared and investigated as described for the corresponding Pr compound (p. 258). The pycnometric density of the octahydrate is 3.489 ± 0.015 g/cm^3. The compound dehydrates in one step at 228°C and the amorphous decomposition product crystallizes at 428°C.

S. S. Antonova, I. V. Shakhno, V. E. Plyushchev (Izv. Vysshikh Uchebn. Zavedenii Khim. Khim. Tekhnol. **15** [1972] 1289/92; C.A. **78** [1973] No. 10950).

6.1.10.3 $\mathbf{Gd(OH)MoO_4 \cdot nH_2O}$ ($=Gd_2O_3 \cdot 2MoO_3 \cdot (2n+1)H_2O$), n = 0 and 1

General data for $M(OH)MoO_4 \cdot nH_2O$ molybdates are given on p. 249.

$Gd(OH)MoO_4 \cdot H_2O$ is prepared from aqueous $Gd(NO_3)_3 + K_2MoO_4 + KOH$ solutions. The monohydrate loses its water of crystallization up to 200°C [1]. Anhydrous $Gd(OH)MoO_4$ can also be prepared by hydrothermal synthesis from Gd_2O_3/MoO_3 mixtures as described for the corresponding Nd compound (p. 259) [2].

The flattened crystals of anhydrous $Gd(OH)MoO_4$ are isomorphous with the monoclinic Nd compound, lattice parameters a = 5.28, b = 12.45, c = 6.69 Å (all ± 0.03 Å), $\beta = 113°$ [2]. ^1H NMR measurements on samples dried in vacuo at 200°C show the line widths ΔH of the Gd to Yb group molybdates to be broader and to have a periodic character, compared with the La to Eu group molybdates with narrow lines which change linearly. Maxima in ΔH are observed for Gd, Dy, Er, and Yb, which contain odd numbers of 4f electrons [3]. The IR spectrum also shows the presence of OH groups in the structure [2].

$Gd(OH)MoO_4$ decomposes at 700°C [2] in accordance with the results of [1] into gadolinium oxomolybdate and H_2O.

References:

[1] A. P. Perepelitsa, V. N. Solomakha (Zh. Neorgan. Khim. **18** [1973] 28/30; Russ. J. Inorg. Chem. **18** [1973] 14/5). – [2] V. I. Protasova, L. Yu. Kharchenko, P. V. Klevtsov (Izv. Akad. Nauk SSSR Neorgan. Materialy **9** [1973] 421/3; Inorg. Materials [USSR] **9** [1973] 374/6). – [3] A. P. Perepelitsa, A. M. Kalinichenko (Ukr. Khim. Zh. **43** [1977] 543/5; Soviet Progr. Chem. **43** No. 5 [1977] 98/100).

6.1.11 Terbium Molybdates

6.1.11.1 $Tb_2(MoO_4)_3 \cdot nH_2O$ (= $Tb_2O_3 \cdot 3MoO_3 \cdot nH_2O$), n = 3 and 4

General data for $M_2(MoO_4)_3 \cdot nH_2O$ molybdates are given on p. 247.

$Tb_2(MoO_4)_3 \cdot nH_2O$ has been prepared by slowly mixing and stirring stoichiometric amounts of aqueous solutions of 0.1 M $Tb(NO_3)_3$ and almost saturated Li_2MoO_4 [1] or 0.1 M Na_2MoO_4 at pH 5 to 6.5 [2] at room temperature. The water content is given as n = 4 after drying the precipitate over $CaCl_2$ or H_2SO_4 [1] or as n = 3 after drying in air [2].

The X-ray powder pattern of the tetrahydrate (for a line diagram see the paper) shows it to be isomorphous with the analogous compounds of the yttrium subgroup (Y, Gd to Lu) [1]. The trihydrate is found to be X-ray amorphous. ^1H NMR spectra of the trihydrate showed three types of chemically different protons, attributed to adsorbed water, water of crystallization, and constitutional water [2]. The density of $Tb_2(MoO_4)_3 \cdot 4H_2O$ is ~4.07 g/cm^3 according to a figure in [1]. In the IR spectrum between 3700 and 400 cm^{-1} of $Tb_2(MoO_4)_3 \cdot 4H_2O$ all absorption bands below 1000 cm^{-1} are ascribed to stretching vibrations of the MoO_4^{2-} groups. Bands due to OH^- groups could not be detected. For a plot see the paper [1].

According to DTA the tetrahydrate loses water in at least two steps, from 80 to 240 and 305 to 335°C [1], while the trihydrate is dehydrated between 150 and 200°C. The minor endothermic effect at ~350°C is attributed to an admixture of a basic salt [2]. The anhydrous decomposition product crystallizes between 560 and 580°C [1]. $Tb_2(MoO_4)_3 \cdot 4H_2O$ is degraded by prolonged storage over P_2O_5 [1]. In aqueous $TbCl_3$ solutions the $[TbMoO_4]^+$ complex forms (see p. 247) [3].

References:

[1] E. M. Avzhieva, I. V. Shakhno, V. E. Plyushchev, G. N. Voronskaya, K. I. Petrov (Zh. Neorgan. Khim. **20** [1975] 1561/5; Russ. J. Inorg. Chem. **20** [1975] 874/6). – [2] A. M. Golub, A. P. Perepelitsa, A. A. Govorov (Zh. Neorgan. Khim. **16** [1971] 660/5; Russ. J. Inorg. Chem. **16** [1971] 352/5). – [3] N. K. Davidenko, G. A. Komashko, K. B. Yatsimirskii (Zh. Neorgan. Khim. **13** [1968] 117/22; Russ. J. Inorg. Chem. **13** [1968] 58/61).

6.1.11.2 $Tb_2Mo_7O_{24} \cdot nH_2O$ (= $Tb_2O_3 \cdot 7MoO_3 \cdot nH_2O$), n = 6 and 8

General data for $M_2Mo_7O_{24} \cdot nH_2O$ molybdates are given on p. 248.

$Tb_2Mo_7O_{24} \cdot nH_2O$ has been prepared and investigated as described for the corresponding Pr compound (p. 258). The pycnometric density of the octahydrate is 3.681 ± 0.015 g/cm^3. The compound dehydrates in two steps at 210 and 234°C, and the amorphous dehydration product crystallizes at 430°C.

S. S. Antonova, I. V. Shakhno, V. E. Plyushchev (Izv. Vysshikh Uchebn. Zavedenii Khim. Khim. Tekhnol. **15** [1972] 1289/92; C.A. **78** [1973] No. 10950).

6.1.11.3 $Tb(OH)MoO_4 \cdot nH_2O$ (= $Tb_2O_3 \cdot 2MoO_3 \cdot (2n+1)H_2O$), n = 0 and 1

General data for $M(OH)MoO_4 \cdot nH_2O$ molybdates are given on p. 249.

$Tb(OH)MoO_4 \cdot H_2O$ is prepared from aqueous $Tb(NO_3)_3 + K_2MoO_4 + KOH$ solutions. The monohydrate loses its water of crystallization up to 200°C [1]. Anhydrous $Tb(OH)MoO_4$ has also been prepared by hydrothermal synthesis from Tb_2O_3/MoO_3 mixtures as described for the corresponding Nd compound (p. 259) [2].

The flattened crystals of anhydrous $Tb(OH)MoO_4$ are isomorphous with the monoclinic Nd compound, lattice parameters a = 5.26, b = 12.42, c = 6.68 Å (all ± 0.03 Å), β = 113° [2]. The ^1H NMR line width has a minimum value among those of the Gd to Yb group molybdates (see p. 263) [3]. The IR spectrum also shows the presence of OH groups in the structure [2].

$Tb(OH)MoO_4$ decomposes at 700°C [2] in accordance with the results of [1] into terbium oxomolybdate and H_2O.

References:

[1] A. P. Perepelitsa, V. N. Solomakha (Zh. Neorgan. Khim. **18** [1973] 28/30; Russ. J. Inorg. Chem. **18** [1973] 14/5). – [2] V. I. Protasova, L. Yu. Kharchenko, P. V. Klevtsov (Izv. Akad. Nauk SSSR Neorgan. Materialy **9** [1973] 421/3; Inorg. Materials [USSR] **9** [1973] 374/6). – [3] A. P. Perepelitsa, A. M. Kalinichenko (Ukr. Khim. Zh. **43** [1977] 543/5; Soviet Progr. Chem. **43** No. 5 [1977] 98/100).

6.1.12 Dysprosium Molybdates

6.1.12.1 $Dy_2(MoO_4)_3 \cdot nH_2O$ (= $Dy_2O_3 \cdot 3MoO_3 \cdot nH_2O$), n = 3 and 4

General data for $M_2(MoO_4)_3 \cdot nH_2O$ molybdates are given on p. 247.

$Dy_2(MoO_4)_3 \cdot 4H_2O$ has been prepared by slowly mixing stoichiometric amounts of aqueous solutions of 0.1 M $Dy(NO_3)_3$ and almost saturated Li_2MoO_4, stirring for ~2 h at room temperature, and subsequent drying of the precipitate over $CaCl_2$ or concentrated H_2SO_4 [1]. $Dy_2(MoO_4)_3 \cdot 3H_2O$ is obtained by hydrating at room temperature over saturated NH_4Cl solution ($p(H_2O) \approx 16$ Torr) the hygroscopic high-temperature modification of $Dy_2(MoO_4)_3$ quenched from between 1160 and 1200°C to room temperature [2, 3].

The X-ray powder patterns show the tetrahydrate to be isomorphous with the molybdate tetrahydrates of the heavier rare earth metals (Gd to Lu) and Y, for a line diagram see the paper [1]. The powder pattern of the trihydrate is different from that of the tetrahydrate and is equivalent to those of the other compounds $M_2(MoO_4)_3 \cdot 3H_2O$ with M = Dy to Lu [2]. Their structure type is called K′ [2] because it does not agree with that of the trihydrates of the K type (M = Y, Ho to Lu) found by [4]. According to a figure in [1] the density of $Dy_2(MoO_4)_3 \cdot 4H_2O$ is ~4.17 g/cm³. In the IR spectrum between 3700 and 400 cm⁻¹ of the tetrahydrate all absorption bands observed below 1000 cm⁻¹ are attributed to MoO_4^{2-} vibrations. For a plot of the spectrum see the paper [1].

At room temperature $Dy_2(MoO_4)_3 \cdot 4H_2O$ is dehydrated by storing over P_2O_5. Thermally, according to DTA, it loses its water in at least two steps, from 100 to 240 and 300 to 355°C [1], while $Dy_2(MoO_4)_3 \cdot 3H_2O$ begins to be degraded at 90°C and is transformed into anhydrous $Dy_2(MoO_4)_3$ at 190°C [2]. The amorphous dehydration products crystallize between 560 and 595°C [1]. In aqueous $DyCl_3$ solutions the soluble $[DyMoO_4]^+$ complex was observed (see p. 247) [5].

References:

[1] E. M. Avzhieva, I. V. Shakhno, V. E. Plyushchev, G. N. Voronskaya, K. I. Petrov (Zh. Neorgan. Khim. **20** [1975] 1561/5; Russ. J. Inorg. Chem. **20** [1975] 874/6). – [2] E. Ya. Rode, G. V. Lysanova, V. G. Kuznetsov, L. Z. Gokhman (Zh. Neorgan. Khim. **13** [1968] 1295/302; Russ. J. Inorg. Chem. **13** [1968] 678/82). – [3] L. A. Drobyshev, I. T. Frolkina (Zh. Fiz. Khim. **44** [1970] 2945; Russ. J. Phys. Chem. **44** [1970] 1681). – [4] K. Nassau, H. J. Levinstein, G. M. Loiacono (J. Phys. Chem. Solids **26** [1965] 1805/16, 1809/11). – [5] N. K. Davidenko, G. A. Komashko, K. B. Yatsimirskii (Zh. Neorgan. Khim. **13** [1968] 117/22; Russ. J. Inorg. Chem. **13** [1968] 58/61).

6.1.12.2 $Dy_2Mo_7O_{24} \cdot nH_2O$ (= $Dy_2O_3 \cdot 7MoO_3 \cdot nH_2O$), n = 6 and 8

General data for $M_2Mo_7O_{24} \cdot nH_2O$ molybdates are given on p. 248.

$Dy_2Mo_7O_{24} \cdot nH_2O$ has been prepared and investigated as described for the corresponding Pr compound (p. 258). The pycnometric density of the octahydrate is 3.682 ± 0.015 g/cm³. The compound dehydrates in one step at 222°C and the amorphous dehydration product crystallizes at 440°C.

S. S. Antonova, I. V. Shakhno, V. E. Plyushchev (Izv. Vysshikh Uchebn. Zavedenii Khim. Khim. Tekhnol. **15** [1972] 1289/92; C.A. **78** [1973] No. 10950).

6.1.12.3 $Dy(OH)MoO_4 \cdot nH_2O$ (= $Dy_2O_3 \cdot 2MoO_3 \cdot (2n+1)H_2O$), n = 0 and 1

General data for $M(OH)MoO_4 \cdot nH_2O$ molybdates are given on p. 249.

$Dy(OH)MoO_4 \cdot H_2O$ has been obtained from aqueous $Dy(NO_3)_3 + K_2MoO_4 + KOH$ solutions. The monohydrate loses its water of crystallization up to 200°C [1]. Anhydrous $Dy(OH)MoO_4$ has also been prepared by hydrothermal synthesis from Dy_2O_3/MoO_3 mixtures as described for the corresponding Nd compound (p. 259) [2].

The flattened crystals of anhydrous $Dy(OH)MoO_4$ are isomorphous with the monoclinic Nd compound, lattice parameters a = 5.23, b = 12.39, c = 6.67 Å (all ± 0.03 Å), β = 113° [2]. The ^1H NMR line width has a maximum value within the Gd to Yb group molybdates (see p. 263) [3]. The IR spectrum also shows the presence of OH groups in the structure [2].

$Dy(OH)MoO_4$ decomposes at 695°C [2] in accordance with the results of [1] into dysprosium oxomolybdate and H_2O.

References:

[1] A. P. Perepelitsa, V. N. Solomakha (Zh. Neorgan. Khim. **18** [1973] 28/30; Russ. J. Inorg. Chem. **18** [1973] 14/5). – [2] V. I. Protasova, L. Yu. Kharchenko, P. V. Klevtsov (Izv. Akad. Nauk SSSR Neorgan. Materialy **9** [1973] 421/3; Inorg. Materials [USSR] **9** [1973] 374/6). – [3] A. P. Perepelitsa, A. M. Kalinichenko (Ukr. Khim. Zh. **43** [1977] 543/5; Soviet Progr. Chem. **43** No. 5 [1977] 98/100).

6.1.13 Holmium Molybdates

6.1.13.1 $Ho_2(MoO_4)_3 \cdot nH_2O$ (= $Ho_2O_3 \cdot 3MoO_3 \cdot nH_2O$), n = 3 and 4

General data for $M_2(MoO_4)_3 \cdot nH_2O$ molybdates are given on p. 247.

$Ho_2(MoO_4)_3 \cdot 4H_2O$ is prepared by slowly mixing stoichiometric amounts of aqueous solutions of 0.1 M $Ho(NO_3)_3$ and almost saturated Li_2MoO_4 [1] or Na_2MoO_4 at pH ≦ 4.68 [2], followed by ~2 h stirring at room temperature and subsequent drying of the precipitate over $CaCl_2$ or concentrated H_2SO_4 [1, 2]. $Ho_2(MoO_4)_3 \cdot 3H_2O$ is obtained by hydrating over saturated NH_4Cl solution ($p(H_2O) \approx 16$ Torr) at room temperature the hygroscopic high-temperature modification of $Ho_2(MoO_4)_3$ quenched from between 1160 and 1200°C to room temperature [3, 4].

According to X-ray powder diagrams $Ho_2(MoO_4)_3 \cdot 4H_2O$ is isomorphous with the other $M_2(MoO_4)_3 \cdot 4H_2O$ molybdates, M = Y and Gd to Lu, for a line diagram see the paper [1]. The d values are listed in [2]. Two different structure types K [3] and K′ [4] are attributed to $Ho_2(MoO_4)_3 \cdot 3H_2O$, see p. 247. As taken from a figure in [1], the density of $Ho_2(MoO_4)_3 \cdot 4H_2O$ is ~4.26 g/cm³. In the IR spectrum between 3700 and 400 cm⁻¹ of $Ho_2(MoO_4)_3 \cdot 4H_2O$ all absorption bands observed below 1000 cm^{-1} are assigned to vibrations of the MoO_4^{2-} groups, for a plot of the spectrum see the paper [1].

At room temperature $Ho_2(MoO_4)_3 \cdot 4H_2O$ is dehydrated by storing over P_2O_5 [1]. According to DTA $Ho_2(MoO_4)_3 \cdot 4H_2O$ is degraded in at least two steps, from 80 to 245 and 320 to 360°C [1], or in three steps at 185, 235, and 375°C [2]. The trihydrate is degraded between 90 and 200°C to anhydrous $Ho_2(MoO_4)_3$ [4]. The amorphous decomposition product crystallizes between 520 and 570°C [1, 2]. In aqueous $HoCl_3$ solutions the soluble $[HoMoO_4]^+$ complex has been observed (see p. 247) [5].

References:

[1] E. M. Avzhieva, I. V. Shakhno, V. E. Plyushchev, G. N. Voronskaya, K. I. Petrov (Zh. Neorgan. Khim. **20** [1975] 1561/5; Russ. J. Inorg. Chem. **20** [1975] 874/6). – [2] E. M. Avzhieva, I. V. Shakhno, V. E. Plyushchev, V. I. Arzhaeva, V. V. Duplitskaya (Izv. Vysshikh Uchebn. Zavedenii Khim. Khim. Tekhnol. **13** [1970] 1237/41; C.A. **74** [1971] No. 60283). – [3] K. Nassau, H. J. Levinstein, G. M. Loiacono (J. Phys. Chem. Solids **26** [1965] 1805/16, 1809/11). – [4] E. Ya. Rode, G. V. Lysanova, V. G. Kuznetsov, L. Z. Gokhman (Zh. Neorgan. Khim. **13** [1968] 1295/302; Russ. J. Inorg. Chem. **13** [1968] 678/82). – [5] N. K. Davidenko, G. A. Komashko, K. B. Yatsimirskii (Zh. Neorgan. Khim. **13** [1968] 117/22; Russ. J. Inorg. Chem. **13** [1968] 58/61).

6.1.13.2 $Ho_2Mo_7O_{24} \cdot nH_2O$ (= $Ho_2O_3 \cdot 7MoO_3 \cdot nH_2O$), n = 6 and 7

General data for $M_2Mo_7O_{24} \cdot nH_2O$ molybdates are given on p. 248.

$Ho_2Mo_7O_{24} \cdot nH_2O$ is formed upon mixing 0.1 M aqueous solutions of $Ho(NO_3)_3$ and ammonium or sodium paramolybdate in the stoichiometric ratio with stirring for 5 h at room temperature at pH ~2.5 adjusted with HNO_3 [1, 2]. After washing with acetone and ether the heptahydrate is obtained over $CaCl_2$ and the hexahydrate over H_2SO_4 or P_2O_5 [1].

X-ray powder diffraction patterns show the compound to be isomorphous with the corresponding compounds of Y and Er to Lu; line diagrams of all these compounds are given in the paper. The pycnometric density of the heptahydrate is 3.688 ± 0.015 g/cm^3. The compound dehydrates in two steps at 224 and 252°C and the amorphous dehydration product crystallizes at 440°C. The DTA curve up to 750°C is given in the paper [1].

References:

[1] S. S. Antonova, I. V. Shakhno, V. E. Plyushchev (Izv. Vysshikh Uchebn. Zavedenii Khim. Khim. Tekhnol. **15** [1972] 1289/92; C.A. **78** [1973] No. 10950). – [2] S. S. Antonova, I. V. Shakhno, V. E. Plyushchev, G. A. Zenkova (Uch. Zap. Mosk. Inst. Tonkoi Khim. Tekhnol. **1** No. 3 [1971] 99/103; C.A. **77** [1972] No. 131389).

6.1.13.3 $Ho(OH)MoO_4 \cdot nH_2O$ (= $Ho_2O_3 \cdot 2MoO_3 \cdot (2n+1)H_2O$), n = 0 and 1

General data for $M(OH)MoO_4 \cdot nH_2O$ molybdates are given on p. 249.

$Ho(OH)MoO_4 \cdot H_2O$ is obtained from aqueous $Ho(NO_3)_3 + K_2MoO_4 +$ KOH solutions. The monohydrate dehydrates up to 200°C [1]. Anhydrous $Ho(OH)MoO_4$ has also been prepared by hydrothermal synthesis from Ho_2O_3/MoO_3 mixtures as described for the corresponding Nd compound (p. 259) [2].

The flattened crystals of anhydrous $Ho(OH)MoO_4$ are isomorphous with the monoclinic $Nd(OH)MoO_4$, lattice parameters a = 5.23, b = 12.37, c = 6.67 Å (all ± 0.03 Å), β = 113° [2]. The ^1H NMR line width has a minimum value within the Gd to Yb group molybdates (see p. 263) [3]. The IR spectrum also shows the presence of OH groups in the structure [2].

$Ho(OH)MoO_4$ decomposes at 690°C [2] in accordance with the results of [1] into holmium oxomolybdate and H_2O.

References:

[1] A. P. Perepelitsa, V. N. Solomakha (Zh. Neorgan. Khim. **18** [1973] 28/30; Russ. J. Inorg. Chem. **18** [1973] 14/5). – [2] V. I. Protasova, L. Yu. Kharchenko, P. V. Klevtsov (Izv. Akad. Nauk SSSR Neorgan. Materialy **9** [1973] 421/3; Inorg. Materials [USSR] **9** [1973] 374/6). – [3] A. P. Perepelitsa, A. M. Kalinichenko (Ukr. Khim. Zh. **43** [1977] 543/5; Soviet Progr. Chem. **43** No. 5 [1977] 98/100).

6.1.14 Erbium Molybdates

6.1.14.1 $Er_2(MoO_4)_3 \cdot nH_2O$ (= $Er_2O_3 \cdot 3MoO_3 \cdot nH_2O$), n = 3 and 4

General data for $M_2(MoO_4)_3 \cdot nH_2O$ molybdates are given on p. 247.

$Er_2(MoO_4)_3 \cdot 4H_2O$ forms on slowly mixing and stirring for ~2 h at room temperature stoichiometric amounts of aqueous solutions of 0.1 M $Er(NO_3)_3$ and almost saturated Li_2MoO_4 and subsequent drying of the precipitate over $CaCl_2$ or concentrated H_2SO_4 [1]. $Er_2(MoO_4)_3 \cdot 3H_2O$ has been obtained either by precipitation from mixtures of 0.1 M aqueous solutions of $Er(NO_3)_3$ and Na_2MoO_4 at pH values from 5 to 7 and then air dried [2] or by hydration over saturated NH_4Cl solution ($p(H_2O) \approx 16$ Torr) of the hygroscopic high-temperature modification of $Er_2(MoO_4)_3$ annealed at 900 to 1000°C, irrespective of the method of cooling [3, 4]. $Er_2(MoO_4)_3$ hydrate of unspecified water content has been obtained from equilibrated mixtures of aqueous $ErCl_3$ and $(NH_4)_6Mo_7O_{24}$ solutions at pH ≈ 5 and a molar ratio $Er_2O_3 : MoO_3 = 1:3$ [5].

According to X-ray powder investigations $Er_2(MoO_4)_3 \cdot 4H_2O$ is isomorphous with the analogous tetrahydrates of Y and Gd to Lu, for a line diagram see the paper [1]. The X-ray powder pattern of $Er_2(MoO_4)_3 \cdot 3H_2O$ shows it to be isomorphous with the trihydrates of Dy to Lu (K′ type series) [4], whereas the agreement with another set of d values (K type series) [3] is limited to just the strongest lines. For the d values see the papers [3, 4]. Other samples of $Er_2(MoO_4)_3 \cdot 3H_2O$ have been found to be X-ray amorphous [2]. As taken from a figure the density of $Er_2(MoO_4)_3 \cdot 4H_2O$ is ~4.30 g/cm³ [1]. The IR spectrum between 3700 and 400 cm^{-1} of the tetrahydrate has been recorded and all absorption bands below 1000 cm^{-1} are attributed to vibrations of the MoO_4^{2-} groups. For a plot of the spectrum see the paper [1]. The IR absorption bands of a sample $Er_2(MoO_4)_3 \cdot nH_2O$ have been assigned to vibrations of the MoO_4^{2-} group of C_{2v} symmetry as follows (values in cm^{-1}): 926 ($\nu_1(A_1)$), 860, 775, 690 ($\nu_3(F_2)$), 390 (ν(Er-O) or $\nu_4(F_2)$) [6].

At room temperature $Er_2(MoO_4)_3 \cdot 4H_2O$ is dehydrated by storing over P_2O_5 and thermally, according to differential thermal analysis, in at least two stages from 80 to 240 and 300 to 390°C [1]. $Er_2(MoO_4)_3 \cdot 3H_2O$ is degraded between 150 and 200°C [2], 90 and 190°C [4], or even below 120°C [3]. A sharp decrease in the intensity of the IR absorption bands characteristic of crystal water (3500 to 3200 and 1660 to 1620 cm^{-1}) has been noted after drying at 100 to 120°C [6]. The amorphous dehydration product crystallizes between 560 and 585°C [1]. In aqueous $ErCl_3$ solutions the soluble $[ErMoO_4]^+$ complex forms (see p. 247) [7].

References:

[1] E. M. Avzhieva, I. V. Shakhno, V. E. Plyushchev, G. N. Voronskaya, K. I. Petrov (Zh. Neorgan. Khim. **20** [1975] 1561/5; Russ. J. Inorg. Chem. **20** [1975] 874/6). – [2] A. M. Golub, A. P. Perepelitsa, A. A. Govorov (Zh. Neorgan. Khim. **16** [1971] 660/5; Russ. J. Inorg. Chem. **16** [1971] 352/5). – [3] K. Nassau, H. J. Levinstein, G. M. Loiacono (J. Phys. Chem. Solids **26** [1965] 1805/16, 1809/11). – [4] E. Ya. Rode, G. V. Lysanova, V. G. Kuznetsov, L. Z. Gokhman (Zh. Neorgan. Khim. **13** [1968] 1295/302; Russ. J. Inorg. Chem. **13** [1968] 678/82). – [5] B. N. Ivanov-Emin, E. I. Zakharikova (Zh. Neorgan. Khim. **14** [1969] 145/9; Russ. J. Inorg. Chem. **14** [1969] 75/8).

[6] B. E. Zaitsev, E. I. Zakharikova, B. N. Ivanov-Emin, G. I. Cherenkova (Zh. Neorgan. Khim. **14** [1969] 1493/6; Russ. J. Inorg. Chem. **14** [1969] 781/3). – [7] N. K. Davidenko, G. A. Komashko, K. B. Yatsimirskii (Zh. Neorgan. Khim. **13** [1968] 117/22; Russ. J. Inorg. Chem. **13** [1968] 58/61).

6.1.14.2 $\mathbf{Er_2Mo_7O_{24} \cdot nH_2O}$ (= $Er_2O_3 \cdot 7MoO_3 \cdot nH_2O$), n = 4, 6, and 7

General data for $M_2Mo_7O_{24} \cdot nH_2O$ molybdates are given on p. 248.

$Er_2Mo_7O_{24} \cdot nH_2O$ precipitates from stoichiometric mixtures of 0.1 M [1, 2] or 0.01 M [2] aqueous solutions of $Er(NO_3)_3$ and ammonium or sodium paramolybdate at pH ≈ 3.6 adjusted by HNO_3 [1 to 3]. After washing with acetone and ether the heptahydrate has been obtained over $CaCl_2$ and the hexahydrate over H_2SO_4 or P_2O_5 [1]. The tetrahydrate has been obtained by [3].

X-ray powder line diagrams are given in [1, 2]. The compound is isomorphous with the corresponding Ho compound (p. 267). The pycnometric density of the heptahydrate is 4.080 ± 0.015 g/cm^3. Dehydration occurs in two steps, at 220 and 254°C [1]. The tetrahydrate loses its water at ~140°C [3]. The amorphous dehydration product crystallizes at 420 [3] to 448°C [1] with decomposition.

References:

[1] S. S. Antonova, I. V. Shakhno, V. E. Plyuŝhchev (Izv. Vysshikh Uchebn. Zavedenii Khim. Khim. Tekhnol. **15** [1972] 1289/92; C.A. **78** [1973] No. 10950). – [2] S. S. Antonova, I. V. Shakhno, V. E. Plyushchev (Izv. Vysshikh Uchebn. Zavedenii Khim. Khim. Tekhnol. **15** [1972] 12/5; C.A. **76** [1972] No. 132181). – [3] A. M. Golub, A. P. Perepelitsa, V. I. Maksin (Izv. Vysshikh Uchebn. Zavedenii Khim. Khim. Tekhnol. **14** [1971] 815/7; C.A. **76** [1972] No. 9894).

6.1.14.3 $\mathbf{Er(OH)MoO_4 \cdot nH_2O}$ (= $Er_2O_3 \cdot 2MoO_3 \cdot (2n+1)H_2O$), n = 0 and 1

General data for $M(OH)MoO_4 \cdot nH_2O$ molybdates are given on p. 249.

$Er(OH)MoO_4 \cdot H_2O$ has been prepared by the same method as $Sc(OH)MoO_4 \cdot H_2O$ (p. 250) from a stoichiometric amount of aqueous $Er(NO_3)_3$ solution which is added to a boiling solution of equimolar amounts of Na_2MoO_4 and NaOH [1]. $Er(OH)MoO_4 \cdot nH_2O$ is also formed by mixing aqueous solutions of $ErCl_3$ with Li_2MoO_4, K_2MoO_4, Cs_2MoO_4, or $(NH_4)_6Mo_7O_{24} \cdot 4H_2O$ [2]. The monohydrate loses its water of crystallization at 139°C [1]. Anhydrous $Er(OH)MoO_4$ has also been prepared by hydrothermal synthesis from Er_2O_3/MoO_3 mixtures like the corresponding Nd compound (p. 259) [3].

The flattened crystals of anhydrous $Er(OH)MoO_4$ are isomorphous with the monoclinic Nd compound, lattice parameters a = 5.15, b = 12.31, c = 6.63 Å (all ± 0.03 Å), β = 113° [3]. The ^1H NMR line width has a maximum value within the Gd to Yb group molybdates (see p. 263) [4]. The IR spectrum also shows the presence of OH groups in the structure [3].

$Er(OH)MoO_4$ decomposes at 654 [1] or 685°C [3] forming erbium oxomolybdate and H_2O.

References:

[1] A. M. Golub, A. P. Perepelitsa, A. A. Govorov (Zh. Neorgan. Khim. **16** [1971] 660/5; Russ. J. Inorg. Chem. **16** [1971] 352/5). – [2] B. N. Ivanov-Emin, E. I. Zakharikova (Zh. Neorgan. Khim. **14** [1969] 145/9; Russ. J. Inorg. Chem. **14** [1969] 75/8). – [3] V. I. Protasova, L. Yu. Kharchenko, P. V. Klevtsov (Izv. Akad. Nauk SSSR Neorgan. Materialy **9** [1973] 421/3; Inorg. Materials [USSR] **9** [1973] 374/6). – [4] A. P. Perepelitsa, A. M. Kalinichenko (Ukr. Khim. Zh. **43** [1977] 543/5; Soviet Progr. Chem. **43** No. 5 [1977] 98/100).

6.1.15 Thulium Molybdates

6.1.15.1 $\mathbf{Tm_2(MoO_4)_3 \cdot nH_2O}$ ($=Tm_2O_3 \cdot 3MoO_3 \cdot nH_2O$), n = 3 and 4

General data for $M_2(MoO_4)_3 \cdot nH_2O$ molybdates are given on p. 247.

$Tm_2(MoO_4)_3 \cdot 4H_2O$ has been prepared by slowly mixing stoichiometric amounts of aqueous solutions of 0.1 M $Tm(NO_3)_3$ and almost saturated Li_2MoO_4, stirring for ~2 h at room temperature, and subsequently drying the precipitate over $CaCl_2$ or concentrated H_2SO_4 [1]. $Tm_2(MoO_4)_3 \cdot 3H_2O$ has been obtained by hydration over saturated NH_4Cl solution ($p(H_2O) \approx$ 16 Torr) of the hygroscopic high-temperature modification of $Tm_2(MoO_4)_3$ after heating at 900 to 1000°C [2, 3].

According to X-ray powder patterns the tetrahydrate (for a line diagram see the paper) [1] is isomorphous with the analogous molybdates of Y and the heavy rare earth metals. For the trihydrates both K [2] and K' [3] structure types (see p. 247) are postulated. As taken from a figure in [1] the density of $Tm_2(MoO_4)_3 \cdot 4H_2O$ is ~4.33 g/cm³. In the IR spectrum of the tetrahydrate between 3700 and 400 cm⁻¹ all absorption bands below 1000 cm⁻¹ are assigned to vibrations of the MoO_4^{2-} groups. For a plot of the spectrum see the paper [1].

At room temperature $Tm_2(MoO_4)_3 \cdot 4H_2O$ is dehydrated by prolonged storage over P_2O_5, and thermally, according to DTA, in two steps from 120 to 240 and from 410 to 435°C [1]. $Tm_2(MoO_4)_3 \cdot 3H_2O$ is reported to be dehydrated between 107 and 190°C [3]. The amorphous dehydration product crystallizes between 520 and 560°C [1].

References:

[1] E. M. Avzhieva, I. V. Shakhno, V. E. Plyushchev, G. N. Voronskaya, K. I. Petrov (Zh. Neorgan. Khim. **20** [1975] 1561/5; Russ. J. Inorg. Chem. **20** [1975] 874/6). – [2] K. Nassau, H. J. Levinstein, G. M. Loiacono (J. Phys. Chem. Solids **26** [1965] 1805/16, 1809/11). – [3] E. Ya. Rode, G. V. Lysanova, V. G. Kuznetsov, L. Z. Gokhman (Zh. Neorgan. Khim. **13** [1968] 1295/302; Russ. J. Inorg. Chem. **13** [1968] 678/82).

6.1.15.2 $\mathbf{Tm_2Mo_7O_{24} \cdot nH_2O}$ ($=Tm_2O_3 \cdot 7MoO_3 \cdot nH_2O$), n = 6 and 7

General data for $M_2Mo_7O_{24} \cdot nH_2O$ molybdates are given on p. 248.

$Tm_2Mo_7O_{24} \cdot nH_2O$ has been prepared and investigated as described for the isomorphous Ho compound (p. 267). The pycnometric density of the heptahydrate is 4.091 ± 0.015 g/cm³. The compound dehydrates in two steps at 214 and 258°C and the dehydration product crystallizes at 436°C.

S. S. Antonova, I. V. Shakhno, V. E. Plyushchev (Izv. Vysshikh Uchebn. Zavedenii Khim. Khim. Tekhnol. **15** [1972] 1289/92; C.A. **78** [1973] No. 10950).

6.1.15.3 $Tm(OH)MoO_4 \cdot nH_2O$ ($=Tm_2O_3 \cdot 2MoO_3 \cdot (2n+1)H_2O$), n = 0 and 2

General data for $M(OH)MoO_4 \cdot nH_2O$ molybdates are given on p. 249.

$Tm(OH)MoO_4 \cdot 2H_2O$ precipitates on addition of aqueous $Tm(NO_3)_3$ to a basic (KOH) aqueous solution of K_2MoO_4. By heating up to 200°C it is dehydrated to $Tm(OH)MoO_4$ [1]. The ^1H NMR spectrum has been recorded on a sample dehydrated in vacuo at 200°C. The line width of the signal has a minimum value in the Gd to Yb group molybdates (see p. 263) [2]. $Tm(OH)MoO_4$ decomposes in the range 680 to 700°C [1].

References:

[1] A. P. Perepelitsa, V. N. Solomakha (Zh. Neorgan. Khim. **18** [1973] 28/30; Russ. J. Inorg. Chem. **18** [1973] 14/5). – [2] A. P. Perepelitsa, A. M. Kalinichenko (Ukr. Khim. Zh. **43** [1977] 543/5; Soviet Progr. Chem. **43** No. 5 [1977] 98/100).

6.1.16 Ytterbium Molybdates

6.1.16.1 $Yb_2(MoO_4)_3 \cdot nH_2O$ ($=Yb_2O_3 \cdot 3MoO_3 \cdot nH_2O$), n = 0.65, 3, and 4

General data for $M_2(MoO_4)_3 \cdot nH_2O$ molybdates are given on p. 247.

$Yb_2(MoO_4)_3 \cdot 4H_2O$ has been obtained by slowly mixing stoichiometric amounts of aqueous solutions of 0.1 M $Yb(NO_3)_3$ and almost saturated Li_2MoO_4 at room temperature, ~2 h stirring, and drying the precipitate over $CaCl_2$ or concentrated H_2SO_4 [1]. $Yb_2(MoO_4)_3 \cdot 3H_2O$ has been prepared either by precipitation from mixtures of 0.1 M aqueous solutions of $Yb(NO_3)_3$ and Na_2MoO_4 at pH values from 5 (preferably) to 7 and air dried [2] or by hydration over saturated NH_4Cl solution ($p(H_2O) \approx 16$ Torr) of the hygroscopic high-temperature modification of $Yb_2(MoO_4)_3$ after heating at 900 to 1000°C [4]. At room air of about 40% relative humidity only $Yb_2(MoO_4)_3 \cdot 0.65H_2O$ forms in several weeks from anhydrous $Yb_2(MoO_4)_3$ [3]. An ytterbium molybdate hydrate of unspecified water content has also been obtained from equilibrated mixtures of aqueous solutions of $YbCl_3$ and $(NH_4)_6Mo_7O_{24}$ at pH ≈ 5 and a molar ratio Yb_2O_3 : MoO_3 = 1 : 3 [5].

According to X-ray powder patterns the tetrahydrate (for a line diagram see the paper) [1] and the trihydrate [4] crystallize differently. Both are isomorphous with the corresponding tetrahydrates and trihydrates, respectively, of Y and the heavy rare earth metals. For the trihydrate the K′ [4] and for the 0.65-hydrate the K type structures [3] are postulated, see p. 247. As taken from a figure the density of $Yb_2(MoO_4)_3 \cdot 4H_2O$ is ~4.38 g/cm³ [1]. In the IR spectrum between 3700 and 400 cm⁻¹ of $Yb_2(MoO_4)_3 \cdot 4H_2O$ all absorption bands below 1000 cm⁻¹ are attributed to vibrations of the MoO_4^{2-} groups. For a plot of the spectrum see the paper [1].

At room temperature $Yb_2(MoO_4)_3 \cdot 4H_2O$ is dehydrated by storing over P_2O_5 and thermally, according to DTA, in at least two steps from 100 to 230 and 270 to 315°C [1]. The trihydrate of $Yb_2(MoO_4)_3$ is degraded between 73 and 170°C [4], the 0.65-hydrate between 60 and 120°C [3]. In $YbCl_3$ solutions the soluble $[YbMoO_4]^+$ complex forms (see p. 247) [6].

References:

[1] E. M. Avzhieva, I. V. Shakhno, V. E. Plyushchev, G. N. Voronskaya, K. I. Petrov (Zh. Neorgan. Khim. **20** [1975] 1561/5; Russ. J. Inorg. Chem. **20** [1975] 874/6). – [2] A. M. Golub, A. P. Perepelitsa, A. A. Govorov (Zh. Neorgan. Khim. **16** [1971] 660/5; Russ. J. Inorg. Chem. **16** [1971] 352/5). – [3] K. Nassau, H. J. Levinstein, G. M. Loiacono (J. Phys. Chem. Solids **26** [1965] 1805/16, 1809/11). – [4] E. Ya. Rode, G. V. Lysanova, V. G. Kuznetsov, L. Z. Gokhman (Zh.

Neorgan. Khim. **13** [1968] 1295/302; Russ. J. Inorg. Chem. **13** [1968] 678/82). – [5] B. N. Ivanov-Emin, E. I. Zakharikova (Zh. Neorgan. Khim. **14** [1969] 145/9; Russ. J. Inorg. Chem. **14** [1969] 75/8).

[6] N. K. Davidenko, G. A. Komashko, K. B. Yatsimirskii (Zh. Neorgan. Khim. **13** [1968] 117/22; Russ. J. Inorg. Chem. **13** [1968] 58/61).

6.1.16.2 $Yb_2Mo_7O_{24}\cdot nH_2O$ ($=Yb_2O_3\cdot 7MoO_3\cdot nH_2O$), n = 6 and 7

General data for $M_2Mo_7O_{24}\cdot nH_2O$ molybdates are given on p. 248.

$Yb_2Mo_7O_{24}\cdot nH_2O$ precipitates from mixtures of aqueous solutions of $Yb(NO_3)_3$ and ammonium paramolybdate at room temperature. The hexahydrate has been obtained by drying at 130°C [1]. The hexa- and heptahydrates have also been prepared and investigated as described for the isomorphous Ho compound (p. 267). The pycnometric density of the heptahydrate is 4.179 ± 0.015 g/cm^3. The compound dehydrates in two steps at 220 and 264°C. The amorphous dehydration product crystallizes at 442°C [2]. The hexahydrate is insoluble in cold and hot water [1].

References:

[1] A. Cleve (Z. Anorg. Allgem. Chem. **32** [1902] 129/63, 151). – [2] S. S. Antonova, I. V. Shakhno, V. E. Plyushchev (Izv. Vysshikh Uchebn. Zavedenii Khim. Khim. Tekhnol. **15** [1972] 1289/92; C.A. **78** [1973] No. 10950).

6.1.16.3 $Yb(OH)MoO_4\cdot nH_2O$ ($=Yb_2O_3\cdot 2MoO_3\cdot(2n+1)H_2O$), n = 0 and 1

General data for $M(OH)MoO_4\cdot nH_2O$ molybdates are given on p. 249.

$Yb(OH)MoO_4\cdot H_2O$ precipitates from mixtures of stoichiometric amounts of aqueous solutions of $Yb(NO_3)_3$, K_2MoO_4, and KOH [1]. The formation of $Yb(OH)MoO_4\cdot nH_2O$ has been observed during the reaction of $YbCl_3$ with ammonium paramolybdate solutions at Yb : Mo = 1 [2]. By drying at 200°C the monohydrate is dehydrated to $Yb(OH)MoO_4$ [1]. The ^1H NMR line width of anhydrous $Yb(OH)MoO_4$ has a maximum value within the Gd to Yb group molybdates (see p. 263) [3].

References:

[1] A. P. Perepelitsa, V. N. Solomakha (Zh. Neorgan. Khim. **18** [1973] 28/30; Russ. J. Inorg. Chem. **18** [1973] 14/5). – [2] B. N. Ivanov-Emin, E. I. Zakharikova (Zh. Neorgan. Khim. **14** [1969] 145/9; Russ. J. Inorg. Chem. **14** [1969] 75/8). – [3] A. P. Perepelitsa, A. M. Kalinichenko (Ukr. Khim. Zh. **43** [1977] 543/5; Soviet Progr. Chem. **43** No. 5 [1977] 98/100).

6.1.17 Lutetium Molybdates

6.1.17.1 $Lu_2(MoO_4)_3\cdot nH_2O$ ($=Lu_2O_3\cdot 3MoO_3\cdot nH_2O$), n = 3 and 4

General data for $M_2(MoO_4)_3\cdot nH_2O$ molybdates are given on p. 247.

$Lu_2(MoO_4)_3\cdot 4H_2O$ is prepared by slowly mixing stoichiometric amounts of aqueous solutions of 0.1 M $Lu(NO_3)_3$ and nearly saturated Li_2MoO_4, ~2 h stirring at room temperature, and drying the precipitate over $CaCl_2$ or concentrated H_2SO_4 [1]. A trihydrate forms on hydration of the hygroscopic high-temperature modification of $Lu_2(MoO_4)_3$ previously heated to 900 to 1000°C [2].

According to X-ray powder patterns the tetrahydrate (for a line diagram see the paper) [1] and the trihydrate [2] crystallize differently. Both are isomorphous with the corresponding tetrahydrates and trihydrates, respectively, of Y and the heavy rare earth metals. The density of $Lu_2(MoO_4)_3 \cdot 4H_2O$ is 4.42 g/cm³ [1]. The IR spectrum between 3700 and 400 cm^{-1} of the tetrahydrate is plotted in the paper. All absorption bands below 1000 cm^{-1} are assigned to vibrations of the MoO_4^{2-} group [1].

At room temperature $Lu_2(MoO_4)_3 \cdot 4H_2O$ is dehydrated by storage over P_2O_5 and thermally, according to DTA, in at least two steps from 150 to 240 and 280 to 310°C [1], while the trihydrate loses its water between 60 and 150°C [2]. The anhydrous dehydration product crystallizes between 530 and 540°C [1].

References:

[1] E. M. Avzhieva, I. V. Shakhno, V. E. Plyushchev, G. N. Voronskaya, K. I. Petrov (Zh. Neorgan. Khim. **20** [1975] 1561/5; Russ. J. Inorg. Chem. **20** [1975] 874/6). – [2] E. Ya. Rode, G. V. Lysanova, V. G. Kuznetsov, L. Z. Gokhman (Zh. Neorgan. Khim. **13** [1968] 1295/302; Russ. J. Inorg. Chem. **13** [1968] 678/82).

6.1.17.2 $Lu_2Mo_7O_{24} \cdot nH_2O$ (= $Lu_2O_3 \cdot 7MoO_3 \cdot nH_2O$), n = 6 and 7

General data for $M_2Mo_7O_{24} \cdot nH_2O$ molybdates are given on p. 248.

$Lu_2Mo_7O_{24} \cdot nH_2O$ has been prepared and investigated as described for the isomorphous Ho compound (p. 267). The pycnometric density of the heptahydrate is 4.320 ± 0.015 g/cm³. The compound dehydrates in two steps at 213 and 262°C, and the amorphous dehydration product crystallizes at 422°C.

S. S. Antonova, I. V. Shakhno, V. E. Plyushchev (Izv. Vysshikh Uchebn. Zavedenii Khim. Khim. Tekhnol. **15** [1972] 1289/92; C.A. **78** [1973] No. 10950).

6.1.17.3 $Lu(OH)MoO_4 \cdot nH_2O$ (= $Lu_2O_3 \cdot 2MoO_3 \cdot (2n+1)H_2O$), n = 0 and 1

General data for $M(OH)MoO_4 \cdot nH_2O$ molybdates are given on p. 249.

$Lu(OH)MoO_4 \cdot H_2O$ is precipitated from mixtures of stoichiometric amounts of aqueous solutions of $Lu(NO_3)_3$, K_2MoO_4, and KOH. It is dehydrated below 200°C to form $Lu(OH)MoO_4$ [1]. The ^1H NMR spectrum of anhydrous $Lu(OH)MoO_4$ shows a narrow spectral line similar to those of the La to Eu group molybdates (see p. 249) characteristic for protons of the OH group [2]. $Lu(OH)MoO_4$ decomposes in the range 680 to 700°C [1].

References:

[1] A. P. Perepelitsa, V. N. Solomakha (Zh. Neorgan. Khim. **18** [1973] 28/30; Russ. J. Inorg. Chem. **18** [1973] 14/5). – [2] A. P. Perepelitsa, A. M. Kalinichenko (Ukr. Khim. Zh. **43** [1977] 543/5; Soviet Progr. Chem. **43** No. 5 [1977] 98/100).

6.2 Metal Rare Earth Molybdates

6.2.1 $LiM(MoO_4)_2 \cdot 2H_2O$ (= $Li_2O \cdot M_2O_3 \cdot 4MoO_3 \cdot 4H_2O$), M = Y, Gd to Lu

These compounds have been obtained from stoichiometric mixtures of an almost saturated aqueous Li_2MoO_4 solution and ~1 M aqueous $M(NO_3)_3$ solutions after equilibration for 6 h at 25°C. At lower $M(NO_3)_3$ concentrations (~0.4 to 0.1 M) an increasing excess of MoO_4^{2-} and longer reaction times are necessary for pure products. Since the precipitates are degraded by water, they have been washed with alcohol and ether. The content of water was found to be independent of the drying agent: $CaCl_2$, H_2SO_4, or P_2O_5 [1]. A previous paper reported a water content of 2.5 H_2O for the Y compound [2].

X-ray powder patterns show all the compounds to be isomorphous, for line diagrams see the paper [1], d values for the Y compound are given in [2]. The pycnometric densities D (in g/cm^3) at 25°C are [1]:

M	Gd	Tb	Dy	Ho	Er	Tm	Yb	Lu
D	3.97	3.99	4.02	4.10	4.12	4.15	4.21	4.27

The compounds can be dehydrated at elevated temperatures and the corresponding thermal effects have been recorded by DTA. The dehydration ranges Δt (in °C) for $CaCl_2$-dried samples are given in the following table [1]:

M	Gd	Tb	Dy	Ho
Δt	100 to 220	100 to 215	90 to 210	80 to 220
M	Er	Tm	Yb	Lu
Δt	80 to 225	140 to 220	90 to 230	120 to 205

For the Y compound the range is 60 to 220°C [2].

References:

[1] E. M. Avzhieva, I. V. Shakhno, V. E. Plyushchev (Izv. Vysshikh Uchebn. Zavedenii Khim. Khim. Tekhnol. **18** [1975] 1685/8; C.A. **84** [1976] No. 83461). – [2] E. M. Avzhieva, I. V. Shakhno, V. E. Plyushchev (Izv. Vysshikh Uchebn. Zavedenii Khim. Khim. Tekhnol. **15** [1972] 1131/5; C.A. **77** [1972] No. 144456).

6.2.2 $NaM(MoO_4)_2 \cdot nH_2O$ (= $Na_2O \cdot M_2O_3 \cdot 4MoO_3 \cdot 2nH_2O$), M = Sc, Y, La to Nd, Gd to Lu; n = 1 to 2.5

The compounds $NaM(MoO_4)_2 \cdot nH_2O$ with M = Sc [1], Y [2, 3], La [4 to 8], Ce and Pr [4, 5], Nd [5], Gd [3, 9], Tb to Lu [3], Ho [9], and Er [10, 11] have been obtained from mixtures of aqueous solutions of rare earth nitrates $M(NO_3)_3$ and Na_2MoO_4, for example with concentrations 0.1 and 1.5 M, respectively, at pH values from 5 to 7, and with molar ratios $M_2O_3:MoO_3 \leqq 1:4$. Equilibration times of 3 h for M = Y, Gd to Lu [3] and 1 to 13 weeks for M = La to Nd [5] have been used in order to redissolve $M_2(MoO_4)_3 \cdot nH_2O$, which may have been formed during precipitation. The formation of the double salts is also favored by precipitation from more concentrated M^{III} solutions. Because of the observed degradation by water the precipitated compounds $NaM(MoO_4)_2 \cdot 2H_2O$ with M = Y, Gd to Lu are washed only with alcohol and ether [2, 3]. The contents of water are n = 2.5 [2] or 2 [3] for M = Y, and n = 2 for La [6], Gd to Lu after drying over $CaCl_2$ [3], Gd, Ho [9], and n = 1.9 for Sc after drying in air [1]. The ^1H NMR data of $NaSc(MoO_4)_2 \cdot 1.9H_2O$ have been interpreted to mean that there is only one mole of water of crystallization, the remainder being adsorbed water [1]. Monohydrates have been obtained for M = La [7, 8], La to Nd after drying to constant weight [5], and also for M = Gd to Lu after drying over P_2O_5 or concentrated H_2SO_4 [3].

The X-ray powder patterns of the monohydrates of the La [7] and Nd [5] compounds have been indexed on the basis of a tetragonal scheelite-type structure with the lattice parameters $a = 5.356 \pm 0.004$, $c = 11.57 \pm 0.01$ Å; Z = 2 [7] and a = 5.302, c = 11.38 Å (from kX) [5], respectively. A list of the d values for the La compound is given in [7]. The structures of both solids are reported to be preserved during dehydration [5, 7]. The calculated density (5.029 g/cm^3) for the La compound [7] appears rather high as compared to the measured densities of the $NaM(MoO_4)_2 \cdot 2H_2O$ compounds with M = Gd to Lu (see below). This latter series of compounds is isomorphous according to the X-ray powder patterns [3]. Line diagrams for Y, Gd to Lu are given in [3], d values for Gd and Ho in [9], and for the 2.5-hydrate of Y in [2].

The pycnometric densities (in g/cm^3) at 25°C for the dihydrates $NaM(MoO_4)_2 \cdot 2H_2O$ increase linearly with the atomic number from 4.09 for M = Gd to 4.75 for M = Lu with the following values for the elements in between: ~4.20 (Tb), 4.29 (Dy), 4.32 (Ho), 4.52 (Er), 4.71 (Yb) (taken from a figure in the paper) [3].

In the IR spectrum between 3600 and 400 cm^{-1} of $NaEr(MoO_4)_2 \cdot nH_2O$ (n not given), bands for the water of crystallization were observed in the 3500 to 3200 and 1660 to 1620 cm^{-1} ranges. Their intensity drops sharply on drying at 100 to 120°C. The intensity and form of the molybdate bands in the 1000 to 400 cm^{-1} range are almost unchanged: 950 ($\nu_1(A_1)$); 865, 770, 690 ($\nu_3(F_2)$); 405 cm^{-1} (ν(Mo-O) or component of the $\nu_4(F_2)$ band) [11].

The thermal dehydration has been investigated by TG and DTA as well as by isothermal dehydration experiments. Some compounds lose their water in more than one step; thus monohydrates have been obtained from the dihydrates with M = Y, Gd to Ho, and Lu [3], and the monohydrates with M = La to Nd are said to form hemihydrates intermediately [5]. While most of the compounds become anhydrous at temperatures between 185 and 200°C [3], it is reported that $NaSc(MoO_4)_2 \cdot 2H_2O$ [1] and the monohydrates of La to Nd [5] lose the remainder of their water of crystallization only above 275°C. The individual temperatures are listed in the following table:

M	n	dehydration temperatures in °C		Ref.
Sc	1.9	205	275	[1]
Y	2	60 to 140	160 to 180	[6]
La	2	below 220		[3]
La	1	180	310	[5]
Ce	1	165	280	[5]
Pr	1	180	315	[5]
Nd	1	200	300	[5]
Gd	2	80 to 130	160 to 185	[6, 7]
Tb	2	65 to 165	190 to 215	[6]
Dy	2	50 to 160	175 to 200	[6]
Ho	2	90 to 150	200	[6, 7]
Er	2	100 to 185		[6]
Tm	2	160 to 192		[6]
Yb	2	80 to 196		[6]
Lu	2	80 to 170	180 to 220	[6]

The 2.5-hydrate of the Y compound is also dehydrated in two steps: up to 166°C (1.5 H_2O) and between 166 and 190°C (1 H_2O) [2].

References:

[1] A. M. Golub, A. P. Perepelitsa, A. M. Kalinichenko (Izv. Vysshikh Uchebn. Zavedenii Khim. Khim. Tekhnol. **15** [1972] 1293/6; C.A. **78** [1973] No. 10949). – [2] E. M. Avzhieva, I. V. Shakhno, V. E. Plyushchev (Izv. Vysshikh Uchebn. Zavedenii Khim. Khim. Tekhnol. **12** [1969] 1307/10; C.A. **72** [1970] No. 85711). – [3] E. M. Avzhieva, I. V. Shakhno, V. E. Plyushchev, O. M. Shiiko (Izv. Vysshikh Uchebn. Zavedenii Khim. Khim. Tekhnol. **14** [1971] 179/82; C.A. **74** [1971] No. 134265). – [4] A. M. Golub, K. S. Aganiyazov (Uch. Zap. Turkmensk. Gos. Ped. Inst. Ser. Estestv. Nauk **34** [1970] 143/8 from C.A. **75** [1971] No. 155366). – [5] A. M. Golub, K. S. Aganiyazov, N. G. Kisel', M. V. Mokhosoev (Izv. Akad. Nauk SSSR Neorgan. Materialy **6** [1970] 170/2; Inorg. Materials [USSR] **6** [1970] 148/50).

[6] S. A. Fedulov, Z. I. Tatarov, L. P. Shklover, N. I. Sergeeva, et al. (Izv. Akad. Nauk SSSR Neorgan. Materialy **2** [1966] 1905; Inorg. Materials [USSR] **2** [1966] 1651/2). – [7] Ts. S. Shugal, I. M. Britan, L. P. Shklover (Kristallografiya **14** [1969] 118/9; Soviet Phys.-Cryst. **14** [1969] 93/4). – [8] G. Carobbi (Gazz. Chim. Ital. **58** [1928] 53/6). – [9] E. M. Avzhieva, I. V. Shakhno, V. E. Plyushchev, V. I. Arzhaeva, V. V. Duplitskaya (Izv. Vysshikh Uchebn. Zavedenii Khim. Khim. Tekhnol. **13** [1970] 1237/41; C.A. **74** [1971] No. 60283). – [10] B. N. Ivanov-Emin, E. I. Zakharikova (Zh. Neorgan. Khim. **14** [1969] 145/9; Russ. J. Inorg. Chem. **14** [1969] 75/8).

[11] B. E. Zaitsev, E. I. Zakharikova, B. N. Ivanov-Emin, G. I. Cherenkova (Zh. Neorgan. Khim. **14** [1969] 1493/6; Russ. J. Inorg. Chem. **14** [1969] 781/3).

6.2.3 $KM(MoO_4)_2 \cdot H_2O$ ($= K_2O \cdot M_2O_3 \cdot 4MoO_3 \cdot 2H_2O$), M = Sc, Y, Gd to Lu

These compounds with M = Y [1] and Gd to Lu [2] are obtained from stoichiometric mixtures of aqueous 0.1M $M(NO_3)_3$ and nearly saturated K_2MoO_4 solutions after equilibration for 3h at 25°C. The precipitates are washed with alcohol and ether. The water content is independent of drying over $CaCl_2$, H_2SO_4, or P_2O_5 [1, 2]. With a slight excess of MoO_4^{2-} and equilibration times of about 1 month, but otherwise comparable conditions, compounds $KM(MoO_4)_2 \cdot nH_2O$ have been precipitated with M = Sc, Y, Tb, Er. The long reaction times are regarded as necessary to redissolve $M_2(MoO_4)_3 \cdot nH_2O$ that may be formed intermediately [3].

According to X-ray powder patterns the compounds with M = Gd to Lu are isomorphous, for the d values see the paper [2]. The d values of $KY(MoO_4)_2 \cdot H_2O$ are given in [1]; if only the strongest lines are considered, there is some agreement between this powder pattern and that reported for anhydrous $KY(MoO_4)_2$ [1]. ^1H NMR data indicate that the water in $KSc(MoO_4)_2 \cdot nH_2O$ is not water of crystallization but adsorbed water [4]. This has also been assumed from DTA for the water in the Y, Tb, and Er compounds [3]. The pycnometric densities of the monohydrates (dried over $CaCl_2$) increase linearly with the atomic number from 4.22 for Gd to 4.56 g/cm³ for Lu [2].

Thermal analyses show all compounds investigated to be dehydrated below 160°C, as follows [2]:

M	Y	Gd	Tb	Dy	Ho	Er	Tm	Yb	Lu
t in °C	125	90	160	90	90	100	140	95	100

The value 154°C is given for $KEr(MoO_4)_2 \cdot nH_2O$ [3].

References:

[1] E. M. Avzhieva, I. V. Shakhno, V. E. Plyushchev (Izv. Vysshikh Uchebn. Zavedenii Khim. Khim. Tekhnol. **13** [1970] 611/4; C.A. **73** [1970] No. 62111). – [2] E. M. Avzhieva, I. V. Shakhno, V. E. Plyushchev, V. V. Arzhaeva, et al. (Izv. Vysshikh Uchebn. Zavedenii Khim. Khim. Tekhnol.

14 [1971] 1457/61; C.A. **76** [1972] No. 30237). – [3] A. M. Golub, A. P. Perepelitsa, V. I. Maksin, K. S. Aganiyazov (Izv. Vysshikh Uchebn. Zavedenii Khim. Khim. Tekhnol. **14** [1971] 328/31; C.A. **75** [1971] No. 11010). – [4] A. M. Golub, A. P. Perepelitsa, A. M. Kalinichenko (Izv. Vysshikh Uchebn. Zavedenii Khim. Khim. Tekhnol. **15** [1972] 1293/6; C.A. **78** [1973] No. 10949).

6.2.4 $NH_4M(MoO_4)_2 \cdot nH_2O$ (= $(NH_4)_2O \cdot M_2O_3 \cdot 4MoO_3 \cdot 2nH_2O$), M = La, Ce, Nd, Sm, Ho, Er, Yb

Yellow $NH_4Ce(MoO_4)_2 \cdot H_2O$ has been precipitated upon dropwise addition of a solution of 8 g $CeCl_3 \cdot 6H_2O$ in 50 mL H_2O to 500 mL of a cold aqueous 10% solution of $(NH_4)_6Mo_7O_{24}$ and subsequent heating of the solution to its boiling point [1]. Compounds $NH_4M(MoO_4)_2 \cdot nH_2O$ with M = La [2], Nd [3], Sm [2, 3], Ho [3], Er [2, 3], and Yb [2] have been obtained from mixtures of aqueous solutions of MCl_3 and $(NH_4)_6Mo_7O_{24}$ with molar ratios M : Mo ≦ 1 : 2. In the cases of M = La and Sm the ratios were less than 1 : 4 and 1 : 3, respectively [2].

The X-ray powder pattern of $NH_4Ce(MoO_4)_2 \cdot H_2O$ (for the d values see the paper) [1] shows some similarity to those of $RbM(MoO_4)_2 \cdot H_2O$ with M = Gd to Lu (cf. p. 278). The IR spectra of the double molybdates have been recorded between 3600 and 400 cm^{-1}. Besides the usual absorption bands due to water of crystallization at 3500 to 3200 and 1660 to 1620 cm^{-1} [1, 3], there are split absorption bands at 1430 and 1405 cm^{-1}, attributed to $\nu_4(F_2)$ of the NH_4^+ ion [1], and five intense bands below 1000 cm^{-1}, assigned to the vibrations of the MoO_4^{2-} group of C_{2v} symmetry [3]. For $NH_4Ce(MoO_4)_2 \cdot H_2O$ the frequencies (in cm^{-1}) are: 930 ($\nu_1(A_1)$); 856, 770, 705 ($\nu_3(F_2)$); 418, 385, 370 ($\nu_4(F_2)$) [1]. The IR spectra of the other double molybdates with M = Nd, Sm, Ho, Er are very similar (for a list of bands see the paper). The distortion of the MoO_4 tetrahedra that causes the splitting of the $\nu_3(F_2)$ vibration is said to be due to an interconnection of the MoO_4 groups by the M^{III} ions [3].

Upon heating, $NH_4Ce(MoO_4)_2 \cdot H_2O$ loses its water between 40 and 120°C. Between 225 and 295°C the compound is further degraded by losing 1 NH_3 and 0.5 H_2O [1]. The reduction in intensity of the water absorption bands in the IR spectra of samples with M = Nd, Sm, Ho, Er after drying at 100 to 120°C indicates these compounds to be dehydrated in this temperature range [3].

References:

[1] M. J. Schwing-Weill (Bull. Soc. Chim. France **1972** 1754/61). – [2] B. N. Ivanov-Emin, E. I. Zakharikova (Zh. Neorgan. Khim. **14** [1969] 145/9; Russ. J. Inorg. Chem. **14** [1969] 75/8). – [3] B. E. Zaitsev, E. I. Zakharikova, B. N. Ivanov-Emin, G. I. Cherenkova (Zh. Neorgan. Khim. **14** [1969] 1493/6; Russ. J. Inorg. Chem. **14** [1969] 781/3).

6.2.5 $(CH_3)_2NH_2M(MoO_4)_2 \cdot nH_2O$ (= $2(CH_3)_2NH \cdot M_2O_3 \cdot 4MoO_3 \cdot (2n+1)H_2O$), M = La, Tm, Lu

$(CH_3)_2NH_2La(MoO_4)_2 \cdot H_2O$, $(CH_3)_2NH_2Tm(MoO_4)_2 \cdot 4H_2O$, and $(CH_3)_2NH_2Lu(MoO_4)_2 \cdot 1.5H_2O$ have been obtained as poorly crystallized precipitates from aqueous solutions of rare earth chlorides or nitrates and dimethylammonium molybdate. In the parent solution molybdate has to be in excess (molar ratio Lu : Mo = 1 : 8) and also excess dimethylammonium chloride has to be present (3 to 5 g per 5 to 10 mL of ~0.2 M rare earth salt solution). Boiling of the reaction mixture for up to 5 h favors crystallinity but also hydrolysis. The La and Lu compounds have been dried at 45 to 50°C; the Tm compound is air dried.

X-ray powder patterns show the La compound to be isomorphous with $CsPr(MoO_4)_2$ and the Tm and Lu compounds with α-$TlEr(MoO_4)_2$. $(CH_3)_2NH_2La(MoO_4)_2 \cdot H_2O$ loses water at 160 and 240°C and starts to decompose at 290°C.

M. V. Mokhosoev, A. D. Gorbalyuk, A. P. Perepelitsa, F. P. Alekseev (Zh. Neorgan. Khim. **25** [1980] 2407/9; Russ. J. Inorg. Chem. **25** [1980] 1331/2).

6.2.6 $RbM(MoO_4)_2 \cdot nH_2O$ (= $Rb_2O \cdot M_2O_3 \cdot 4MoO_3 \cdot 2nH_2O$), M = Sc, Y, Nd, Sm, Gd to Lu; n = 0.6, 1

$RbSc(MoO_4)_2 \cdot 0.6H_2O$ has been prepared from mixtures of aqueous solutions of Rb_2MoO_4 and $Sc(NO_3)_3$ at 25°C and pH 5 to 7. Equilibration times of up to 1 month and excess MoO_4^{2-} are necessary to redissolve $Sc_2(MoO_4)_3 \cdot nH_2O$ that may have formed intermediately. The precipitate is air dried [1]. Monohydrates $RbM(MoO_4)_2 \cdot H_2O$ with M = Y, Gd to Lu have been precipitated from stoichiometric mixtures of nearly saturated aqueous Rb_2MoO_4 and 0.1M aqueous $M(NO_3)_3$ solutions (for $Y(NO_3)_3$ also 0.01M solution) at 25°C. As the precipitates are degraded by water, they are washed with acetone and ether [2]. Monohydrates with M = Gd to Lu are also formed by hydrating in moist air the hygroscopic anhydrous compounds $RbM(MoO_4)_2$. The rate of hydration increases from Gd to Lu [3, 4]. Samples $RbM(MoO_4)_2 \cdot nH_2O$, M = Y, Tb, Er, with unspecified water content have been obtained under the conditions described for M = Sc [5]. $RbNd(MoO_4)_2 \cdot nH_2O$ has been precipitated from an aqueous $NdCl_3$ solution with Rb_2MoO_4 solution [6].

The X-ray powder patterns of $RbM(MoO_4)_2 \cdot H_2O$ with M = Er, Tm, Yb, Lu prepared from the anhydrous salts have been indexed on the basis of a monoclinic unit cell; Z = 4. The lattice parameters are given in the following table [3]:

compound	a in Å	b in Å	c in Å	β
$RbEr(MoO_4)_2 \cdot H_2O$	7.805	20.86	5.055	91°15′
$RbTm(MoO_4)_2 \cdot H_2O$	7.782	20.85	5.045	91°19′
$RbYb(MoO_4)_2 \cdot H_2O$	7.771	20.86	5.046	91°23′
$RbLu(MoO_4)_2 \cdot H_2O$	7.749	20.89	5.039	91°29′

For the indexed powder pattern of the Tm compound see the paper [3]; that of the Yb compound is given unindexed in [4]. The precipitated monohydrates $RbM(MoO_4)_2 \cdot H_2O$ with M = Y, Gd to Lu are isomorphous with each other according to their X-ray powder patterns. For the d values and a line diagram of the Er compound see the paper [2]. There is no good agreement between the sets of d values of the differently prepared monohydrates, yet the intensity distribution is similar and the misfit in the d values might reasonably be due to poor resolution and absorption effects during generation of the sets. While different unit cells for the monohydrates and their anhydrous forms have been reported [3], nearly identical powder patterns for both have also been published [2]. The ^1H NMR spectra of the $RbM(MoO_4)_2 \cdot nH_2O$ compounds with M = Sc, Y, Tb, Er indicate the water to be not water of crystallization but occluded [1, 5].

In the range below 1000 cm^{-1} the IR spectrum of $RbNd(MoO_4)_2 \cdot nH_2O$ shows the typical bands of the molybdate vibrations (see for example p. 277). The intensity of the IR absorption bands due to water of crystallization (3500 to 3200 and 1660 to 1620 cm^{-1}) decreases abruptly after drying the sample at 100 to 120°C [6]. Dehydration temperatures from DTA for the monohydrates are [2]:

M	Y	Gd	Tb	Dy	Ho	Er	Tm	Yb	Lu
t in °C	160	150	150	140	130	120	140	150	130

$RbSc(MoO_4)_2 \cdot 0.6H_2O$ loses its water below 200°C [1]. The thermogram of $RbEr(MoO_4)_2 \cdot nH_2O$ shows dehydration at 160°C [5]. $RbYb(MoO_4)_2 \cdot H_2O$ is reported to lose water in two steps of $0.6H_2O$ from 80 to 160°C and $0.4H_2O$ from 160 to 360°C [3].

References:

[1] A. M. Golub, A. P. Perepelitsa, A. M. Kalinichenko (Izv. Vysshikh Uchebn. Zavedenii Khim. Khim. Tekhnol. **15** [1972] 1293/6; C.A. **78** [1973] No. 10949). – [2] M. V. Bobkova, I. V. Shakhno, V. E. Plyushchev (Izv. Vysshikh Uchebn. Zavedenii Khim. Khim. Tekhnol. **14** [1971] 1625/9; C.A. **76** [1972] No. 67521). – [3] V. K. Rybakov, V. K. Trunov, V. I. Spitsyn (Dokl. Akad. Nauk SSSR **192** [1970] 369/71; Dokl. Phys. Chem. Proc. Acad. Sci. USSR **190/195** [1970] 393/5). – [4] V. K. Rybakov, V. K. Trunov (Zh. Neorgan. Khim. **16** [1971] 1320/5; Russ. J. Inorg. Chem. **16** [1971] 698/701). – [5] A. M. Golub, A. P. Perepelitsa, V. I. Maksin, K. S. Aganiyazov (Izv. Vysshikh Uchebn. Zavedenii Khim. Khim. Tekhnol. **14** [1971] 328/31; C.A. **75** [1971] No. 11010).

[6] B. E. Zaitsev, E. I. Zakharikova, B. N. Ivanov-Emin, G. I. Cherenkova (Zh. Neorgan. Khim. **14** [1969] 1493/6; Russ. J. Inorg. Chem. **14** [1969] 781/3).

6.2.7 $\mathbf{CsM(MoO_4)_2 \cdot nH_2O}$ (= $Cs_2O \cdot M_2O_3 \cdot 4MoO_3 \cdot 2nH_2O$), M = Sc, Y, Gd to Lu; n = 1, 1.8, 2, 2.4

$CsSc(MoO_4)_2 \cdot 1.8H_2O$, $CsY(MoO_4)_2 \cdot 2H_2O$, $CsEr(MoO_4)_2 \cdot 2.4H_2O$ have been obtained from mixtures of aqueous solutions of rare earth nitrates and excess Cs_2MoO_4 at pH 5 to 7, with equilibration at 25°C for about 1 month. The precipitates are air dried [1]. The monohydrates $CsM(MoO_4)_2 \cdot H_2O$ with M = Y, Gd to Lu have been precipitated from 0.1 M (also 0.01 and 0.3 M for Y, Gd, Er) aqueous $M(NO_3)_3$ solutions by addition of the stoichiometric amount of almost saturated aqueous Cs_2MoO_4 solution, with equilibration at 25°C for 5 d. Since the precipitates are degraded by water, they are washed with acetone and ether. The compounds contain 1 mol H_2O after drying at 40°C [2]. $CsEr(MoO_4)_2 \cdot nH_2O$ forms upon mixing aqueous solutions of $ErCl_3$ and Cs_2MoO_4 with molar ratios Er:Mo = 1:2 to 1:4 [3].

X-ray powder patterns show the monohydrates with M = Y, Gd to Lu to be isomorphous, for the d values see the paper [2]. ^1H NMR spectra indicate that the water in $CsSc(MoO_4)_2 \cdot 1.8H_2O$, $CsY(MoO_4)_2 \cdot 2H_2O$, and $CsEr(MoO_4)_2 \cdot 2.4H_2O$ is not water of crystallization but occluded [1]. The IR spectrum of $CsEr(MoO_4)_2 \cdot nH_2O$ between 3600 and 400 cm^{-1} shows the same characteristics as those of the corresponding $AM(MoO_4)_2 \cdot nH_2O$ compounds, A = alkali metal (see for example p. 277) [4].

Thermal degradation temperatures from DTA for the compounds with M = Y, Gd to Lu are [2]:

M	Y	Gd	Tb	Dy	Ho	Er	Tm	Yb	Lu
t in °C	140	150	140	140	140	140	150	150	160

The water-rich Sc, Y, and Er compounds lose their water up to 200°C [1].

References:

[1] A. M. Golub, A. P. Perepelitsa, A. M. Kalinichenko (Izv. Vysshikh Uchebn. Zavedenii Khim. Khim. Tekhnol. **15** [1972] 1293/6; C.A. **78** [1973] No. 10949). – [2] M. V. Bobkova, I. V. Shakhno, V. E. Plyushchev, O. I. Smirnova (Zh. Neorgan. Khim. **17** [1972] 1263/7; Russ. J. Inorg. Chem. **17** [1972] 657/9). – [3] B. N. Ivanov-Emin, E. I. Zakharikova (Zh. Neorgan. Khim. **14** [1969] 145/9; Russ. J. Inorg. Chem. **14** [1969] 75/8). – [4] B. E. Zaitsev, E. I. Zakharikova, B. N. Ivanov-Emin, G. I. Cherenkova (Zh. Neorgan. Khim. **14** [1969] 1493/6; Russ. J. Inorg. Chem. **14** [1969] 781/3).

6.3 Molybdolanthanoates

6.3.1 $(NH_4)_6[La_2Mo_{14}O_{48}]\cdot 24H_2O$ (= $6(NH_4)_2O\cdot 2La_2O_3\cdot 28MoO_3\cdot 48H_2O$)

This compound has been prepared by dropping an aqueous $La(NO_3)_3$ solution into a $(NH_4)_6Mo_7O_{24}$ solution. After 1 d the compound is precipitated in the form of light yellow crystals by the addition of NH_4Cl or NH_4NO_3. It is regarded as a molybdolanthanate. However, the formula of the corresponding molybdocerate (see below) has been slightly revised by more recent investigations. Treatment with K, Rb, or Cs salts yields the less soluble K, Rb, or Cs molybdolanthanates.

G. A. Barbieri (Atti Reale Accad. Lincei [5] **17** I [1908] 540/5).

6.3.2 $(NH_4)_7[Ce_2^{III}(OH)Mo_{14}O_{48}]\cdot 18H_2O$ (= $7(NH_4)_2O\cdot 2Ce_2O_3\cdot 28MoO_3\cdot 37H_2O$)

This compound precipitates from a mixture generated by dropwise addition of a solution of 8 g $CeCl_3\cdot 6H_2O$ in 50 mL H_2O to 500 mL of an aqueous 10% solution of $(NH_4)_6Mo_7O_{24}$ at room temperature, after cooling to 5°C. From analytical results the compound is best described as $(NH_4)_7[Ce_2(OH)Mo_{14}O_{48}]\cdot nH_2O$ with $10 \leqq n \leqq 22$ [1]. This compound was obtained earlier from solutions of $Ce(NO_3)_3$ and $(NH_4)_6Mo_7O_{24}$, but the formula given, $(NH_4)_6[Ce_2Mo_{14}O_{48}]\cdot 24H_2O$, differs from that of the recent investigation [2]. The formation of a heteropoly anion is established by the red-orange color of the compound [2] and by electrophoretic measurements [1].

The compound crystallizes as needles or parallelepipeds based on rhombs. A single crystal, kept in contact with its mother liquor, was used to determine the unit cell, which proved to be triclinic with the lattice parameters a = 20.13, b = 30.15, c = 17.28 Å, α = 103.95°, β = 115.75°, γ = 112.25°; Z = 5. The pycnometric density is 2.99, the calculated density 3.11 g/cm³. The IR spectrum between 1800 and 300 cm^{-1} resembles very much that of $(NH_4)_6Mo_7O_{24}\cdot 4H_2O$, so it is probable that the paramolybdate group is present in the heteropoly anion (for a figure of the spectrum see the paper) [1].

According to thermogravimetry $(NH_4)_7[Ce_2(OH)Mo_{14}O_{48}]\cdot 18H_2O$ is dehydrated in several scarcely resolved steps, starting at 40°C. Above 110°C H_2O and also NH_3 are lost and the last mole of H_2O is freed in a well-resolved step at 430 to 435°C. The total weight loss of 17.9% corresponds to $7NH_3$ and $22H_2O$ per 2Ce [1].

The crystals of $(NH_4)_7[Ce_2(OH)Mo_{14}O_{48}]\cdot 18H_2O$ are stable only in contact with their mother liquor. In air at ambient temperature they soon become dull and crumble [1]. The NH_4 compound yields the less soluble K, Rb, or Cs molybdocerates on treatment with K, Rb, or Cs salts [2].

References:

[1] M. J. Schwing-Weill (Bull. Soc. Chim. France **1972** 1754/61). – [2] G. A. Barbieri (Atti Reale Accad. Lincei [5] **17** I [1908] 540/5).

6.3.3 $(NH_4)_{2.5}Ce_{1.5}^{III}[H_2Ce^{III}Mo_{12}O_{42}]\cdot 9H_2O$ (= $5(NH_4)_2O\cdot 5Ce_2O_3\cdot 48MoO_3\cdot 40H_2O$) and Other Molybdocerates(III)

Solutions of molybdocerates(III) can easily be obtained from molybdocerates(IV) (or the acid) by electrochemical or chemical reduction (see p. 285). However, the reduced solutions are not stable and decay in a few hours. Only the ammonium compound given in the heading and a caesium compound (see below) have been precipitated and investigated.

If reduced $(NH_4)_6[H_2CeMo_{12}O_{42}]\cdot 9H_2O$ solutions (see p. 285) are allowed to stand for 24 to 30 h, they occasionally deposit small quantities of well-formed dark rectangular blocks. The yield never exceeds 25% of the solute in the original solution, and the colors of different products vary from maroon to brown, although they are all isomorphous by X-ray powder diffraction (but not with the unreduced ammonium 12-molybdocerate(IV)). Solutions of these crystals in 1.0M H_2SO_4 when freshly prepared gave optical spectra and voltammograms characteristic of the fully reduced electrolyzed solution, and the IR spectra of solid samples were identical with that of the oxidized material. The crystals therefore undoubtedly contain the $[Ce^{III}Mo_{12}O_{42}]^{9-}$ ion, but also, according to chemical analysis, extra cerium, presumably cationic. The composition of one sample is given in the heading. It is likely that differently colored products contained different amounts of Ce^{III}. EPR measurements on freshly reduced solution and on solids, at ambient and liquid nitrogen temperatures, revealed no signals of Mo^V (apart from small traces of an impurity). Thus the reduced complex should be formulated as $[H_nCe^{III}Mo_{12}O_{42}]^{(9-n)-}$ rather than as $[H_nCe^{IV}Mo^VMo^{VI}_{11}O_{42}]^{(9-n)-}$. Other evidence for the Ce^{III} formulation is (1) the value of the reduction potential (+0.45V) and (2) the absence of an intense $Mo^V \rightarrow Mo^{VI}$ intervalence transition (observed in all mixed-valence molybdates at 800 to 1000 nm). The observed transition at 450 nm in the spectrum of the freshly reduced solution can be assigned to a $Ce^{III} \rightarrow Mo^{VI}$ charge transfer.

Addition of Cs^+ to a freshly reduced solution yields a pale pink-brown precipitate which could not be redissolved for recrystallization. The IR spectrum of this material was essentially identical with that of ammonium 12-molybdocerate(IV) in the metal-oxygen vibrational region (1000 to 400 cm^{-1}).

L. McKean, M. T. Pope (Inorg. Chem. **13** [1974] 747/9).

6.3.4 $\mathbf{H_8Ce^{IV}Mo_8O_{30}\cdot 17H_2O}$ $(=CeO_2\cdot 8MoO_3\cdot 21H_2O)$

The bright yellow acid is prepared from its aqueous solutions by mixing with an equal volume of 60% H_2SO_4 as quickly as possible [1]. Solutions of the acid are prepared either by prolonged heating of $H_8CeMo_{12}O_{42}$ in 8N H_2SO_4 [2] or by ion exchange from a solution of $(NH_4)_4[H_4CeMo_8O_{30}]$ (see below). Titration curves show the acid to be octabasic [1].

The ^1H NMR spectrum suggests that the solid compound contains H_3O^+ and OH^-. The IR spectra of the 17-, 9.5-, and 6-hydrates and the anhydrous decomposition product (CeO_2 + 8MoO_3) in the region of the deformation vibrations of H_2O (1800 to 1400 cm^{-1}) are plotted in the paper. It is concluded that H_2O molecules are in the inner sphere of the heteropoly anion, and can be given off only with decomposition [2].

The solid acid loses water of crystallization on standing in the air. It is soluble in H_2O, acetone, and alcohols but not in ether [1].

The spectrum of an aqueous 2×10^{-5} M solution has been recorded between 220 and 1100 nm. The acid absorbs in the 220 to 520 nm region. The absorption increases with decreasing wavelength. When the acidity decreases, a plateau between 230 and 250 nm appears (see plot in [2]). In alkaline solutions there is a maximum at 243 ± 5 nm which shifts to longer wavelengths with increasing pH. It is assumed that there are two anions in equilibrium in the solutions. The solutions of the acid are not stable [1, 2].

References:

[1] Nguyen D'eu, E. A. Torchenkova, V. I. Spitsyn (Dokl. Akad. Nauk SSSR **198** [1971] 1350/3; Dokl. Chem. Proc. Acad. Sci. USSR **196/201** [1971] 536/8). – [2] V. I. Spitsyn, E. A. Torchenkova, Nguyen D'eu, V. F. Chuvaev (Allgem. Prakt. Chem. **22** [1971] 55/9).

6.3.5 $(NH_4)_4[H_4Ce^{IV}Mo_8O_{30}]\cdot 5.5\,H_2O$ ($=2(NH_4)_2O\cdot CeO_2\cdot 8\,MoO_3\cdot 7.5\,H_2O$)

This microcrystalline, dark red compound has been precipitated by the additon of a saturated aqueous NH_4NO_3 solution to a solution of $(NH_4)_8[CeMo_{12}O_{42}]\cdot 8\,H_2O$ which had been treated with 8N HNO_3 for 1d at room temperature. After washing with NH_4NO_3 solution and alcohol the compound was air dried (yield 73.2%).

The spectrum of the freshly prepared solution (2×10^{-5} M) between 220 and 320 nm shows the absorption maximum at 243 ± 5 nm and the intensity ($\varepsilon=26000$ at 245 nm and pH 7) of the color to decrease relative to the $(NH_4)_8[CeMo_{12}O_{42}]$ solution.

Hot water rapidly dissolves the salt, giving a dark red solution. Heteropoly compounds can be precipitated by K, Ba, Hg, and other heavy metal ions.

Nguyen D'eu, E. A. Torchenkova, V. I. Spitsyn (Dokl. Akad. Nauk SSSR **198** [1971] 1350/3; Dokl. Chem. Proc. Acad. Sci. USSR **196/201** [1971] 536/8).

6.3.6 $H_8Ce^{IV}Mo_{10}O_{36}\cdot 22\,H_2O$ ($=CeO_2\cdot 10\,MoO_3\cdot 26\,H_2O$)

Yellow crystals of this acid are precipitated from its aqueous solutions by mixing with an equal volume of 60% H_2SO_4 as fast as possible [1]. The solutions of the acid itself are prepared from solutions of $H_8CeMo_{12}O_{42}$ by prolonged heating in 2N HNO_3 [2] or from solutions of $(NH_4)_6[H_2CeMo_{10}O_{36}]\cdot 7\,H_2O$ (see below) by ion exchange. Titration curves show the acid to be octabasic [1].

From the ^{1}H NMR spectrum it is assumed that the solid compound contains H_3O^+ and OH^-. The IR spectra of the 18-, 11.5-, and 8-hydrates and the anhydrous decomposition product ($CeO_2+10\,MoO_3$) in the region of the deformation vibrations of H_2O (1800 to 1400 cm^{-1}) are plotted in the paper. It is concluded that H_2O molecules, which can be given off only with decomposition, are in the inner sphere of the heteropoly anion [2].

The crystals lose water when standing in air [1]. The acid is soluble in H_2O, alcohols, and acetone but not in ether [1].

Spectra of 2×10^{-5}M solutions between 220 and 520 nm are very similar to those of $H_8Ce^{IV}Mo_8O_{30}$ (p. 281). They show increasing absorption at shorter wavelengths with a plateau in the range from 230 to 250 nm. When the solution of the acid is neutralized and brought to pH 11, a maximum at 243 ± 5 nm emerges. There is a well defined isosbestic point, see plots in the papers [1, 2].

References:

[1] Nguyen D'eu, E. A. Torchenkova, V. I. Spitsyn (Dokl. Akad. Nauk SSSR **198** [1971] 1350/3; Dokl. Chem. Proc. Acad. Sci. USSR **196/201** [1971] 536/8). – [2] V. I. Spitsyn, E. A. Torchenkova, Nguyen D'eu, V. F. Chuvaev (Allgem. Prakt. Chem. **22** [1971] 55/9).

6.3.7 $(NH_4)_6[H_2Ce^{IV}Mo_{10}O_{36}]\cdot 7\,H_2O$ ($=3(NH_4)_2O\cdot CeO_2\cdot 10\,MoO_3\cdot 8\,H_2O$)

This microcrystalline, bright yellow compound is precipitated by addition of a saturated aqueous NH_4NO_3 solution to a solution of $(NH_4)_8[CeMo_{12}O_{42}]\cdot 8\,H_2O$ (p. 289) which had been reacted for 1d with 2N HNO_3 at room temperature. After washing with HNO_3 solution and alcohol, the compound was air dried (yield 88.7%).

The spectrum between 220 and 320 nm of the freshly prepared aqueous 2×10^{-5}M solution shows the absorption maximum at 243 ± 5nm and the intensity of the color ($\varepsilon=37000$ at 245 nm and pH 7) to decrease relative to the $(NH_4)_8[CeMo_{12}O_{42}]$ solution.

The salt dissolves rapidly in hot water, giving a dark red solution. Heteropoly compounds can be precipitated by K, Ba, Hg, and other heavy metal ions.

Nguyen D'eu, E. A. Torchenkova, V. I. Spitsyn (Dokl. Akad. Nauk SSSR **198** [1971] 1350/3; Dokl. Chem. Proc. Acad. Sci. USSR **196/201** [1971] 536/8).

6.3.8 $H_8Ce^{IV}Mo_{12}O_{42}\cdot nH_2O$ ($=CeO_2\cdot 12MoO_3\cdot(n+4)H_2O$), n = 0, 2, 4.5, 6, 8.7, 10, 18

The crystalline, yellow $H_8CeMo_{12}O_{42}\cdot 18H_2O$ precipitates from its 2 to 4% aqueous solutions (see below), when an equal volume of 60% H_2SO_4 is added quickly enough to prevent the heteropoly ion from degradation. The filtered precipitate is carefully washed six to eight times with concentrated HNO_3 and then six to eight times with ether. Yields are 97 to 98% of theoretical [1]. Potentiometric titrations indicate the acid to be octabasic [3]. The 18-hydrate cannot be dried in vacuo without decomposition [2]. By drying the 18-hydrate over P_2O_5 10- and 8.7-hydrates are obtained. Drying at elevated temperatures leads to further loss of water: n = 6 at 125°C, 4.5 at 140°C, 2 at 150°C, and 0 at 175°C [3]. A sample of the acid with unknown water content was obtained as a yellow glass-like mass by prolonged evaporation of aqueous solutions in air [4].

The only information about the crystal structure of $H_8CeMo_{12}O_{42}$ related in the literature is a remark saying that there is a rhombohedral unit cell with the hexagonal lattice parameters $a\approx12$ and $c\approx16$ Å [5]. A table of d values is given for the 18-hydrate, which is isomorphous with $H_8ThMo_{12}O_{42}\cdot 18H_2O$ (p. 302) [1]. IR and Raman spectra (see below) indicate that except for small distortions the $CeMo_{12}O_{42}$ group has the same structure as for $(NH_4)_8[CeMo_{12}O_{42}]\cdot 12H_2O$ (p. 289) and $(NH_4)_2[H_6CeMo_{12}O_{42}]\cdot 12H_2O$ [6]. The crystal structure of the latter has been determined by single crystal X-ray data, see p. 287. Solid state ^{1}H NMR spectra of all hydrates at 80 K have been interpreted by assuming three differently bound states for hydrogen: OH^-, H_2O, and H_3O^+. Accordingly, the 18-hydrate has been formulated tentatively as $(H_3O)_{8-x}[CeMo_{12}O_{42-x}(OH)_x]\cdot(10+x)H_2O$ with $x\approx2$ to 3. During dehydration the number of H_3O^+ is reduced, leading to a formulation of the ~9-hydrate as $(H_3O)_5[CeMo_{12}O_{39}(OH)_3]\cdot 4H_2O$. Narrow lines present in all ^{1}H NMR spectra of the different hydrates are explained as due to protons free from dipole interaction with H_2O molecules [3], see also [7]. According to X-ray photoelectron spectra binding energies in $H_8CeMo_{12}O_{42}$ are Ce 4d = 112.7, Mo $3d_{5/2}$ = 233.3, Mo $3d_{3/2}$ = 236.4, O 1s = 531.3 eV (reference C 1s = 285 eV) [5].

IR and Raman spectra of $H_8CeMo_{12}O_{42}\cdot 18H_2O$ between 1710 and 180 cm^{-1} are found to be in good agreement with those of $(NH_4)_8[CeMo_{12}O_{42}]\cdot 12H_2O$ in the range attributed to the frequencies of the $CeMo_{12}O_{42}$ groups [6, 8]; frequencies (in cm^{-1}) and assignments (the solution was 0.1M) [8]:

IR (solid)	R (solid)	R (solution)	assignment
	973	973	MoO_2, $\nu_s(A_g)$
965	957	957	MoO_2, $\nu_s(T_u,T_g)$
932	932	932	MoO_2, $\nu_{as}(T_u,T_g)$
	900	900	MoO_2, $\nu_s(E_g)$
718	720	720	
608			

IR (solid)	R (solid)	R (solution)	assignment
560	560	557	MoOMo, ν_{as}, ν_s (A_g, E_g, T_u, T_g)
485	500	505	
462	470	470	
408	400	408	
380	350	350	MoO_2, δ(A_g, T_g)
371	250	250	
220	183		

The IR spectra in the range of the H_2O bending vibrations (1800 to 1400 cm^{-1}) for the 18-hydrate, a 9.5-hydrate, and anhydrous $CeO_2+12MoO_3$ are plotted in [7].

References:

[1] P. Bajdala, E. A. Torchenkova, V. I. Spitsyn (Dokl. Akad. Nauk SSSR **196** [1971] 1344/5; Dokl. Chem. Proc. Acad. Sci. USSR **196/201** [1971] 137/8). – [2] M. Filowitz, R. K. C. Ho, W. G. Klemperer, W. Shum (Inorg. Chem. **18** [1979] 93/103, 95). – [3] V. F. Chuvaev, P. Bajdala, E. A. Torchenkova, V. I. Spitsyn (Dokl. Akad. Nauk SSSR **196** [1971] 1097/100; Dokl. Chem. Proc. Acad. Sci. USSR **196/201** [1971] 131/4). – [4] Z. F. Shakhova, S. A. Gavrilova (Zh. Neorgan. Khim. **3** [1958] 1370/3; Russ. J. Inorg. Chem. **3** No. 6 [1958] 138/42). – [5] I. V. Tat'yanina, E. A. Torchenkova, L. P. Kazanskii, V. I. Spitsyn (Dokl. Akad. Nauk SSSR **234** [1977] 1136/9; Dokl. Phys. Chem. Proc. Acad. Sci. USSR **232/237** [1977] 597/600).

[6] L. P. Kazanskii, E. A. Torchenkova, V. I. Spitsyn (Dokl. Akad. Nauk SSSR **209** [1973] 141/3; Dokl. Phys. Chem. Proc. Acad. Sci. USSR **208/213** [1973] 209/11). – [7] V. I. Spitsyn, E. A. Torchenkova, Nguyen D'eu, V. F. Chuvaev (Allgem. Prakt. Chem. **22** [1971] 55/9). – [8] V. I. Spitsyn, E. A. Torchenkova, L. P. Kazanskii, P. Bajdala (Z. Chem. [Leipzig] **14** [1974] 1/8).

6.3.9 Aqueous $H_8Ce^{IV}Mo_{12}O_{42}$ Solutions

Preparation. Solutions of $H_8CeMo_{12}O_{42}$ are prepared from the neutral $(NH_4)_8[CeMo_{12}O_{42}]\cdot nH_2O$ salt (p. 289) which is transformed into the acid salt by dissolving in 2% H_2SO_4 and adding a saturated aqueous solution of NH_4NO_3. Solutions of this are then converted to solutions of the pure acid using an acidic ion exchange resin [1]. $H_8CeMo_{12}O_{42}$ solutions can also be obtained directly from aqueous $(NH_4)_8[CeMo_{12}O_{42}]\cdot nH_2O$ suspensions by shaking them with the exchanger for ~2h [2].

The ^{17}O-enriched acid was obtained by dissolving the unenriched compound in ^{17}O-enriched water and storing this solution for 24 h at 25°C. After cooling to 0°C the acid was precipitated by concentrated H_2SO_4. From the washed precipitate a 14% ^{17}O-enriched solution was obtained for the NMR measurements (see below) [3].

Properties. The main structural feature of solid $H_8CeMo_{12}O_{42}\cdot nH_2O$, the $CeMo_{12}O_{42}$ unit, is preserved in the yellow aqueous solutions. NMR spectra of aqueous solutions of ^{17}O-enriched $H_8CeMo_{12}O_{42}$ showed resonance lines at 898 ppm with a line width of 650 Hz, assigned to the terminal O atoms, and at 214 ppm with a line width of 690 Hz, attributed to one of the oxygen atoms of the shared face (cf. p. 288) [3].

The Raman spectrum of an aqueous 0.1M $H_8CeMo_{12}O_{42}$ solution is in good agreement with that of solid $H_8CeMo_{12}O_{42}\cdot 18H_2O$, see the table above [4, 5].

UV spectra ($\lambda > 220$ nm) of aqueous 2×10^{-5} M solutions vary with the pH of the solution. All spectra show strongly increasing absorption at $\lambda < 230$ nm. A shoulder at ~230 to 245 nm and pH < 2 transforms into a relative maximum at 245 ± 5 nm [6] ($\varepsilon = 44000$ [7]) and pH values between 4.5 and 11.3 [4 to 8]. For plots of the spectra see [6 to 8]. There is a well defined isosbestic point in the spectra at pH values between these regions, which points to an equilibrium between two differently protonated species ($[H_2CeMo_{12}O_{42}]^{6-}$ and $[CeMo_{12}O_{42}]^{8-}$) in solution [8]. The absorption at 205 nm has been attributed to a $p_\pi \rightarrow d_\pi$ charge transfer, and the band at 247 nm to a transfer from a bonding three center orbital [4, 5]. The spectra have been interpreted in more detail on the basis of an MO scheme for the cis-$MoO_2O'_4$ octahedron of C_{2v} symmetry. From this model one can estimate the energies of four transitions at 30200, 37500, 38600, and 47900 cm^{-1}, which compare to the positions of three Gaussian curves at 31000, 41000, and 50900 cm^{-1}, into which the observed spectrum can be resolved. (This interpretation refers to the presumably very similar spectra of the heteropoly acids with the hetero atoms Ce, Th, and U.) From the same MO model the authors postulate a direct d-d overlap between the two Mo atoms of the Mo_2O_9 group, to account for the remarkable stability of the $[CeMo_{12}O_{42}]^{8-}$ ion. The d orbitals, usually empty in Mo^{VI}, are said to be partly filled by formation of π bonds to oxygen [9].

In its aqueous solutions $H_8CeMo_{12}O_{42}$ acts as an octabasic acid [1, 10, 11]. The dissociation constants of the first 6 protons are in the range of 6×10^{-3} to 1×10^{-3}, while the last two are somewhat less acidic [1]. The acid is fully dissociated only at concentrations less than 4×10^{-6} M [12, 13] or at pH > 4.5 [6, 8]. Overall protonation constants have been calculated for the different steps of protonation by fitting the calculated neutralization curve to the experimental one at 20°C and ionic strength ~0.1. The first four protons dissociate completely and $[H_4CeMo_{12}O_{42}]^{4-}$ then dissociates as a weak acid. Overall protonation constants log K of $[H_nCeMo_{12}O_{42}]^{(8-n)-}$ with n = 1 to 4 are 4.16(1), 7.15(1), 9.13(9), and 11.25(4), respectively [14].

Electrochemical and Chemical Reactions. Electrolytic reduction in neutral media [15] gives dark brown solutions, which show an intense absorption at 18000 cm^{-1} (≙ 555 nm) [17]. Unlike hydrated Ce^{4+} ions tetravalent cerium in aqueous solutions of $H_8CeMo_{12}O_{42}$ (pH 0 to 1) is not reduced by Fe^{2+}, but it is reduced in a one-electron step by $SnCl_2$ or $TiCl_3$ solutions. The reduced solutions are brown. The molybdocerate(III) can easily be reoxidized by permanganate or cerium(IV) salts to the molybdocerate(IV) [18]. The normal potential vs. the standard hydrogen electrode (SHE) of this first reduction by $TiCl_3$ in 1M HCl at 25°C is +0.79 V, corresponding to +0.56 V vs. the saturated calomel electrode (SCE). It is dependent on the HCl concentration: +0.72 V at 5×10^{-2} M HCl and +0.97 V at 8 M HCl (vs. SHE) [11]. Titration with $SnCl_2$ (in 1M HCl) gave the redoxpotential as +0.7 V [17]. After reduction by one equivalent ($Ce^{IV} \rightarrow Ce^{III}$) the heteropoly ion can be further reduced by $TiCl_3$ [11, 18]. A transient brown color is then observed, the optical spectrum of which shows an absorption band at 440 nm. In 8 M HCl at 45°C all the 12 Mo^{VI} can be reduced to Mo^V (for the potential of this reduction see the figure in the paper). No comment on the structure of the reduced species is made [11]. From theoretical considerations the $[CeMo_{12}O_{42}]^{8-}$ ion has been predicted not to be reducible to an isostructural mixed valence "heteropoly blue" [19]. By cyclic voltammometry on solutions of $H_8CeMo_{12}O_{42}$ in 1 M H_2SO_4 a reduction peak of +0.49 V vs. SCE was found, which was attributed to the reduction of Ce^{IV} to Ce^{III}. No further reduction and no evidence of Mo^V could be observed down to −0.1 V, at which point a complex multi-electron process occurred. The reduction is reversible only in 1 M H_2SO_4, not in 0.1 M HNO_3 or buffers up to pH 5 (see also [6]). Small reduction peaks occasionally observed at +0.3 and +0.15 V could be traced to isopolymolybdate impurities. Optical spectra of the brown-orange reduced solution (in 1 M H_2SO_4) show a broad shoulder at ~450 nm assigned to a $Ce^{III} \rightarrow Mo^{VI}$ charge transfer. (The discrepancy between this spectrum and that of [17] (see above) is said to be due to irreversible reduction conditions.) The reduced solutions after 24 to 30 h precipitate a solid molybdocerate(III) (see

p. 280) [16]. Other polarographic investigations on 5×10^{-4} M $H_8CeMo_{12}O_{42}$ in 0.5 M aqueous H_2SO_4 give three reduction waves with $E_{1/2} = +0.33$, $+0.06$, and -0.34 V vs. SCE, the first two values measured by a Pt microelectrode, the last one by a dropping mercury electrode. The solution after the first reduction did not give an EPR signal over a broad range of temperature [9].

The $[CeMo_{12}O_{42}]^{8-}$ ion is remarkably stable towards bases. Ceric hydroxide precipitates only after addition of 10 equivalents of NaOH, corresponding to pH $\approx$ 11 [1, 6], or at pH $\geqq$ 10 [8]. $H_8CeMo_{12}O_{42}$ is only slightly soluble in concentrated mineral acids [5]. In strongly acidic solutions the heteropoly acid is degraded. Aging for 1 d at room temperature [7] or heating [8] in 2 N HNO_3 forms a new anion $[H_2CeMo_{10}O_{36}]^{6-}$. Using 8 N HNO_3 yields $[H_4CeMo_8O_{30}]^{4-}$ [7, 8].

The formation of soluble complexes $[CeM^{II}_2Mo_{12}O_{42}]^{4-}$ is reported with M^{II} = Mn, Fe, Co, Ni, Cu, Zn, Cd [5]. For reactions with trivalent rare earth ions see p. 291. In weakly acidic solutions the molybdoceric acid is decomposed by excess $Th(ClO_4)_4$, forming a precipitate (which has not been investigated). In strong acids the complex $[CeTh_2Mo_{12}O_{42}]$ is formed [5]. With NbO_3^- [8] or $Ta_6O_{19}^{8-}$ ions [20] in aqueous solution, $[H_2CeMo_{12}O_{42}]^{6-}$ forms the complexes $[CeNb_2Mo_{12}O_{46}]^{6-}$ and $[CeTa_2Mo_{12}O_{49}]^{12-}$, respectively.

$H_8CeMo_{12}O_{42}$ may be extracted from aqueous solutions to a slight extent by ketones and ethylacetoacetate but not by alcohols, esters, ethers, benzene, gasoline, and 1,2-dichloroethane [10].

References:

[1] L. C. W. Baker, G. A. Gallagher, T. P. McCutcheon (J. Am. Chem. Soc. **75** [1953] 2493/5). – [2] Z. F. Shakhova, S. A. Gavrilova (Zh. Neorgan. Khim. **3** [1958] 1370/3; Russ. J. Inorg. Chem. **3** No. 6 [1958] 138/42). – [3] M. Filowitz, R. K. C. Ho, W. G. Klemperer, W. Shum (Inorg. Chem. **18** [1979] 93/103). – [4] L. P. Kazanskii, E. A. Torchenkova, V. I. Spitsyn (Dokl. Akad. Nauk SSSR **209** [1973] 141/3; Dokl. Phys. Chem. Proc. Acad. Sci. USSR **208/213** [1973] 209/11). – [5] V. I. Spitsyn, E. A. Torchenkova, L. P. Kazanskii, P. Bajdala (Z. Chem. [Leipzig] **14** [1974] 1/8).

[6] V. I. Spitsyn, E. A. Torchenkova, G. G. Stepanova (Rec. Chem. Progr. **31** [1970] 89/101). – [7] Nguyen D'eu, E. A. Torchenkova, V. I. Spitsyn (Dokl. Akad. Nauk SSSR **198** [1971] 1350/3; Dokl. Chem. Proc. Acad. Sci. USSR **196/201** [1971] 536/8). – [8] V. I. Spitsyn, E. A. Torchenkova, Nguyen D'eu, V. F. Chuvaev (Allgem. Prakt. Chem. **22** [1971] 55/9). – [9] L. P. Kazanskii, E. A. Torchenkova, V. I. Spitsyn (Dokl. Akad. Nauk SSSR **213** [1973] 118/21; Dokl. Phys. Chem. Proc. Acad. Sci. USSR **208/213** [1973] 932/4). – [10] Yu. F. Shkaravskii (Ukr. Khim. Zh. **29** [1963] 356/9; C.A. **59** [1963] 2366).

[11] É. Piémont, J. P. Schwing (Bull. Soc. Chim. France **1975** 1476/8). – [12] E. Matijević, M. Kerker (J. Am. Chem. Soc. **81** [1959] 5560/6). – [13] J. R. Keller, E. Matijević, M. Kerker (J. Phys. Chem. **65** [1961] 56/8). – [14] I. V. Tat'yanina, A. P. Borisova, E. A. Torchenkova, V. I. Spitsyn (Dokl. Akad. Nauk SSSR **256** [1981] 612/5; Dokl. Chem. Proc. Acad. Sci. USSR **256/261** [1981] 33/5). – [15] R. D. Peacock (Private communication to McKean and Pope according to [16]).

[16] L. McKean, M. T. Pope (Inorg. Chem. **13** [1974] 747/9). – [17] R. D. Peacock, T. J. R. Weakley (J. Chem. Soc. A **1971** 1937/40). – [18] A. Barbier, L. Malaprade (Compt. Rend. **256** [1963] 168/9). – [19] M. T. Pope (Inorg. Chem. **11** [1972] 1973/4). – [20] E. A. Torchenkova, Nguyen D'eu, L. P. Kazanskii, V. I. Spitsyn (Izv. Akad. Nauk SSSR Ser. Khim. **1973** 734/8; Bull. Acad. Sci. USSR Div. Chem. Sci. **22** [1973] 715/8).

6.3.10 $\mathbf{Na_8[Ce^{IV}Mo_{12}O_{42}]}$ (= $4Na_2O \cdot CeO_2 \cdot 12MoO_3$), $\mathbf{K_8[Ce^{IV}Mo_{12}O_{42}]}$ (= $4K_2O \cdot CeO_2 \cdot 12MoO_3$)

These compounds have been prepared by a procedure analogous to the preparation of $H_8CeMo_{12}O_{42} \cdot 18H_2O$ (cf. p. 283) as described in [1]. The water contents of the samples have not been specified, so it remains unclear whether the samples still contained water of crystallization or had been dehydrated under XPS conditions. The binding energies (in eV relative to C 1s = 285 eV) referred to $K_8[CeMo_{12}O_{42}]$ are Ce 4d = 112.1, Mo $3d_{5/2}$ = 232.8, Mo $3d_{3/2}$ = 236.0, O 1s = 530.6, and K 1p = 293.1 ± 0.1 [2, 3]. Compared to $H_8CeMo_{12}O_{42}$ the binding energies of Mo 3d and O 1s are lower in the potassium salt. This is regarded as due to the smaller electronegativity of K^+ relative to H^+, which gives rise to less positive charge on the Mo atoms [3].

The replacement of K by Na has practically no influence on the binding energy of the Mo $3d_{5/2}$ and O 1s electrons (232.7 and 530.6 eV [4]), probably because the crystal lattice is determined chiefly by the large heteropoly anion [2].

References:

[1] P. Bajdala, E. A. Torchenkova, V. I. Spitsyn (Dokl. Akad. Nauk SSSR **196** [1971] 1344/5; Dokl. Chem. Proc. Acad. Sci. USSR **196/201** [1971] 137/8). – [2] L. P. Kazanskii, V. I. Spitsyn (Dokl. Akad. Nauk SSSR **227** [1976] 140/3; Dokl. Phys. Chem. Proc. Acad. Sci. USSR **226/231** [1976] 225/7). – [3] I. V. Tat'yanina, E. A. Torchenkova, L. P. Kazanskii, V. I. Spitsyn (Dokl. Akad. Nauk SSSR **234** [1977] 1136/9; Dokl. Phys. Chem. Proc. Acad. Sci. USSR **232/237** [1977] 597/600). – [4] V. N. Molchanov, L. P. Kazanskii, E. A. Torchenkova, V. I. Spitsyn (Izv. Akad. Nauk SSSR Ser. Khim. **1978** 1248/51; Bull. Acad. Sci. USSR Div. Chem. Sci. **27** [1978] 1085/7).

6.3.11 $\mathbf{(NH_4)_2[H_6Ce^{IV}Mo_{12}O_{42}] \cdot 12H_2O}$ (= $(NH_4)_2O \cdot CeO_2 \cdot 12MoO_3 \cdot 15H_2O$)

A small amount of crystalline $(NH_4)_2[H_6CeMo_{12}O_{42}] \cdot 12H_2O$ has been obtained during the precipitation of $(NH_4)_6[H_2CeMo_{12}O_{42}] \cdot 10H_2O$ (see below), which indeed constituted the major fraction of the product [1].

The lattice parameters of the rhombohedral crystals are a = 10.589 ± 0.004 Å, α = 78.108° ± 0.004°, Z = 1; space group $R\bar{3}$-C_{3i}^2 (No. 148) [1]. This corresponds to a hexagonal cell with a = 13.34, c = 21.80 Å; Z = 3. The structure has been determined by single crystal methods. Atomic parameters (hexagonal cell) [2]:

atom	position	x	y	z
Ce	3a	0	0	0
Mo(1)	18f	0.28996(8)	0.09992(8)	0.04160(4)
Mo(2)	18f	0.14709(8)	0.18088(8)	0.12470(4)
O(1)	18f	0.4084(7)	0.2000(8)	0.0020(4)
O(2)	18f	0.1841(6)	0.1868(6)	0.0230(3)
O(3)	18f	0.3337(7)	0.0124(8)	0.0728(4)
O(4)	18f	0.1177(7)	0.0070(6)	0.0909(3)
O(5)	18f	0.3009(7)	0.1960(7)	0.1145(4)
O(6)	18f	0.1883(8)	0.3234(8)	0.1284(4)
O(7)	18f	0.1205(7)	0.1366(8)	0.1990(4)

atom	position	x	y	z
O(8)	18f	0.4634(12)	0.4409(11)	0.0604(7)
O(9)	18f	0.1896(12)	0.5507(12)	0.1034(7)
N(1)	6c	0	0	0.3117(9)

The characteristic structural unit, the $[CeMo_{12}O_{42}]^{8-}$ ion, is shown in **Fig. 79** (from [3]). It is formed by six Mo_2O_9 units composed of two face-sharing MoO_6 octahedra. By corner sharing (and thus matching the stoichiometric requirement: $Mo_{12}O_{42} \triangleq 6\,Mo_2O_7$) these units form a nearly undistorted icosahedral cage around the central Ce atom [1]. So in addition to its point symmetry $\overline{3}$ in the crystal structure, the actual symmetry of the heteropoly anion is nearly cubic (m3) [4]. All twelve MoO_6 octahedra are in fact chemically equivalent and there are just three different types of oxygen: Two cis located, terminal O^c are bonded to only one Mo with distances Mo-O^c = 1.68 Å. One oxygen atom (O^b) is coordinated to both of the Mo atoms of an Mo_2O_9 unit with Mo-O^b = 1.98 Å. The other two "shared face" oxygen atoms (O^a) have distances Mo-O^a = 2.28 Å to two Mo atoms, plus a third Mo-O^a = 1.94 Å as the corner-sharing apices of the octahedra of the next Mo_2O_9 units. The O^a atoms also form the icosahedron around Ce with Ce-O^a = 2.50 ± 0.01 Å. The H_2O molecules and NH_4^+ ions appear to fill the space between the anions and to connect the structure in a hydrogen-bonded network. The acidic hydrogen atoms were not located and their positions are open to conjecture [1]. The calculated density is 3.30 g/cm³ [2].

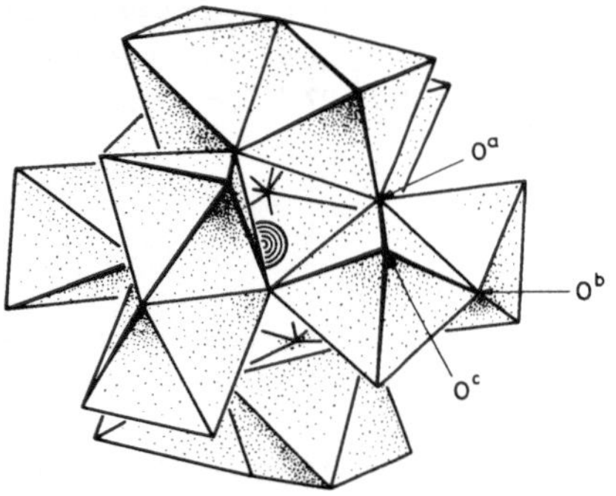

Fig. 79. The structure of the $[CeMo_{12}O_{42}]^{8-}$ ion [3].

References:

[1] D. D. Dexter, J. V. Silverton (J. Am. Chem. Soc. **90** [1968] 3589/90). – [2] D. D. Dexter (Diss. Georgetown Univ. 1968, pp. 1/184, 52/71; Diss. Abstr. B **29** [1968] 2825). – [3] M. T. Pope (Heteropoly and Isopoly Oxometalates, Springer, Berlin 1983, p. 29). – [4] H. T. Evans (Perspectives Struct. Chem. **4** [1971] 1/59, 31).

6.3.12 $(NH_4)_6[H_2Ce^{IV}Mo_{12}O_{42}] \cdot n\,H_2O$ (= $3(NH_4)_2O \cdot CeO_2 \cdot 12\,MoO_3 \cdot (n+1)H_2O$), n = 0, 1, 2.5, 7.3, 10

The light yellow $(NH_4)_6[H_2CeMo_{12}O_{42}] \cdot 10\,H_2O$ is obtained from a solution of $(NH_4)_8[CeMo_{12}O_{42}] \cdot 8\,H_2O$ (see below) in 2% H_2SO_4 on adding a saturated solution of NH_4NO_3 [1 to 3]. The compound has been recrystallized from hot water and washed with methanol [2]. Drying at elevated temperatures gives reduced water contents: n = 7.3 at 60°C, 2.5 at 80°C, 1 at 110°C, and 0 from 140 to 200°C [4].

The monoclinic crystals proved unsuitable for accurate X-ray structural investigations [5]. The solid state ^1H NMR signals consist of a broad and a narrow component. The intensity of the first parallels the degree of hydration and yields an interproton distance of 1.60 Å for the water of crystallization. The narrow component is due to the acid protons and those of the NH_4^+ ion. As the shape of the narrow line is independent of the sample temperature in the range of 4.2 to 300 K, the rotation barrier between different possible orientations of the NH_4 ion is regarded as remarkably low. The IR spectra show the NH_4 vibrations $F\delta(Q)=1410$ and $F\nu(Q)=3170$ to $3165\ cm^{-1}$, which are reduced ($F\delta(Q)$) and increased ($F\nu(Q)$) values compared to the corresponding frequencies of NH_4F and thus also reflect the weak bonds of NH_4^+ to the rigid lattice. The removal of the $10H_2O$ of crystallization does not affect the NH_4^+ lines [4].

Anhydrous $(NH_4)_6[H_2CeMo_{12}O_{42}]$ decomposes at temperatures above 200°C, releasing water and ammonia [4].

References:

[1] G. A. Barbieri (Atti Reale Accad. Lincei [5] **23** I [1914] 805/12). – [2] L. C. W. Baker, G. A. Gallagher, T. P. McCutcheon (J. Am. Chem. Soc. **75** [1953] 2493/5). – [3] E. Matijević, M. Kerker (J. Am. Chem. Soc. **81** [1959] 5560/6). – [4] V. F. Chuvaev, G. G. Stepanova, E. A. Torchenkova, V. I. Spitsyn (Dokl. Akad. Nauk SSSR **166** [1966] 406/9; Dokl. Phys. Chem. Proc. Acad. Sci. USSR **166/171** [1966] 26/9). – [5] D. D. Dexter, J. V. Silverton (J. Am. Chem. Soc. **90** [1968] 3589/90).

6.3.13 $(NH_4)_8[Ce^{IV}Mo_{12}O_{42}]\cdot 8H_2O$ ($=4(NH_4)_2O\cdot CeO_2\cdot 12MoO_3\cdot 8H_2O$)

Yellow $(NH_4)_8[CeMo_{12}O_{42}]\cdot 8H_2O$ precipitates when 50 mL of an aqueous 5% solution of $(NH_4)_2[Ce(NO_3)_6]$ is added dropwise to a boiling solution of 30 g $(NH_4)_6Mo_7O_{24}$ in 100 mL H_2O [1 to 4]. Different contents of water of crystallization, n=12 [5, 6] and 9 [7], have also been reported.

The 9-hydrate has been investigated by ^1H NMR and IR spectroscopy. The results are interpreted as described for $(NH_4)_6[H_2CeMo_{12}O_{42}]\cdot nH_2O$ (see above) [7]. IR and Raman spectra of the solid 12-hydrate (for tables of the frequencies see the papers) are comparable to those of $H_8CeMo_{12}O_{42}\cdot 18H_2O$ (see p. 283) and have been assigned in the same way assuming the same structure for the $CeMo_{12}O_{42}$ unit as in $(NH_4)_2[H_6CeMo_{12}O_{42}]\cdot 12H_2O$ (cf. p. 287) [5, 6]. $(NH_4)_8[CeMo_{12}O_{42}]\cdot 8H_2O$ is insoluble in H_2O and soluble in dilute inorganic acids [1].

References:

[1] G. A. Barbieri (Atti Reale Accad. Lincei [5] **23** I [1914] 805/12). – [2] L. C. W. Baker, G. A. Gallagher, T. P. McCutcheon (J. Am. Chem. Soc. **75** [1953] 2493/5). – [3] E. Matijević, M. Kerker (J. Am. Chem. Soc. **81** [1959] 5560/6). – [4] Z. F. Shakhova, S. A. Gavrilova (Zh. Neorgan. Khim. **3** [1958] 1370/3; Russ. J. Inorg. Chem. **3** No. 6 [1958] 138/42). – [5] L. P. Kazanskii, E. A. Torchenkova, V. I. Spitsyn (Dokl. Akad. Nauk SSSR **209** [1973] 141/3; Dokl. Phys. Chem. Proc. Acad. Sci. USSR **208/213** [1973] 209/11).

[6] V. I. Spitsyn, E. A. Torchenkova, L. P. Kazanskii, P. Bajdala (Z. Chem. [Leipzig] **14** [1974] 1/8). – [7] V. F. Chuvaev, G. G. Stepanova, E. A. Torchenkova, V. I. Spitsyn (Dokl. Akad. Nauk SSSR **166** [1966] 406/9; Dokl. Phys. Chem. Proc. Acad. Sci. USSR **166/171** [1966] 26/9).

6.3.14 Molybdocerates(IV) of Organic N-Bases

$[(CH_3)_2NH_2]_8[CeMo_{12}O_{42}]\cdot 20H_2O$ was prepared by adding molybdoceric(IV) acid to a solution containing excess dimethylammonium chloride. The precipitate was dissolved in warm water and added to another solution containing excess $(CH_3)_2NH_2Cl$. The precipitated yellow crystals can be recrystallized from warm water [1]. Other authors add excess concentrated $(CH_3)_2NH_2Cl$ solution to a molybdoceric acid solution. The precipitate was filtered after 12 h. The compound dehydrates at 140°C. The anhydrous compound is stable up to 180°C [2]. The 20-hydrate is very slightly soluble in cold water [1].

The compounds in the table below have been prepared from the molybdoceric acid plus acidified solutions of excesses of the bases. The precipitates are filtered after 12 h, washed with acidified H_2O and methanol, and dried. The stability ranges have been obtained by thermogravimetric analyses [2].

base	formula	stability range of the anhydrous compound in °C
methyl amine	$(CH_3NH_3)_8[CeMo_{12}O_{42}]\cdot nH_2O$	140 to 200
guanidine	$(CH_5N_3H)_7[HCeMo_{12}O_{42}]\cdot nH_2O$	140 to 200
pyridine	$(C_5H_5NH)_7[HCeMo_{12}O_{42}]\cdot nH_2O$	120 to 140
quinoline	$(C_9H_7NH)_7[HCeMo_{12}O_{42}]$	100 to 120
urotropine	$(C_6H_{12}N_4H)_6[H_2CeMo_{12}O_{42}]\cdot nH_2O$	140 to 180
hydroxyquinoline	$(C_9H_7NOH)_6[H_2CeMo_{12}O_{42}]\cdot nH_2O$	200 to 250
acridine	$(C_{13}H_9NH)_6[H_2CeMo_{12}O_{42}]\cdot nH_2O$	140 to 180
rivanol	$(C_{15}H_{15}ON_3H)_6[H_2CeMo_{12}O_{42}]\cdot nH_2O$	200 to 220
quinacrine	"$(C_{23}H_{30}ON_3H)_{3.5}[H_4CeMo_{12}O_{42}]$"	120 to 160

Yellow microcrystalline diazonium salts of the type $(RC_6H_4N_2)_4[H_4CeMo_{12}O_{42}]$ with R = H, p-CH_3, p-CH_3O, and p-NO_2 were produced by pouring together chilled aqueous solutions of the molybdoceric acid and an HCl solution of the corresponding diazonium salt. The precipitates were washed with water, alcohol, and ether and dried under vacuum over P_2O_5.

The IR spectra between 1700 and 400 cm^{-1} of the p-CH_3O and p-NO_2 compounds are plotted in the paper. The spectra of all four compounds show absorption bands in the region 2300 to 2200 cm^{-1} corresponding to the valence vibrations of the N≡N bond. The compounds are practically insoluble in water, alcohol, and ether, but soluble in dimethylformamide. They do not decompose when stored in a dark desiccator for 2 to 3 months [3].

References:

[1] L. C. W. Baker, G. A. Gallagher, T. P. McCutcheon (J. Am. Chem. Soc. **75** [1953] 2493/5). – [2] Z. F. Shakhova, S. A. Gavrilova (Vestn. Mosk. Univ. Ser. Mat. Mekhan. Astron. Fiz. Khim. **14** No. 2 [1959] 179/83; C. A. **1960** 9715). – [3] V. P. Sagalovich, S. A. Gavrilova, V. V. Kozlov, Yu. M. Kulikov, Z. F. Shakhova (Dokl. Akad. Nauk SSSR **168** [1966] 832/5; Dokl. Chem. Proc. Acad. Sci. USSR **166/171** [1966] 752/4).

6.3.15 $Cs_4[H_4Ce^{IV}Mo_{12}O_{42}]\cdot 9H_2O$ ($=2Cs_2O\cdot CeO_2\cdot 12MoO_3\cdot 11H_2O$)

This compound has been precipitated from mixtures of aqueous solutions of 12-molybdoceric acid and a Cs salt [1]. The compound has also been erroneously referred to as a 12-hydrate. The solid state ^1H NMR signal can be separated into two parts of different line widths.

The intensity ratio matches the ratio of hydrous to acidic protons. Allowing also for thermal motion an H-H distance of 1.63 Å in the H_2O has been calculated from the line width and the second moment [2].

References:

[1] V. I. Spitsyn, E. A. Torchenkova, G. G. Stepanova (At. Energiya **15** [1963] 519/20; Soviet At. Energy **15** [1963] 1316/7). – [2] V. F. Chuvaev, G. G. Stepanova, E. A. Torchenkova, V. I. Spitsyn (Dokl. Akad. Nauk SSSR **166** [1966] 406/9; Dokl. Phys. Chem. Proc. Acad. Sci. USSR **166/171** [1966] 26/9).

6.3.16 $(NH_4)_2Cs_4[H_2Ce^{IV}Mo_{12}O_{42}]\cdot 7H_2O$ (= $(NH_4)_2O\cdot 2Cs_2O\cdot CeO_2\cdot 12MoO_3\cdot 8H_2O$)

This bright yellow compound was obtained from mixtures of aqueous solutions of $(NH_4)_6[H_2CeMo_{12}O_{42}]\cdot 10H_2O$ and a Cs salt. The precipitate was washed with water and methanol and dried in air. The solubility (related to anhydrous $(NH_4)_2O\cdot 2Cs_2O\cdot CeO_2\cdot 12MoO_3$) increases from 0.176 g in 100 mL H_2O to 0.200 g in 100 mL 0.05 M HNO_3. The compound has been used for the colorimetric determination of Cs.

V. I. Spitsyn, E. A. Torchenkova, G. G. Stepanova (At. Energiya **15** [1963] 519/20; Soviet At. Energy **15** [1963] 1316/7).

6.3.17 $M_6^{III}Ce_2^{IV}Mo_{24}O_{85}\cdot nH_2O$ (= $3M_2O_3\cdot 2CeO_2\cdot 24MoO_3\cdot nH_2O$), M^{III} = La, Ce, Pr, Sm; n = 24 to 47

The 0.1 and 0.01M HCl solutions of $H_8CeMo_{12}O_{42}$ or $(NH_4)_6[H_2CeMo_{12}O_{42}]$ and the chlorides or nitrates of La, Ce, Pr, and Sm yield soluble complexes with a molar ratio $M^{III}:Ce^{IV} = 2:1$. Upon further addition of M^{3+} a solid of composition $M_6Ce_2Mo_{24}O_{85}\cdot nH_2O$ precipitates, with n = 24 (or 36 [2]) for M = La, 29 for Ce, 46 for Pr, and 47 for Sm [1]. The presence of nitric acid increases the water content with increasing HNO_3 concentration, which results in a change of the color of the salts from yellow to brown. The results of dehydration of the La and Sm salts show that their water is all bound the same [2]. Compounds with M = La and Ce are X-ray amorphous, while they are crystalline for Pr and Sm [1, 2]. The solubility in water at 20°C increases from 0.09 g/L for M = La to 0.73 g/L for Sm [1].

References:

[1] E. A. Torchenkova, G. G. Stepanova, V. I. Spitsyn (Dokl. Akad. Nauk SSSR **157** [1964] 1167/70; Dokl. Chem. Proc. Acad. Sci. USSR **154/159** [1964] 791/4). – [2] V. I. Spitsyn, E. A. Torchenkova, G. G. Stepanova (Rec. Chem. Progr. **31** [1970] 89/101).

6.3.18 $M_{20}^{I}M_6^{III}Ce_4^{IV}Mo_{48}O_{171}\cdot nH_2O$ (= $10M_2O\cdot 3M_2O_3\cdot 4CeO_2\cdot 48MoO_3\cdot nH_2O$), $M^I = NH_4$, Cs; M^{III} = Y, Er, Yb

$H_8CeMo_{12}O_{42}$ (or $(NH_4)_6[H_2CeMo_{12}O_{42}]$) and the chlorides or nitrates of Y, Er, and Yb in aqueous solution form a heteropoly anion with a mole ratio $M^{III}:Ce^{IV} = 2:1$. In contrast to M^{III} = La, Ce, Pr, and Sm (see the preceding section), the yttrium group elements do not form a precipitate at any $M^{III}:Ce^{IV}$ ratio. The absorption spectrum between 220 and 400 nm of the solution with the 2:1 complex shows no fundamental changes relative to that of $H_8CeMo_{12}O_{42}$ [1, 2]. Attempts to isolate NH_4 and Cs salts gave precipitates of the composition $M^{I}_{20}M^{III}_{6}Ce^{IV}_{4}Mo_{48}O_{171} \cdot nH_2O$. These are readily recrystallized from water [2].

References:

[1] E. A. Torchenkova, G. G. Stepanova, V. I. Spitsyn (Dokl. Akad. Nauk SSSR **157** [1964] 1167/70; Dokl. Chem. Proc. Acad. Sci. USSR **154/159** [1964] 791/4). – [2] V. I. Spitsyn, E. A. Torchenkova, G. G. Stepanova (Rec. Chem. Progr. **31** [1970] 89/101).

6.3.19 $Ag_8[Ce^{IV}Mo_{12}O_{42}]$ ($=4Ag_2O \cdot CeO_2 \cdot 12MoO_3$)

In accordance with the Gmelin system this compound has been described in "Silber" B 4, 1974, p. 344.

7 Molybdate Hydrates with Subgroup 4 Metals and Thorium

7.1 Titanium Molybdate Hydrates

7.1.1 Undefined Titanium Molybdates

A sample of composition $TiO_2 \cdot MoO_3 \cdot 3.5\,H_2O$ has presumably been prepared similarly to the $ZrO_2 \cdot MoO_3 \cdot 3.7\,H_2O$, see p. 298. ^1H NMR and IR spectroscopy indicate the presence of three types of deformed water molecules with different bond energies [1].

Samples with Ti : Mo ratios of 2 : 1, 1 : 1, and 1 : 2 have been prepared from aqueous solutions of titanium chloride or sulfate and sodium molybdate. Dried at 40 or 100°C they are X-ray amorphous. All samples are cation exchangers [2, 3]. For $Ti(OH)_3$-$Mo(OH)_3$ solid solutions see p. 346.

References:

[1] M. V. Mokhosoev, L. A. Pozharskaya, V. P. Pitsyuga (Dokl. Akad. Nauk SSSR **251** [1980] 392/6; Dokl. Phys. Chem. Proc. Acad. Sci. USSR **250/255** [1980] 203/6). – [2] M. Qureshi, H. S. Rathore (J. Chem. Soc. A **1969** 2515/8). – [3] M. Qureshi, H. S. Rathore, R. Kumar (J. Therm. Anal. **3** [1971] 371/8).

7.2 Molybdotitanates

Older data are given in "Molybdän", 1935, p. 374.

7.2.1 $H_4TiMo_{12}O_{40} \cdot 10\,H_2O$ (= $TiO_2 \cdot 12\,MoO_3 \cdot 12\,H_2O$)

This heteropoly acid has been obtained as dark brown deliquescent crystals by crystallization from its aqueous solutions in a vacuum desiccator over sulfuric acid [1]. Aqueous solutions have been obtained by ion exchange from the ammonium salt [2] (a method that has also been reported as unsuccessful in this case [1]), or by the ether method. In this last method, a mixture of aqueous solutions of $(NH_4)_2TiF_6$ (10%) and ammonium paramolybdate (saturated hot) was acidified to pH 1 and heated in a platinum dish. The reaction mixture was then shaken 3 to 4 times with ether, while 5 to 10 mL of a 1 : 1 diluted sulfuric acid were added each time. The yellow lower layer, which contained the etherate of the 12-molybdotitanic acid, was separated and the etherate decomposed with water. The yield is 30% [1]. In aqueous solution $H_4TiMo_{12}O_{40} \cdot 10\,H_2O$ acts as a tetrabasic acid [2, 3], dissociation constant $\sim 6.9 \times 10^{-4}$ [2].

References:

[1] Z. F. Shakhova, E. N. Semenovskaya (Zh. Neorgan. Khim. **7** [1962] 1084/6; Russ. J. Inorg. Chem. **7** [1962] 556/8). – [2] A. Liberti, G. Giombini, E. Cervone (Ric. Sci. **25** [1955] 883/90). – [3] Z. F. Shakhova, E. N. Semenovskaya (Zh. Neorgan. Khim. **13** [1968] 1887/9; Russ. J. Inorg. Chem. **13** [1968] 983/4).

7.2.2 $(NH_4)_{4-x}[H_xTiMo_{12}O_{40}] \cdot n\,H_2O$ (= $(2 - x/2)(NH_4)_2O \cdot TiO_2 \cdot 12\,MoO_3 \cdot (n + x/2)H_2O$), x = 0, 1; n = 5, 10

This salt has been obtained by reaction of concentrated aqueous solutions of $(NH_4)_2TiF_6$ and $(NH_4)_6Mo_7O_{24}$. Stoichiometric amounts were mixed, acidified slightly, and heated in a platinum crucible to 60°C. After further acidification by HCl and a few drops of nitric acid golden yellow crystals of $(NH_4)_{4-x}[H_xTiMo_{12}O_{40}]\cdot nH_2O$ precipitated on addition of concentrated NH_4Cl solution [1, 2]. They were washed with aqueous NH_4Cl solution and alcohol and then dried in air [1]. Alternatively, up to a fivefold excess of paramolybdate has been used in 0.1 N HCl solutions [3]. The compound can be recrystallized by dissolution in dilute H_2SO_4 and precipitation by an NH_4 salt [1].

While the value of x is not specified by [1, 3], it is given as 1 by [2] and as zero by [4, 5]. Yet from the precipitation conditions and the powder patterns the compounds presumably are identical. Apparently the content of water of crystallization has not been redetermined by chemical analyses. A formulation as the 10-hydrate is cited from older papers (see "Molybdän", 1935, p. 374), while the structure type to which the crystals are shown to belong is a pentahydrate structure for $(NH_4)_3[HTiMo_{12}O_{40}]$ [2].

If during the formation of the 12-molybdotitanate anion soluble phosphates [6, 7] or silicates [8] are also present, ternary heteropoly complexes are formed, see also [2].

X-ray powder patterns [2] show $(NH_4)_{4-x}[H_xTiMo_{12}O_{40}]\cdot nH_2O$ to be isomorphous with cubic $H_3PW_{12}O_{40}\cdot 5H_2O$ of the Keggin-type structure [9]; space group $Pn\bar{3}m$-O_h^4 (No. 224), lattice parameter a = 11.695 Å; Z = 2. Indexed $\sin^2\vartheta$ values are given in the paper. The measured density is 3.5 g/cm^3 [5].

The IR spectrum between 1200 and 600 cm^{-1} and the integrated apparent molar absorptivities of the five individual absorption bands plus their tentative assignments are given in the paper [10]. The absorption spectrum between 300 and 460 nm of aqueous molybdotitanate solution is similar to that of molybdate solution with a steep increase at wavelengths below 360 nm, for a plot see the paper [11].

$(NH_4)_4[TiMo_{12}O_{40}]\cdot nH_2O$ is sparingly [4, 10] soluble in H_2O, giving a yellow solution [1, 3]. Increasing the pH decomposes the heteropoly anion and the solution becomes colorless at pH 4.5 [1]. Titrating a suspension of the NH_4 salt with 4 equivalents of NaOH forms a soluble sodium molybdotitanate. On addition of a further 20 equivalents the heteropoly anion decomposes, precipitating titanium hydroxide [4]. In acid solutions decomposition begins in 0.45 N HCl (at a Ti:Mo mole ratio of 1:60) with precipitation of molybdenum trioxide [3]. In solution the molybdotitanate anion is reducible by $SnCl_2$ [11 to 13], hydrazine, ferrous ammonium sulfate, and other reducing agents [13] to a heteropoly blue. Reduced solutions show additional absorption bands at 660 to 740 nm (14250 cm^{-1} [12]) in the 500 to 900 nm range [11]. The reduced heteropoly blue can be used to determine titanium colorimetrically at 755 nm [13].

References:

[1] A. Liberti, G. Giombini, E. Cervone (Ric. Sci. **25** [1955] 883/90). – [2] J. W. Illingworth, J. F. Keggin (J. Chem. Soc. **1935** 575/80). – [3] Z. F. Shakhova, E. N. Semenovskaya (Zh. Neorgan. Khim. **7** [1962] 1084/6; Russ. J. Inorg. Chem. **7** [1962] 556/8). – [4] Z. F. Shakhova, E. N. Semenovskaya (Zh. Neorgan. Khim. **13** [1968] 1887/9; Russ. J. Inorg. Chem. **13** [1968] 983/4). – [5] A. Liberti, A. Santoro (Ric. Sci. **24** [1954] 2079/82).

[6] Yu. F. Shkaravskii (Zh. Neorgan. Khim. **11** [1966] 120/7; Russ. J. Inorg. Chem. **11** [1966] 64/9). – [7] Yu. F. Shkaravskii (Zh. Neorgan. Khim. **11** [1966] 797/802; Russ. J. Inorg. Chem. **11** [1966] 433/6). – [8] V. F. Barkovskii, T. L. Radovskaya (Zh. Neorgan. Khim. **12** [1967] 911/5; Russ. J. Inorg. Chem. **12** [1967] 522/5). – [9] J. F. Keggin (Proc. Roy. Soc. [London] A **144** [1934] 75/100). – [10] N. E. Sharpless, J. S. Munday (Anal. Chem. **29** [1957] 1619/22).

[11] T. L. Radovskaya, V. F. Barkovskii (Zh. Neorgan. Khim. **11** [1966] 1972/3; Russ. J. Inorg. Chem. **11** [1966] 1052/3). – [12] N. A. Alikina, V. F. Barkovskii, T. L. Radovskaya, V. S. Shvarev (Zh. Neorgan. Khim. **13** [1968] 1880/6; Russ. J. Inorg. Chem. **13** [1968] 979/83). – [13] J. C. Guyon, M. G. Mellon (Anal. Chem. **34** [1962] 856/9).

7.2.3 Molybdotitanates of Organic N-Bases

The compounds given in the table below have been precipitated from a solution of the molybdotitanic acid by an excess of HCl solutions of the bases. The precipitates were filtered after 12 h, washed with dilute HCl, methanol, and ether, and dried in air. The yellow crystalline compounds were stored in a desiccator over $CaCl_2$. Thermogravimetric analyses show the anhydrous compounds to be stable over small temperature ranges. Total decomposition to $TiO_2+12MoO_3$ takes place at temperatures >500°C.

base	formula	stability range of the anhydrous compound in °C
dimethylamine	$[(CH_3)_2NH_2]_4[TiMo_{12}O_{40}]\cdot nH_2O$	80 to 100
guanidine	$(CH_5N_3H)_4[TiMo_{12}O_{40}]\cdot nH_2O$	60 to 100
pyridine	$(C_5H_5NH)_4[TiMo_{12}O_{40}]\cdot nH_2O$	80 to 100
quinoline	$(C_9H_7NH)_4[TiMo_{12}O_{40}]\cdot nH_2O$	80 to 100
urotropine	$(C_6H_{12}N_4H)_4[TiMo_{12}O_{40}]\cdot nH_2O$	80 to 100
hydroxyquinoline	$(C_9H_7NOH)_4[TiMo_{12}O_{40}]\cdot nH_2O$	60 to 100
5,7-dibromo-8-hydroxy-quinoline	$(C_9H_5NOBr_2H)_4[TiMo_{12}O_{40}]\cdot nH_2O$	60 to 100
pyramidon	$(C_{13}H_{17}ON_3H)_3[HTiMo_{12}O_{40}]\cdot nH_2O$	80 to 120

All the compounds are slightly soluble in H_2O; only the dimethylamine compound dissolves easily. In strong bases and NH_3 solution the compounds decompose with precipitation of titanium oxide hydrate.

Z. F. Shakhova, E. N. Semenovskaya, O. V. Kuznetsova (Vestn. Mosk. Univ. Khim. **17** No. 4 [1962] 61/5; C.A. **58** [1963] 4517).

7.3 Zirconium Molybdate Hydrates

Older data are given in "Molybdän", 1935, p. 306.

7.3.1 $Zr(OH)_2Mo_2O_7\cdot 2H_2O$ ($=ZrO_2\cdot 2MoO_3\cdot 3H_2O$)

For the preparation of this compound aqueous reactant solutions, 200 mL of 0.5 M $ZrOCl_2\cdot 8H_2O$ and 200 mL of 1M $Na_2MoO_4\cdot 2H_2O$, were mixed by simultaneous dropwise addition to 100 mL of H_2O with continuous vigorous stirring. The slurry of precipitated gel and mother liquor was stirred overnight, then acidified by dropwise addition with continuous stirring of 500 mL of 6 to 8 M HCl, and refluxed for several days. Partial or sometimes complete dissolution of the gel occurred, and the solution turned greenish yellow. On continued heating a new precipitate appeared, a microcrystalline solid, which after cooling was separated and washed free of sodium by 1M HCl and free of chloride by H_2O. Clear, colorless single crystals were grown from 1g of this product in 10 mL of 4 M HCl during one to two months at 175°C in a sealed Pyrex glass tube [1].

A precipitate with Zr:Mo=1:2 also has been obtained by precipitation from hot homogeneous solutions. Gelatinous precipitates from Zr:Mo=1:2 mixtures of $ZrOCl_2$ and ammonium paramolybdate solutions were redissolved in 30% H_2O_2 solution, acidified to 0.5 M with HCl. Hot decomposition of the soluble peroxomolybdate gave a precipitate of good crystallinity [2].

The X-ray powder pattern of pulverized single crystals was identical to that of the microcrystalline product and agreed with the single crystal data [1]. Apparently there is also an agreement with the X-ray diffraction pattern of the product obtained by crystallization from homogeneous solution [2].

Single crystals of $Zr(OH)_2Mo_2O_7 \cdot 2H_2O$ are tetragonal. Their habit is characterized by an elongation along [001] and the forms {011}, {111}, and {100}. The lattice parameters at 25°C are $a = 11.45 \pm 0.01$, $c = 12.49 \pm 0.01$ Å; Z=8. Space group $I4_1cd$-C_{4v}^{12} (No. 110). The positional parameters of the nonhydrogen atoms are as follows (R=0.037):

atom	position	x	y	z
Mo(1)	16b	0.02471(5)	0.16384(5)	0.23016(11)
Zr(2)	8a	0	0	0.0
O(3)	16b	0.17488(47)	0.16151(69)	0.22254(71)
O(4)	16b	−0.17523(52)	0.15977(64)	0.20593(66)
O(5)	16b	0.18209(43)	−0.00294(56)	−0.00897(62)
O(6)	16b	0.00311(64)	0.11293(46)	−0.13980(52)
O(7)	16b	0.00680(56)	0.17731(47)	0.06225(54)
O(8)	8a	0	0	0.17140(64)

The basic structural feature is a unit of a pentagonal bipyramidal ZrO_7 coordination polyhedron, which shares two neighboring equatorial edges (O(8)-O(7)) with two slightly distorted MoO_6 octahedra, see **Fig. 80**. These units are linked by two common O atoms into chains along the twofold axes parallel to [001]. The three-dimensional structure is built up from the chains by corner sharing (O(5)) and by hydrogen bonds. The Zr-O distances range from 2.088 to 2.175 Å about a mean value of 2.145 Å. The Mo atom is displaced from the center of the surrounding oxygen octahedron approximately along the pseudo threefold axis. This gives three shorter and three longer Mo-O distances: 1.722(6), 1.755(7), and 1.797(5) Å to O(3), O(6), and O(5) plus 2.034(3), 2.113(7), and 2.310(6) to O(8), O(7), and O(4), respectively. From chemical and geometric arguments the two O atoms with the longest bonds to Mo are considered to be water (O(4)) and hydroxyl oxygen atoms (O(7)). This assignment establishes zig-zag chains of hydrogen bonds O(3)···H-O(4)-H···O(7)-H···O(3)··· parallel to [100]. The O-O distances along the hydrogen bonds are O(4)-O(3)=2.68, O(4)-O(7)=2.77, and O(7)-O(3)=2.80 Å. Thus the crystal chemical formula can be written as $ZrO_5(OH)_2[MoO(H_2O)]_2$ in accordance with the IR spectrum (see below) [1].

The measured density is 4.02 ± 0.14 at 25°C, the calculated density 3.774 g/cm³ [1].

In the IR absorption spectrum the very sharp and strong band at 3340 cm^{-1} is assigned to the OH stretching vibration, the broad band at 3320 cm^{-1} to the H_2O stretching vibration, and the sharp medium band at 1665 cm^{-1} to the H_2O bending vibration. The five bands between 1035 and 725 cm^{-1} are regarded as superpositions of metal-oxygen and various H_2O vibrations [1]. The absorption spectra of [1] and [2] differ significantly only in an absorption band at about 1400 cm^{-1} in the spectrum of [2] (for a plot see the paper).

^{1}H NMR spectra at temperatures between 93 and 423 K of air-dried and partially dehydrated samples showed that all protons are in a rigid lattice below 110 K. The height of the rotational barrier between different equilibrium positions of H_2O at higher temperatures was calculated to be 7.5 ± 0.4 kcal/mol. From the second moments distances H-H $= 1.45 \pm 0.02$ Å have been calculated, which are anomalously short for crystal hydrates. This is thought to be due to interaction with other O atoms which are close to the H_2O molecule. The distance between singular protons (not associated with H_2O) is 2.23±0.02 Å, from secondary moments [3]. The ^{1}H NMR results agree with those of the crystallographic and IR investigations of [1].

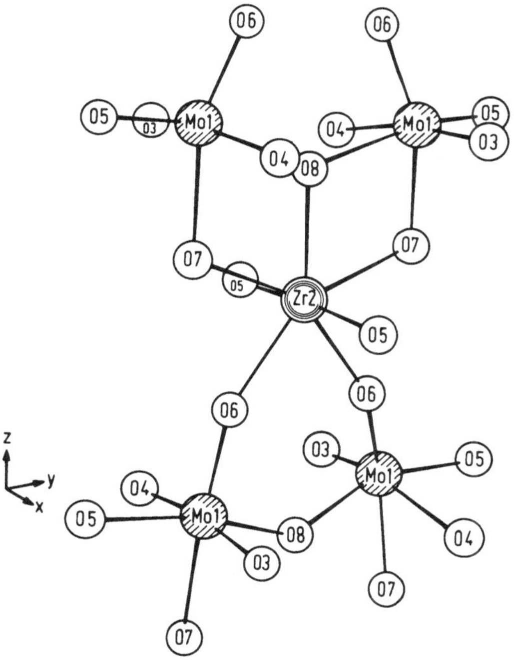

Fig. 80. Characteristic structural unit of $Zr(OH)_2Mo_2O_7 \cdot 2H_2O$ [1].

On heating, the compound is dehydrated. In thermogravimetric experiments weight loss starts at about 200°C with nearly complete dehydration at about 350°C. Dehydration was also observed on drying to constant weight at 100°C for some 3 d [1, 3]. Hydroxyl groups are not degraded below 150°C [3]. Decomposition products are $ZrMo_2O_8$ and at least one other as yet unidentified compound [1].

Solid $Zr(OH)_2Mo_2O_7 \cdot 2H_2O$ has been titrated with sodium hydroxide solution at 25°C. Endpoints were reached only after several days at 9.72 mequiv. OH^-/g by direct titration and 8.97 mequiv. OH^-/g on back titration. At this point the material was X-ray amorphous. A chemical reaction according to $Zr(OH)_2Mo_2O_7 \cdot 2H_2O(s) + 4OH^- + (x+y-5)\ H_2O \rightarrow ZrO_{2-x}(OH)_{2x}(H_2O)_y(s) + 2MoO_4^{2-}$ would correspond to a neutralization capacity of 8.6 mequiv. OH^-/g [1], see also [2]. Partition coefficients for Ca, Sr, Ba, Zn, and Cd were determined and plotted against pH [2].

References:

[1] A. Clearfield, R. H. Blessing (J. Inorg. Nucl. Chem. **34** [1972] 2643/63). – [2] M. I. Gómez del Rio, P. Sánchez-Batanero, F. Burriel Marti (Quim. Anal. **29** [1975] 147/52). – [3] M. V. Mokhosoev, L. A. Pozharskaya, V. G. Pitsyuga, S. S. Palitsyna, T. I. Loboda (Deposited Doc. VINITI-3391-76 [1976] 1/13; C.A. **89** [1978] No. 155846).

7.3.2 Undefined Zirconium Molybdates

Formation of amorphous precipitates with composition $ZrO_2 \cdot MoO_3 \cdot 3.7\,H_2O$ after drying in air has been reported on mixing equimolar amounts of aqueous $ZrOCl_2$ and Na_2MoO_4 solutions [1]. ^{1}HNMR and IR spectroscopy indicate the presence of three types of deformed water molecules with different bond energies [2]. A refractive index of n=1.767 is given for the optically isotropic crystals. TG and DTA curves show an endothermal effect at 160°C due to loss of H_2O [1].

Zirconium molybdates of unknown composition have been obtained by precipitation from acidic aqueous solutions of $ZrOCl_2$ (or $ZrO(NO_3)_2$) and ammonium paramolybdate, $(NH_4)_6Mo_7O_{24}$. The precipitates have been used as ion exchangers, see for example [3 to 6].

From the influence of both zirconium salts ($ZrOCl_2$) and molybdate ($(NH_4)_2MoO_4$) on the rate of the oxidation of I^- by H_2O_2 in an acidic medium at 25°C the formation of a soluble complex with the supposed composition $[ZrMoO_4]^{2+}$ was postulated at low concentrations of ZrO^{2+} (~10^{-6} M) and MoO_4^{2-} (~10^{-8} M). The stability constant is calculated from optical density measurements as 1.3×10^9 [7].

References:

[1] T. I. Loboda, V. I. Krivobok, T. T. Got'manova, M. V. Mokhosoev (Khim. Tekhnol. Molibdena Vol'frama **1978** No. 4, pp. 161/7; Ref. Zh. Khim. **1979** No. 5 B 884). – [2] M. V. Mokhosoev, L. A. Pozharskaya, V. P. Pitsyuga (Dokl. Akad. Nauk SSSR **251** [1980] 392/6; Dokl. Phys. Chem. Proc. Acad. Sci. USSR **250/255** [1980] 203/6). – [3] N. E. Denisova, E. S. Boichinova, A. A. Peregudova (Zh. Prikl. Khim. **39** [1966] 1235/41; J. Appl. Chem. [USSR] **39** [1966] 1159/64). – [4] N. E. Denisova, E. S. Boichinova (Izv. Akad. Nauk SSSR Neorgan. Materialy **3** [1967] 1049/54; Inorg. Materials [USSR] **3** [1967] 933/7). – [5] S. S. Rodin, A. K. Lavrukhina (Radiokhimiya **4** [1962] 623/4; Soviet Radiochem. **4** [1962] 548/9).

[6] H. J. Riedel (Nukleonik **5** [1963] 48/54). – [7] K. B. Yatsimirskii, L. P. Raizman (Zh. Neorgan. Khim. **8** [1963] 1107/11; Russ. J. Inorg. Chem. **8** [1963] 574/6).

7.4 Molybdozirconates

Older data are given in "Molybdän", 1935, p. 375.

7.4.1 $H_4ZrMo_{12}O_{40} \cdot 10\,H_2O$ (= $ZrO_2 \cdot 12\,MoO_3 \cdot 12\,H_2O$)

Clear yellow solutions of $H_4ZrMo_{12}O_{40}$ have been prepared from suspensions of the ammonium salt (see below) by ion exchange [1, 2]. On drying in a vacuum desiccator the solid free acid precipitates as extremely hygroscopic brown crystals [2]. In aqueous solution the acid is tetrabasic, dissociation constant ~9.8×10^{-4} [1].

Thermal analyses show the solid acid to contain 10 mol H_2O. The anhydrous acid can be obtained between 20 and 40°C. With increasing temperature the constitutional water is lost and above 360°C decomposition to ZrO_2 and MoO_3 appears [3].

References:

[1] A. Liberti, G. Giombini, E. Cervone (Ric. Sci. **25** [1955] 883/90). – [2] Z. F. Shakhova, E. N. Semenovskaya, E. N. Timofeeva (Zh. Neorgan. Khim. **6** [1961] 330/3; Russ. J. Inorg. Chem. **6** [1961] 168/70). – [3] Z. F. Shakhova, E. N. Semenovskaya, E. N. Timofeeva (Vestn. Mosk. Univ. Khim. **17** No. 1 [1962] 55/9; C.A. **57** [1962] 1829).

7.4.2 $(NH_4)_{4-x}[H_xZrMo_{12}O_{40}]\cdot nH_2O$ ($=(2-x/2)(NH_4)_2O\cdot ZrO_2\cdot 12MoO_3\cdot(n+x/2)H_2O$), x = 0, 1; n = 5, 10

This yellow compound has been obtained in acidic solution by reaction of $(NH_4)_2ZrF_6$ and ammonium molybdate and subsequent precipitation with concentrated NH_4Cl solution, analogous to the preparation of the 12-molybdotitanate salt (p. 293) [1 to 3]. Yields of 46% have been reported [3]. The compound can be recrystallized from solution in dilute H_2SO_4 by addition of NH_4Cl [4]. The salt is formulated as the tetraammonium salt (x = 0) [2 to 4] and as containing 10 mol H_2O, determined by thermogravimetric analysis [4]. There is also the formulation as the pentahydrate of the acidic triammonium salt, which is not a result of chemical analysis but deduced from the observed isomorphism with $H_3PW_{12}O_{40}\cdot 5H_2O$ and the salts $M^I_3[PM^{VI}_{12}O_{40}]\cdot nH_2O$, M^{VI} = Mo, W [1]. Ternary heteropoly compounds are easily formed in the presence of phosphate [5, 6], see also [1].

The X-ray powder diffraction pattern shows $(NH_4)_4[ZrMo_{12}O_{40}]\cdot nH_2O$ to be isomorphous with the corresponding Ti compound (p. 294), and thus also with cubic $H_3PW_{12}O_{40}\cdot 5H_2O$ of the Keggin type; space group $Pn\bar{3}m\text{-}O_h^4$ (No. 224) [1, 7]. The lattice parameter a = 11.694 Å; Z = 2. The measured density is 3.6 g/cm³ [7].

The decahydrate can be thermally dehydrated. The anhydrous compound is stable between 20 and 80°C. The decomposition to the oxides takes place above 500°C [4].

Aqueous solutions of $(NH_4)_4[ZrMo_{12}O_{40}]\cdot nH_2O$ have been studied by potentiometry, polarography, and photometry at different pH values. The 12-molybdozirconate anion is stable in acidic solutions with pH values up to 4.7, and decomposes at higher basicities [2]. It is also decomposed by HCl of concentrations >0.22 N with separation of molybdenum oxide hydrate [3]. In solution $(NH_4)_4[ZrMo_{12}O_{40}]$ is reducible by ascorbic acid to a blue product. The reaction kinetics has been studied between 3 and 45°C and the influence of the presence of the metal ions Fe^{3+}, Sb^{3+}, Bi^{3+}, and Cu^{2+} on the reduction rate and reduction orders has been determined. The reaction rate increases in the presence of these ions. From Arrhenius plots the activation energy has been found to be 13.9 ± 0.8 kcal/mol without Fe^{3+} and remains unchanged within the given standard deviation in the presence of Fe^{3+} [8].

References:

[1] J. W. Illingworth, J. F. Keggin (J. Chem. Soc. **1935** 575/80). – [2] A. Liberti, G. Giombini, E. Cervone (Ric. Sci. **25** [1955] 883/90). – [3] Z. F. Shakhova, E. N. Semenovskaya, E. N. Timofeeva (Zh. Neorgan. Khim. **6** [1961] 330/3; Russ. J. Inorg. Chem. **6** [1961] 168/70). – [4] Z. F. Shakhova, E. N. Semenovskaya, E. N. Timofeeva (Vestn. Mosk. Univ. Khim. **17** No. 1 [1962] 55/9; C. A. **57** [1962] 1829). – [5] Yu. F. Shkaravskii (Zh. Neorgan. Khim. **11** [1966] 120/7; Russ. J. Inorg. Chem. **11** [1966] 64/9).

[6] Yu. F. Shkaravskii (Zh. Neorgan. Khim. **11** [1966] 797/802; Russ. J. Inorg. Chem. **11** [1966] 433/6). – [7] A. Liberti, A. Santoro (Ric. Sci. **24** [1954] 2079/82). – [8] N. K. Sokovikova, E. N. Semenovskaya, I. P. Alimarin (Zh. Analit. Khim. **30** [1975] 2365/71; J. Anal. Chem. [USSR] **30** [1975] 1988/92).

7.4.3 Molybdozirconates of Organic N-Bases

The compounds given in the table below have been precipitated from a solution of the molybdozirconic acid by addition of acidic solutions of the bases. The yellow crystalline precipitates were filtered off after 12 h, washed with cold water, dilute HCl, and methanol, and dried in air. The methylamine compound could not be prepared by this method because it is very easily soluble in aqueous solutions. Thermal analyses show that the compounds can be dehydrated. Decomposition to $ZrO_2+12MoO_3$ occurs at temperatures >500°C.

base	formula	stability range of the anhydrous compound in °C
guanidine	$(CH_5N_3H)_4[ZrMo_{12}O_{40}]\cdot nH_2O$	120 to 200
pyridine	$(C_5H_5NH)_4[ZrMo_{12}O_{40}]\cdot nH_2O$	120 to 180
quinoline	$(C_9H_7NH)_4[ZrMo_{12}O_{40}]\cdot nH_2O$	160 to 220
urotropine	$(C_6H_{12}N_4H)_4[ZrMo_{12}O_{40}]\cdot nH_2O$	120 to 150
8-hydroxyquinoline	$(C_9H_7NOH)_4[ZrMo_{12}O_{40}]\cdot nH_2O$	200 to 260
5,7-dibromo-8-hydroxyquinoline	$(C_9H_5NOBr_2H)_4[ZrMo_{12}O_{40}]\cdot nH_2O$	200 to 240

All the compounds are slightly soluble in H_2O. Strong bases and ammonia decompose them, precipitating zirconium oxide hydrate.

Z. F. Shakhova, E. N. Semenovskaya, E. N. Timofeeva (Vestn. Mosk. Univ. Khim. **17** No. 1 [1962] 55/9; C.A. **57** [1962] 1829).

7.4.4 $Cs_4[ZrMo_{12}O_{40}]\cdot 10H_2O$ (=$2Cs_2O\cdot ZrO_2\cdot 12MoO_3\cdot 10H_2O$)

This Cs salt has been prepared by a method analogous to that used for the corresponding NH_4 salt (see p. 299). It forms a greenish yellow microcrystalline powder [1].

The compound can be thermally dehydrated. The anhydrous form is stable between 20 and 80°C. It decomposes into the oxides above 300°C [2].

References:

[1] Z. F. Shakhova, E. N. Semenovskaya, E. N. Timofeeva (Zh. Neorgan. Khim. **6** [1961] 330/3; Russ. J. Inorg. Chem. **6** [1961] 168/70). – [2] Z. F. Shakhova, E. N. Semenovskaya, E. N. Timofeeva (Vestn. Mosk. Univ. Khim. **17** No. 1 [1962] 55/9; C.A. **57** [1962] 1829).

7.5 Hafnium Molybdate Hydrates

7.5.1 $Hf(OH)_2Mo_2O_7\cdot 2H_2O$ (=$HfO_2\cdot 2MoO_3\cdot 3H_2O$)

This compound has been obtained by mixing and stirring at 25°C for 12 h solutions of 0.5 M $HfOCl_2$ and 1M Na_2MoO_4 and then refluxing the acidified (4 to 6 M HCl) suspension for several days following the procedure described for the analogous Zr compound (p. 295). Chemical analysis gives the content of water after drying over $CaCl_2$ as higher than that shown by the formula, probably due to adsorbed water (4.2 to 5.7 H_2O).

Assuming isomorphism with the tetragonal $Zr(OH)_2Mo_2O_7\cdot 2H_2O$ the X-ray powder pattern of $Hf(OH)_2Mo_2O_7\cdot 2H_2O$ could be indexed with the lattice parameters a=11.45, c=12.42 Å. For d values see the paper, for details of the crystal structure see the Zr compound.

The ^1H NMR spectrum, recorded at 93 K, confirmed the existence of two different types of protons. The IR spectrum recorded between 3700 and 400 cm^{-1} contains weak and poorly resolved absorption bands between 1200 and 1000 cm^{-1} attributed to deformation vibrations of cation-coordinated hydroxyl groups. The water of crystallization gives a broad band in the 3600 to 3000 cm^{-1} range having maxima at 3210 and 3400 cm^{-1} assigned to the symmetric and antisymmetric stretching vibrations. The H_2O bending mode is observed at 1655 cm^{-1}. For a plot of the spectrum see the paper. The crystals are optically isotropic with a refractive index n = 1.820.

Upon heating, H_2O is lost in two steps at 170 and 240°C, which are poorly resolved in DTA curves. The anhydrous product crystallizes at 480°C, forming $Hf(MoO_4)_2$.

T. I. Loboda, V. I. Krivobok, M. V. Mokhosoev, T. T. Got'manova (Zh. Neorgan. Khim. **23** [1978] 3006/9; Russ. J. Inorg. Chem. **23** [1978] 1669/71).

7.5.2 $HfO_2 \cdot MoO_3 \cdot 4.6\,H_2O$

A sample of this composition has presumably been prepared similarly to the $ZrO_2 \cdot MoO_3 \cdot 3.7\,H_2O$, see p. 298. ^1H NMR and IR spectroscopy indicate the presence of three types of deformed water molecules with different bond energies.

M. V. Mokhosoev, L. A. Pozharskaya, V. G. Pitsyuga (Dokl. Akad. Nauk SSSR **251** [1980] 392/6; Dokl. Phys. Chem. Proc. Acad. Sci. USSR **250/255** [1980] 203/6).

7.6 Molybdohafnates in Solutions

Neither a molybdohafnic acid nor one of its salts has so far been isolated as a pure solid phase. Until now they have only been obtained in solutions.

Optimum conditions for the formation of 12-molybdohafnate from $(NH_4)_2HfF_6$ in aqueous solution (13.5 g Hf/L) are a 14-fold excess of $(NH_4)_6Mo_7O_{24}$ at pH 0.8 and 0.4 mL of 0.28 M H_3BO_3 solution per 13.5 mg Hf for the fixation of fluoride ions. The reaction is complete after 10 minutes at room temperature. The molar ratio Zr : Mo = 1 : 12 was established by chemical analysis (after extraction of the reduced form by n-butanol). The absorption spectrum shows a maximum at 360 nm in the 330 to 450 nm range. The heteropoly acid is most stable in H_2SO_4 (up to ~15 N) and least stable in $HClO_4$ (up to ~2 N) compared to hydrochloric, nitric, and acetic acid [1].

The 12-molybdohafnate ion is reducible to a blue-green form by aqueous solutions of Sn^{II}, Mo^{V} compounds, ascorbic acid, or metallic Mo [1 to 3]. From very acidic solutions (between 0.78 and 15 N in H_2SO_4) the reduced 12-molybdohafnic acid can be extracted by O-containing organic solvents such as n-butanol, isoamyl alcohol, methylethylketone, and various mixtures of these with benzene [1, 2]. The absorption spectrum of the reduced form has a maximum at 720 to 740 nm [2, 3] in aqueous solution, which varies between 750 and 820 nm in organic solvents [2]. The reduced 12-molybdohafnate has therefore been used for the photometric determination of Hf down to concentrations of 0.4 mg Hf/L [2, 3]. The reduction rate increases in the presence of Fe^{3+}, Cu^{2+}, Sb^{3+}, and Bi^{3+}. The different influence of Fe^{3+} on the reduction rate of 12-molybdozirconate and 12-molybdohafnate allows the photometric determination of Hf in the presence of a fourfold excess of Zr [7]. The formation of ternary heteropoly complexes in the presence of soluble phosphates [3 to 5], silicates [1, 3], Al^{3+} [3, 6, 7], or Fe^{3+} [3, 7, 8] has been postulated.

References:

[1] Z. F. Shakhova, E. N. Semenovskaya, N. K. Sokovikova (Zh. Analit. Khim. **25** [1970] 485/9; J. Anal. Chem. [USSR] **25** [1970] 416/9). – [2] Z. F. Shakhova, E. N. Semenovskaya, N. K. Sokovikova, V. A. Koval'chuk (Zh. Analit. Khim. **25** [1970] 490/4; J. Anal. Chem. [USSR] **25** [1970]

420/3). – [3] Z. F. Shakhova, E. N. Semenovskaya, N. K. Sokovikova (Zh. Analit. Khim. **23** [1968] 1164/8; J. Anal. Chem. [USSR] **23** [1968] 1023/6). – [4] Yu. F. Shkaravskii, Ya. F. Ometsinskaya (Ukr. Khim. Zh. **32** [1966] 1023/6; Soviet Progr. Chem. **32** [1966] 768/71). – [5] Yu. F. Shkaravskii (Zh. Neorgan. Khim. **11** [1966] 797/802; Russ. J. Inorg. Chem. **11** [1966] 433/6).

[6] E. N. Semenovskaya, N. K. Sokovikova, O. K. Primachek (Vestn. Mosk. Univ. Khim. **25** [1970] 367/8; Moscow Univ. Chem. Bull. **25** No. 3 [1970] 69/70). – [7] N. K. Sokovikova, E. N. Semenovskaya, I. P. Alimarin (Zh. Analit. Khim. **30** [1975] 2365/71; J. Anal. Chem. [USSR] **30** [1975] 1988/92). – [8] I. P. Alimarin, E. N. Semenovskaya, N. K. Sokovikova (Zh. Analit. Khim. **26** [1971] 126/30; J. Anal. Chem. [USSR] **26** [1971] 108/12).

7.7 Thorium Molybdate Hydrates

Older data for $Th(MoO_4)_2 \cdot nH_2O$ are given in "Molybdän", 1935, pp. 306/7. More recent data for $Th(MoO_4)_2 \cdot nH_2O$ and $Th_3(Mo_7O_{24})_2 \cdot nH_2O$ are given in "Thorium" Erg.-Bd. C 2, 1976, p. 133.

7.8 Molybdothorates

Older data are given in "Molybdän", 1935, p. 375.

7.8.1 $H_8ThMo_{12}O_{42} \cdot nH_2O$ ($=ThO_2 \cdot 12MoO_3 \cdot (n+4)H_2O$), n = 0, 2, 6, 8, 11, 18

Crystals of $H_8ThMo_{12}O_{42} \cdot 18H_2O$ have been precipitated from 2 to 4% solutions of the acid by rapid addition with stirring of an equal volume of 60% H_2SO_4. The change in the acidity has to be as fast as possible, as $H_8ThMo_{12}O_{42}$ decomposes in 3% H_2SO_4 solution. It can be precipitated from 30% H_2SO_4 with no appreciable decomposition, even when heated to 90°C. The precipitate is washed with concentrated HNO_3 and ether. The yield is nearly quantitative [1, 2]. Aqueous solutions of $H_8ThMo_{12}O_{42}$ are obtained by ion exchange from a suspension of the ammonium salt $(NH_4)_8[ThMo_{12}O_{42}] \cdot nH_2O$ (p. 305) [1, 3]. The 18-hydrate is dehydrated to an 11-hydrate by storing over P_2O_5 and further by heating at 100, 120, 140, and 175°C to samples with n = 8, 6, 2, and 0, respectively [4].

From the similarity of the X-ray powder patterns [1] and of other physical properties [2, 4 to 6], the structure of $H_8ThMo_{12}O_{42} \cdot 18H_2O$ is regarded as isomorphous with the analogous Ce compound (cf. p. 283). For d values see the paper [1]. There is also a rhombohedral unit cell with the hexagonal lattice parameters a = 12, c = 16 Å mentioned for $H_8ThMo_{12}O_{42}$ and the analogous Ce and U compounds [6]. (The powder pattern and the unit cell are not consistent.) The structure of the $[MMo_{12}O_{42}]^{8-}$ anion, M = Ce, Th, is of the Dexter-Silverton type and is described for the Ce compound.

^1H NMR spectra of $H_8ThMo_{12}O_{42} \cdot 18H_2O$ and of partially dehydrated samples ($n \approx 8$ to 10) have been recorded at 80 K. The resonance signal can be split into three components. The narrow part with a line width of $\Delta H = 2$ G has been attributed to protons free of dipole interaction with H_2O. From the second moment ($\Delta H^2 = 1.5$ G^2) shortest interproton distances of 2.5 Å have been calculated. The broader components correspond to protons of H_2O ($\Delta H = 11$ G) and H_3O^+ ($\Delta H = 19$ G). From their relative intensities the 18-hydrate has been formulated as $(H_3O)_6[ThMo_{12}O_{40}(OH)_2] \cdot 12H_2O$ [4, 7]. In partially dehydrated samples the intensity of the H_2O line is increased, indicating a growing number of H_2O protons and a decrease in H_3O^+ protons. The NMR signal of a sample of $H_8ThMo_{12}O_{42}$ consists only of an H_2O line and a narrow line of low intensity [4].

XPS spectra yield binding energies (given in eV relative to C 1s = 285 eV): Th $4f_{7/2}$ = 335.7, Mo $3d_{5/2}$ = 233.5, Mo $3d_{3/2}$ = 236.6, and O 1s = 531.3. There is a small satellite mentioned with ΔE = 5.9 eV at the high energy side of the Th $4f_{7/2}$ line. The differences relative to the binding energies in other heteropoly compounds are discussed. Due to the UHV conditions (10^{-8} Torr) during the measurement the state of hydration of the sample is not defined [6].

IR and Raman spectra are reported for $H_8MMo_{12}O_{42} \cdot 18H_2O$ with M = Th. The frequencies of the following table are valid also for M = Ce and U; assignments have been made assuming the Dexter-Silverton structure [5]:

Raman	IR	assignments
—	1710	$\delta(H_3O^+)$
—	1670	$\delta(H_2O)$ of two types of water
—	1631	
976	970	ν_1 and ν_3 (cis-MoO_2)
957	—	
933	934	
883	—	ν_3(Mo-O-Mo)
863	860	
—	718	
—	620	Mo-O vibration
563	—	ν_1 (Mo-O-Mo)
513	—	
488	480	various Mo-O vibrations
—	462	
—	410	
388	—	
353	—	cis-MoO_2 bending vibration
285	—	ν_2 (Mo-O-Mo)
243	—	
180	—	

Raman spectra of 0.1M aqueous solutions are similar to those of the solid 18-hydrate [5].

The absorption of light in the UV range depends on the pH of the solution. There is strong absorption at shorter wavelengths, and a shoulder at λ = 248 nm in solutions of pH 3.6 and lower becomes a relative maximum at pH 4.8 to 10.3 [7]. The observation of an isosbestic point (at λ = 238 nm) indicates that there is a pH-dependent equilibrium between two species, $[ThMo_{12}O_{40}(OH)_2]^{6-}$ and $[ThMo_{12}O_{42}]^{8-}$ [2, 7]. The UV spectrum has been interpreted on the basis of an MO scheme for the cis-MoO_2O_4' octahedron (cf. p. 285) of C_{2v} symmetry [8], see also [9].

Crystals of the 18-hydrate decompose after several months of storage. In DTA experiments loss of H_2O starts at 60°C with maximum rate at 175°C. Weight becomes constant at 445°C. Prolonged heating at 125°C leads to an insoluble solid with the composition $ThO_2 \cdot 12MoO_3 \cdot 4H_2O$, while brief heating to 200°C gives the same composition in a water-soluble form, which probably is no longer a heteropoly acid [7].

$H_8ThMo_{12}O_{42}$ acts as an octabasic acid in aqueous solutions [1, 7]. A pH of 2 was measured for 0.004 M solutions [3]. The anion is remarkably stable towards bases and decomposes at pH

14 only after 2 h at room temperature [7]. Dilute acid solutions decompose it with precipitation of thorium oxide hydrate and partial reduction of molybdenum [3]. In 0.002 M aqueous solution $H_8ThMo_{12}O_{42}$ decomposes within a week with formation of $H_4Mo_8O_{26}$ and precipitation of polynuclear $Th_3Mo_{12}O_{42} \cdot 29H_2O$ (p. 306) [7]. Soluble complexes of 12-molybdothorate form with such metal ions as Mn, Fe, Co, Ni, Cu, Zn, Cd, Y, Er, Yb, Nb, and Ta [2].

References:

[1] P. Bajdala, E. A. Torchenkova, V. I. Spitsyn (Dokl. Akad. Nauk SSSR **196** [1971] 1344/5; Dokl. Chem. Proc. Acad. Sci. USSR **196/201** [1971] 137/8). – [2] V. I. Spitsyn, E. A. Torchenkova, L. P. Kazanskii, P. Bajdala (Z. Chem. [Leipzig] **14** [1974] 1/8). – [3] Z. F. Shakhova, S. A. Gavrilova, V. F. Zakharova (Zh. Neorgan. Khim. **7** [1962] 1752/3; Russ. J. Inorg. Chem. **7** [1962] 904/5). – [4] V. F. Chuvaev, P. Bajdala, E. A. Torchenkova, V. I. Spitsyn (Dokl. Akad. Nauk SSSR **196** [1971] 1097/100; Dokl. Chem. Proc. Acad. Sci. USSR **196/201** [1971] 131/4). – [5] L. P. Kazanskii, E. A. Torchenkova, V. I. Spitsyn (Dokl. Akad. Nauk SSSR **209** [1973] 141/3; Dokl. Phys. Chem. Proc. Acad. Sci. USSR **208/213** [1973] 209/11).

[6] I. V. Tat'yanina, E. A. Torchenkova, L. P. Kazanskii, V. I. Spitsyn (Dokl. Akad. Nauk SSSR **234** [1977] 1136/9; Dokl. Phys. Chem. Proc. Acad. Sci. USSR **232/237** [1977] 597/600). – [7] E. A. Torchenkova, P. Bajdala, V. S. Smurova, V. I. Spitsyn (Dokl. Akad. Nauk SSSR **199** [1971] 120/3; Dokl. Chem. Proc. Acad. Sci. USSR **196/201** [1971] 568/70). – [8] L. P. Kazanskii, E. A. Torchenkova, V. I. Spitsyn (Dokl. Akad. Nauk SSSR **213** [1973] 118/21; Dokl. Phys. Chem. Proc. Acad. Sci. USSR **208/213** [1973] 932/4). – [9] A. L. Vil'dt, L. P. Tsyganok (Zh. Neorgan. Khim. **21** [1976] 1835/9; Russ. J. Inorg. Chem. **21** [1976] 1005/8).

7.8.2 $Na_8[ThMo_{12}O_{42}] \cdot nH_2O$ ($=4Na_2O \cdot ThO_2 \cdot 12MoO_3 \cdot nH_2O$)

This sodium salt precipitates on neutralizing an $H_8ThMo_{12}O_{42}$ solution with excess NaOH. In the solution at pH $\approx$14 the compound begins to decompose with liberation of thorium oxide hydrate after 2 h. The water content of the sodium salt has not been determined [1].

XPS gives the binding energies of Mo $3d_{5/2}$ and O 1s electrons in $Na_8[ThMo_{12}O_{42}]$ as 232.7 and 530.6 eV, respectively (standard C 1s = 285.0 eV). The values are compared with other values of normal, isopoly, and heteropoly molybdates [2].

References:

[1] E. A. Torchenkova, P. Bajdala, V. S. Smurova, V. I. Spitsyn (Dokl. Akad. Nauk SSSR **199** [1971] 120/3; Dokl. Chem. Proc. Acad. Sci. USSR **196/201** [1971] 568/70). – [2] V. N. Molchanov, L. P. Kazanskii, E. A. Torchenkova, V. I. Spitsyn (Izv. Akad. Nauk SSSR Ser. Khim. **1978** 1248/51; Bull. Acad. Sci. USSR Div. Chem. Sci. **27** [1978] 1085/7).

7.8.3 $K_8[ThMo_{12}O_{42}] \cdot nH_2O$ ($=4K_2O \cdot ThO_2 \cdot 12MoO_3 \cdot nH_2O$)

This potassium salt has been prepared from the free acid [1]. Anhydrous $K_8[ThMo_{12}O_{42}]$ has been investigated by XPS at a pressure of 10^{-8} Torr in the analyzer. The binding energies (in eV relative to C 1s = 285.0 eV) are given as: Th $4f_{7/2}$ = 335.2, Mo $3d_{5/2}$ = 232.9, Mo $3d_{3/2}$ = 235.9, O 1s = 530.6, and K 1p = 293.1 ± 0.1. A satellite line at ΔE = 5.9 eV was observed at the high energy side of the Th $4f_{7/2}$ peak [2].

References:

[1] P. Bajdala, E. A. Torchenkova, V. I. Spitsyn (Dokl. Akad. Nauk SSSR **196** [1971] 1344/5; Dokl. Chem. Proc. Acad. Sci. USSR **196/201** [1971] 137/8). – [2] I. V. Tat'yanina, E. A. Torchenkova, L. P. Kazanskii, V. I. Spitsyn (Dokl. Akad. Nauk SSSR **234** [1977] 1136/9; Dokl. Phys. Chem. Proc. Acad. Sci. USSR **232/237** [1977] 597/600).

7.8.4 $(NH_4)_5[H_3ThMo_{12}O_{42}] \cdot 11\,H_2O$ $(=5(NH_4)_2O \cdot 2ThO_2 \cdot 24MoO_3 \cdot 25H_2O)$

This acidic ammonium salt has been prepared by stirring a slurry of $(NH_4)_8[ThMo_{12}O_{42}] \cdot 8H_2O$ in a 2.5% aqueous solution of $H_8ThMo_{12}O_{42}$. The presence of three acidic protons was confirmed by analysis of the neutralization curve. A line diagram of the X-ray powder pattern is given in the paper. Prolonged standing (~5 h) of a saturated solution of $(NH_4)_5[H_3ThMo_{12}O_{42}]$ at 50 to 60°C gave $(NH_4)_4[Th_2Mo_{12}O_{42}] \cdot 14H_2O$ (see p. 306).

E. A. Torchenkova, P. Bajdala, V. S. Smurova, V. I. Spitsyn (Dokl. Akad. Nauk SSSR **199** [1971] 120/3; Dokl. Chem. Proc. Acad. Sci. USSR **196/201** [1971] 568/70).

7.8.5 $(NH_4)_8[ThMo_{12}O_{42}] \cdot nH_2O$ $(=4(NH_4)_2O \cdot ThO_2 \cdot 12MoO_3 \cdot nH_2O)$, n = 7, 8

$(NH_4)_8[ThMo_{12}O_{42}] \cdot 8H_2O$ has been obtained by dropwise addition with stirring of 50 mL of 2.5% $Th(SO_4)_2$ solution to a solution of 9 g $(NH_4)_6Mo_7O_{24}$ in 30 mL of H_2O. The amorphous precipitate was crystallized by boiling the mixture and then holding it on a water bath for 30 to 40 min. After cooling, the mixture was left for 12 h and filtered. The precipitate was washed with aqueous 1% NH_4NO_3, ethanol, and dried in air; yield 44% [1]. Instead of thorium sulfate also the nitrate can be used. The hexagonal crystals obtained from the mother liquor contain $7H_2O$ as water of crystallization. Needles of an octahydrate are obtained by supersaturation with NH_4NO_3 of an aqueous $H_8ThMo_{12}O_{42}$ solution neutralized with aqueous ammonia. The yield is nearly quantitative [2]. The compound can be recrystallized from 0.1 N HCl by the addition of saturated $(NH_4)_2SO_4$ solution [1].

For a line diagram of the X-ray powder pattern of the heptahydrate see the paper [2]. IR and Raman spectra are reported for $(NH_4)_8[MMo_{12}O_{42}] \cdot nH_2O$ with M = Ce, Th, and U (one set of frequencies for all three compounds). The spectra are similar to those of the solid acid (cf. p. 303). The differences in the spectra are not discussed. Assignments have been made on the basis of the structure of the Dexter-Silverton type. For a table of the frequencies see the paper [3]. The EPR spectrum of γ-irradiated $(NH_4)_8[ThMo_{12}O_{42}] \cdot 8H_2O$ is different from those with reduced Keggin-type ions $[MM'_{12}O_{40}]^{n-8}$ [4].

References:

[1] Z. F. Shakhova, S. A. Gavrilova, V. F. Zakharova (Zh. Neorgan. Khim. **7** [1962] 1752/3; Russ. J. Inorg. Chem. **7** [1962] 904/5). – [2] E. A. Torchenkova, P. Bajdala, V. S. Smurova, V. I. Spitsyn (Dokl. Akad. Nauk SSSR **199** [1971] 120/3; Dokl. Chem. Proc. Acad. Sci. USSR **196/201** [1971] 568/70). – [3] L. P. Kazanskii, E. A. Torchenkova, V. I. Spitsyn (Dokl. Akad. Nauk SSSR **209** [1973] 141/3; Dokl. Phys. Chem. Proc. Acad. Sci. USSR **208/213** [1973] 209/11). – [4] I. V. Potapova, L. P. Kazanskii, V. I. Spitsyn (Issled. Svoistv Primenenie Geteropolikislot Katal., Mater. Vses. Soveshch., Novosibirsk 1978, pp. 135/9; Ref. Zh. Khim. **1979** No. 1V86).

7.8.6 Molybdothorates of Organic N-Bases

The compounds given in the table below have been precipitated from a solution of the molybdothoric acid by acidified solutions of the bases. The precipitates were filtered after 12 h, washed with dilute H_2SO_4 and methanol, and dried in air. The white compounds are microcrystalline aside from the piperazine and hydroxyquinoline compounds, which are amorphous. Thermogravimetric analyses show that the compounds can be dehydrated. At higher temperatures (>340°C for the pyridine and >660°C for the piperazine compound) the compounds decompose giving $ThO_2+12MoO_3$.

base	formula	stability range of the anhydrous compound in °C
guanidine	$(CH_5N_3H)_8[ThMo_{12}O_{42}]\cdot nH_2O$	140 to 240
pyridine	$(C_5H_5NH)_6[H_2ThMo_{12}O_{42}]\cdot nH_2O$	100 to 140
quinoline	$(C_9H_7NH)_6[H_2ThMo_{12}O_{42}]\cdot nH_2O$	120 to 160
hydroxyquinoline	$(C_9H_7NOH)_7[HThMo_{12}O_{42}]\cdot nH_2O$	200 to 260
ethylenediamine	$(C_2H_8N_2H_2)_4[ThMo_{12}O_{42}]\cdot nH_2O$	120 to 160
piperazine	$(C_4H_{10}N_2H_2)_4[ThMo_{12}O_{42}]\cdot nH_2O$	300 to 340
hexamethylenediamine	$(C_6H_{12}N_2H_6)_3[H_2ThMo_{12}O_{42}]\cdot nH_2O$	200 to 260

S. A. Gavrilova, Z. F. Shakhova, G. M. Petrachkova (Vestn. Mosk. Univ. Khim. **19** No. 2 [1964] 54/8; C.A. **61** [1964] 9461).

7.8.7 $Th_3Mo_{12}O_{42}\cdot 29H_2O$ (= $3ThO_2\cdot 12MoO_3\cdot 29H_2O$)

Crystals of this compound have been obtained from aqueous solutions of 0.002 M $H_8ThMo_{12}O_{42}$ by standing for two weeks or by addition of an excess of $Th(NO_3)_4$ solution to a freshly prepared 10% solution of $H_8ThMo_{12}O_{42}$. The compound is assumed to be a polynuclear heteropoly complex of the type $Th[Th_2Mo_{12}O_{42}]\cdot 29H_2O$.

For a line diagram of the X-ray powder pattern see the paper. According to broad line ^1H NMR spectra all protons belong to water of crystallization. On passing through an acidic cation exchanger the compound is decomposed and only $H_8ThMo_{12}O_{42}$ is present in the eluate.

E. A. Torchenkova, P. Bajdala, V. S. Smurova, V. I. Spitsyn (Dokl. Akad. Nauk SSSR **199** [1971] 120/3; Dokl. Chem. Proc. Acad. Sci. USSR **196/201** [1971] 568/70).

7.8.8 $(NH_4)_4[Th_2Mo_{12}O_{42}]\cdot 14H_2O$ (= $(NH_4)_2O\cdot ThO_2\cdot 6MoO_3\cdot 7H_2O$)

This polynuclear heteropoly compound is obtained in crystalline form from a saturated solution of $(NH_4)_5[H_3ThMo_{12}O_{42}]$ (p. 305) at 50 to 60°C or on dissolving over a 3 h period 12.8 g of $(NH_4)_8[ThMo_{12}O_{42}]\cdot nH_2O$ and 13.7 g of $H_8ThMo_{12}O_{42}\cdot 18H_2O$ in 1L of H_2O at 50°C. After a further 3 h at 70°C, the product precipitates on cooling. On acidic ion exchangers the compound decomposes and only 12-molybdothoric acid is eluted.

E. A. Torchenkova, P. Bajdala, V. S. Smurova, V. I. Spitsyn (Dokl. Akad. Nauk SSSR **199** [1971] 120/3; Dokl. Chem. Proc. Acad. Sci. USSR **196/201** [1971] 568/70).

7.8.9 $CeTh_2Mo_{12}O_{42}$ ($=CeO_2 \cdot 2ThO_2 \cdot 12MoO_3$)

This soluble heteropoly complex is formed by the reaction of thorium perchlorate and 12-molybdoceric(IV) acid (p. 283) in 1% H_2SO_4 solution. A precipitate does not form even with a great excess of Th^{4+}, whereas in weakly acidic solutions (pH 2.6) a precipitate is obtained after several seconds at the mole ratio Th^{4+}: $[CeMo_{12}O_{42}]^{8-} = 2:1$.

If a solution with the ratio 1.33:1 is passed through a cation exchanger, the Th^{4+} cations are retained quantitatively. This indicates an appreciable dissociation of the heteropoly complex. Potentiometric titrations show the complex to be stable up to pH≈6 (acidic Th^{4+} solutions are stable only up to pH 3.6).

V. I. Spitsyn, E. A. Torchenkova, L. P. Kazanskii, P. Bajdala (Z. Chem. [Leipzig] **14** [1974] 1/8, 5/6).

8 Molybdate Hydrates with Main Group 4 Metals

8.1 Molybdogermanates

Remarks. Germanium molybdenum oxide hydrates are not known but heteropoly compounds are easily formed. The molybdogermanates often adopt the "Keggin type" structure first reported by Keggin for the 12-tungstophosphoric acid [1]. A modern introduction to the structure and chemistry of heteropoly compounds is given by Pope [2]. Older data are given in "Molybdän", 1935, pp. 375/6.

References:

[1] J. F. Keggin (Nature **131** [1933] 908/9; Proc. Roy. Soc. [London] A **144** [1934] 75/100). – [2] M. T. Pope (Heteropoly and Isopoly Oxometalates, Springer, Berlin 1983).

8.1.1 $(NH_4)_6[H_4GeMo_9O_{34}]$ ($=3(NH_4)_2O \cdot GeO_2 \cdot 9MoO_3 \cdot 2H_2O$)

This compound forms on reaction at 0°C of a freshly prepared metagermanate solution and an ammonium paramolybdate suspension in 4 M acetic acid. After 12 h at 0°C ammonium nitrate is added, and the crystallized product is separated by filtration. The isomeric form (α or β) could not be determined.

The compound decomposes on attempts at recrystallization. Acidification of a solution in a 1:1 mixture of water and dioxane yields the β-$[GeMo_{12}O_{40}]^{4-}$ ion.

M. Fournier, R. Massart (Compt. Rend. C **279** [1974] 875/7).

8.1.2 The 11-Molybdogermanate Ion $[H_2GeMo_{11}O_{39}]^{6-}$

While the two known salts of the 11-molybdogermanate ion are derived from the octabasic and the monoprotonated heptabasic forms (see below), the ion present in the very pale yellow aqueous solutions has been formulated as $[H_2GeMo_{11}O_{39}]^{6-}$. It is obtained either according to $GeO_2 + 11\,MoO_4^{2-} + 16\,H^+ \rightarrow [H_2GeMo_{11}O_{39}]^{6-} + 7\,H_2O$ by acidification with HCl of a mixture of GeO_2 and Na_2MoO_4 in the stoichiometric ratio [1], or by neutralization of an $H_4GeMo_{12}O_{40}$ solution according to $7[GeMo_{12}O_{40}]^{4-} + 20\,OH^- \rightarrow 7[H_2GeMo_{11}O_{39}]^{6-} + Mo_7O_{24}^{6-} + 3\,H_2O$ [1, 2]. The formation has been followed by potentiometry as well as photometry [1]. It has been proposed that its structure is derived from the Keggin-type structure by the loss of an Mo atom together with its unshared O atom [3].

On acidification the α isomer of the $GeMo_{12}$ species is obtained [4], see also [1, 2]. The lacunary heteropoly ion easily reacts with di- and trivalent ions of the transition elements $Z = Mn^{2+}$, Co^{2+}, Ni^{2+}, Cu^{2+} [3], Fe^{3+} [5]. The mixed heteropoly ions of the form $[GeMo_{11}ZO_{39}(OH_2)]^{n-}$ [3, 4] are less stable than their tungsten analogues and exist only in a narrow pH range, in the case of $Z = Co^{2+}$ from 4.3 to 5 [3]. A heteropoly ion of composition $[GeMo_8Fe_2O_{31}]^{4-}$ is said to form during the reaction of 11-molybdogermanate with Fe^{3+} [5].

References:

[1] M. Biquard, M. Lamache (Bull. Soc. Chim. France **1971** 32/7). – [2] P. Souchay, A. Tchakirian (Ann. Chim. [Paris] [12] **1** [1946] 249/61, 250/3). – [3] C. Tourné, G. Tourné (Bull. Soc. Chim. France **1969** 1124/36). – [4] M. Fournier, R. Massart (Compt. Rend. C **279** [1974] 875/7). – [5] P. Souchay, M. Lamache, M. Petit, F. Brunet (Bull. Soc. Chim. France **1971** 37/42).

8.1.3 The Na Salt of the $GeMo_{11}$ Species

A solution of this salt can be obtained from boiling aqueous Na_2GeO_3 to which 8 equivalents of MoO_3 are added. After the complete dissolution of the MoO_3 an aqueous solution of Na_2MoO_4 corresponding to 3 equivalents Mo is added dropwise to the hot solution, which is then heated for 1d without stirring. On evaporation first the solution becomes syrupy and then the salt solidifies. The exact composition of the salt has not been determined.

C. Tourné, G. Tourné (Bull. Soc. Chim. France **1969** 1124/36, 1131).

8.1.4 $K_8[GeMo_{11}O_{39}]\cdot 14H_2O$ ($=4K_2O\cdot GeO_2\cdot 11MoO_3\cdot 14H_2O$)

The whitish powdery compound precipitates on addition of potassium acetate to a solution of the sodium salt [1]. This latter solution is obtained either by decomposition of $H_4GeMo_{12}O_{40}$ by the stoichiometric amount of NaOH [1 to 3], or by acidification (HCl, HNO_3, etc.) to pH $\approx$ 4.5 of an aqueous solution containing germanate and molybdate in the molar ratio Ge:Mo = 1:11 [1, 3].

References:

[1] M. Biquard, M. Lamache (Bull. Soc. Chim. France **1971** 32/7). – [2] P. Souchay, A. Tchakirian (Ann. Chim. [Paris] [12] **1** [1946] 249/61, 250/3). – [3] C. Tourné, G. Tourné (Bull. Soc. Chim. France **1969** 1124/36, 1130/1).

8.1.5 α-$(NH_4)_6Na[HGeMo_{11}O_{39}]$ ($=6(NH_4)_2O\cdot Na_2O\cdot 2GeO_2\cdot 22MoO_3\cdot H_2O$)

This compound forms on reaction at 0°C of sodium metagermanate solution with ammonium paramolybdate in a molar ratio of Ge:Mo = 1:11 in 4M monochloroacetic acid. After washing the oily product with a concentrated ethanolic acetic acid solution, the product crystallizes at −20°C as a fine powder.

M. Fournier, R. Massart (Compt. Rend. C **279** [1974] 875/7).

8.1.6 Redox Reactions of the $GeMo_{12}$ Species in Solutions

To simplify the description of the reduction reactions the unreduced $H_4GeMo_{12}O_{40}$ is designated as the 0 species and the reduced products as I, II, etc. according to the total count of transferred electrons, see, for example [1].

In solution $H_4GeMo_{12}O_{40}$ is easily reduced by agents as H_2S, $SnCl_2$, or hydrazine [2] to the IV species and by $CrSO_4$ [3], $(NH_4)_2Fe(SO_4)_2$, or diphenylcarbazone [4] to the III species. With excess $CrSO_4$ the heteropoly anion decomposes [3]. The IV species is easily reoxidized to the 0 species even by air [2]. The characteristically blue reduction products ("heteropoly blues") are extractable into organic solvents and have been isolated [2 to 4], see p. 311. Also the observed acceleration of the reduction rate of molybdate (e.g. by I^- [5] or hydroquinone [6]) in the presence of germanate is in fact due to the formation of the more easily reducible 12-molybdogermanate. The kinetics of these reactions have been studied [5, 6].

The reduction behavior has been studied by potentiometry and by polarography with both the rotating platinum and the dropping mercury electrodes. For the formation of the I species $[GeMo^VMo^{VI}_{11}O_{40}]^{5-}$ see p. 312. The 0$\leftrightarrow$I reduction has also been found in di-n-octylamine solution [7].

Polarographic half-wave potentials of the α and β isomers in 0.5 M HCl vs. saturated calomel electrode (SCE) are [8]:

anion	$E_{1/2}$ in V (number of electrons)		
α-$GeMo_{12}$	+0.36 (2)	+0.24 (2)	+0.06 (2)
β-$GeMo_{12}$	+0.50 (2)	+0.40 (2)	0.00 (2)

In 0.2 N H_2SO_4 solution the α-0↔α-II potential is reported as +0.6 V vs. the standard hydrogen electrode, which corresponds to 0.352 V vs. SCE. The potential is strongly dependent on the acidity [3]. The 0↔II reduction has also been found in ethanol and di-n-octylamine solutions [7]. The number of reduced Mo atoms per $[GeMo_{12}O_{40}]^{n-}$ ion is also given as 3 at the first equivalence point [3]. The α-II species is stable up to pH 6. At higher pH values it decomposes. The absorption spectrum between 600 and 1200 nm has a maximum at ~800 nm. The absorption coefficient at 800 nm during the reduction is plotted in the paper. The β-II species shows two maxima in its absorption spectrum, at 700 and 1020 nm. In acid solutions it disproportionates into the β-0 and the β-IV species. The reaction is second-order relative to the β-II species. Rate constants for various pH values are given in the paper [8]. The EPR spectrum of a frozen (77 K) $[GeMo_{12}O_{40}]^{6-}$ solution has been measured; g = 1.948. The results are compared with those of other reduced $[XMo_{12}O_{40}]^{n-}$ species [9].

The IV species has been described from polarographic experiments by various authors [7, 10, 11]. While for $[GeMo_{12}O_{40}]^{4-}$ (0 species) the β isomer transforms spontaneously into the α isomer, the β isomer of the IV species is the more stable one [8, 12]. The β-IV species is stable in the pH range 0 to 14 [13]. The transformation α-IV→β-IV has a half decay period of ~3 h at 40°C and 1 h at 60°C in 0.5 N HCl and is first-order relative to α-IV. The reaction is slower at reduced acidity. This reversed stability relation can be used to obtain the unreduced β isomer by electrochemical reoxidation [8]. The blue color of the aqueous solution is due to an intense absorption in the spectrum (measured between 600 and 1200 nm) of both of the isomers at 800 to 825 nm in 0.5 N HCl solution [8] or at ~920 nm (β isomer) at pH 6.9 [13]. The intensity of the absorption of the β isomer is about twice of that of the α isomer [8], see the papers [8, 13] for the spectra. The acid of the β-IV species can be prepared from stoichiometric amounts of Na_2GeO_3, Mo^V (sic), and Na_2MoO_4 in HCl solution (no other details are given) [8].

The α-VI and β-VI species have been obtained by procedures similar to those used for the corresponding molybdosilicate species [12]. Reduction waves in polarograms of neutral solutions of the β-IV species have been attributed to the 6 and 8 electron reduction products. The $E_{1/2}$ values vs. SCE are −0.45 and −0.65 V, respectively. The α-IV species could be reduced at pH 3 to 6 only to the α-VI species. The absorption coefficients (at 920 nm) and the absorption spectrum between 600 and 1200 nm (for the β-VI species) are plotted. The spectrum has a maximum at ~700 nm. The K salts of the α-VI and β-VI species have been isolated but no description of them is given [13].

References:

[1] M. T. Pope (Heteropoly and Isopoly Oxometalates, Springer, Berlin 1983, p. 106). – [2] H. Hahn, F. Hahn (Naturwissenschaften **49** [1962] 539/40). – [3] Z. F. Shakhova, R. K. Motorkina (Zh. Obshch. Khim. **26** [1956] 2663/73; J. Gen. Chem. [USSR] **26** [1956] 2969/76). – [4] Z. F. Shakhova, R. K. Motorkina (Tr. Komis. Analit. Khim. Akad. Nauk SSSR Inst. Geokhim. Analit. Khim. **8** [1958] 100/9; C.A. **1958** 19444). – [5] I. I. Alekseeva, I. I. Nemzer, A. P. Rysev (Izv. Vysshikh Uchebn. Zavedenii Khim. Khim. Tekhnol. **13** [1970] 1423/7; C.A. **74** [1971] No. 57698).

[6] V. R. Rudenko (Zh. Neorgan. Khim. **24** [1979] 73/6; Russ. J. Inorg. Chem. **24** [1979] 40/3). – [7] M. L. Plöger, F. Pottkamp, F. Umland (Z. Anorg. Allgem. Chem. **407** [1974] 211/26). – [8]

M. Biquard, P. Souchay (Bull. Soc. Chim. France **1971** 437/44). – [9] L. P. Kazanskii (Izv. Akad. Nauk SSSR Ser. Khim. **1978** 274/8; Bull. Acad. Sci. USSR Div. Chem. Sci. **27** [1978] 235/8). – [10] S. V. Lugovoi, Z. Z. Odud (Nov. Polyarogr. Tezisy Dokl. 6th Vses. Soveshch. Polarogr., Riga 1975, p. 232 from C.A. **86** [1977] No. 23388).

[11] K. Grasshoff, H. Hahn (Z. Anal. Chem. **180** [1961] 18/31). – [12] R. Massart, P. Souchay (Compt. Rend. **257** [1963] 1297/9). – [13] M. Biquard (Compt. Rend. C **281** [1975] 309/12).

8.1.7 $H_8GeMo_4^VMo_8^{VI}O_{40}\cdot12H_2O$ ($=GeO_2\cdot2$"Mo_2O_5"$\cdot8MoO_3\cdot16H_2O$)

The reactions with bases show this "germanium molybdenum blue IV" to be an acid in which 4 or 8 hydrogen ions can be replaced by metals [1]. ^{1}H NMR investigations indicate the H to be present in the compound in OH groups, giving the formula $GeMo_{12}O_{32}(OH)_8\cdot12H_2O$ [2].

The compound has been prepared by reduction of 12-molybdogermanic acid, $H_4GeMo_{12}O_{40}$, with hydrazine. The crude product was extracted by methylethyl ketone and purified by ion exchange. The product was then isolated by evaporation of the solution in a rotary evaporator [2]. Reduction can also be effected by H_2S in weakly acidic solutions, by $SnCl_2$, or by a Zn reductor. The compound can be recrystallized from aqueous solution as blue-black platelets [1].

Broadline ^{1}H NMR spectra between −113 and +24°C show at low temperatures a signal corresponding to the sum of a broad contribution due to protons of H_2O and a narrow part due to hydroxyl protons. At room temperature there is a rapid exchange among all proton positions [2].

$GeMo_{12}O_{32}(OH)_8\cdot12H_2O$ is readily soluble in water and polar organic solvents but insoluble in nonpolar solvents. In strongly alkaline solutions the compound decomposes. In solution it is easily oxidized even by air to the fully oxidized molybdogermanic acid [1].

References:

[1] H. Hahn, F. Hahn (Naturwissenschaften **49** [1962] 539/40). – [2] H. Marsmann, H. Hahn (Z. Naturforsch. **21b** [1966] 188).

8.1.8 $(NH_4)_8[GeMo_{12}O_{40}]\cdot6.9H_2O$, $(N_2H_5)_8[GeMo_{12}O_{40}]\cdot9.8H_2O$, $[Zn(NH_3)_4]_4[GeMo_{12}O_{40}]\cdot4.7H_2O$, $[Cd(NH_3)_4]_4[GeMo_{12}O_{40}]\cdot4.9H_2O$, $[Cd(en)_2]_4[GeMo_{12}O_{40}]\cdot3.8H_2O$, $Tl_4[GeMo_{12}O_{36}(OH)_4]\cdot4.0H_2O$, $(nitron)_4[GeMo_{12}O_{36}(OH)_4]\cdot3.1H_2O$

The Tl and the nitron salts are derived from the $[GeMo_{12}O_{36}(OH)_4]^{4-}$ anion; all the others from the $[GeMo_{12}O_{40}]^{8-}$ anion. In addition the salts $[Ni(NH_3)_4]_4[GeMo_{12}O_{40}]\cdot7.3H_2O$ and $[Cu(en)_2]_4[GeMo_{12}O_{40}]\cdot6.7H_2O$ have been prepared (en = ethylenediamine). The water contents depend on the preparation conditions. The ammine salts easily lose NH_3.

H. Hahn, F. Hahn (Naturwissenschaften **49** [1962] 539/40).

8.1.9 The Acid of the $GeMo_3^VMo_9^{VI}$ Species

Crystals of this heteropoly blue have been prepared on a celluloid support by evaporation from a 0.25 to 0.5% aqueous 12-molybdogermanic acid solution reduced by Cr^{II} [1, 2] or Fe^{II} [1, 3, 4]. Electron diffraction shows them to be isomorphous with crystals of the unreduced molybdogermanic acid (cf. p. 313) with a cubic unit cell, lattice parameter a = 23.16 Å; Z = 8 [2, 4], the d values are listed in [4].

Chemical analyses show a quarter of the molybdenum in these samples to be reduced to Mo^{V} [1, 2], which corresponds to a three-electron reduction. The absorption spectrum of the blue reduced solution of this compound has an intense absorption at 820 to 825 nm, while at other wavelengths the absorption does not differ significantly from that of the unreduced species [2, 3]. Attempts to achieve further reduction by $CrSO_4$ apparently decomposed the heteropoly ion [1].

References:

[1] Z. F. Shakhova, R. K. Motorkina (Zh. Obshch. Khim. **26** [1956] 2663/73; J. Gen. Chem. [USSR] **26** [1956] 2969/76). – [2] I. P. Alimarin, Z. F. Shakhova, R. K. Motorkina (Dokl. Akad. Nauk SSSR **106** [1956] 61/4; Proc. Acad. Sci. USSR Div. Chem. Sci. **106/111** [1956] 1/4). – [3] Z. F. Shakhova, R. K. Motorkina (Tr. Komis. Analit. Khim. Akad. Nauk SSSR Inst. Geokhim. Analit. Khim. **8** [1958] 100/9; C.A. **1958** 19444). – [4] Z. F. Shakhova, G. N. Tishchenko, R. K. Motorkina (Zh. Obshch. Khim. **27** [1957] 1118/24; J. Gen. Chem. [USSR] **27** [1957] 1200/6).

8.1.10 The α-$[GeMo^{V}Mo^{VI}_{11}O_{40}]^{5-}$ Ion

Solutions of this one-electron reduced polyanion (I species) have been prepared by electrochemical reduction at constant potential on a mercury or platinum electrode of 10^{-2}M $[(C_4H_9)_4N]_4[GeMo_{12}O_{40}]$ solutions in DMF containing 0.2 to 0.5 mol/L $(C_4H_9)_4NBF_4$ as a supporting electrolyte. The progress of the reaction was followed by coulometry, potentiometry, and polarography.

The reduced ion has the structure of the α-Keggin type like the α-$[(C_4H_9)_4N]_4[GeMo_{12}O_{40}]$ (p. 319). At low temperatures the unpaired electron is trapped at a single Mo atom and thus polarizes its surrounding. EPR results suggest that the Mo^{V} atoms do not occupy exactly the same positions as the Mo^{VI} atoms, but presumably are slightly shifted towards the centers of the octahedra.

EPR studies on $[GeMo_{12}O_{40}]^{5-}$ in the frozen electrolyzed solution showed resonance signals at all temperatures between 10 and 200 K, indicating the unpaired electron to be in an orbitally nondegenerate ground state, i.e., in a $b_2(d_{xy})$ orbital. Below 40 K a hyperfine structure is observed. A computer simulation of the spectrum near 10 K showed the electron trapped on a single molybdenum atom in an only axially distorted ligand field and yielded the following EPR parameters: $g_{\parallel}(g_z) = 1.935$, $g_{\perp}(g_x, g_y) = 1.951$, $A_{\parallel} = 68.5$, $A_{\perp} = 33.6$ G. The anisotropy of the g tensor is large compared to that of the isostructural reduced polyanion with other heteroatoms (Si, As, P). Above 40 K the signal broadens, and above 45 K the hyperfine components vanish, see the paper for plots of the spectra. The ground state delocalization coefficient $\beta = 0.79$ and the Fermi contact term K = 0.69 have been derived from the hyperfine coupling parameters, $A_{\parallel}$ and $A_{\perp}$. The ground state is more delocalized in true Mo^{V}/Mo^{VI} mixed valence systems than, for example, in $[XMo^{V}W_{11}O_{40}]^{n-}$ anions. The peak-to-peak line width of the EPR signal increases with temperature, but also contains temperature independent components. The temperature dependent part yields the activation energy of the thermal electron hopping, $E_{th} = 0.045$ eV.

Electronic spectra show a broad absorption with a maximum at 13700 cm^{-1} (molar extinction coefficient 1160) and shoulders at 11600 and 7300 cm^{-1}. Owing to the very short time scale of UV-visible spectroscopy, the mobile electron appears trapped even at room temperature. Thus both the crystal field (d-d) bands of Mo^{V} and intervalence bands due to optical excitation $Mo^{V} \cdots O \cdots Mo^{VI} \rightarrow Mo^{VI} \cdots O \cdots Mo^{V}$ are observed. Accordingly the shoulder at 7300 cm^{-1} is thought to result from intervalence transitions, both "intra Mo_3O_{13} group" and "extra group" (between two corner sharing MoO_6 octahedra of different Mo_3O_{13} groups). From this assignment the optical activation energy $E_{opt} = 0.913$ eV has been calculated. One of the components at 11600 and 13700 cm^{-1} was tentatively assigned to a third intervalence transition which arises from the antibonding interaction of two d_{xy} orbitals of two corner linked, "extra group" Mo^{VI} atoms. This level is separated by approximately 2J (twice the transfer integral) from the less excited bonding state of the previously mentioned "extra group" intervalence transition. The other component is regarded as the d-d transition $^2B_2 \rightarrow {}^2E$. These assignments seem more likely than attribution of either absorption to a transition into a ^2E level which is split due to low site symmetry. A splitting of the ^2E level of about 2000 cm^{-1} should also give rise to a clearly orthorhombic EPR spectrum, which is not observed. From the former assignment the transfer integral is calculated as $J_{opt} = 0.26$ eV, which compares reasonably well to $J = 0.25$ eV derived from E_{opt} and E_{th}. For comparisons with other one-electron reduced heteropoly anions see the paper.

C. Sanchez, J. Livage, J. P. Launay, M. Fournier, Y. Jeannin (J. Am. Chem. Soc. **104** [1982] 3194/202).

8.1.11 α-$H_4GeMo_{12}O_{40} \cdot nH_2O$ (= $GeO_2 \cdot 12MoO_3 \cdot (2+n)H_2O$), $0 \leqq n \leqq 28$

Preparation. Yellow crystallized 12-molybdogermanic acid has been prepared by decomposing ether extracts of acidic (0.2 to 0.5 N) aqueous solutions of the acid or its salts by addition of a small amount of H_2O and evaporation of the ether [1, 2]. The aqueous solutions are obtained from mixtures of sodium germanate [2] or GeO_2 [1, 3] and paramolybdates [1, 3] or MoO_3 [2] with excess molybdate [1, 2] at acidities of about 0.3 to 0.4 N H_2SO_4 [1] or pH 1.5 [2]. Stoichiometric amounts are used in [3], but polarographic measurements show that an excess of 4 to 5 equivalents of molybdate is necessary for the complete formation of the heteropoly acid (1 equivalent $\triangleq$ 12 Mo per 1 Ge) [2]. Crystallized acid yields of 92% with respect to Ge have been achieved. Recrystallization from aqueous solution is not recommended because reduced heteropoly acids may form [1]. The mother liquor yields the 28-hydrate, which effloresces rapidly. After drying over silica gel the water content is $9H_2O$ [2], over P_2O_5 only $\sim 6H_2O$ [4].

Structure. Electron diffraction patterns have been indexed assuming cubic symmetry with the lattice parameter $a = 23.05$ Å; $Z = 8$. From the similarity of the lattice parameter the structure is thought to be analogous to that of $H_3PW_{12}O_{40} \cdot 29H_2O$. An alternative indexing with $a' = a/2$ would leave just one weak line unindexed, the other lines obeying the extinction rules of a body-centered cubic lattice (see the paper for the indexed d values). The acid is isomorphous with a reduced derivative, referred to as 12-molybdogermanic heteropoly blue (see p. 311), and with 10-molybdo-2-vanadogermanic acid ($GeMo_{10}V_2$ species, p. 338) [5]. There are no published crystal structure determinations for the acid, yet the crystal structure has been determined for the sodium and guanidinium salts, which contain the α-$[GeMo_{12}O_{40}]^{4-}$ anion of the Keggin type, see pp. 318/9.

To investigate the state of the water, ^1H NMR spectra have been recorded at 80 K of several partially or fully dehydrated samples of $H_4GeMo_{12}O_{40} \cdot nH_2O$ with $27 \geqq n \geqq 0$. The spectra of the water-rich samples show broad absorption lines due to H_2O and $\sim 4H_3O^+$, while the spectra of nearly dehydrated samples are characterized by a narrow line, most intense for $n = 2$. A change in the relative intensity of the narrow line is observed at $n = 8$. This narrow line indicates the presence of isolated protons H^+, which are free from dipole interactions with other protons [6]. A careful analysis of the line shapes suggests that during dehydration H_3O^+ and H_2O are increasingly distorted, leading to a partial separation of H^+ from its H_2O and increased interproton distances, H-H = 1.67 to 1.75 Å. The larger H-H distances of the water molecules are said to be due to polarizing effects of the heteropoly anion. These polarizing effects are less in salts of the heteropoly acids and larger in heteropoly acids of tungsten [7]. This transition $H_3O^+ \rightarrow H^+ + H_2O$ during dehydration allows estimation of stability and proton accepting properties relative to the analogous compounds of Si and P. These properties of the Ge, Si, and P compounds can be correlated with their redox potentials [6].

Optical Properties. IR spectra of the 12-molybdogermanic acid have been recorded on samples identified as the α isomer with approximately 6 H_2O [4, 8] and with unspecified water content (dried in flowing N_2) [9]. Assignments have been made on the basis of symmetry considerations and by comparison with IR and Raman spectra of other $[XM_{12}O_{40}]^{n-}$ ions. With the ideal T_d symmetry all possible vibrations of an isolated Keggin-type ion may be represented by $\Gamma_{vib} = 9A_1 + 4A_2 + 13E + 16F_1 + 22F_2$, of which the following are Raman or IR active: $\Gamma_{Raman} = 9A_1 + 13E + 22F_2$, $\Gamma_{IR} = 22F_2$ [4, 8]. The following table gives the wave numbers of the absorption bands (in cm^{-1}), the relative intensities, and the assignments according to [4], which agree reasonably with the assignments of [9] (the frequencies of the most intense absorption maxima in the spectra of [9] appear to be shifted to higher energies by 20 to 30 cm^{-1}); the typical Raman frequencies of the acid in aqueous solution are given for comparison [4]:

IR	975	955	875	802	765	515
intensity	w	s	m	s	vs	w
Raman	—	961	882	—	—	—
assignment	—	ν_{as}(Mo-O_d)	ν(Mo-O-Mo)	ν(Ge-O)	ν(Mo-O-Mo)	—
IR	460	440	365	325	242	222
intensity	s	m	s	m	w	w
Raman	—	—	358	326	—	207
assignment	δ(O-Ge-O)	—	—	—	—	—

O_d designates the O atom that is coordinated to just one Mo atom [4]. All absorption bands below 450 cm^{-1} are regarded as due to deformation vibrations of the MoO_6 groups [9]. Compared with 12-tungstoheteropoly acids the vibrations of the MoO_6 groups are shifted to lower energies, due to reduced π bonding [4, 9]. Valence and deformation vibrations of the crystal water are observed at 3400 and 1610 cm^{-1}, respectively [9]. Because the solid acid is decomposed by laser radiation the Raman spectrum has been determined only for solutions of the acid [4], see p. 316.

Chemical Properties. While ^{1}H NMR spectra of a sample designated as anhydrous $H_4GeMo_{12}O_{40}$ have been discussed in terms of heteropoly acids [6], the same authors (in part) state that the compound decomposes spontaneously even at a low content of water of crystallization [10]. Samples with higher water content (~6 H_2O) are thermally labile and decompose above 100°C (forming MoO_3) or during Raman experiments with laser radiation [4]. The acid is easily soluble in water [2].

References:

[1] R. K. Motorkina (Zh. Neorgan. Khim. **2** [1957] 92/105; Russ. J. Inorg. Chem. **2** No. 1 [1957] 142/61, 148/52). – [2] K. Grasshoff, H. Hahn (Z. Anal. Chem. **180** [1961] 18/31, 19/20). – [3] P. Souchay, A. Tchakirian (Ann. Chim. [Paris] [12] **1** [1946] 249/61, 250/3). – [4] C. Rocchiccioli-Deltcheff, R. Thouvenot, R. Franck (Spectrochim. Acta A **32** [1976] 587/97). – [5] Z. F. Shakhova, G. N. Tishchenko, R. K. Motorkina (Zh. Obshch. Khim. **27** [1957] 1114/24; J. Gen. Chem. [USSR] **27** [1957] 1200/6).

[6] V. F. Chuvaev, E. V. Vanchikova, L. I. Lebedeva, V. I. Spitsyn (Dokl. Akad. Nauk SSSR **210** [1973] 370/3; Dokl. Chem. Proc. Acad. Sci. USSR **208/213** [1973] 411/4). – [7] V. F. Chuvaev, V. I. Spitsyn (Dokl. Akad. Nauk SSSR **232** [1977] 145/7; Dokl. Phys. Chem. Proc. Acad. Sci. USSR

232/237 [1977] 35/7). – [8] C. Rocchiccioli-Deltcheff, R. Thouvenot, R. Franck (Compt. Rend. C **280** [1975] 751/4). – [9] G. Lange, H. Hahn, K. Dehnicke (Z. Naturforsch. **24b** [1969] 1498/507). – [10] V. F. Chuvaev, V. I. Spitsyn (Dokl. Akad. Nauk SSSR **232** [1977] 1124/6; Dokl. Phys. Chem. Proc. Acad. Sci. USSR **232/237** [1977] 191/3).

8.1.12 $H_4GeMo_{12}O_{40}$ Solutions

Solutions of 12-molybdogermanic acid, both aqueous and nonaqueous, have been extensively studied because of the importance of this compound for the analytical chemistry of Ge.

Aqueous Solutions. In acidic aqueous solutions the $GeMo_{12}$ species is formed spontaneously from soluble germanates and molybdates. Formation and stability depend on the acidity: The pH range has been limited to values between ~1 and 3.6 (e.g., 0.8 to 3.6 in H_2SO_4, 1.0 to 3.4 in HNO_3, and 0.9 to 3.6 in $HClO_4$) [1]. Optimal conditions are pH values of about 1.5 to 2 and up to 100% excess molybdate [2 to 4]. The reaction is complete after 30 min [2]. The instability constant is $K=([H_4GeO_4][H_2Mo_3O_{10}]^4)/[H_4GeMo_{12}O_{40}]=1.38\times10^{-13}$ [1] and the formation constant is 1.6×10^{11}, assuming the reaction $H_4GeO_4+2H_6Mo_6O_{21}\rightleftharpoons H_4GeMo_{12}O_{40}+6H_2O$ [2]. In microanalytical situations a 200-fold or greater excesses of molybdate have been used [5, 6]. Another work questions the conclusion that at pH 2.2 with various molar ratios Ge:Mo in the reaction mixture the 12-molybdogermanate ion is the predominant species and suggests that the Ge:Mo ratio varies from 1:1 to 1:48 [7]. Yet a molar ratio Ge:Mo = 1:12 has been confirmed for the heteropoly species formed in acidic mixtures of germanate and molybdate solutions with Ge:Mo = 1:8 to 1:15 at pH 2.4 [1] and in 0.64 M perchloric acid [8].

On titration of 4.55×10^{-3} M solutions of $H_4GeMo_{12}O_{40}$ in 1 M NaCl with NaOH a first equivalence point is observed at pH 2 to 2.5 after addition of 4 equivalents of OH^-, which indicates the formation of the $[GeMo_{12}O_{40}]^{4-}$ ion [9]. A next equivalence point is observed after a further 2.86 equivalents of NaOH at a pH of about 4. At this point the 12-molybdogermanate ion decomposes according to $7[GeMo_{12}O_{40}]^{4-}+20OH^-\rightarrow7[H_2GeMo_{11}O_{39}]^{6-}+Mo_7O_{24}^{6-}+3H_2O$ [9, 10], cf. also p. 308.

Also, in strongly acidic solutions the 12-molybdogermanic acid reversibly decomposes [8, 11]. The decomposition in 0.3 to 1 M $HClO_4$ solutions with a constant $[GeMo_{12}O_{40}]^{4-}$ concentration of 5×10^{-3} M (related to Mo) has been formulated as $H_4GeMo_{12}O_{40}+nH_2O \underset{k_2}{\overset{k_1}{\rightleftharpoons}} GeO_2+12/x(Mo_xO_y)^{6x-2y}+4(6y/x-18)H^+$. The reaction order is 1 with respect to the 12-molybdogermanate, and the rate constants k_1 and k_2 are given as 0.304 and $0.0434\,h^{-1}$, respectively. The value of x is approximately 4 for Mo concentrations between 0.01 and 0.2 M. The reaction order is unchanged at perchloric acid concentrations up to 11 M [8]. In contrast, decomposition into molybdenyl cations, molybdenum oxide chlorides, or molybdenum oxide sulfates has been proposed to be produced by perchloric or hydrochloric acids of concentrations > 2 M or sulfuric acid concentrations > 1 M [11]. Stability against acids also depends on the total Mo concentration [8], and it has been proposed that the formation and decomposition of the 12-molybdogermanate is governed by the ratio $[H^+]/[Mo]$ rather than by the absolute value of the acid concentration [4]. (Because of the stability relationships between the two known isomers, in the absence of any specifications, one may reasonably assume that the decomposition data apply to the α isomer.)

Organic Solvents. $H_4GeMo_{12}O_{40}$ is soluble in various O- or N-containing organic solvents such as ethers, alcohols, and ketones. Extraction with ether has been used for the preparation of the free acid from the acidic aqueous reaction mixture [2, 3, 12]. For analytical purposes 5:1 ether/pentanol mixtures or methyl ethyl ketone [13], n-butyl acetate [6, 13, 14], 1-butanol

[14, 15], butanol/chloroform mixtures [7, 14, 15], and mixtures of amine derivatives (di- and tri-n-octylamine) with chloroform [5, 6] or 1,2-dichlorethane [16] have been applied. The extractability by amines is better than that by O-containing solvents.

The distribution coefficients between the organic and the aqueous phase depend on the pH and differ among the various 12-molybdohetero acids. This has been used for selective extractions [5, 13]. Thus for the 12-molybdogermanic acid the distribution coefficient $D = c_{org}/c_{aq}$ is less than 0.08 for n-butyl acetate or diethyl ether and 2N HCl. For a 5:1 diethyl ether/1-pentanol mixture D = 100 with 1 to 1.4N HCl and for methylisobutyl ketone D = 620 with 0.1 to 2N HCl [13].

Structure. The vibrational and electronic spectra of solutions of the acid and the 12-molybdogermanate ion have been interpreted assuming the Keggin-type structure analogous to that in the solid state.

There are five possible isomers of the $[XM_{12}O_{40}]^{n-}$ ion. The electrostatically more favorable forms termed α and β have so far been found. In the β isomer one of the Mo_3O_{13} subunits is rotated by 60° from its orientation in the α isomer. As the O atoms concerned form in good approximation a planar hexagon, the rotation causes only minor distortions. The α isomer appears to be thermodynamically more stable [17]. The α-$[GeMo_{12}O_{40}]^{4-}$ ion is always described as the final reaction product [6, 18, 19]. The β isomer is intermediately formed [5, 6], especially in highly acidic reaction mixtures [18]. It transforms spontaneously into the α isomer within 2h at room temperature [4] or within 10min at 80°C [6]. Solutions of the two isomers differ in their optical and polarographic properties. The β isomer can be stabilized [5] or the transformation rate decreased [19] by addition of amines or alcohols. The transformation is first-order with respect to the β isomer [6, 19]. At pH 1.3 in aqueous solution the transformation parameters are: activation enthalpy 13.4 kcal/mol, free activation energy 20.7 kcal/mol, and activation entropy $-25\ cal \cdot mol^{-1} \cdot K^{-1}$. For rate constants and half-lives between 0 and 40°C see the paper [6]. In a 50vol% H_2O/CH_3OH solution, 0.5M in HCl, the half-lives are 73 and ~20min at 40 and 60°C, respectively [19].

The protons of the $H_4GeMo_{12}O_{40}$-etherate are attached to the $(C_2H_5)_2O$ molecules, forming $(C_2H_5)_2OH^+$, and not to the H_2O molecules also present, according to the analysis of the IR and Raman spectra [12].

Spectra. The Raman spectrum of the aqueous solution [20] and the IR spectrum of the etherate [12] have been assigned assuming the structure of the α isomer of the Keggin type and by comparison with IR and Raman spectra of other 12-molybdo- or 12-tungstohetero species. The typical bands are similar to those of the solid acid (see p. 314). Lists of the wave numbers are given in the papers [12, 20].

IR bands of the etherate at 3420 and 2990 cm^{-1} are attributed to OH stretching vibrations of H_2O and $(C_2H_5)_2OH^+$, respectively, while those at 1750 and 1656 cm^{-1} are assigned to deformation vibrations of H_2O and those at 1107 cm^{-1} to δ(OH) of the protonated ether [12].

Electronic spectra of the aqueous solution from 220 to 1100nm are reported to show no maxima but only increasing extinction at wavelengths below 300nm [21, 22]. Observations of an absorption maximum at 370nm [23] (see also [4]) may be due to the use of different reference media. In 20vol% 1-butanol/$CHCl_3$ solution a shoulder at ~300nm is found [14]. Spectra in organic solvents are said not to differ from those in aqueous solutions [22].

Chemical Reactions. While reactions of the $H_4GeMo_{12}O_{40}$ with ether lead to the etherate [12] of only limited stability, the reactions with the more basic organic amines lead to soluble ion associates or scarcely soluble salts. Di-n-octylamine [6] or bis (2-ethylhexyl)amine [16] associates have been studied; the latter has the formula $[(C_4H_9\text{-}CH(C_2H_5)\text{-}CH_2)_2NH_2]_4GeMo_{12}O_{40}$ [16].

For the analytical determination by gravimetry [24, 25], potentiometry [26], or colorimetry the 12-molybdogermanic acid has been precipitated by 8-hydroxyquinoline and derivatives [24, 25], nitron [26], crystal violet [27], and antipyrine dyes [28 to 30]. The precipitates have been formulated as neutral [24, 25, 27, 28] or acid salts [27, 30] with a ratio of protonated amine to Ge = 4:1 and (3 or 2):1, respectively, yet nonintegral ratios have also been obtained [29]. Caution has been advised with respect to the significance of formulas given of salts of cations that are liable also to precipitate other iso- or heteropoly compounds [31]. Some salts are described on p. 318ff.

If during the formation of the 12-molybdogermanic acid metavanadate ions are present, ternary heteropoly compounds such as $H_6GeMo_{10}V_2O_{40} \cdot nH_2O$ are obtained (see p. 338) [3, 32, 33]. The existence of ternary heteropoly species with Cr^{3+} (see p. 357) [34, 35] and Fe^{3+} [36, 37] has been proposed. The redox reactions are described on p. 309.

References:

[1] W. Kemula, S. Rosołowski (Roczniki Chem. **34** [1960] 835/42; C.A. **1961** 9137). – [2] K. Grasshoff, H. Hahn (Z. Anal. Chem. **180** [1961] 18/31). – [3] R. K. Motorkina (Zh. Neorgan. Khim. **2** [1957] 92/105; Russ. J. Inorg. Chem. **2** No. 1 [1957] 142/61, 148/52). – [4] A. Halász, E. Pungor (Talanta **18** [1971] 569/75). – [5] F. Alt, F. Umland (Z. Anal. Chem. **274** [1975] 103/8).

[6] M. L. Plöger, F. Pottkamp, F. Umland (Z. Anorg. Allgem. Chem. **407** [1974] 211/26). – [7] L. I. Lebedeva, F. A. Muradova, V. K. Potrokhov (Zh. Neorgan. Khim. **16** [1971] 2743/7; Russ. J. Inorg. Chem. **16** [1971] 1461/4). – [8] M. Biquard, P. Souchay (Ann. Chim. [Paris] [14] **10** [1975] 163/7). – [9] M. Biquard, M. Lamache-Duhameaux (Bull. Soc. Chim. France **1971** 32/7). – [10] P. Souchay, A. Tchakirian (Ann. Chim. [Paris] [12] **1** [1946] 249/61, 250/3).

[11] F. Chaveau, P. Souchay, R. Schaal (Bull. Soc. Chim. France **1959** 1190/6). – [12] G. Lange, H. Hahn, K. Dehnike (Z. Naturforsch. **24b** [1969] 1498/507). – [13] S. J. Simon, D. F. Boltz (Anal. Chem. **47** [1975] 1758/63). – [14] C. Wadelin, M. G. Mellon (Anal. Chem. **25** [1953] 1668/73). – [15] L. I. Lebedeva, E. V. Vanchikova (Izv. Vysshikh Uchebn. Zavedenii Khim. Khim. Tekhnol. **19** [1976] 476/8; C.A. **85** [1976] No. 25976).

[16] N. Ivanov, D. Boikova (Dokl. Bolg. Akad. Nauk **31** [1978] 873/6; C.A. **90** [1979] No. 179568). – [17] M. T. Pope (Inorg. Chem. **15** [1976] 2008/10). – [18] A. Halász, E. Pungor (Talanta **18** [1971] 557/67). – [19] M. Biquard, P. Souchay (Bull. Soc. Chim. France **1971** 437/44). – [20] C. Rocchiccioli-Deltcheff, R. Thouvenot, R. Franck (Spectrochim. Acta A **32** [1976] 587/97).

[21] I. P. Alimarin, Z. F. Shakhova, R. K. Motorkina (Dokl. Akad. Nauk SSSR **106** [1956] 61/4; Proc. Acad. Sci. USSR Div. Chem. Sci. **106/111** [1956] 1/4). – [22] Z. F. Shakhova, R. K. Motorkina (Tr. Komis. Analit. Khim. Akad. Nauk SSSR Inst. Geokhim. Anal. Khim. **8** [1958] 100/9; C.A. **1958** 19444). – [23] L. I. Lebedeva (Zh. Neorgan. Khim. **12** [1967] 1287/92; Russ. J. Inorg. Chem. **12** [1967] 681/4). – [24] I. P. Alimarin, O. A. Alekseeva (Zh. Prikl. Khim. SSSR **12** [1939] 1900/6 from C.A. **1940** 7777). – [25] T. Dupuis (Mikrochem. Ver. Mikrochim. Acta **35** [1950] 449/65).

[26] H. Hahn, R. Wagenknecht (Z. Anal. Chem. **182** [1961] 343/57). – [27] F. V. Mirzoyan, V. M. Tarayan, E. Kh. Airiyan, N. A. Grigoryan (Talanta **27** [1980] 1055/9). – [28] V. J. Dick, A. Maurer (Rev. Roumaine Chim. **14** [1969] 1603/11; C.A. **73** [1970] No. 31112). – [29] A. P. Kreshkov, I. F. Kolosova, M. B. Ogareva, Z. P. Dobronevskaya (Zh. Analit. Khim. **26** [1971] 1322/6; J. Anal. Chem. [USSR] **26** [1971] 1178/82). – [30] V. P. Zhivopistsev, T. B. Cherepanova (Zh. Analit. Khim. **32** [1977] 977/80; J. Anal. Chem. [USSR] **32** [1977] 767/9).

[31] G. A. Tsigdinos (Top. Current Chem. **76** [1978] 1/64, 9). – [32] B. N. Ivanov-Emin (Zh. Obshch. Khim. **10** [1940] 826/30 from C.A. **1941** 2434). – [33] Z. F. Shakhova, G. N.

Tishchenko, R. K. Motorkina (Zh. Obshch. Khim. **27** [1957] 1118/24; J. Gen. Chem. [USSR] **27** [1957] 1200/6). – [34] L. I. Lebedeva, F. A. Muradova (Zh. Neorgan. Khim. **16** [1971] 1056/8; Russ. J. Inorg. Chem. **16** [1971] 561/3). – [35] L. I. Lebedeva, F. A. Muradova, G. N. Prokof'eva (Zh. Neorgan. Khim. **16** [1971] 2025/6; Russ. J. Inorg. Chem. **16** [1971] 1081/2).

[36] Z. F. Shakhova, E. N. Dorokhova (Zh. Neorgan. Khim. **10** [1965] 2060/4; Russ. J. Inorg. Chem. **10** [1965] 1121/4). – [37] F. A. Muradova, L. I. Lebedeva, G. V. Dement'eva (Zh. Neorgan. Khim. **16** [1971] 2877/8; Russ. J. Inorg. Chem. **16** [1971] 1531/2).

8.1.13 α-$Na_4[GeMo_{12}O_{40}]\cdot 8H_2O$ (= $2Na_2O\cdot GeO_2\cdot 12MoO_3\cdot 8H_2O$)

This compound has been prepared from a solution of GeO_2, $Na_2MoO_4\cdot 2H_2O$, and $NaClO_4$ in 0.206 M $HClO_4$, with 0.12 M MoO_4^{2-}, 0.02 M $Ge(OH)_4$, and 3.0 M $NaClO_4$, which was allowed to evaporate at room temperature. After a few days air-stable, sea green, prismatic crystals were obtained [1].

Three-dimensional X-ray diffraction data showed the symmetry of the structure to be triclinic, space group $P\bar{1}$-C_i^1 (No. 2), lattice parameters $a=14.421(1)$, $b=13.187(1)$, $c=11.596(1)$ Å, $\alpha=114.31(1)°$, $\beta=103.88(1)°$, $\gamma=76.45(1)°$; $Z=2$. $D_x=3.62$ g/cm^3 [1].

The crystal structure was refined to $R=0.029$. The positional parameters of the atoms are listed in the paper. In the structure $[GeMo_{12}O_{40}]^{4-}$ ions are connected in a three-dimensional frame work by O-Na-O and O-Na-H_2O-Na-O bridges and by hydrogen bonds. The $GeMo_{12}O_{40}$ unit represents the α isomer of the Keggin-type structure. The GeO_4 tetrahedron is almost regular with Ge-O distances from 1.72 to 1.74 Å, O-O from 2.82 to 2.85 Å, and angles O-Ge-O from 108.8° to 110.2°. The distortion of the Mo coordination by O is more due to the noncentral position of Mo in the O_6 octahedron than to large differences in the O-O distances, which range from 2.53 to 2.91 Å. There are four different groups of Mo-O distances with the following ranges: (1) 1.67 to 1.70 Å, O atoms coordinated to one Mo atom only, (2) 1.79 to 1.86 Å, coordinated to two Mo atoms, (3) 1.99 to 2.08 Å, coordinated to two Mo atoms, and (4) 2.26 to 2.32 Å, coordinated to Ge as well as to three Mo atoms. The O atoms in group (1) are always in a trans position relative to those in group (4), while those in group (2) are in trans position relative to those in group (3) [1]. The separation of groups (2) and (3) corresponds to the reduction of the (pseudo)symmetry of the Keggin type from $\bar{4}3m$-T_d ((2) and (3) are symmetry related) to 23-T [2, 3] and is not correlated to the participation of the O atoms in corner sharing or edge sharing between the MoO_6 groups. This differentiation has been attributed to varying π bond contributions to the Mo-O bonds. The Mo arrangement corresponds to the corners of a slightly distorted cubo-octahedron. Within the Mo_3O_{13} groups the Mo-Mo distances range from 3.32 to 3.38 Å, between the groups from 3.70 to 3.78 Å. For the coordination of the Na^+ ions see the paper. The Na-O distances range from 2.25 to 2.77 Å [1].

References:

[1] R. Strandberg (Acta Cryst. B **33** [1977] 3090/6). – [2] H. D'Amour, R. Allmann (Z. Krist. **143** [1976] 1/13, 8). – [3] H. Ichida, A. Kobayashi, Y. Sasaki (Acta Cryst. B **36** [1980] 1382/7).

8.1.14 $K_4[GeMo_{12}O_{40}]$ (= $2K_2O\cdot GeO_2\cdot 12MoO_3$)

This compound has been prepared by familiar methods as given in [1]. XPS measurements give the binding energy of the O 1s electron as 531.04 eV relative to C 1s = 285.0 eV. From comparison with other heteropoly species $[XM_{12}O_{40}]^{n-}$ the force constants of the M-O bonds are thought to decrease in the order M = W, Mo, V [2].

References:

[1] P. Souchay (Ions Minéraux Condensées, Masson, Paris 1969). – [2] L. P. Kazanskii, V. I. Spitsyn (Dokl. Akad. Nauk SSSR **227** [1976] 140/3; Dokl. Phys. Chem. Proc. Acad. Sci. USSR **226/231** [1976] 225/7).

8.1.15 α-$[(C_4H_9)_4N]_4[GeMo_{12}O_{40}]$ (= $2[(C_4H_9)_4N]_2O\cdot GeO_2\cdot 12MoO_3$)

Preparation can be achieved from an HNO_3 solution of Na_2MoO_4 to which a 0.38 M Na_2GeO_3 solution in 0.52 M NaOH is added dropwise. The solution turns yellow and after 30 min at 80°C the β→α isomerization has taken place. Precipitation can be performed by adding an aqueous solution of $(C_4H_9)_4NBr$. Recrystallization in acetone yields small yellow crystals. The IR spectrum contains 16 bands between 985 and 340 cm^{-1}.

C. Sanchez, J. Livage, J. P. Launay, M. Fournier, Y. Jeannin (J. Am. Chem. Soc. **104** [1982] 3194/202).

8.1.16 α-$[C(NH_2)_3]_4[GeMo_{12}O_{40}]$ (= $4HNC(NH_2)_2\cdot GeO_2\cdot 12MoO_3\cdot 2H_2O$)

This guanidinium salt of the 12-molybdogermanic acid was prepared by addition of 21 mg of $C(NH_2)_3Cl$ to 7.5 mL of a mixture containing 0.6 g $Na_2MoO_4\cdot 2H_2O$ and 29 mg GeO_2 in 0.51 M $HClO_4$. The solution was set out for evaporation at room temperature and within a few days light yellow, prismatic crystals were obtained.

The crystal structure determined by single crystal X-ray methods was refined to R = 0.050. The symmetry is triclinic, space group $P\bar{1}$-C_i^1 (No. 2), lattice parameters a = 12.123(2), b = 12.159(2), c = 16.655(3) Å, α = 76.35(1)°, β = 78.46(1)°, γ = 66.99(1)°; Z = 2. The calculated density is 3.21 g/cm^3. For tables of positional parameters and interatomic distances see the paper.

The crystal structure consists of $[GeMo_{12}O_{40}]^{4-}$ ions in a three-dimensional frame work, linked by numerous hydrogen bonds to the guanidinium cations. The heteropoly ion represents the α isomer of the Keggin-type structure. The central GeO_4 tetrahedron is rather regular with Ge-O distances from 1.729 to 1.745 Å and angles O-Ge-O between 108.8° and 110.0°. The Mo-O distances are split into three groups ranging from 1.66 to 1.71, from 1.79 to 2.09, and from 2.26 to 2.31 Å. This is correlated to the coordination of the O atoms by 1 Mo, by 2 Mo, or by 1 Ge and 3 Mo atoms, respectively. The Mo-Mo distances within the Mo_3O_{13} groups range from 3.32 to 3.36 Å, and those between the groups from 3.70 to 3.76 Å. The guanidinium ions are approximately planar with a mean C-N distance of 1.33 Å. A proposal regarding the hydrogen bond system is given in the paper.

R. Strandberg, B. Hedman (Acta Cryst. B **38** [1982] 773/8).

8.1.17 $M_{4-x}[H_xGeMo_{12}O_{40}]\cdot nH_2O$ (= $(2-x/2)M_2O\cdot GeO_2\cdot 12MoO_3\cdot (n+x/2)H_2O$), M = NH_4, Rb, Cs; x < 1

Hydrated alkali salts of the 12-molybdogermanic acid have been precipitated by addition of NH_4^+, Rb^+, or Cs^+ ions to a concentrated solution of the acid. This solution was prepared from stoichiometric mixtures of solutions of sodium germanate and molybdenum trioxide hydrate which then were concentrated by evaporation and acidified to 0.2 to 0.5 N H_2SO_4. The precipitated salts are powders of small yellow crystals, with a greenish tinge in the case of Cs.

According to chemical analysis the composition of the rubidium compound is $Rb_4[GeMo_{12}O_{40}]\cdot 9H_2O$, whereas for NH_4 and Cs it is $1.85(NH_4)_2O\cdot GeO_2\cdot 12MoO_3\cdot 4.65H_2O$ and $1.82Cs_2O\cdot GeO_2\cdot 11.5MoO_3\cdot 6H_2O$.

X-ray powder patterns show the three compounds to be isomorphous. The saturated solutions of the salts in water have pH values of 2.86, 4.19, and 3.97 for NH_4, Rb, and Cs, respectively. In water solution the Rb and Cs salts hydrolyze slightly. The solubilities in H_2O at 25°C are 8.78, 0.9, and 0.057 wt% for the NH_4, Rb, and Cs salts, respectively. The solubilities in up to 40 wt% sulfuric acid at 25°C have been determined, see tables and plots in the paper. The solubilities have maxima at about 15 wt% H_2SO_4. Upon diluting the colorless solution at this concentration, the NH_4 compound decomposes forming a blue-green solution. The solubility of the Cs compound has also been determined at 25°C in 2 and 5.3 wt% HNO_3 and 2 to 9 wt% oxalic acid. In the latter decomposition by reduction takes place. The solubilities of 12-molybdogermanates are usually greater than those of the silicon analogues.

Stored in a closed box for 5 months the Cs compound becomes slightly blue as a result of partial reduction.

F. M. Perel'man, A. Ya. Zvorykin, T. N. Yakubovskaya (Zh. Neorgan. Khim. **3** [1958] 1374/80; Russ. J. Inorg. Chem. **3** No. 6 [1958] 143/53).

8.2 Compounds with Tin

Older data are given in "Molybdän", 1935, pp. 307, 376.

8.2.1 $(NH_4)_2[Mo^V_2O_4(OH)_4\cdot 4Sn^{II}(OH)_2]\cdot 8H_2O$

A mixture of 0.5 g $(NH_4)_6Mo_7O_{24}\cdot 4H_2O$ in 50 vol% alcohol and 0.65 g $SnCl_2\cdot 2H_2O$ in concentrated HCl reacts in 1 d to give a red solution of a $Mo_2O_3^{4+}$ species. At pH >7 NH_3 precipitates the compound $(NH_4)_2[Mo_2O_4(OH)_4\cdot 4Sn(OH)_2]\cdot 8H_2O$. (With NaOH, NH_3, and pyridine also similar compounds, but containing Cl^-, have been precipitated.) The color and the diamagnetic character prove the existence of a dimeric Mo^V unit with two oxygen bridges, which were found in many other Mo^V compounds. The IR spectrum supports this structure.

O. Zdrafcu, M. Brezeanu (Bul. Inst. Politeh. Gheorghe Gheorghiu-Dej Bucuresti Ser. Chim. Metal. **43** [1981] 37/43; C.A. **96** [1982] No. 227933).

8.2.2 Stannic Molybdate

From aqueous mixtures of $SnCl_4$ and Na_2MoO_4 (or $(NH_4)_2MoO_4$), gels precipitate which can be used as cation exchangers. After standing for several hours, the gel is washed with water, filtered, and dried at 40°C. The product consists of small light yellow particles. The acid form is obtained by immersing the gel in 2 M HNO_3 for 24 h. Titration of the acid form shows that the gels contain a monobasic group. X-ray diffraction proves the amorphous character.

The properties of the gels depend on the composition and concentration of the reaction mixture. Products from mixtures with Sn:Mo$\geqq$1 have a composition of Sn:Mo$\approx$2, from mixtures with Sn:Mo<1 products with Sn:Mo$\approx$1 are obtained. A decreasing Sn:Mo ratio of the reaction mixture gives products with increasing ion exchange capacity (Sn:Mo = 0.5, 1 mequiv/g; Sn:Mo = 2, 0.54 mequiv/g) but with decreasing stability. The products from mixtures Sn:Mo = 0.66 or 0.5 hydrolyze appreciably in water. Products from concentrated mixtures

(0.5 M) are not gels or else dissolve in water upon washing [1, 2]. The reproducibility of the properties of the gels is improved remarkably if the precipitate is refluxed in the mother liquor for 24 h. These yellowish green products have a higher exchange capacity and a better thermal stability [3]. The distribution coefficients for 25 metal ions on different gels in water, NH_4Cl solutions, and at different pH values (0 to 4) are tabulated, also the separation conditions for 10 ion pairs [1]. Many inorganic and organic reductants colorize the exchanger beads. The different colors may be used as spot tests for the reductants (e.g. Fe^{2+}). The color reactions of 19 cations and 8 anions are tabulated [4].

The thermogravimetric curves from 50 to 900°C are plotted. They show 3 to 4 steps. Above 100°C the color darkens, becoming black at 600°C. Heating the products at 100°C diminishes the exchange capacity considerably [1, 2]. Above 200°C the preparations begin to crystallize [1].

Samples with Sn:Mo=1:1 are stable against HNO_3, HCl, $HClO_3$, H_2SO_4, formic, acetic, and oxalic acid, aqueous NaOH and NH_3 solutions [2].

References:

[1] M. Qureshi, K. Husain, J. P. Gupta (J. Chem. Soc. A **1971** 29/32). – [2] M. Qureshi, J. P. Rawat (J. Inorg. Nucl. Chem. **30** [1968] 305/11). – [3] M. Qureshi, R. Kumar, V. Sharma (Anal. Chem. **46** [1974] 1855/8). – [4] M. Qureshi, J. P. Rawat (Chemist-Analyst **56** [1967] 89/90).

8.2.3 The $SnMo_{12}$ Heteropoly Species in Solution

The optimal conditions for the formation of the $SnMo_{12}$ species in solution have been elucidated photometrically at 250 nm. A mixture of $SnCl_4$ and $(NH_4)_2MoO_4$ should contain Mo in a molar excess of 1:14. The solution is stable for 30 min only in the small pH range 3 to 4 at room temperature. Heating on the water bath for several min decomposes the complex. The composition Sn:Mo=1:12 was proved by the isomolar series method. For the equilibrium $SnMo_{mn} \rightleftharpoons Sn_{free} + n\,Mo_m$ with mn=12, it was found that n=3 and m=4. Therefore it is assumed that the complex is formed in a reaction between Sn^{IV} and $Mo_4O_{13}^{2-}$. The stability constant β is $(0.7\ to\ 5.0) \times 10^{12}$ (log β=11.9 to 12.7). The absorption spectrum (plotted from 220 to 300 nm) has a maximum at 250 nm (ε=12000±100).

S. A. Morosanova, V. N. Muzykantova, L. A. Shkatova, I. P. Alimarin (Izv. Akad. Nauk SSSR Ser. Khim. **1975** 1919/22; Bull. Acad. Sci. USSR Div. Chem. Sci. **1975** 1801/3); see also S. A. Morosanova, V. N. Muzykantova, L. A. Shkatova (VINITI No. 2786-74 [1974]; Ref. Zh. Khim. **1975** No. 4V84).

8.3 Lead Molybdate Hydrates

8.3.1 Aqueous Lead Molybdate Solutions and Suspensions

There exist several papers about lead molybdates in aqueous solution at different pH ranges without describing the isolation or detailed characterization of the individual compounds.

$PbMoO_4 \cdot nH_2O$ is formed by mixing the solutions of $Pb(NO_3)_2$ and Na_2MoO_4 in equimolar amounts at pH 4.2 to 5.5 and 20°C. At pH<5.3 the precipitation is not complete because of the appreciable solubility of the compound in acidic solutions. The water content n is about 0.7 and depends on the temperature of the mixture. The salt is dehydrated at 105°C and melts at 1070°C [1].

Equilibrium mixtures of a 0.5 M solution of $Pb(NO_3)_2$ and 0.0125 M solutions of the compounds $MoO_3 \cdot 2H_2O$, $Na_2MoO_4 \cdot 2H_2O$, $Na_2Mo_2O_7 \cdot 6H_2O$, $Na_2Mo_3O_{10} \cdot 7H_2O$, or $Na_2Mo_4O_{13} \cdot 7H_2O$ have been analyzed conductometrically in the molar range Pb : Mo = 0 to 1.4 at room temperature. The titration curves show only one inflection point at the molar ratio Pb : Mo = 1, indicating the formation of $PbMoO_4$. Basic or acidic lead molybdates cannot be identified, because the curves are nonlinear in the molar range Pb : Mo < 1. The addition of HCl or $PbCl_2$ to the mixtures does not affect the yield of $PbMoO_4$ [2]. To suspensions of $PbMoO_4$ generated by mixing solutions of $Pb(NO_3)_2$ and Na_2MoO_4 varying amounts of NaOH have been added. The pH measurements indicate the exclusive formation of $PbMoO_4$ up to pH > 12 and temperatures up to 85°C; basic compounds cannot be detected [3].

In a coprecipitation reaction various amounts of a 1M solution of $Pb(NO_3)_2$ were added to a 0.5 M solution of Na_2MoO_4 in a molar ratio Pb : Mo ≧ 2 and the pH of the mixture was adjusted to > 10 by adding 25% ammonia. The X-ray diffraction patterns of the air-dried precipitates show lines of $Pb(OH)_2$ and very weakly of $PbMoO_4$. Heating them to 290 to 320°C results in crystallization of Pb_2MoO_5, simultaneously removing water present in the structure as OH groups, which is proved by DTA, diffraction patterns, and IR spectra (Pb_2MoO_5 has an intense band at 820 to 825 cm^{-1}) [4].

Phototurbidimetric measurements at 22°C determining the relationship between pH and optical density of suspensions of $PbMoO_4$ in $HClO_4$-acidic aqueous solution show the characteristic curves with a maximum caused by pH-dependent changes in the size of the suspended particles. They also show that the position of the maximum depends on the age of the suspension, shifting about 0.1 units to higher pH in 4 d. The most probable reason for this effect is the hydrolysis of the dilute solutions of the lead salt [5]. (This method may be used for the determination of the solubility product of $PbMoO_4$, see "Molybdän" Erg.-Bd. B 2, 1976, p. 256.)

The conductometric titration curves of 0.1M solutions of Na_2MoO_4 acidified by 3N HNO_3 to the ratios 6Mo : 6H^+, 6Mo : 8H^+, 6Mo : 9H^+, and 6Mo : 12H^+, with a 0.1N solution of $Pb(NO_3)_2$ show 2 or 3 inflection points interpreted as the formation of the condensed molybdates $Pb_3Mo_6O_{21}$, PbH_4MoO_6, and Pb_3MoO_6 [6]. PbH_4MoO_6 is said to belong also to the only inflection point of the conductometric titration of a 50% alcoholic 0.0059 M solution of MoO_3 with 0.5 M $Pb(NO_3)_2$ [7]. The conductometric titration curve of 0.0014 N Na_2MoO_4 with 0.1N $Pb(NO_3)_2$ has one inflection point belonging to $PbMoO_4$, whereas the conductometric inflection point of the titration of a 0.0018 N aqueous solution of MoO_3 with 0.1N $Pb(NO_3)_2$ is interpreted as formation of PbH_4MoO_6. The two inflection points of the conductometric titration of 0.0033 N $(NH_4)_2Mo_2O_7$ with 0.1N $Pb(NO_3)_2$ indicate the formation of $PbMo_2O_7$ and $PbMoO_4$ [8].

References:

[1] A. N. Zobnina, I. P. Kislyakov (Izv. Vysshikh Uchebn. Zavedenii Khim. Khim. Tekhnol. **13** [1970] 143/7; C.A. **73** [1970] No. 41306). – [2] J. Byé (Ann. Chim. [Paris] [11] **20** [1945] 463/550, 535). – [3] B. Charreton (Bull. Soc. Chim. France **1956** 337/47). – [4] Zh. G. Bazarova, M. V. Mokhosoev, E. A. Kirillov, K. N. Fedorov (Izv. Akad. Nauk SSSR Neorgan. Materialy **14** [1978] 1504/6; Inorg. Materials [USSR] **14** [1978] 1175/7). – [5] M. L. Chepelevetskii, K. F. Kharitonovich (Zh. Analit. Khim. **18** [1963] 357/9; J. Anal. Chem. [USSR] **18** [1963] 314/6).

[6] R. Ripan, A. Duca (Acad. Rep. Populare Romine Bul. Stiint. Sect. Stiint. Techn. Chim. **6** [1954] 215/41, 228; C.A. **1956** 15314). – [7] R. Ripan, M. Puscasu (Acad. Rep. Populare Romine Fil. Cluj Studii Cercetari Chim. **12** [1961] 47/53; C.A. **57** [1962] 11918). – [8] R. Ripan, A. Duca (Acad. Rep. Populare Romine Bul. Stiint. Sect. Stiint. Techn. Chim. **6** [1954] 251/75; C.A. **1956** 15310).

8.3.2 $Pb_4(OH)_2(MoO_4)_3 \cdot nH_2O$ and $Pb_{10}(OH)_2(MoO_4)_9 \cdot nH_2O$ (= $4PbO \cdot 3MoO_3 \cdot (n+1)H_2O$ and $10PbO \cdot 9MoO_3 \cdot (n+1)H_2O$)

The precipitates of mixtures of aqueous solutions of $Pb(NO_3)_2$ and Na_2MoO_4, stirred for 48 h at 25°C and pH>5.6, consist of $PbMoO_4$ and basic lead molybdates whose compositions depend on the molar ratio Pb:Mo of the mixture. At Pb:Mo≦1.1, $Pb_{10}(OH)_2(MoO_4)_9 \cdot nH_2O$ (Pb:Mo=1.11) and at Pb:Mo>1.1, $Pb_4(OH)_2(MoO_4)_3 \cdot nH_2O$ (Pb:Mo=1.33) is precipitated. The formation of the two compounds, which are stable up to pH≧3, has also been proved by potentiometric and conductometric methods.

The ranges of pH for the formation of the neutral and basic lead molybdates are close together, therefore the composition of the precipitate depends on the order of mixing the reaction solutions of $Pb(NO_3)_2$ and Na_2MoO_4: adding the Na_2MoO_4 solution to $Pb(NO_3)_2$, $PbMoO_4 \cdot nH_2O$ is precipitated (n has the same characteristics as in $Pb_4(OH)_2(MoO_4)_3 \cdot nH_2O$). In the reverse mixing procedure $Pb_{10}(OH)_2(MoO_4)_9 \cdot nH_2O$ is formed first; upon further addition of $Pb(NO_3)_2$, $Pb_4(OH)_2(MoO_4)_3 \cdot nH_2O$ is formed.

The water content n of the air-dried $Pb_4(OH)_2(MoO_4)_3 \cdot nH_2O$ is about 0.7 depending on the temperature of the precipitation. The dehydration takes place at 75 to 116°C in one step. At 337°C the compound decays to $PbMoO_4$ and PbO.

A. N. Zobnina, I. P. Kislyakov (Izv. Vysshikh Uchebn. Zavedenii Khim. Khim. Tekhnol. **13** [1970] 143/7; C.A. **73** [1970] No. 41306).

8.3.3 Lead Peroxomolybdates $PbMoO_8 \cdot nH_2O$, $PbMoO_7 \cdot nH_2O$, $PbMoO_6 \cdot nH_2O$, and $PbMoO_5 \cdot nH_2O$

The lead peroxomolybdates are prepared by the reaction of various amounts of cold H_2O_2 with mixtures of solutions of equivalent quantities of $Pb(NO_3)_2$ and Na_2MoO_4. The crystalline products can be dehydrated completely over P_2O_5 in vacuum.

The DTA results in the table show that the thermal stability increases both with decreasing peroxo and water content:

compound	temperature of beginning O_2 evolution in °C (all ±1)	final temperature of dehydration in °C (all ±1)
$PbMoO_8 \cdot nH_2O$	0	106
$PbMoO_7 \cdot 5H_2O$	15	118
$PbMoO_7 \cdot 2H_2O$	20	109
$PbMoO_6 \cdot 2H_2O$ (as prepared)	32	74
$PbMoO_6 \cdot 2H_2O$ (after 10 d)	44	118
$PbMoO_6 \cdot 0.2H_2O$	53	126
$PbMoO_5 \cdot 2H_2O$	60	124
$PbMoO_5 \cdot 0.2H_2O$	65	130

Freshly prepared samples are more unstable and can decompose explosively to $PbMoO_4$. The observed exothermic effects make probable the following reaction scheme for the degradation of $PbMoO_8$: $PbMoO_8 \cdot nH_2O \rightarrow PbMoO_6 \cdot nH_2O + O_2 \rightarrow PbMoO_5 \cdot nH_2O + 1.5O_2 \rightarrow PbMoO_4 + 2O_2 + nH_2O$.

The dehydration of $PbMoO_7 \cdot 5H_2O$, $PbMoO_6 \cdot 2H_2O$, and $PbMoO_5 \cdot 2H_2O$ has been examined thermogravimetrically (the method is described in [1]). More than 2 molecules of H_2O in $PbMoO_7 \cdot 5H_2O$ are adsorbed; these molecules are lost first. Then the dihydrate $PbMoO_7 \cdot 2H_2O$ passes to the monohydrate. The last H_2O is bound more strongly and removed very slowly. $PbMoO_6 \cdot 2H_2O$ and $PbMoO_5 \cdot 2H_2O$ have similar behavior. It is therefore concluded that one molecule of H_2O in each of the dihydrates is water of crystallization, the other one has zeolitic character.

The enthalpies of decomposition (ΔH_d) of $PbMoO_7 \cdot 2H_2O$, $PbMoO_6 \cdot 2H_2O$, and $PbMoO_5 \cdot 2H_2O$ have been determined calorimetrically, and from these the binding energies of the peroxo group related to the MoO_4^{2-} ion are calculated. Because of its thermal instability the ΔH_d of $PbMoO_8 \cdot 2H_2O$ was derived from DTA data:

compound	ΔH_d in kcal/mol	binding energy in kcal
$PbMoO_8 \cdot 2H_2O$	88.0	35.9
$PbMoO_7 \cdot 2H_2O$	81.3	31.1
$PbMoO_6 \cdot 2H_2O$	76.4	20.3
$PbMoO_5 \cdot 2H_2O$	23.0	27.1

The kinetic constants K of thermal decomposition, according to the generalized equation of Erofeev and Kolmogorov ($\log[-\log(1-\alpha)] = \log K + n \log t$; α = fraction of the substance having reacted, t = time in min, and n = a constant), have been determined for $PbMoO_6$ ($K = 21.5 \times 10^{-3}$ at 52°C) and $PbMoO_5$ ($K = 48.9 \times 10^{-3}$ at 65°C) [2].

References:

[1] G. A. Bogdanov, I. K. Prokhorova, T. M. Kurokhtina, A. S. Chernyshev (Zh. Fiz. Khim. **40** [1966] 1724/8; Russ. J. Phys. Chem. **40** [1966] 932/4). – [2] T. M. Kurokhtina, G. A. Bogdanov, G. L. Smorgonskaya (Izv. Vysshikh Uchebn. Zavedenii Khim. Khim. Tekhnol. **18** [1975] 1022/4; C.A. **83** [1975] No. 187555).

9 Molybdate Hydrates with Subgroup 5 Metals

9.1 Compounds with Vanadium

Older data are given in "Molybdän", 1935, pp. 311, 376/80.

9.1.1 Molybdovanadates(IV)

9.1.1.1 The V^{IV}-Mo^{VI}-H_2O System

Depending on the concentrations, pH, and temperature, aqueous mixtures of MoO_4^{2-} and VO^{2+} form three different species containing V^{IV} and Mo^{VI} in the ratios V : Mo = 1 : 2, 1 : 1, and 2 : 1. In order to avoid the oxidation of V^{IV} to V^{V} the solutions must be free of oxygen (deaerated by bubbling Ar), especially if pH > 2. Reaction times of several days are necessary. In accordance with the group of hexametalates $M_6O_{19}^{x-}$ with the Lindquist type structure ($Nb_6O_{19}^{8-}$, see [1]) the compounds are formulated as $V_mMo_nO_{19}^{x-}$ (m + n = 6), because this structure was proved by many physical data for the W-containing anions $[V^{IV}W_5^{VI}O_{19}]^{4-}$ and $[V_2^{IV}W_4^{VI}O_{19}]^{6-}$ (see [2]). But detailed X-ray structure determinations do not exist. Therefore, other formulations like $V_mMo_nO_{16}^{(x-6)-}$ or $V_mMo_nO_{18}^{(x-2)-}$ can be found in the literature, see, e.g., [3].

The photometric titration curve of a 0.02 M Na_2MoO_4 solution at pH 5 and 60°C with a $VOSO_4$ solution under equilibrium conditions, shows two inflection points at the molar ratios V : Mo = 1 : 2 and 2 : 1 corresponding to $V_2Mo_4O_{19}^{6-}$ (I) and $V_4Mo_2O_{19}^{10-}$ (II). The acidimetric titration of a mixture of 0.005 M $VOSO_4$ and 0.01 M Na_2MoO_4 with NaOH and HCl, respectively, in the pH range 1 to 13 demonstrates the different pH-dependent equilibria of I. Each step of the titration must be heated to 45°C for 50 h. The curve shows at pH 4 the formation of I as a violet solution according to $2VO^{2+} + 4MoO_4^{2-} + H_2O + H^+ \rightarrow H_3V_2Mo_4O_{19}^{3-}$. At pH 8 to 9 the destruction of I occurs: $2H_3V_2Mo_4O_{19}^{3-} + 12OH^- \rightarrow V_4O_9^{2-} + 8MoO_4^{2-} + 9H_2O$. The acidic destruction is observed at pH 2.4 following the scheme $3H_3V_2Mo_4O_{19}^{3-} + 15H^+ \rightarrow 6VO^{2+} + 3Mo_4O_{13}^{2-} + 12H_2O$. At pH 6 the curve shows the formation of small amounts of II which forms a blue-black solution. The acidimetric titration curve of a mixture of 0.04 M $VOSO_4$ and 0.02 M Na_2MoO_4 (each step kept at 70°C for 15 d) shows at pH 5 the formation of II according to $4VO^{2+} + 2MoO_4^{2-} + 7OH^- \rightarrow H_7V_4Mo_2O_{19}^{3-}$. At pH 8, II is decomposed: $H_7V_4Mo_2O_{19}^{3-} + 3OH^- \rightarrow V_4O_9^{2-} + 2MoO_4^{2-} + 5H_2O$. In the pH range 5 to 3.9, an equilibrium between I and II exists according to $2H_7V_4Mo_2O_{19}^{3-} + 3H^+ \rightleftharpoons H_3V_2Mo_4O_{19}^{3-} + 6VO(OH)_2 + H_2O$. At pH 3.2 the redissolution of $VO(OH)_2$ occurs, and at pH 10 to 11 a condensation equilibrium of V^{IV} is observed [3, 4].

The $H_6V_3Mo_3O_{19}^{2-}$ species (III) which gives a red solution is formed at low temperatures (0°C) and higher concentrations. The photometric titration curve at 560 nm of a 0.05 M Li_2MoO_4 solution with VO^{2+} at pH 4.75 and 0°C under equilibrium conditions shows two inflection points at the molar ratios V: Mo = 1 : 2 and 1 : 1 corresponding to the anions $H_3V_2Mo_4O_{19}^{3-}$ (I) and III. (If Na_2MoO_4 is used, the salt $Na_2H_4V_2Mo_4O_{19}$ is precipitated in the pH range 2.7 to 8.) The acidimetric titration of a mixture of 0.05 M Li_2MoO_4 and 0.05 M $VOCl_2$ with HCl and LiOH, respectively, at 0°C shows at pH 4.75 the formation of the 1 : 1 species according to $3VO^{2+} + 3MoO_4^{2-} + 2H_2O + 2OH^- \rightarrow H_6V_3Mo_3O_{19}^{2-}$. At pH 10.5 the anion is decomposed: $4H_6V_3Mo_3O_{19}^{2-} + 22OH^- \rightarrow 3V_4O_9^{2-} + 12MoO_4^{2-} + 23H_2O$. At pH 3.1, III is in equilibrium with I: $4H_6V_3Mo_3O_{19}^{2-} + 14H^+ \rightleftharpoons 3H_4V_2Mo_4O_{19}^{2-} + 6VO^{2+} + 13H_2O$. Therefore, in the pH range 2.7 to 4.2 $Li_2H_4V_2Mo_4O_{19}$ is precipitated. At pH < 3.1 the very soluble free acid $H_8V_3Mo_3O_{19}$ is formed as a red solution. At pH 1.75, III is decomposed forming $Mo_4O_{13}^{2-}$ or $HMo_6O_{20}^{3-}$ and VO^{2+} [4], see also [5]. The absorption spectra in the range 350 to 800 nm of I, II, III, and the free acid of III are plotted in [4]. From the three anions defined salts have been isolated (see below).

The oxidation of VO^{2+} solutions by oxygen at pH > 2 is catalyzed by small amounts of MoO_4^{2-}. Kinetic measurements at pH 3.4, $p(O_2) = 730$ Torr, and 30°C in solutions of $VOSO_4$ and Na_2MoO_4 with V : Mo = 1 : 1/60 have shown that the activating species is the complex III, because the ammonium salt $[(C_2H_5)_4N]_2H_4V_3Mo_3O_{18}$ can be isolated from the reaction mixture. Two molecules of III react with one O_2 molecule, which is reduced in two steps to H_2O according to the scheme

$$3VO^{2+} + 3MoO_4^{2-} + 3H_2O \rightleftharpoons H_6V_3^{IV}Mo_3O_{18}$$
$$2H_6V_3^{IV}Mo_3O_{18} + O_2 \rightarrow 2H_5V_2^{IV}V^{V}Mo_3O_{18} + H_2O_2$$
$$H_2O_2 + 2V^{IV} + 2H^+ \rightarrow 2H_2O + 2V^{V}$$
$$H_5V_2^{IV}V^{V}Mo_3O_{18} + V^{IV} + H^+ \rightleftharpoons H_6V_3^{IV}Mo_3O_{18} + V^{V}$$

This mechanism is reasonable because the oxidation rate is independent of VO^{2+}, second-order in Na_2MoO_4, and first-order in the O_2 concentration. V^V and Mo^V inhibit the reaction [6]. The same complex III seems to be the O_2 activating species in the oxidation of reduced heteropoly acids of the type $H_{3+n}[PMo_{12-n}V_nO_{40}]$, n = 1 to 4, because they dissociate liberating VO^{2+} which reacts with molybdate in the above manner [7]. The higher stability of V^{IV}-Mo^{VI} complexes in relation to $[PMo_{12-n}V_nO_{40}]^{(3+n)-}$ with n > 1 is also seen in photometric titration curves of aqueous mixtures of VO^{2+} and PO_4^{3-} titrated with MoO_4^{2-}, which only show the formation of $[PMo_{11}VO_{40}]^{4-}$ and the V_2Mo_4 species (I), but not phosphomolybdates with more than one V^{IV} atom [8].

References:

[1] I. Lindquist (Arkiv Kemi **5** [1953] 247/50). – [2] C. M. Flynn Jr., M. T. Pope (Inorg. Chem. **12** [1973] 1626/34). – [3] D. Labonnette, S. Ostrowetsky (Compt. Rend. C **282** [1976] 341/4). – [4] D. Labonnette (J. Chem. Res. S **1979** 252/3; J. Chem. Res. M **1979** 2801/31). – [5] S. Ostrowetsky, D. Labonnette (Compt. Rend. C **282** [1976] 169/72).

[6] L. I. Kuznetsova, K. I. Matveev (React. Kinet. Catal. Letters **3** [1975] 305/10; C.A. **85** [1976] No. 10711). – [7] L. I. Kuznetsova, E. N. Yurchenko, K. I. Matveev (Tezisy Dokl. 12th Vses. Chugaevskoe Soveshch. Khim. Kompleksn. Soedin., Novosibirsk 1975, Vol. 2, p. 188; C.A. **85** [1976] No. 167200). – [8] P. Souchay, G. Bertho (Compt. Rend. C **262** [1966] 42/5).

9.1.1.2 $K_3H_7V_4Mo_2O_{19} \cdot 7$ to $9H_2O$ (= $3K_2O \cdot 8VO_2 \cdot 4MoO_3 \cdot 21$ to $25H_2O$)

Stoichiometric quantities of 1 M aqueous, oxygen-free solutions of K_2MoO_4 and $VOCl_2$ are mixed and adjusted to pH 5.5 with KOH. The mixture is heated to 70°C for 15 d. After the separation of precipitated $VO(OH)_2$ by filtration of the hot solution, black octahedral crystals precipitate slowly at room temperature. They can be washed with ethanol or ether [1]. In an earlier publication the compound was precipitated from an Na_2MoO_4-$VOSO_4$ solution by KCl [2].

The solid compound is stable in the air, but the solution in contact with oxygen is decomposed rapidly to decavanadate and molybdate [1].

References:

[1] D. Labonnette (J. Chem. Res. S **1979** 252/3; J. Chem. Res. M **1979** 2801/31, 2819). – [2] D. Labonnette, S. Ostrowetsky (Compt. Rend. C **282** [1976] 341/4).

9.1.1.3 $[(C_2H_5)_4N]_4H_6V_4Mo_2O_{19} \cdot 7\,H_2O$ ($= 2\,[(C_2H_5)_4N]_2O \cdot 4\,VO_2 \cdot 2\,MoO_3 \cdot 10\,H_2O$)

In a procedure analogous to the preparation of the potassium salt (see above), this tetraethyl ammonium salt is obtained as a blue-black powder which is oxidized rapidly in the air.

D. Labonnette (J. Chem. Res. S **1979** 252/3; J. Chem. Res. M **1979** 2801/31, 2819).

9.1.1.4 $H_8V_3Mo_3O_{19}$ ($= 3\,VO_2 \cdot 3\,MoO_3 \cdot 4\,H_2O$)

In the V^{IV}-Mo^{VI}-H_2O system (p. 325) the acid is obtained in the pH range 2.2 to 2.7. The absorption spectrum of the red solution shows a maximum at 500 nm. The acid has not been isolated and investigated in detail [1, 2].

References:

[1] S. Ostrowetsky, D. Labonnette (Compt. Rend. C **282** [1976] 169/72). – [2] D. Labonnette (J. Chem. Res. S **1979** 252/3; J. Chem. Res. M **1979** 2801/31, 2822).

9.1.1.5 $K_2H_6V_3Mo_3O_{19} \cdot 7\,H_2O$ ($= K_2O \cdot 3\,VO_2 \cdot 3\,MoO_3 \cdot 10\,H_2O$)

This compound was prepared from 1 M aqueous solutions of Li_2MoO_4 and $VOCl_2$ which are mixed in stoichiometric quantities, the pH adjusted to 4.75. The mixture is cooled to 0°C for several days. Adding KCl the brown-red potassium salt precipitates.

D. Labonnette (J. Chem. Res. S **1979** 252/3; J. Chem. Res. M **1979** 2801/31, 2825).

9.1.1.6 Alkylammonium Salts of $H_8V_3Mo_3O_{19}$

The tetramethylammonium salts were prepared from 1 M aqueous, oxygen-free solutions of Na_2MoO_4 and $VOSO_4$ which are mixed with a slight excess of VO^{2+}. The pH is adjusted to 5 and the mixture heated to 60°C for several hours. The precipitated $VO(OH)_2$ is filtered and solid $(CH_3)_4NBr$ added to the filtrate. After 1 d at room temperature brown-black crystals precipitate which can be washed with water, ethanol, and ether. They have the composition $[(CH_3)_4N]_2H_6V_3Mo_3O_{19} \cdot 5\,H_2O$. From the same mixture precipitates at pH 2.6 and 0°C immediately after the addition of $(CH_3)_4NBr$ the salt $[(CH_3)_4N]H_7V_3Mo_3O_{19} \cdot 14\,H_2O$. The two ammonium compounds oxidize rapidly in the air [1, 2].

From $VOSO_4$-Na_2MoO_4 solutions a compound of composition $[(C_2H_5)_4N]_2H_4V_3Mo_3O_{18}$ (or possibly $[(C_2H_5)_4N]_2H_6V_3Mo_3O_{19}$) has been precipitated (see als p. 326). No details of precipitation and properties of the compound are given [3].

References:

[1] S. Ostrowetsky, D. Labonnette (Compt. Rend. C **282** [1976] 169/72). – [2] D. Labonnette (J. Chem. Res. S **1979** 252/3; J. Chem. Res. M **1979** 2801/31, 2825). – [3] L. I. Kuznetsova, K. I. Matveev (React. Kinet. Catal. Letters **3** [1975] 305/10).

9.1.1.7 $Li_2H_4V_2Mo_4O_{19}$ ($=Li_2O \cdot 2VO_2 \cdot 4MoO_3 \cdot 2H_2O$)

This compound precipitates in the V^{IV}-Mo^{VI}-H_2O system (p. 325) in the presence of LiCl between pH 2.7 and 4.2. The compound has not been isolated and investigated in detail.

D. Labonnette (J. Chem. Res. M **1979** 2801/31, 2820).

9.1.1.8 Sodium Salts of $H_6V_2Mo_4O_{19}$

$Na_3H_3V_2Mo_4O_{19} \cdot 6H_2O$. Stoichiometric quantities of 1 M aqueous, oxygen-free solutions of Na_2MoO_4 and $VOSO_4$ are mixed and acidified with HCl to pH 4. An excess of NaCl is added and the mixture heated to 45°C for 2 d. After several days blue-violet crystals precipitate from the filtered solution. They can be washed with water, ethanol, and petroleum ether.

$Na_2H_4V_2Mo_4O_{19}$ precipitates in the V^{IV}-Mo^{VI}-H_2O system (p. 325) between pH 2.7 and 8. No other details are given for this compound.

D. Labonnette (J. Chem. Res. S **1979** 252/3; J. Chem. Res. M **1979** 2801/31, 2808, 2820).

9.1.1.9 Potassium Salts of $H_6V_2Mo_4O_{19}$

Stoichiometric quantities of 1 M aqueous, oxygen-free solutions of Na_2MoO_4 and $VOCl_2$ are mixed and acidified with HCl. An excess of KCl is added and the mixture heated to 45°C for 2 d. From the filtered solution precipitate crystals after some days: $K_4H_2V_2Mo_4O_{19} \cdot 9H_2O$ as violet octahedra at pH 4.5, $K_3H_3V_2Mo_4O_{19} \cdot 3$ to $5H_2O$ at pH 4, and $K_2H_4V_2Mo_4O_{19}$ at pH 3 [1, 2].

References:

[1] D. Labonnette, S. Ostrowetsky (Compt. Rend. C **282** [1976] 341/4). – [2] D. Labonnette (J. Chem. Res. S **1979** 252/3; J. Chem. Res. M **1979** 2801/31, 2811).

9.1.2 Molybdovanadates(IV, V)

9.1.2.1 $K_6V^{IV}V^{V}_2Mo_{10}O_{40} \cdot 13H_2O$ ($=3K_2O \cdot VO_2 \cdot V_2O_5 \cdot 10MoO_3 \cdot 13H_2O$)

From the mother liquor of the preparation of $K_8[V_8Mo_4O_{36}] \cdot 12H_2O$ (see p. 331), the compound precipitates as dark brown cube-shaped crystals after several months, besides a yellow phase of other molybdovanadates. The yield of this reduced species, which is stable in the air, is very small. The crystals are rapidly decayed by X-rays. For the structure investigation they were enclosed together with some mother liquor in a sealed capillary.

The cubic unit cell of the crystals has the parameter a = 10.6124(5) Å; Z = 1. The single crystal X-ray structure determination (R = 0.061) gives the space group $P\bar{4}3m$-T_d^1 (No. 215). Atomic positions:

atom	position	x	y	z
K(1)	3c	0	0.5	0.5
0.5 K(2)	6f	0.3661(10)	0	0
Mo/V	12i	0.26736(8)	0.26736(8)	0.49981(40)
V	1b	0.5	0.5	0.5

atom	position	x	y	z
O(1)	4e	0.5895(10)	0.5895(10)	0.5895(10)
O(2)	12i	0.3928(8)	0.3928(8)	0.2214(12)
O(3)	12i	0.3604(8)	0.3604(8)	0.8201(10)
O(4)	12i	0.1599(7)	0.1599(7)	0.5093(38)
Aq(1)	12i	0.2229(14)	0.2229(14)	0.0057(35)
Aq(2)	1a	0	0	0

The anion has the well-known α-Keggin-type structure. A very regular V^VO_4 tetrahedron with V-O distances of 1.64(2) Å is surrounded by 4 M_3O_{13} units (M = Mo, V^{IV}, or V^V). The 10 Mo atoms and 2 V atoms are in a random distribution. The 3 MO_6 octahedra of the M_3O_{13} unit have one O atom common with the VO_4 tetrahedron, with an M-O distance of 2.350(10) Å. The M-O distances of bridging O atoms are 1.816(4) to 2.009(6), those of terminal O atoms 1.617(11) Å. These values are typical for the Keggin-type structure. The M-M distances between the octahedra are 3.49 Å, and the V-M distances from the central V atom are also 3.49 Å. The 6 K^+ ions are 8-coordinated in square antiprisms; three of them link only with O atoms of the anions, the other three with 4 anion O atoms and 4 water molecules. H bonds are very weak. There is a rotational disorder in the structure with the anion distributed between two positions interrelated by a 90° rotation. Interatomic distances are tabulated in the paper.

The EPR spectrum at 77 K (plotted in the paper) shows that the reduced metal atom of this heteropoly blue compound is V^{IV}. In analogy to all other known heteropoly blue compounds, it is supposed that the reduced atom, V^{IV}, is coordinated octahedrally, so that the anion may be formulated as $[(V^{IV}, V^V, Mo^{VI}_{10})V^VO_{40}]^{6-}$.

The density was found to be strongly dependent on the crystal size, the larger crystals being less dense probably due to water inclusions. The smallest crystals were found to have a pycnometric density of 2.98 g/cm³. The calculated density is 3.085 g/cm³.

A. Björnberg, B. Hedman (Acta Cryst. B **36** [1980] 1018/22).

9.1.2.2 $K_3V_6MoO_{19}\cdot15H_2O$ (= $3K_2O\cdot2VO_2\cdot5V_2O_5\cdot2MoO_3\cdot30H_2O$)

This compound is obtained from an aqueous mixture of "molybdic acid" and potassium metavanadate in a molar ratio of 1:12, boiled for ~4 h. Small red needles precipitate under vacuum, which can be recrystallized from hot water. (In the paper the formula $K_3V_6MoO_{19}\cdot11H_2O$ is given, but the molecular weight and the calculated values for elemental analysis given coincide only with $K_3V_6MoO_{19}\cdot15H_2O$.)

From single crystal investigations the parameters of the monoclinic unit cell are a = 9.42, b = 15.38, c = 10.12 Å, β = 105°; Z = 2. Space group $P2_1/c-C^5_{2h}$ (No. 14). The experimental density is 2.53 g/cm³.

S. K. Roy, G. C. Bhattacharya (J. Indian Chem. Soc. **53** [1967] 527).

9.1.3 Molybdovanadates(V)

Remarks. In aqueous solutions containing molybdates and vanadates numerous complex equilibria exist not only between several polymolybdates and polyvanadates but also between different heteropoly compounds containing the two elements, which can form for their part different redox equilibria. Therefore different species exist together which explains the difficulties in isolating well-defined compounds from these solutions. Of the many examples found in the literature not all may be pure, well-defined V-Mo complexes or they may have another composition.

9.1.3.1 The V^{V}-Mo^{VI}-H_2O System

In this system anions with V:Mo ratios of 9:1, 1:1, 1:2, 1:3, and 1:5 have been found. Compounds have been isolated with the ratio 1:2 (p. 333), 1:3 (p. 333), and 1:5 (p. 335).

In aqueous 0.04 M solutions of $NaVO_3$ at pH 4 mixed with varying amounts of Na_2MoO_4 (0 to 0.12 M) one can find after 2 d by the photometric method at 350 nm one inflection point at Mo:V = 2, corresponding to the $[V_2Mo_4O_{19}]^{4-}$ species, which may be derived from the group of hexametalates $M_6O_{19}^{x-}$ such as $Mo_6O_{19}^{2-}$. The isolation of salts of this polyacid is complicated by the decavanadate formation in the same pH region [1, 2]. The ^{51}V NMR spectrum of $[V_2Mo_4O_{19}]^{4-}$ at room temperature and pH 4.5 shows a chemical shift of 502 ± 2 ppm (line width 0.07 G) related to $VOCl_3$ [3].

In more dilute and more acidic solutions $[V_2Mo_4O_{19}]^{4-}$ is in equilibrium with another hexametalate ion, $[VMo_5O_{19}]^{3-}$, according to $[HV_2Mo_4O_{19}]^{3-} + HMo_6O_{21}^{5-} + 2H^+ \rightleftharpoons 2[VMo_5O_{19}]^{3-} + 2H_2O$. The existence of this species depends mainly on the concentration of vanadate and the pH of the solution. Vanadate solutions above 0.04 M only show $[V_2Mo_4O_{19}]^{4-}$, but the spectrophotometric method of 2×10^{-4} M vanadate solutions with varying molybdate concentrations at pH 3 show only the formation of $[VMo_5O_{19}]^{3-}$, whereas a 4×10^{-3} M vanadate solution at pH 3 has two inflection points at Mo:V = 2 and 5. The influence of the absolute concentration of V on the equilibrium may be an effect of the ionic strength of the mixture, because the addition of NaCl (or $NaClO_4$) also shifts the equilibrium to the left. In contrast to $[V_2Mo_4O_{19}]^{4-}$ this polyanion is formed instantaneously. In solutions with pH < 2 it is decomposed [4]. At pH 3 the intense yellow solutions are stable for at least 24 h, but contact with glass must be avoided because of the formation of 12-molybdosilicic acid. The yellow species can be reduced by 1-amino-2-naphthol-4-sulfonic acid to a purple species [5]. The absorption spectrum from 250 to 450 nm is plotted in [4] showing a shoulder at 340 nm. The acid and some salts have been isolated (see p. 335).

In molybdate/vanadate mixtures of higher vanadate content a $[V_9MoO_{28}]^{5-}$ ion is formed at pH 4 in a slow reaction (5 d). It can be derived from decavanadate $V_{10}O_{28}^{6-}$, which exists in this pH region. When the decavanadate concentration of equilibrated mixtures of 0.05 M molybdate solutions and varying vanadate contents at pH 4 is plotted logarithmically the curve shows two equivalent points at V:Mo = 1 and 9. By a detailed analysis of the curve the 1:1 species can be formulated as $[HV_3Mo_3O_{19}]^{4-}$ [1, 2] (the correct formula is given in [4]). The absorption spectra of $[HV_3Mo_3O_{19}]^{4-}$, $[V_2Mo_4O_{19}]^{4-}$, and $[V_9MoO_{28}]^{5-}$ have no characteristic maxima in the region 250 to 450 nm, but they show an isosbestic point at 360 nm. They are plotted in [2].

V:Mo = 1:1 and 1:5 species were also found besides a 1:3 species by the spectrophotometric method at 400 nm in solutions of 0.002 M vanadate and 0.069 M $HClO_4$ with varying amounts of molybdate [6]. From the 1:3 species well defined Na salts are known (see p. 333).

References:

[1] P. Souchay, F. Chauveau (Compt. Rend. **247** [1958] 1619/23). – [2] F. Chauveau (Bull. Soc. Chim. France **1960** 834/48, 845). – [3] L. P. Kazanskii, V. I. Spitsyn (Dokl. Akad. Nauk SSSR **223** [1957] 381/4; Dokl. Phys. Chem. Proc. Acad. Sci. USSR **220/225** [1957] 721/4). – [4]

F. Chauveau, P. Souchay (Bull. Soc. Chim. France **1963** 561/5). – [5] G. W. Wallace, M. G. Mellon (Anal. Chim. Acta **23** [1960] 355/62), G. W. Wallace (Diss. Purdue Univ. 1957; Diss. Abstr. **20** [1960] 2546).

[6] R. W. Hunt, Jr., L. G. Hargis (Anal. Chem. **49** [1977] 779/84).

9.1.3.2 $(NH_4)_2[V_6MoO_{19}]\cdot 12H_2O$ (= $(NH_4)_2O\cdot 3V_2O_5\cdot MoO_3\cdot 12H_2O$)

From an aqueous mixture of 0.05 M "molybdic acid" and 0.2 M ammonium metavanadate, boiled for 4 h, small silky yellowish crystals are obtained under vacuum. The lattice parameters of the orthorhombic unit cell are a = 9.62, b = 16.25, c = 6.24 Å (all ± 0.02 Å); Z = 2. Space group Pnnn-D_{2h}^2 (No. 48). The pycnometric density is 3.240, the calculated 3.259 g/cm³.

S. K. Roy (Indian J. Chem. A **15** [1977] 358).

9.1.3.3 $K_8[V_8Mo_4O_{36}]\cdot 12H_2O$ (= $4K_2O\cdot 4V_2O_5\cdot 4MoO_3\cdot 12H_2O$)

This compound is obtained as yellow acicular crystals by recrystallizing $K_7[V_5Mo_8O_{40}]\cdot \sim 8H_2O$ (see p. 332) from water. The product is unstable in air. During the X-ray exposures the crystals were enclosed together with part of the mother liquor in a sealed glass capillary. They show a very good cleavage ||(100).

The monoclinic crystals have the lattice parameters a = 23.255(2), b = 11.708(1), c = 18.453(8) Å, β = 114.11(2)°; Z = 4. The space group from the three-dimensional X-ray structure determination (R = 0.055) is C2/c-C_{2h}^6 (No. 15). The atomic coordinates are listed in the paper.

In the crystal the anions $[V_8Mo_4O_{36}]^{8-}$ are arranged in two different layers ||(100) of mutual sequence, which are held together by O-K^+-O and H bonds. The layers differ in the spatial orientation of the anions (see figure in the paper). The shortest O-O distance between two anions is 2.814(9) Å. The anion consists of 12 edge-sharing polyhedra paired by an inversion center. Two MoO_6 and four VO_6 octahedra form a nearly planar ring, which is capped on both sides by a VO_5 square pyramid. Two other MoO_6 octahedra outside the ring share an edge with the VO_6 octahedra of the ring, and two VO_5 trigonal bipyramids are linked to VO_6 and MoO_6 octahedra outside the ring, see **Fig. 81**, p. 332 (from [2]). There is no direct contact between the MoO_6 octahedra. They are distorted according to the different binding situation of their O atoms. The VO_6 octahedra and VO_5 trigonal bipyramids show extremely long V-O distances to the quadruply and triply coordinated O atoms. The following table shows the M-O distances:

polyhedron	terminal O	bridging O	three-coordinated O	four-coordinated O
MoO_6 octahedra	1.689(6) to 1.740(4)	1.856(4) to 1.973(4)	2.012(4) to 2.241(4)	2.376(4)
VO_6 octahedra	1.605(5) to 1.629(5)	1.769(4) to 1.919(4)	1.713(4) to 1.991(4)	2.638(4) to 2.761(4)
VO_5 square pyramids	1.612(5)	—	1.861(4) to 2.190(4)	1.707(4) to 3.032(5)
VO_5 trigonal bipyramids	1.625(6) to 1.651(5)	1.836(5)	1.836(4) to 2.670(4)	—

The Mo-Mo distance is 5.270(1) Å, the Mo-V distances range from 3.094(1) to 3.464(2) Å, and the V-V distances from 3.011(2) to 3.371(1) Å. A complete list of the interatomic distances and angles is given in the paper.

The pycnometric density is 2.744(3), the calculated 2.746 g/cm^3 [1].

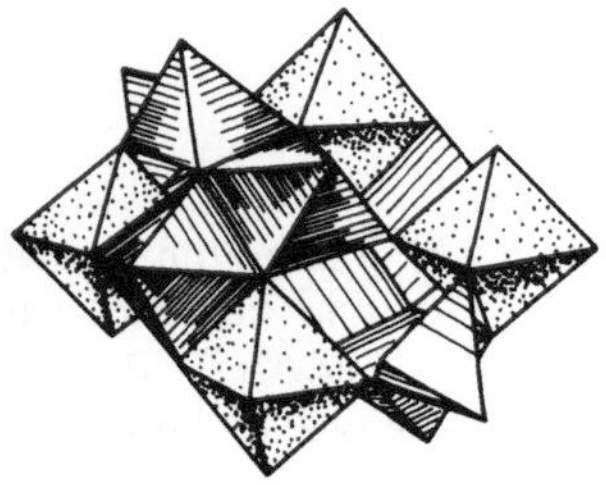

Fig. 81. The structure of the $[V_8Mo_4O_{36}]^{8-}$ anion [2].

References:

[1] A. Björnberg (Acta Cryst. B **35** [1979] 1989/95). – [2] M. T. Pope (Heteropoly and Isopoly Oxometalates, Springer, Berlin 1983, p. 55).

9.1.3.4 $K_7[V_5Mo_8O_{40}]\cdot{\sim}8H_2O$ ($=7K_2O\cdot5V_2O_5\cdot16MoO_3\cdot{\sim}16H_2O$)

This compound is obtained from a mixture of 0.025 mol KOH, 0.05 mol NH_4VO_3, 0.025 mol MoO_3, and 0.1 mol KCl in 110 mL water, acidified with 12.5 mL 2 M HCl. The yellow acicular crystals precipitate overnight. They are unstable in air. Recrystallization from water gives $K_8[V_8Mo_4O_{36}]\cdot12H_2O$ (see above). During the X-ray exposures the crystals were sealed together with part of the mother liquor in a glass capillary.

The monoclinic unit cell has the lattice parameters a = 19.435(4), b = 20.237(5), c = 12.769(3) Å, β = 108.29(2)°; Z = 4. Space group $P2_1/n\text{-}C^5_{2h}$ (No. 14). The atomic coordinates are listed in the paper.

The three-dimensional X-ray structure determination (R = 0.070) shows the $[V_5Mo_8O_{40}]^{7-}$ anion to consist of a ring of eight edge-sharing MoO_6 octahedra, which are puckered in a two-up-two-down order. To these octahedra four VO_5 trigonal bipyramids are linked sharing two edges with the ring. The remaining V atom is the center of the anion in a VO_4 tetrahedron sharing each corner with two MoO_6 octahedra. The anion contains 24 terminal and 16 three-coordinated but no bridging O atoms. The MoO_6 octahedra are considerably distorted, the Mo-O distances range from 1.680(11) to 1.709(10) Å for terminal O atoms and from 1.901(10) to 2.386(10) Å for the three-coordinated O atoms. The V-O distances of the very regular VO_4 tetrahedron are 1.693(9) to 1.734(10) Å; the angles differ from the ideal tetrahedral value by about 3°. The VO_5 trigonal bipyramids all have one extremely long V-O bond (2.712(9) to 2.857(10) Å), the other 4 tetrahedron-like V-O distances range from 1.607(16) to 1.650(12) Å for terminal O atoms and from 1.826(10) to 1.875(11) Å for three-coordinated O atoms. For edge-sharing polyhedra the Mo-Mo distances range from 3.302(2) to 3.377(2) Å and the Mo-V distances from 3.426(3) to 3.524(3) Å. Two of the K^+ ions are situated in the center of the cavities on the two sides of the anion, coordinating with a six-membered ring, which is formed by the axial terminal O atoms of the anion, see **Fig. 82** (in [2] all the V atoms are shown occupying tetrahedra instead of trigonal bipyramids). This arrangement stabilizes the unusual anion structure, avoiding the formation of the Keggin-type structure (as in $[V_5W_8O_{40}]^{7-}$). The other K^+ ions coordinate 7 to 11 O atoms of the anion or the water molecules in a rather irregular geometry with distances of about 3.25 Å. The anion rings are sandwiched on top of each other with a slight mutual twist. The layers thus formed are parallel to the (010) plane. A complete list of the interatomic distances and angles is given in the paper.

Fig. 82. The structure of the anion in $K_7[V_5Mo_8O_{40}]\cdot8\,H_2O$ showing the position of one of the K^+ ions [1]

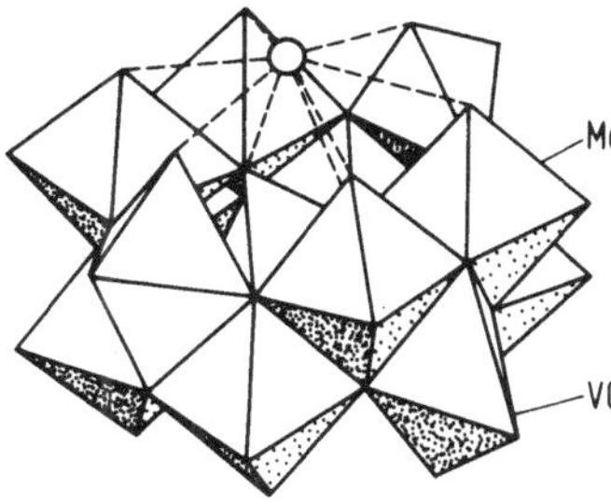

The pycnometric density is 2.89(1), the calculated 2.896 g/cm^3 [1].

References:

[1] A. Björnberg (Acta Cryst. B **36** [1980] 1530/6). – [2] M. T. Pope (Heteropoly and Isopoly Oxometalates, Springer, Berlin 1983, p. 55).

9.1.3.5 $\mathbf{(C_{17}H_{20}NO_3)_2H_2V_2Mo_4O_{19}\cdot3\,H_2O}$ $(=(C_{17}H_{20}NO_3)_2O\cdot V_2O_5\cdot4\,MoO_3\cdot4\,H_2O)$?

This morphinium salt is possibly the only separated salt of the $[V_2Mo_4O_{19}]^{4-}$ species. Adding a morphine solution to a slightly acidified mixture of 1 volume of a 10% (weight?) ammonium molybdate and 4 volumes of a saturated ammonium vanadate solution a yellow product precipitates with the composition $C_{17}H_{19}NO_3\cdot V(OH)_5\cdot2\,MoO_3$, which may be formulated as $(C_{17}H_{20}NO_3)_2H_2V_2Mo_4O_{19}\cdot3\,H_2O$.

W. Deckert (Z. Anal. Chem. **112** [1938] 241/57, 253).

9.1.3.6 $\mathbf{Na_6[V_2Mo_6O_{26}]\cdot16\,H_2O}$ $(=3\,Na_2O\cdot V_2O_5\cdot6\,MoO_3\cdot16\,H_2O)$, $\mathbf{Na_4H_2V_2Mo_6O_{26}\cdot16\,H_2O}$ $(=2\,Na_2O\cdot V_2O_5\cdot6\,MoO_3\cdot17\,H_2O)$?

$Na_6[V_2Mo_6O_{26}]\cdot16\,H_2O$ dominates the pH range 5.5 to 6.5 in mixtures of sodium paramolybdate and metavanadate. From the orange solutions containing the two components in the molar ratio V : Mo = 1:1, 1:2, 1:3, 1:6, or 1:12 only this salt is obtained. It crystallizes as yellow prisms when the reaction volume is reduced slowly by evaporation to 1/3. The best yield is obtained from the 1:3 and 1:6 mixtures. The precipitate can be recrystallized from water and dried in the air [1].

To get crystals for X-ray studies, hot solutions of 1.94 g $NaVO_3\cdot4\,H_2O$ in 30 mL H_2O and of 7.36 g $Na_2MoO_4\cdot2\,H_2O$ in 30 mL H_2O were mixed. 13.33 mL 3M HCl were added dropwise followed by 5 g NaCl. The solution was filtered at 313 K. After 10 d of slow evaporation at room temperature yellow prisms had precipitated. The crystals were found to be unstable in air. For X-ray investigation they were sealed with part of the mother liquor in a glass capillary [2].

The lattice parameters of the triclinic unit cell are a = 10.176(2), b = 10.416(4), c = 10.292(2) Å, α = 113.19(2)°, β = 95.54(2)°, γ = 101.73(2)°; Z = 1. The single crystal structure determination (R = 0.036) gives the space group $P\bar{1}$-C_i^1 (No. 2). The atomic coordinates are listed in the paper [2].

The $[V_2Mo_6O_{26}]^{6-}$ anion is isostructural with α-$[Mo_8O_{26}]^{4-}$ (see Fig. 13, p. 17), $[As_2Mo_6O_{26}]^{6-}$, and $[(CH_3As)_2Mo_6O_{26}]^{8-}$. It consists of a planar ring of six edge-sharing MoO_6 octahedra capped by two VO_4 tetrahedra. In the crystal the anions are connected by Na^+ ions and H bonds. The shortest O-O distances between two anions are very large with 3.152(4) Å. The Na^+ ions are coordinated octahedrally by 1 to 3 anionic O atoms and by 5 to 3 water molecules. The MoO_6 octahedra show some distortions, whereas the VO_4 tetrahedra are very regular, reaching the ideal tetrahedral angle within 1°. The Mo-O distances in the MoO_6 octahedra are 1.706(4) to 1.726(3) Å for terminal O, 1.898(3) to 1.925(3) Å for bridging O, and 2.313(3) to 2.383(3) Å for three-coordinated O. In the VO_4 tetrahedra the V-O distances are 1.640(3) Å for the terminal O and 1.775(3) to 1.783(3) Å for the three-coordinated O. The short Mo-Mo distances are 3.270(1) to 3.284(1) Å, the V-Mo distances 3.712(1) to 3.770(1) Å. The interatomic distances and angles are given in the paper [2].

The pycnometric density is 2.627(10), the calculated 2.621 g/cm³ [2]. The anisotropic crystals have the refractive indices n_1 = 1.718 and n_2 = 1.685. The IR spectrum from 3600 to 400 cm^{-1} is plotted [1].

The crystals of the 16-hydrate are not stable in the air. They lose water to form a product with 11 to 12 H_2O. The thermogram of this hydrate from 20 to 840°C (plotted in the paper) shows that the remaining water is lost in several steps at 150, 230, and 290°C. At 230°C a product with 0.7 H_2O results which is darker and less soluble, but its IR spectrum is identical with that of the 11.5-hydrate. At temperatures > 230°C the product is completely dehydrated, which is accompanied by the destruction of the compound [1].

The crystals of $Na_6[V_2Mo_6O_{26}] \cdot 16H_2O$ dissolve readily in water to give a light yellow solution, which darkens to orange when heated. The solution is stable against acids. Bases decompose the anion to molybdate and vanadate [1].

$Na_4H_2V_2Mo_6O_{26} \cdot 10H_2O$. From the mixtures of molybdate and vanadate in the molar ratio V : Mo = 1:1, 1:2, 1:3, and 1:6 precipitate in the more acidic pH range 4.0 to 5.0 yellow crystals, also with a ratio V : Mo = 1:3 but with a lower base content than at pH 5.5 to 6.5. They can be recrystallized from water and have the composition $2Na_2O \cdot V_2O_5 \cdot 6MoO_3 \cdot 17H_2O$, which may be $Na_4V_2Mo_6O_{25} \cdot 17H_2O$ or probably $Na_4H_2V_2Mo_6O_{26} \cdot 16H_2O$ (the authors propose $Na_6V_3Mo_9O_{37.5}$ but they also formulate the proved structure of $Na_6[V_2Mo_6O_{26}]$ as $Na_9V_3Mo_9O_{39}$). The anisotropic crystals have the refractive indices n_1 = 1.686 and n_2 = 1.754. The IR spectrum from 3200 to 400 cm^{-1} and the X-ray powder diagram are plotted. The plotted thermogram from 20 to 700°C shows that 16 of the 17 water molecules are lost in the temperature range 0 to 220°C accompanied by two endothermic effects. The monohydrate has the structure of the initial compound, which is proved by the IR spectra. The remaining water molecule leaves at 220 to 270°C causing the destruction of the compound [3].

References:

[1] A. A. Amirbekova, A. K. Il'yasova (Izv. Akad. Nauk Kaz. SSR Ser. Khim. **19** [1969] 81/4; C. A. **71** [1969] No. 131140). – [2] A. Björnberg (Acta Cryst. B **35** [1979] 1995/9). – [3] A. B. Bekturov, A. K. Il'yasova, A. A. Amirbekova (Zh. Neorgan. Khim. **15** [1970] 2781/5; Russ. J. Inorg. Chem. **15** [1970] 1446/8).

9.1.3.7 $H_3VMo_5O_{19}$ (= $V_2O_5 \cdot 10\,MoO_3 \cdot 3\,H_2O$) and its Salts

The free acid may be isolated as an ether addition compound from the same aqueous mixtures which form $H_3VMo_6O_{22}$ (see below), but the conditions to get the VMo_5 compound are not clear. It is only mentioned that sometimes the VMo_5 instead of the VMo_6 compound is formed [1]. Furthermore, $[VMo_5O_{19}]^{3-}$ is decomposed in the pH range 1.1 to 1.3 [2] which is the range of the mixtures of molybdate and metavanadate for the formation of $[VMo_6O_{22}]^{3-}$ (see below). It is possible to extract the yellow acid from more basic aqueous solutions (2×10^{-4} M) with pentanol by adding 8-hydroxyquinoline (10^{-3} M). So at pH 2.0 to 2.5 extraction yields up to 75% can be reached with pentanolic oxine solutions in contrast to only 40% with pentanol only. Oxine (B) and the acid (A) form B_2A salts which are soluble in pentanol [3]. The acid is slightly more stable in methanolic solution than in aqueous solutions. Titration with methanolic KOH solution shows the basicity of the acid to be two [1].

$[(n\text{-}C_4H_9)_4N]_3[VMo_5O_{19}]$ has been isolated from the stoichiometric mixture of a methanolic solution of $(n\text{-}C_4H_9)_4NOH$ and a solution of V_2O_5 and $\alpha\text{-}[(n\text{-}C_4H_9)_4N]_4[Mo_8O_{26}]$ in acetonitrile. ^{17}O NMR and IR spectroscopy (1000 to 650 cm^{-1}) show the compound to contain the $[VMo_5O_{19}]^{3-}$ anion which is isostructural with $[VW_5O_{19}]^{3-}$ and $[Mo_6O_{19}]^{2-}$ (see Fig. 16, p. 18) [4].

By the addition of a 10 (wt?)% solution of urotropine (hexamethylendiamine), a 10% solution of guanidine, $(NH_2)_2CNH$, or a 5% solution of pyramidone (N-dimethyl-4-amino-antipyrine) in 0.1 M HCl to a 0.02 M solution of $H_3VMo_5O_{19}$ the corresponding salts precipitate. After 24 h the crystalline products can be filtered off. They are washed with 0.1 M HCl and then with ethanol, and dried at 50 to 60°C. They contain 2 to 3 moles of the base per mole acid. It seems difficult to obtain defined salts [1].

References:

[1] E. F. Tkach, N. A. Polotebnova (Zh. Neorgan. Khim. **14** [1969] 3354/7; Russ. J. Inorg. Chem. **14** [1969] 1768/70). – [2] E. Chauveau, P. Souchay (Bull. Soc. Chim. France **1963** 561/5). – [3] E. F. Tkach (Khim. Koord. Neorg. Org. Soedin. **1978** 32/5; Ref. Zh. Khim. **1978** 19G337). – [4] W. G. Klemperer, W. Shum (J. Chem. Soc. Chem. Commun. **1979** 60/1).

9.1.3.8 $H_3VMo_6O_{22} \cdot \sim 6\,H_2O$ (= $V_2O_5 \cdot 12\,MoO_3 \cdot \sim 15\,H_2O$)

The acid is obtained from its ether addition compound. A stoichiometric aqueous mixture of $Na_2MoO_4 \cdot 2\,H_2O$ and $NaVO_3 \cdot 2\,H_2O$ at pH 1.1 to 1.3 (HCl) is boiled for 1 h. After 24 h the orange solution is filtered, then ether and concentrated HCl are added, and the solution cooled with ice. The separated ether addition compound is washed with ether/HCl and decomposed with water. After 24 h the free acid can be separated as a glassy mass with a yield of 53 to 55%, which depends on the acid used in the reaction: HCl and HNO_3 give the same yield, H_2SO_4 lessens it considerably. The compound can also be extracted by a mixture of pentanol and ethyl methyl ketone in the presence of HNO_3 (0.02 to 0.08 N) with a yield of 50 to 52% [1].

The solid acid is stable for a long time; after one year storage it showed only a slight loss of water (1 to 2%). It is sparingly soluble in cold water but easily soluble in hot water. In water, hydrolysis is noted after 3 to 4 h, which explains the difficulties in determining the basicity by titration with KOH. The acid obtained by crystallization of the ether addition compound without addition of water has a basicity of six, that which is crystallized with the addition of water has basicity seven. Therefore a dimeric formula is ascribed to the acid, $H_6V_2Mo_{12}O_{44} \cdot 11$ to $13\,H_2O$ [1]. However, a molecular mass determination of the acid neutralized with NaOH in fused

$Na_2SO_4 \cdot 10H_2O$ (Glauber's salt) shows that the anion is monomeric having the formula $[VMo_6O_{22}]^{3-}$ [2]. The absorption spectrum from 220 to 340 nm in aqueous and butanolic solution is plotted in [1].

Several reductants like ascorbic acid, $N_2H_4 \cdot HCl$, or $SnCl_2$ reduce the acid in 2 h to a species of the same V : Mo = 1 : 6 ratio as the yellow oxidized form. To the hot solution of the acid at pH 1.1 to 1.3 an aqueous solution of an equimolar amount of ascorbic acid is added. From the cooled mixture the product is extracted by ether/HCl. The separated ether addition compound is decomposed with water, from which blue acicular crystals are obtained. The acidity of the reaction mixture affects the stability of the product. The maximum stability within a period of 16 to 18 h is reached in 0.004 to 0.01 M H_2SO_4 for ascorbic acid and 0.02 to 0.2 M H_2SO_4 for hydrazine. The titration of the reduced solution with KOH shows a basicity of 6, i.e., the blue compound is derived from the dimeric acid $H_6V_2Mo_{12}O_{44}$. Potentiometric titrations of $H_3VMo_6O_{22}$ with the reductants show that the acid takes up two electrons to reduce 2 of the 12 Mo^{VI} to Mo^{V}. With an excess of the reductant more Mo atoms can be reduced, but the products have other compositions than the oxidized form, i.e., V : Mo = 1 : 7, 1 : 8, or 1 : 9. The absorption spectrum has a maximum at 780 to 800 nm; in the UV range the spectra of the yellow and the blue form of the 1:6 compound are identical [3].

References:

[1] N. A. Polotebnova, E. F. Tkach (Zh. Neorgan. Khim. **14** [1969] 1040/2; Russ. J. Inorg. Chem. **14** [1969] 542/3). – [2] E. F. Tkach, N. A. Polotebnova (Zh. Neorgan. Khim. **16** [1971] 1913/6; Russ. J. Inorg. Chem. **16** [1971] 1016/8). – [3] E. F. Tkach, N. A. Polotebnova (Zh. Neorgan. Khim. **16** [1971] 210/3; Russ. J. Inorg. Chem. **16** [1971] 109/11).

9.1.3.9 $5Na_2O \cdot V_2O_5 \cdot 22MoO_3 \cdot 58H_2O$

From aqueous mixtures of molybdate and vanadate at pH 1.5 to 2 containing the components in the molar ratio V : Mo = 1 : 6, yellow crystals can be isolated on evaporation. They dissolve slowly in water and can be recrystallized from 0.01 M $HClO_4$. Their analytical composition is $5Na_2O \cdot V_2O_5 \cdot 22MoO_3 \cdot 58H_2O$, for which the formula $Na_5VMo_{11}O_{38} \cdot 29H_2O$ is proposed.

The refractive indices of the anisotropic crystals are $n_1 = 1.730$ and $n_2 > 1.780$. The IR spectrum is very similar to that of $7Na_2O \cdot V_2O_5 \cdot 46MoO_3 \cdot 80H_2O$ (p. 337). The thermoanalysis shows that the compound has lost most of its water at 200 to 210°C. The remaining 1 to 2 water molecules leave at 200 to 250°C. Above 700°C it is decomposed separating MoO_3. Line diagrams of the X-ray powder patterns of the compound and its dehydration products are plotted.

A. B. Bekturov, A. K. Il'yasova, A. A. Amirbekova (Zh. Neorgan. Khim. **15** [1970] 2781/5; Russ. J. Inorg. Chem. **15** [1970] 1446/8).

9.1.3.10 $H_3VMo_{12}O_{40} \cdot nH_2O$ ($= V_2O_5 \cdot 24MoO_3 \cdot (2n+3)H_2O$)

The acid is obtained from an aqueous mixture of $Na_2MoO_4 \cdot 2H_2O$ and $Na_3VO_4 \cdot nH_2O$ in a molar ratio of about 12:1. The mixture is passed through a cationic ion exchanger in the H^+ form. From the very acidic effluent (pH 1) which contains less than 0.2% of the Na^+ ions, the acid crystallizes by evaporation.

V. Chiola, J. G. Lawrence (U.S. 3446575 [1969]; C.A. **71** [1969] No. 23320).

9.1.3.11 $7\,Na_2O\cdot V_2O_5\cdot 46\,MoO_3\cdot 80\,H_2O$

Under the same conditions as described for $5\,Na_2O\cdot V_2O_5\cdot 22\,MoO_3\cdot 58\,H_2O$ (see p. 336), mixtures of the ratio V:Mo = 1:12 give lemon-yellow crystals of composition $7\,Na_2O\cdot V_2O_5\cdot 46\,MoO_3\cdot 80\,H_2O$. Their recrystallization from 0.01 M $HClO_4$ is difficult because of hydrolysis.

The refractive indices are $n_1 = 1.760$ and $n_2 > 1.780$. The IR spectrum is very similar to that of $5\,Na_2O\cdot V_2O_5\cdot 22\,MoO_3\cdot 58\,H_2O$. The heating curve shows most of the water to be lost between 200 and 210°C. The remaining 1 to 2 H_2O are lost at 200 to 250°C. Above 700°C the anhydrous compound decomposes, separating MoO_3.

A. B. Bekturov, A. K. Il'yasova, A. A. Amirbekova (Zh. Neorgan. Khim. **15** [1970] 2781/5; Russ. J. Inorg. Chem. **15** [1970] 1446/8).

9.1.4 Mixed Metal Heteropoly Compounds

9.1.4.1 Preparations Containing Zr, Mo, and V

Well-defined ternary heteropoly compounds containing Zr, Mo, and V have not been isolated. But some gels with varying composition obtained from aqueous mixtures of salts of the three components have been described, because they can be used as redox and ion exchangers. They are tentatively designated as zirconium molybdovanadates. So aqueous solutions of 0.4 M $(NH_4)_2MoO_4$ and 0.04 M NH_4VO_3 were added to 0.17 M $Zr(NO_3)_4$ in 1 N HNO_3 in the molar ratio Zr:Mo:V = 1:1:1 (I) [1] or 3:1:1 (II) [2]. For the mixture Zr:Mo:V = 6:1:1 (III) a 0.94 M $ZrO(NO_3)_2$ solution in 6 N HNO_3 was used. The pH was adjusted to 9 with 3 M NH_3. The resulting gel was dried at 100°C, granulated in H_2O, and dried again [1]. The analytical composition of the three products is Zr:Mo:V = 1.68:0.088:1.00 (I), 3.33:0.285:1.00 (II), and 6.76:0.815:1.00 (III). The bulk densities are 1.57, 0.92, and 0.75 g/cm^3, respectively [2]. If the reaction mixture has a pH <5 the products are appreciably soluble, at pH 5 to 8 their chemical stability in acidic media is less than at a precipitation pH of 9. The color of the gels is yellow to orange in the oxidized state and dark gray in the reduced state (reduction is possible with Sn^{2+}, I^-, or Fe^{2+}). The latter can be reoxidized by acidic MnO_4^- solutions. The static electron-exchange capacity is 0.32 for I, and 0.52 mequiv/g for III [1].

The ion exchange capacities for Li^+ are 4.20 (I), 3.80 (II), and 3.70 mequiv/g (III), for Na^+ 2.98 (I), 2.19 (II), and 2.50 mequiv/g (III) [2]. The ion exchangers also take up Ca^{2+}, Cu^{2+}, Pb^{2+}, and Fe^{3+} from solutions [1]. Potentiometric titrations with LiOH and NaOH show that the gels are weak acids with the pK values 9.13 ± 0.11 (I), 9.04 ± 0.09 (II), and 8.56 ± 0.15 (III). Probably the acidic character is caused by the equilibrium $H_2VO_4^- \rightleftharpoons HVO_4^{2-} + H^+$ which has pK values between 7.88 and 9.36 under various conditions. MoO_4^{2-} and ZrO^{2+} groups form the framework of the polymer [2].

The chemical stability of the exchangers was tested at different pH values. Appreciable amounts of dissolved Zr were found in acidic media because the acids depolymerize the gels and form soluble Zr complexes. Increasing the pH from 0.8 to 2.9 diminishes the dissolved amounts of Zr, Mo, and V [3]. At pH >10 mainly V is dissolved [2]. Also the composition of the gels affects the stability. The removal of H_2O from the framework by heating at 100 to 250°C increases the stability but decreases the ion exchange capacity. The electron-exchange capacity is constant in products heated up to 350°C. At temperatures >350°C the color changes from yellow to green and the chemical stability decreases [3].

References:

[1] R. G. Safina, E. S. Boichinova, N. E. Denisova (Zh. Prikl. Khim. **44** [1971] 2337/9; J. Appl. Chem. [USSR] **44** [1971] 2397/8). – [2] R. G. Safina, N. E. Denisova, E. S. Boichinova (Zh. Prikl. Khim. **46** [1973] 2432/5; J. Appl. Chem. [USSR] **46** [1973] 2584/7). – [3] E. S. Boichinova, N. E. Denisova, R. G. Safina (Zh. Prikl. Khim. **49** [1976] 468/70; J. Appl. Chem. [USSR] **49** [1976] 470/2).

9.1.4.2 The $HfMo_{12}V$ Type Heteropoly Anion

A defined compound of this type has not been isolated, but the reduced form of an Hf-Mo-V complex was used to determine Hf photometrically in alloys with a high V content. For example, aqueous mixtures of a 1.1×10^{-4} M solution of $HfO(NO_3)_2\cdot2H_2O$, a 0.1 M solution of NH_4VO_3, and a 14% Na_2MoO_4 solution were adjusted to pH 1 by 1.8 N H_2SO_4, boiled for 10 min, and reduced with a 2% ascorbic acid solution. The resulting mixture has an absorption maximum at 730 nm; $\bar{\varepsilon}=3400+30$. (The absorption spectrum from 500 to 1000 nm is plotted.) The composition of the species was found by the photometric molar ratio method to be Hf:Mo:V=1:12:1.

E. N. Semenovskaya, N. K. Sokovikova, L. K. Smirnova (Vestn. Mosk. Univ. Khim. **28** [1973] 358/62; Moscow Univ. Chem. Bull. **28** No. 3 [1973] 77/80).

9.1.4.3 $H_6GeMo_{10}V_2O_{40}\cdot nH_2O$ ($=GeO_2\cdot10MoO_3\cdot V_2O_5\cdot(3+n)H_2O$) and its Solutions

The free acid was isolated for the first time from a mixture of GeO_2, NaOH, ammonium paramolybdate, and ammonium vanadate in the molar ratio $GeO_2:MoO_3:V_2O_5=1:12:12$. The mixture was boiled to eliminate NH_4^+ and acidified with H_2SO_4. Then ether was added. Three layers were formed from which the (lowest) oily orange layer contained most of the compound as etherate, which was separated and decomposed with small amounts of water. Evaporating the ether octahedral crystals of orange-red color were obtained. They contain about 28 H_2O, but effloresce in the air losing H_2O; the water content diminishes from 25 to 9 to 10 wt% [1]. Better and less ambiguous preparative results are obtained with an aqueous mixture of GeO_2, sodium paramolybdate, and a 10- to 15-fold excess of sodium metavanadate in 0.1 M H_2SO_4. The mixture is boiled for 1 h and then extracted with ether/H_2SO_4, which must be free of alcohol because the latter reduces the compound. Washing the extract with water, contact of the ether phase with a too high H^+ concentration, or heating it may lower the yield. After adding water ($^1/_2$ to $^2/_3$ of the volume) to the ether solution, well-formed transparent dark red octahedrons crystallize on standing in the air. They are readily soluble in water [2]. Similar procedures are described in [3 to 5].

For the analytical composition Ge:Mo:V=1:10:2 the formulas $H_4GeMo_{10}V_2O_{39}\cdot nH_2O$, $n\approx28$ [1] and $H_8GeMo_{10}V_2O_{41}\cdot nH_2O$ [2] were proposed. An electron diffraction study showed the acid to be isomorphous with $H_4GeMo_{12}O_{40}$. Therefore the formula $H_6GeMo_{10}V_2O_{40}\cdot nH_2O$ is supposed to be the most probable. The unit cells of the two cubic compounds are nearly identical: a = 23.05 Å for $H_4GeMo_{12}O_{40}\cdot nH_2O$ and 23.10 Å for $H_6GeMo_{10}V_2O_{40}\cdot nH_2O$; Z = 8. The measured electron diffraction intensities can be indexed with a = 23.10 Å but also with a' = a/2 except one very weak line, which is caused by water of crystallization. It is assumed that the lattice parameter a belongs to the ~29-hydrate and the unit cell with a' = 11.50 to 11.60 Å to the pentahydrate of the compound. The indexed d values are tabulated; the intensity curves of the electron diffraction patterns are plotted [6], see also [10].

The compound melts at 50 to 60°C. The measured density at 25°C is 2.60 to 2.65 [1], the calculated 2.48 g/cm³ [6].

Aqueous and nonaqueous solutions of the heteropoly acid have been investigated because they are used in analytical chemistry for the spectrophotometric determination of Ge. The solutions are prepared by mixing GeO_2, Na_2MoO_4 (or $(NH_4)_2MoO_4$), and $NaVO_3$ in 0.2 N H_2SO_4 and boiling for 2 min. The precipitated $V_2O_5 \cdot nH_2O$ is filtered. Another method adds a solution of the vanadomolybdic acid to the dissolved GeO_2 [1]. These orange solutions are most stable in the small H^+ concentration range of 0.20 to 0.28 N [2, 7].

The absorption spectrum of the aqueous solution from 220 to 1100 nm shows high absorption in the UV region dropping rapidly with increasing wavelength. It is very similar to the spectrum of the lemon-yellow solution of $H_4GeMo_{12}O_{40}$, but the latter absorbs less in the region 320 to 600 nm. The same spectrum is obtained when the heteropoly acid is dissolved in isoamyl or butyl alcohol. A plot is given in [7, 8].

The potentiometric titration curve of an aqueous solution of the acid with 0.2 N KOH has two steps. The first one at a molar ratio of acid:KOH = 1:5 corresponds to the neutralization indicating that the acid is pentabasic (formulated as $K_5H_3GeMo_{10}V_2O_{41}$). The second step at a ratio 1:>22 corresponds to the decomposition of the acid to its three components [9].

The dissolved heteropoly acid was reduced by $CrSO_4$, $(NH_4)_2Fe(SO_4)_2 \cdot 6H_2O$, ascorbic acid, Al, and Sb^{III}. The latter reaction can be used for the colorimetric determination of Sb, because the intensity of the formed blue color depends linearly on the Sb content in the range 10 to 70 mg/L using a 5% solution of $H_6GeMo_{10}V_2O_{40}$ [5]. The absorption spectrum of the blue solution formed by $CrSO_4$ or $(NH_4)_2FeSO_4 \cdot 6H_2O$ as reducing agent is very similar to the spectra of the unreduced $GeMo_{12}$ and $GeMo_{10}V_2$ acids in the UV region. But this part is not too characteristic because defined maxima fail in the UV spectra of all Ge-Mo and Ge-Mo-V compounds in contrast to the analogous W compounds. In the visible region a maximum at about 825 nm can be observed according to the blue color. The spectrum from 220 to 1100 nm is plotted [8, 10]. (The legends of the two plots give different assignments.)

More information about the reduction reactions of $H_6GeMo_{10}V_2O_{40}$ were obtained from redox titrations and polarographic measurements. The potentiometric titration curve of an aqueous solution of the acid with $CrSO_4$ at 90°C under CO_2 atmosphere shows three jumps in potential corresponding to the reduction reactions $V^V \rightarrow V^{IV}$ (a), $Mo^{VI} \rightarrow Mo^V$ (b), and $Mo^V \rightarrow Mo^{III}$ (c). (The jump corresponding to $V^{IV} \rightarrow V^{III}$ coincides with that of $Mo^{VI} \rightarrow Mo^V$.) The first two equivalence points are only slightly separated, but it is assumed that V is reduced first. To measure the redox potentials for the three reactions half the amount of $CrSO_4$ necessary for the respective complete reduction was added to a solution of the acid in 6 N HCl. The potential was determined at 22°C and after heating the solution to 90°C. The values relative to the hydrogen electrode were E = 0.900 (a), 0.616 (b), and 0.260 V (c) at 90°C and 1.036 (a), 0.561 (b), and 0.232 V (c) at 22°C. These data show that the V-containing compound is a considerably stronger oxidizing agent than $H_4GeMo_{12}O_{40}$ which has at 22°C the two redox potentials E = 0.550 and 0.159 V. During the titration with Cr^{II} the formation of a heteropoly blue containing Ge, Mo, and V was not observed. Also, the potential of the blue reaction could not be determined because of the low stability of the blue compound in the air. Obviously the synchronous reduction of Mo and V is not possible without changing the structure severely [11]. This is proved by investigations of the reduction of $H_6GeMo_{10}V_2O_{40}$ with ascorbic acid and Al. The potentiometric titration curve of a solution of the acid in 1 N H_2SO_4 with ascorbic acid has one jump in potential causing a blue solution. This reaction is complete in 24 h at room temperature or in 30 min on heating. The analyses of the solution and the isolated blue compound (extracted with ether) show that the blue compound is free of V; all the V content of the heteropoly acid is found as VO^{2+} in solution. The blue species contains not more than 3 Mo^V from a total of about 12 Mo atoms in the molecule. The polarogram of $H_6GeMo_{10}V_2O_{40}$ in 1 N H_2SO_4 shows two irreversible reduction waves with $E_{1/2} = -0.20$ and -0.40 V (relative to the saturated calomel electrode). The diffusion controlled current of the second wave is proportional to the concentration of the heteropoly acid and has its maximal value at pH 2.20. In alcoholic solutions the polarogram has only one irreversible, diffusion controlled wave at $E_{1/2} = -0.27$ V. In both solvents the polarogram of the reduction of the blue species has characteristics similar to the unreduced acid. Coulometric measurements have shown that $H_6GeMo_{10}V_2O_{40}$ is electrochemically reduced by 4 electrons. Analysis of the products of an electrolysis of the heteropoly acid at −0.6 V shows that probably 2 of the electrons reduce 2 Mo^{VI} to 2 Mo^{V} and the other 2 react with 2 V^{V} to 2 V^{IV}, so that the products of the blue reaction of $H_6GeMo_{10}V_2O_{40}$ are $[GeMo^{VI}_{10}Mo^{V}_{2}O_{40}]^{6-}$ and VO^{2+} [12].

References:

[1] B. N. Ivanov-Emin (Zh. Obshch. Khim. **10** [1940] 826/30; C.A. **1941** 2434). – [2] R. K. Motorkina (Zh. Neorgan. Khim. **2** [1957] 92/105; Russ. J. Inorg. Chem. **2** No. 1 [1957] 142/61, 153). – [3] A. I. Kokorin, M. B. Bardin, N. A. Polotebnova (Uch. Zap. Kishinev. Univ. **7** [1953] 59/62 from C.A. **1957** 930). – [4] A. I. Kokorin, N. A. Polotebnova (Uch. Zap. Kishinev. Univ. **7** [1953] 63/7 from Ref. Zh. Khim. **1955** 11462). – [5] A. I. Kokorin, N. A. Polotebnova (Tr. Komis. Analit. Khim. **7** [1956] 205/10; C.A. **1956** 15328).

[6] Z. F. Shakhova, G. N. Tishchenko, R. K. Motorkina (Zh. Obshch. Khim. **27** [1957] 1118/24; J. Gen. Chem. [USSR] **27** [1957] 1200/6). – [7] Z. F. Shakhova, R. K. Motorkina (Zh. Analit. Khim. **11** [1956] 698/703; J. Anal. Chem. [USSR] **11** [1956] 749/53). – [8] Z. F. Shakhova, R. K. Motorkina (Tr. Komis. Analit. Khim. **8** [1958] 100/9; C.A. **1958** 19444). – [9] A. I. Kokorin, N. A. Polotebnova (Zh. Obshch. Khim. **27** [1957] 304/10; J. Gen. Chem. [USSR] **27** [1957] 339/44). – [10] I. P. Alimarin, Z. F. Shakhova, R. K. Motorkina (Dokl. Akad. Nauk SSSR **106** [1956] 61/4; Proc. Acad. Sci. USSR Div. Chem. Sci. **106/111** [1956] 1/4).

[11] Z. F. Shakhova, R. K. Motorkina (Zh. Obshch. Khim. **26** [1956] 2667/73; J. Gen. Chem. [USSR] **26** [1956] 2969/76). – [12] L. A. Furtune, N. A. Polotebnova, A. A. Kozlenko (Zh. Neorgan. Khim. **18** [1973] 2185/8; Russ. J. Inorg. Chem. **18** [1973] 1155/7).

9.1.4.4 $M_3[HGeMo_{10}V_2O_{39}]\cdot 1.5H_2O$ ($=3M_2O\cdot 2GeO_2\cdot 20MoO_3\cdot 2V_2O_5\cdot 4H_2O$), M = Rb, Cs

The insoluble yellow-orange salts were obtained from aqueous solutions of the heteropoly acid (see above) on adding an M^+ salt. The precipitates were dried at 105 to 110°C.

B. N. Ivanov-Emin (Zh. Obshch. Khim. **10** [1940] 826/30; C.A. **1941** 2434).

9.1.4.5 $(C_5H_5NH)_4[GeMo_{10}V_2O_{39}]$ and $[C(NH_2)_3]_4[GeMo_{10}V_2O_{39}]$

The insoluble pyridinium and guanidinium salts were obtained from aqueous solutions of the heteropoly acid (see above) on adding the organic bases. The precipitates were dried at 105 to 110°C.

B. N. Ivanov-Emin (Zh. Obshch. Khim. **10** [1940] 826/30; C.A. **1941** 2434).

9.2 Compounds with Niobium

9.2.1 The Nb_2O_5-MoO_3-H_2O System

Solutions of potassium niobate acidified with $HClO_4$ to pH 2 to 2.5 are colloids. The addition of small amounts of molybdate causes a rapid precipitation. Phototurbidimetric measurements at 536 nm determining the relation between optical density and the molar ratio Mo : Nb show a maximum of the particle size of the suspension at Mo:Nb≈2, and the influence of ageing, which enlarges the particle size and shifts the maximum to a higher Mo:Nb ratio. At ratios ≧8 metastable solutions are formed (a mixture of Mo:Nb=8 remains clear for 2 d) whose stability increases with increasing MoO_4^{2-} content. On heating these solutions $Nb(OH)_5$ precipitates, absorbing MoO_4^{2-} in the constant ratio Mo:Nb=4:1. This is interpreted as the formation of a basic salt of the tetramolybdic acid, $NbO(OH)Mo_4O_{13}$ or $Nb(OH)_3Mo_4O_{13}$.

Yu. F. Shkaravskii (Zh. Neorgan. Khim. **8** [1963] 2668/74; Russ. J. Inorg. Chem. **8** [1963] 1399/403).

9.2.2 Nb_xMo_y Type Heteropoly Compounds

A well-defined Nb-Mo compound is unknown, but solutions of a molybdoniobic acid and its reduced blue form have been described because they can be used for the spectrophotometric determination of Nb.

In aqueous mixtures of potassium niobate and a large excess of sodium molybdate a heteropoly compound is formed in at least 10 min. The optimal pH is 1.5. Temperature variations from 0 to 100°C have little effect on the reaction. The absorption spectrum of the solution shows high absorption at 200 nm, falling rapidly with increasing wavelength (plotted from 200 to 500 nm).

The reduction was made in strongly acidic media to avoid the synchronous blue reaction of the MoO_4^{2-} excess. Intense blue solutions were obtained with ferrous ammonium sulfate or chlorostannous acid. Ascorbic acid, sulfite, or hydrazine hydrochloride gave weakly colored products. The maximum blue intensity was reached 3 min after adding the reductant; 2 min later the intensity began to decrease. The absorption spectrum of the blue solution, plotted from 400 to 1000 nm, shows a maximum at 725 nm.

J. C. Guyon, G. W. Wallace, Jr., M. G. Mellon (Anal. Chem. **34** [1962] 640/3), J. C. Guyon (Diss. Purdue Univ. 1961; Diss. Abstr. **22** [1962] 3364).

9.2.3 The $[Ce^{IV}Mo_8Nb_4O_{40}]^{8-}$ Ion in Solution

The absorption spectra of aqueous mixtures of $(NH_4)_4[H_4Ce^{IV}Mo_8O_{30}]\cdot 5.5H_2O$ (I), see p. 282, and $KNbO_3\cdot 2H_2O$ have a clearly higher absorption in the UV region (with a maximum at about 243 nm) than a solution of I only. Their pH stability range is 1.0 to 11.8, which is smaller in the acidic part than in the case of I. The molar ratio method shows that the complex formed in the mixture contains Nb in the ratio Nb:I=4:1, i.e., Ce:Mo:Nb=1:8:4, so that the mixture reacts according to $H_8CeMo_8O_{30}+4KNbO_3\rightarrow K_4[H_4CeMo_8Nb_4O_{40}]+2H_2O$. Titration with NaOH gives a basicity of 8. From a comparison of the chemical and spectroscopic properties it is assumed that Ce is located in the center of the anion coordinated by 12 O atoms of the MoO_6 and Nb_2O_5 groups, as in the proved structure of the $[CeMo_{12}O_{42}]^{8-}$ ion (see p. 287). A salt with the $[Ce^{IV}Mo_8Nb_4O_{40}]^{8-}$ anion has not been isolated.

V. I. Spitsyn, E. A. Torchenkova, Nguyen D'eu, V. F. Chuvaev (Allgem. Prakt. Chem. **22** [1971] 55/9).

9.2.4 $(NH_4)_8[Ce^{IV}Mo_{10}Nb_2O_{41}]\cdot 10H_2O$ (=$4(NH_4)_2O\cdot CeO_2\cdot 10MoO_3\cdot Nb_2O_5\cdot 10H_2O$)

In aqueous mixtures of $(NH_4)_6[H_2Ce^{IV}Mo_{10}O_{36}]\cdot 7H_2O$ (I), see p. 282, and $KNbO_3\cdot 2H_2O$, which have a weaker light absorption in the UV region than I only, a ternary heteropoly complex is formed according to $H_8CeMo_{10}O_{36} + 2KNbO_3 \rightarrow K_2[H_6CeMo_{10}Nb_2O_{41}]$. The solution is stable in the range pH 11 to 3 N H_2SO_4. The molar ratio Nb : I = 2 : 1 and the basicity of 8 were proved experimentally. The structure is assumed to contain a central CeO_{12} unit like the $[CeMo_{12}O_{42}]^{8-}$ ion (see p. 287). The ammonium salt $(NH_4)_8[Ce^{IV}Mo_{10}Nb_2O_{41}]\cdot10H_2O$ has been isolated from the mixture as a very soluble product, which is decomposed totally by acidic cation exchangers.

V. I. Spitsyn, E. A. Torchenkova, Nguyen D'eu, V. F. Chuvaev (Allgem. Prakt. Chem. **22** [1971] 55/9).

9.2.5 $H_{11}Ce^{IV}Mo_{12}NbO_{46}\cdot nH_2O$ (= $2CeO_2\cdot24MoO_3\cdot Nb_2O_5(2n+11)H_2O$), n = 0.5 to 26.5

An aqueous solution of $(NH_4)_6[Ce^{IV}Mo_{12}Nb_2O_{46}]\cdot28H_2O$ (see below) in contact with an acidic cation exchanger reacts to give the title compound through precipitation of $Nb_2O_5\cdot nH_2O$, according to $H_6CeMo_{12}Nb_2O_{46} + mH_2O \rightarrow H_{11}CeMo_{12}NbO_{46} + 0.5Nb_2O_5\cdot nH_2O$. The acid (with n = 22.5) precipitates if 60% H_2SO_4 is added to the eluate. It is quite soluble in water, alcohol, and acetone, but insoluble in ether [1]. In another preparation the acid prepared contains $26.5H_2O$. On drying above P_2O_5 the 17.5- and the 6-hydrates are obtained. The 4.5-hydrate forms at 90°C and the 0.5-hydrate at 105°C [2].

The ^1H NMR spectra of the several hydrates at 80 K show that the protons are not equivalent. The first derivative of the signal of the highly hydrated species has a narrow line (width 2 G) in the center of the spectrum and a broad one with two maxima at $\Delta H = \pm10$ and ±6 G. The broad component is the sum of the signals of the protons of H_2O molecules and H_3O^+ ions ($\Delta H = \pm10$ G belongs more to H_3O^+, ±6 G more to H_2O). The narrow line is typical for hydroxyl protons. The proton distribution derived from the integrated line intensities gives 17.5 H_2O, $9H_3O^+$, and 2 OH^-, so that the formula of the 26.5-hydrate is $(H_3O)_{11-x}[CeMo_{12}NbO_{46-x}(OH)_x]\cdot(15.5+x)H_2O$, $x \approx 2$. On dehydration the amount of H_3O^+ decreases; the narrow line changes little. Therefore it is assumed that the OH groups are part of the inner anion structure. From the van Vleck equation their distance is calculated to be about 2.5 Å. The dehydration of the hexahydrate increases the relative intensity of the H_2O line and decreases the two other lines because the H_3O^+ ions form H_2O with O atoms of the anion. The second moments of the spectra are tabulated and the spectra are plotted in [2].

References:

[1] V. I. Spitsyn, E. A. Torchenkova, Nguyen D'eu, V. F. Chuvaev (Allgem. Prakt. Chem. **22** [1971] 55/9). – [2] V. F. Chuvaev, Nguyen D'eu, E. A. Torchenkova, V. I. Spitsyn (Dokl. Akad. Nauk SSSR **209** [1973] 635/8; Dokl. Phys. Chem. Proc. Acad. Sci. USSR **205/210** [1973] 279/82).

9.2.6 $Na_6[Ce^{IV}Mo_{12}Nb_2O_{46}]\cdot20H_2O$ and $(NH_4)_6[Ce^{IV}Mo_{12}Nb_2O_{46}]\cdot28H_2O$ (= $3Na_2O\cdot CeO_2\cdot12MoO_3\cdot Nb_2O_5\cdot20H_2O$ and $3(NH_4)_2O\cdot CeO_2\cdot12MoO_3\cdot Nb_2O_5\cdot28H_2O$)

In aqueous mixtures of $(NH_4)_6[H_2Ce^{IV}Mo_{12}O_{42}]\cdot8H_2O$ (I), see p. 288, and $KNbO_3\cdot2H_2O$ (II) a ternary heteropoly complex is formed according to $[H_2CeMo_{12}O_{42}]^{6-} + 2NbO_3^- \rightleftharpoons [CeMo_{12}Nb_2O_{46}]^{6-} + 2OH^-$. The constant of this equilibrium was calculated to be $K = (4.6 \pm 1.1) \times 10^{-19}$. The complex is stable in the pH range 1.8 to 11.9. The photometric molar ratio method at 380 nm gives the composition Nb : I = 2 : 1. The absorption spectrum of the solution from 210 to 280 nm is very similar to that of I with high absorption in the UV region, a shoulder at 235 nm, and a decreasing intensity with increasing wavelength (a plot is given).

The ammonium salt precipitates from a hot solution containing an excess of II (3.56 g I, 0.97 g II). The product is washed with a 10% NH_4NO_3 solution and methanol. It can be recrystallized from warm water. In a similar way the sodium salt was isolated.

V. I. Spitsyn, E. A. Torchenkova, Nguyen D'eu, V. F. Chuvaev (Allgem. Prakt. Chem. **22** [1971] 55/9).

9.2.7 $M_2[Ce^{IV}Mo_{12}Nb_2O_{46}]\cdot nH_2O$ (= $M_2O_3\cdot CeO_2\cdot 12MoO_3\cdot Nb_2O_5\cdot nH_2O$), M = Y, Er, Yb

These compounds precipitate from an aqueous solution containing $KNbO_3$ and $(NH_4)_6[H_2Ce^{IV}Mo_{12}O_{42}]$ in the molar ratio of about 3:1 when a solution of the respective rare earth salt is added. The analytical compositions are $Y_2[CeMo_{12}Nb_2O_{46}]\cdot 26H_2O$, $Er_2[CeMo_{12}Nb_2O_{46}]\cdot 25H_2O$, and $Yb_2[CeMo_{12}Nb_2O_{46}]\cdot 24H_2O$.

V. I. Spitsyn, E. A. Torchenkova, Nguyen D'eu, V. F. Chuvaev (Allgem. Prakt. Chem. **22** [1971] 55/9).

9.3 Compounds with Tantalum

9.3.1 The Ta_2O_5-MoO_3-H_2O System

This system behaves very much as the analogous Nb system (see p. 340). The maximum of the phototurbidimetric curve lies at the ratio Mo:Ta ≈ 0.5. The metastable tantalate-molybdate solution remains clear for at least 10 d. The formation of the corresponding basic tantalum salt, $TaO(OH)Mo_4O_{13}$ or $Ta(OH)_3Mo_4O_{13}$, is assumed.

Yu. F. Shkaravskii (Zh. Neorgan. Khim. **8** [1963] 2668/74; Russ. J. Inorg. Chem. **8** [1963] 1399/403).

9.3.2 $K_8[H_2MoTa_{12}O_{38}]\cdot 17H_2O$ (= $4K_2O\cdot MoO_3\cdot 6Ta_2O_5\cdot 18H_2O$)

An aqueous mixture of 30 mL 0.02 M potassium hexatantalate and of 60 mL 0.05 M "molybdic acid" was refluxed for 4 h with exclusion of CO_2 from air. After 72 h in vacuum a white amorphous powder precipitated having the given analytical composition. The molecular weight was determined cryoscopically in fused $Na_2SO_4\cdot 10H_2O$ to be 3395 (calc. 3494). Thermogravimetry under reduced pressure showed that in a first step $6H_2O$ at 75°C and in a second step $5H_2O$ at 160°C were eliminated. The remaining water was lost above 340°C. Near 720°C further decomposition occurred.

S. K. Roy, H. C. Mishra (J. Indian Chem. Soc. **55** [1978] 612/3).

9.3.3 $TaMo_{12}$ Type Heteropoly Compounds

A well-defined compound has not been isolated but was obtained in solution together with the reduced blue form because they can be used for the spectrophotometric determination of Ta.

Aqueous mixtures of K_2TaF_7 and an excess of Na_2MoO_4, adjusted to pH 1.35 to 1.45 by sulfuric acid, react immediately to give a yellow solution which is stable for 10 min even in strongly acidic media, in contrast to the analogous Nb solution which decays in a few seconds. The absorption spectrum from 340 to 500 nm is plotted. It shows high absorption at 340 nm falling with increasing wavelength.

The reduction is made in strongly acidic media (H_2SO_4) to avoid the synchronous blue reaction of the MoO_4^{2-} excess. Ascorbic acid and Sn^{II} give intense blue solutions, hydrazine hydrochloride a pale blue, and Fe^{II} a pale green product. The reaction with chlorostannous acid is complete in less than 1 min. The blue solution is stable for 1 h. The methods of photometric molar ratio and of continuous variations show that the blue complex has the composition Ta:Mo = 1:12. The absorption spectrum from 600 to 900 nm is plotted, showing a maximum at 820 nm.

J. C. Guyon (Anal. Chim. Acta **30** [1964] 395/400).

9.3.4 $(NH_4)_8[Ce^{IV}Mo_{10}Ta_2O_{41}]\cdot14H_2O$ ($=4(NH_4)_2O\cdot CeO_2\cdot10MoO_3\cdot Ta_2O_5\cdot14H_2O$)

In aqueous mixtures of $(NH_4)_6[H_2CeMo_{10}O_{36}]\cdot7H_2O$ (I), see p. 282, and $K_8Ta_6O_{19}\cdot16H_2O$ a ternary heteropoly complex is formed according to $3H_8CeMo_{10}O_{36}+K_8Ta_6O_{19}\rightarrow 3K_2[H_6CeMo_{10}Ta_2O_{41}]+2KOH+2H_2O$. The basicity of 8 and the molar ratio Ta:I = 2:1 were proved by titration and the photometric molar ratio method. The absorption spectrum of the solution is nearly identical with I. The ammonium salt can be isolated from the mixture (but no details are given). In contact with an acidic cation exchanger it is decomposed to $Ta_2O_5\cdot nH_2O$ and I. From the comparison of the chemical and spectral properties it is assumed that the anion has the same central CeO_{12} unit as $[CeMo_{12}O_{42}]^{8-}$ or $[CeMo_8Nb_4O_{40}]^{8-}$ (see pp. 287/8 and 341).

E. A. Torchenkova, Nguyen D'eu, L. P. Kazanskii, V. I. Spitsyn (Izv. Akad. Nauk SSSR Ser. Khim. **1973** 734/8; Bull. Acad. Sci. USSR Div. Chem. Sci. **1973** 715/8).

9.3.5 The $[Ce^{IV}Mo_{12}Ta_2O_{49}]^{12-}$ Ion in Solution

In aqueous mixtures of $(NH_4)_6[H_2CeMo_{12}O_{42}]\cdot8H_2O$ (I), see p. 288, and $K_8Ta_6O_{19}\cdot16H_2O$ a ternary heteropoly complex is formed according to $3H_2CeMo_{12}O_{42}^{6-}+Ta_6O_{19}^{8-}+2H_2O\rightarrow 3H_2CeMo_{12}Ta_2O_{49}^{10-}+4H^+$. The molar ratio method proves the composition Ta:I = 2:1. The solution is stable in the range pH 12 to 4 N H_2SO_4 for one week. The pH dependence of the optical density shows that the complex exists in two forms in the solution. The absorption spectrum from 220 to 320 nm is nearly identical to that of I (a plot is given) with a somewhat reduced intensity.

E. A. Torchenkova, Nguyen D'eu, L. P. Kazanskii, V. I. Spitsyn (Izv. Akad. Nauk SSSR Ser. Khim. **1973** 734/8; Bull. Acad. Sci. USSR Div. Chem. Sci. **1973** 715/8).

9.3.6 $H_{12}Ce^{IV}Mo_{12}Ta_2O_{49}\cdot nH_2O$ ($=CeO_2\cdot12MoO_3\cdot Ta_2O_5\cdot(6+n)H_2O$), n = 0 to 30

The free acid is obtained by passing a solution of the ammonium salt (see below) through a column of an acidic cation exchanger. On adding 60% H_2SO_4 to the eluate the acid precipitates as the 30-hydrate. It is readily soluble in water, dilute acids, alcohol, and acetone, but insoluble in ether. The titration with NaOH proves the basicity of 12. The acid protons are nonequivalent. Titration of the last two protons is accompanied by a more rapid increase in the pH value [1]. The 30-hydrate dehydrates to the 27-hydrate at 40°C. The 16.5- and the 6-hydrate can be obtained by drying over P_2O_5. The 4.5-hydrate forms at 90°C and the anhydrous acid at 105°C [2].

The ^1H NMR spectra of the several hydrates at 80 K are nearly identical to those of $H_{11}Ce^{IV}Mo_{12}NbO_{46} \cdot n\,H_2O$ (see p. 342). The proton distribution according to the integrated line intensities of the 27-hydrate gives $17\,H_2O$, $10\,H_3O^+$, and $2\,OH^-$, so that the acid can be written as $(H_3O)_{12-x}[CeMo_{12}Ta_2O_{49-x}(OH)_x] \cdot (15+x)H_2O$, $x \approx 2$. The 16.5-hydrate has only 6 to 7 H_3O^+, and the proton distribution of the hexahydrate gives the formula $(H_3O)_4[CeMo_{12}Ta_2O_{43}(OH)_4] \cdot 4.5\,H_2O$. The behavior of the OH line and the influence of further dehydration of the hexahydrate on the spectrum are in accord with the spectra of the Nb compound. The second moments of the spectra are tabulated and the spectra are plotted in [2].

References:

[1] E. A. Torchenkova, Nguyen D'eu, L. P. Kazanskii, V. I. Spitsyn (Izv. Akad. Nauk SSSR Ser. Khim. **1973** 734/8; Bull. Acad. Sci. USSR Div. Chem. Sci. **1973** 715/8). – [2] V. F. Chuvaev, Nguyen D'eu, E. A. Torchenkova, V. I. Spitsyn (Dokl. Akad. Nauk SSSR **209** [1973] 635/8; Dokl. Phys. Chem. Proc. Acad. Sci. USSR **205/210** [1973] 279/82).

9.3.7 $(NH_4)_{12}[Ce^{IV}Mo_{12}Ta_2O_{49}] \cdot 22.5\,H_2O$ $(= 6(NH_4)_2O \cdot CeO_2 \cdot 12\,MoO_3 \cdot Ta_2O_5 \cdot 22.5\,H_2O)$

To a hot solution (~60°C) of 10 g $(NH_4)_6[H_2CeMo_{12}O_{42}] \cdot 8\,H_2O$ (I), see p. 288, in 300 mL H_2O a 4.5% solution of $K_8Ta_6O_{19} \cdot 16\,H_2O$ is added gradually up to a molar ratio Ta : I = 3 : 1. The mixture is boiled for 2 min and filtered hot. To the filtrate a saturated solution of NH_4NO_3 is added. The crystalline lemon-yellow precipitate is washed with 20% NH_4NO_3 solution and methanol. It is soluble in hot water but insoluble in organic solvents.

E. A. Torchenkova, Nguyen D'eu, L. P. Kazanskii, V. I. Spitsyn (Izv. Akad. Nauk SSSR Ser. Khim. **1973** 734/8; Bull. Acad. Sci. USSR Div. Chem. Sci. **1973** 715/8).

10 Molybdate Hydrates with Chromium

10.1 Chromium Molybdenum Oxides

10.1.1 $Cr(OH)_2$-$Mo(OH)_3$ Solid Solutions

The coprecipitate of the two hydroxides is formed by reducing a methanolic mixture of $CrCl_3$ and $MoCl_5$ with zinc amalgam and adding methanolic KOH. The Mo concentration reaches up to Cr : Mo = 10.

The product is able to reduce molecular nitrogen if small amounts of water are added; the optimum of the reduction to hydrazine is reached in a solution with 2% water and 0.4 M KOH. Mechanical mixtures of the two hydroxides show a much smaller reductive power. Most of the reduction product is ammonia formed directly without the intermediate step of hydrazine formation. The oxidative consumption of Cr^{2+} during the nitrogen reaction at 333 K is caused by the reduction of nitrogen (25%) and the formation of hydrogen (75%) according to $Cr(OH)_2 \rightarrow CrOOH + 0.5\,H_2$. The Mo component reduces hydrazine to ammonia. The $Cr(OH)_2$-$Mo(OH)_3$ solid solutions have been compared with the nitrogen-reducing capability of $Ti(OH)_3$-$Mo(OH)_3$ solid solutions.

E. M. Burbo, N. T. Denisov, S. I. Kobeleva (Kinetika Kataliz **22** [1981] 1401/6; Kinet. Catal. [USSR] **22** [1981] 1108/13).

10.1.2 $Cr_2O_3 \cdot 4\,MoO_3 \cdot 12\,H_2O$

By mixing 0.1M solutions of $CrCl_3 \cdot 6\,H_2O$ and $Na_2MoO_4 \cdot 2\,H_2O$ in the molar ratio Cr : Mo = 1 : 2 at pH 6.1 and room temperature a dark green gel is precipitated, which when washed with water and dried at 40°C has the above composition. In water, it has an equilibrium pH of 3.5, and dissolves completely in 4 M HCl or 4 M HNO_3, but may be purified with 1M acids.

The amorphous compound shows IR bands at about 3300, 1610, and 1380 cm^{-1}, which are typical for the stretching vibrations of interstitial water and OH groups ($\nu_1(H_2O$ or $OH)$), the deformation vibration of interstitial water ($\delta_1(H_2O)$), and the M-OH deformation vibration ($\delta_2(OH)$), respectively. The band at about 600 to 480 cm^{-1} corresponds to the stretching M-O vibration ($\nu_2(Cr\text{-}O)$), and the deformation vibration of the anion ($\delta_3(MoO_4)$).

At pH 6 to 7 the product possesses cation exchange capabilities with an exchange capacity for K^+ of 0.34 mequiv/g. It has been used to separate Pb^{2+} from numerous metal ions. TG and DTA curves up to 1000°C are given in the paper. They indicate several dehydration and oxide forming reactions. Constant weight is reached at >500°C.

M. Qureshi, R. Kumar, H. S. Rathore (Talanta **19** [1972] 1377/86).

10.2 Metal Chromate Molybdate Hydrates

10.2.1 The Na_2CrO_4-Na_2MoO_4-H_2O System

The sodium chromate hydrates are described in "Chrom" B, 1962, pp. 468/72; the sodium molybdate hydrates are in this volume, pp. 51/74.

The solubility characteristics of this system have been analyzed at 0, 8, 15, 19, 22, and 25°C. At 0 and 8°C the solid phase contains decahydrate solid solutions over the whole concentration range but shows a discontinuity in the solid solution series, see **Fig. 83a** (for a diagram at 0°C see the paper). Crystals rich in chromate are formed in liquid solutions along the curve AC and crystals rich in molybdate along the curve CB; the curve CD corresponds to metastable solid solutions rich in molybdate which are observed only at 8°C. At point C of the diagram, the liquid solution is in equilibrium with two solid solutions of compositions E′ and F′. The line EF is the calculated composition of the decahydrate solid solutions free from adhering solution.

At the higher temperatures the solid phase consists of different solid solutions depending on the composition of the liquid solution and the temperature. So at 15°C solid solutions rich in molybdate crystallize as dihydrates and ones rich in chromate as decahydrates, at 19°C solid solutions of the deca-, hexa-, (metastable) tetra-, and dihydrates can be found, at 22°C hexa- and tetrahydrates, and at 25°C hexa-, tetra-, and dihydrates. Fig. 83b shows the solubility curve at 22°C. Point A represents the solubility of $Na_2CrO_4 \cdot 6H_2O$, the stable compound at this temperature. Along AC there is an equilibrium between liquid solutions and solid solutions of the hexahydrates. Their calculated composition is JK. At C, liquid solution, and hexa- and tetrahydrate solid solutions are in equilibrium. CD is the equilibrium between liquid solution and the tetrahydrate solid solutions whose calculated compositions correspond to LM. DB is the equilibrium between liquid solution and dihydrate solid solutions. Diagrams for 15, 19, and 25°C are given in the paper.

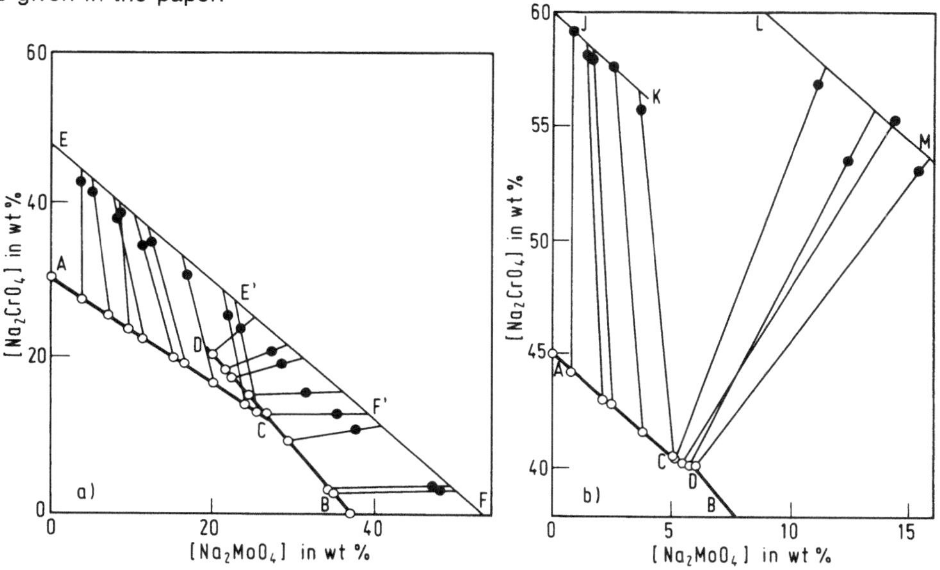

Fig. 83. Solubility curves of the Na_2CrO_4-Na_2MoO_4-H_2O system, a) at 8°C, b) at 22°C. Open circles represent compositions of the saturated solutions, solid circles those of the moist solids in equilibrium with the corresponding solutions. For the letters see the text.

Fig. 84, p. 348, shows the relation between the composition of the liquid solution and the temperature at which this solution coexists in equilibrium with decahydrate solid solutions and another solid phase. Points A (at 10.27±0.05°C) and D are the transition temperatures of $Na_2MoO_4 \cdot 10H_2O$ to the dihydrate and of $Na_2CrO_4 \cdot 10H_2O$ to the hexahydrate, respectively. Along AB the solution is in equilibrium with dihydrate solid solutions containing very little Na_2CrO_4, and with decahydrate solid solutions rich in molybdate. Along CB and BE there are dihydrate and decahydrate solid solutions rich in chromate, and along CD decahydrate and hexahydrate solid solutions both rich in chromate are found. F is the metastable transition of $Na_2CrO_4 \cdot 10H_2O$ into $Na_2CrO_4 \cdot 4H_2O$ from which the curve FG can be obtained in the absence of the hexahydrate corresponding to a metastable equilibrium of the liquid solution with deca- and tetrahydrate solid solutions.

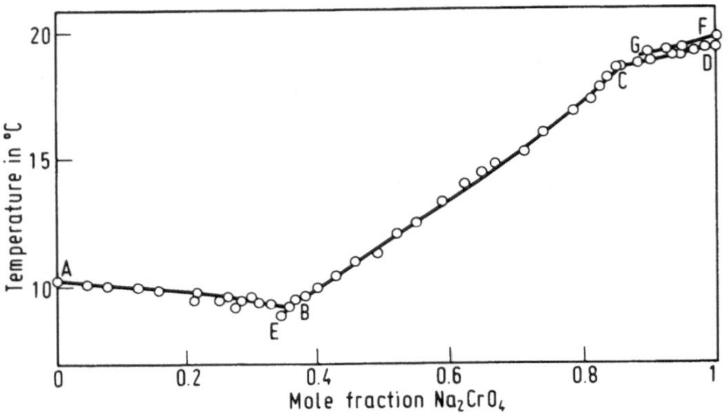

Fig. 84. Temperatures at which saturated liquid solutions of the Na_2CrO_4-Na_2MoO_4-H_2O system are in equilibrium with two or more solid phases. For the letters see the text.

W. E. Cadbury, Jr. (J. Am. Chem. Soc. **67** [1945] 262/8).

10.3 Molybdochromates

Older data are given in "Molybdän", 1935, pp. 386/7.

10.3.1 The 6-Molybdochromate(III) Ion $[Cr(OH)_6Mo_6O_{18}]^{3-}$

This heteropoly anion is one of the group of the isostructural 6-molybdometalates with the general formula $[H_6X^{n+}Mo_6O_{24}]^{(6-n)-}$, X = Al^{III} (p. 231), Ga^{III} (p. 237), Fe^{III}, Co^{III}, Rh^{III}, and Ni^{II} [1]. The structure of this ion was uncertain until the late sixties because different coordination concepts [2], the role of constitutional water [3, 4], and the monomeric or dimeric character [5 to 7] were disputed. Formulas like $[Cr(MoO_4)_6]^{9-}$, $[CrMo_6O_{21}]^{3-}$, $[CrO_3(MoO_3)_6]^{3-}$, or $[(CrO_6Mo_6O_{15})_2]^{6-}$ can be found in the older literature. But cryoscopic measurements in fused $Na_2SO_4 \cdot 10H_2O$ have confirmed the monomeric character [4, 8]. The thermoanalyses of several 6-molybdochromate salts have shown that three molecules of water are constitutional and if they are removed the anion decomposes [9, 10]. The crystal structure determination of $Na_3[Cr(OH)_6Mo_6O_{18}] \cdot 8H_2O$ (p. 351) [11] has proved that the anion is of the Anderson type [12], i.e., a central CrO_6 octahedron is surrounded by a planar ring formed by six MoO_6 octahedra which share edges with one another. Indirect conclusions of H bond arrangements make it very probable that the six protons are associated with the six oxygen atoms of the CrO_6 octahedron. This is supported by IR- and ^1H NMR-spectroscopic data [7, 13]. Therefore the anion is now written as $[Cr(OH)_6Mo_6O_{18}]^{3-}$.

The conductometric curve of the titration of a freshly prepared 0.02 M solution of $[Cr(H_2O)_6](NO_3)_3$ with 0.1 M $(NH_4)_6Mo_7O_{24}$ has one inflection point at the molar ratio Mo : Cr = 6 according to the formation of the 6-molybdochromate anion. The same titration with a solution of $[Cr(H_2O)_6](NO_3)_3$ or $[Cr(H_2O)_6]Cl_3$ heated for 60 h shows inflection points at Mo : Cr < 6 (nitrate 4.78, chloride 4.50), possibly according to the partial formation of bridged complexes. The reverse titration of the paramolybdate with fresh chromic nitrate has one to three inflection points depending on the pH of the solution [14].

The absorption spectrum of the aqueous solution has three bands, at $\nu = 55.1 \times 10^{13}$ (log $\varepsilon = 0.84$), 75.0×10^{13} (0.91), and 125×10^{13} Hz (4.6). The first two bands are nearly identical with those of $[Cr(H_2O)_6]^{3-}$. The strong absorption at 125×10^{13} Hz due to the heteropoly molybdate ion is found in all the salts of the anions $[X^{III}(OH)_6Mo_6O_{18}]^{3-}$, X = Al, Cr, Fe, Co [15].

The self-diffusion coefficient D of the anion was measured at 30°C by the open capillary method to be $(7.4 \pm 0.2) \times 10^{-6}$ cm²/s at infinite dilution. This value is identical with D of the analogous Co^{3+} heteropoly anion [16].

Isotope exchange experiments with ^{51}Cr, ^{99}Mo, and ^{18}O have shown that Cr exchanges between $[Cr(OH)_6Mo_6O_{18}]^{3-}$ and $[Cr(H_2O)_6]^{3+}$ as a CrO_6 unit without breaking Cr-O bonds and that the Cr exchange is remarkably more rapid than in other chromium complexes. Half times range from 4.3 to 45 min at 29.5°C, total ionic strength 0.446, and pH 1 to 2, obeying the kinetic equation $R = k_1([Cr(OH)_6Mo_6O_{18}]^{3-}) + k_2([Cr(H_2O)_6]^{3+} \cdot [Cr(OH)_6Mo_6O_{18}]^{3-})$ with $k_1 = 1.4 \times 10^{-2}$ min⁻¹ and $k_2 = 0.4$ L·mol⁻¹·min⁻¹ at pH 1.06. The kinetics of the oxygen exchange between $[Cr(H_2O)_6]^{3+}$, $[Cr(OH)_6Mo_6O_{18}]^{3-}$, and the solvent revealed three kinds of differently bound oxygen atoms in the heteropoly anion. The Mo and O exchanges between the paramolybdate $[Mo_7O_{24}]^{6-}$ and the heteropoly anion show that MoO_4^{2-} units dissociate accompanied by a synchronous attack of two solvent oxygen atoms on the heteropoly complex [4, 17].

The stability of a heteropoly molybdate may depend more on the nature of the bond between the central ion (e.g. Cr^{3+}) and oxygen than on its coordination number or ionic radius. This is concluded from the comparison of thermodynamic data of several binary molybdenum-oxygen and central ion-oxygen compounds with MO considerations and spectroscopic data of heteropoly compounds and proved for the 6-molybdochromate(III) ion [18]. The pK value of the dissociation equilibrium $[Cr(OH)_6Mo_6O_{18}]^{3-} \rightleftharpoons Cr^{3+} + 6\,MoO_4^{2-} + 6\,H^+$ has been determined in 1.0 M and 3.0 M $NaClO_4$ at 25.0 ± 0.1°C to be $pK = 54 \pm 1$ for the Na salt, what is considerably higher than the values of the corresponding Ga^{3+} and Ni^{2+} complexes (pK = 48 and 31, respectively) [19].

The potentiometric titration curve of an aqueous solution of the potassium salt of $[Cr(OH)_6Mo_6O_{18}]^{3-}$ with NaOH shows one step at the molar ratio Cr : NaOH = 1 : 9 corresponding to the decomposition of the anion according to $[Cr(OH)_6Mo_6O_{18}]^{3-} + 9\,OH^- \rightarrow 6\,MoO_4^{2-} + Cr(OH)_3 + 6\,H_2O$ [5].

The cyclic voltammogram of 0.001 M $(NH_4)_3[Cr(OH)_6Mo_6O_{18}]$ in 1 M $NaClO_4$ at sweep rates of 2 V/s and lower is typical for a system with a rapid electron transfer at the interface followed by a slow chemical reaction. The anodic peak is considerably smaller than the cathodic one suggesting that the reduced species decompose. The analogous molybdo heteropoly compounds with Al^{III}, Fe^{III}, Co^{III}, and I^{VII} have similar voltammograms indicating that the isolation of any heteropoly blues from these compounds is precluded [20].

References:

[1] G. A. Tsigdinos (Top. Current Chem. **76** [1978] 1/64, 38). – [2] A. F. Kapustinskii, A. A. Shidlovskii (Izv. Sekt. Platiny SSSR **30** [1955] 44/66, 60; C.A. **50** [1956] 10459). – [3] M. T. Pope, L. C. W. Baker (J. Phys. Chem. **63** [1959] 2083/4). – [4] G. A. Tsigdinos (Diss. Boston Univ. 1961; Diss. Abstr. **22** [1961] 732). – [5] L. C. W. Baker, G. Foster, W. Tan, F. Scholnick, T. P. McCutcheon (J. Am. Chem. Soc. **77** [1955] 2136/42).

[6] C. W. Wolfe, M. L. Block, L. C. W. Baker (J. Am. Chem. Soc. **77** [1955] 2200). – [7] T. Wada (Compt. Rend. B **263** [1966] 51/4). – [8] L. C. W. Baker (Intern. Conf. Coord. Chem. **6** [1961] 604/12, 610). – [9] A. La Ginestra, R. Cerri (Gazz. Chim. Ital. **95** [1965] 26/32). – [10] A. La Ginestra, R. Cerri, F. Giannetta, P. Fiorucci (J. Therm. Anal. **2** [1970] 107/17).

[11] A. Perloff (Inorg. Chem. **9** [1970] 2228/39). – [12] J. S. Anderson (Nature **140** [1937] 850). – [13] L. P. Kazanskii, S. Holguin Quinones, B. N. Ivanov-Emin, L. A. Filatenko (Koord. Khim. **4** [1978] 1676/83; Soviet Coord. Chem. **4** [1978] 1279/86). – [14] H. T. Hall, H. Eyring (J. Am. Chem. Soc. **72** [1950] 782/90). – [15] Y. Shimura, H. Ito, R. Tsuchida (J. Chem. Soc. Japan Pure Chem. Sect. **75** [1954] 560/2; C.A. **48** [1954] 13422).

[16] L. C. W. Baker, M. T. Pope (J. Am. Chem. Soc. **82** [1960] 4176/9). – [17] K. H. Lee (Diss. Georgetown Univ. 1970; Diss. Abstr. B **31** [1970] 5240). – [18] L. I. Lebedeva (Zh. Neorgan. Khim. **12** [1967] 1287/92; Russ. J. Inorg. Chem. **12** [1967] 681/4). – [19] O. W. Rollins (J. Inorg. Nucl. Chem. **33** [1971] 75/80). – [20] G. A. Tsigdinos, C. J. Hallada (J. Less-Common Metals **36** [1974] 79/93, 88).

10.3.2 The 6-Molybdochromic Acid

For the solid acid no exact formula is given in [1 to 3]. Because of its tribasicity it is formulated as $H_3CrMo_6O_{21}$ in [4, 5]. Possibly it should be written $H_3[Cr(OH)_6Mo_6O_{18}]$ in analogy to the alkali salts.

The pink solution of the free acid is obtained by passing a solution of the ammonium salt (p. 354) through a column of a cation exchange resin of the sulfonic acid type. The solid compound, a green, very soluble powder, is obtained as the residue by evaporating the effluent. Long contact times with the resin (overnight) completely destroy the heteropoly compound [1]. A slurry of $MoO_3 \cdot 2H_2O$ in a 1M solution of $CrCl_3$ is dissolved completely at 50°C apparently forming the heteropoly acid [2].

The acid is tribasic according to potentiometric titrations with NaOH. The curves show two steps at three and twelve moles NaOH per mole acid. The first step corresponds to the neutralization of the three acidic H^+, the second one is caused by the decomposition of the heteropoly anion to the precipitating $Cr(OH)_3$ and MoO_4^{2-}. The three dissociation constants of the moderately strong acid (pK 2 to 3) have nearly the same value, so that only one dissociation step is observed [3]. The tribasic character of the acid has also been proved by a spectrophotometric method measuring the optical density of a solution of methyl orange or other indicators such as bromocresol green in the presence of various amounts of the heteropoly acid [4, 5].

Besides the salts of the acid described below, there have been synthesized five compounds with cationic cobalt complexes, cis- and trans-$[Co(NO_2)_2(NH_3)_4]_3X$, cis- and trans-$[Co(NO_2)_2en_2]_3X$, and $[CoCl_2py_4]_3X$, $X=[Cr(OH)_6Mo_6O_{18}]^{3-}$, en = ethylene diamine, py = pyridine [6]. For the cation complexes see "Kobalt" Erg.-Bd. B 2, 1964, pp. 480, 505, and 568, respectively.

References:

[1] L. C. W. Baker, B. Loev, T. P. McCutcheon (J. Am. Chem. Soc. **72** [1950] 2374/7). – [2] M. L. Freedman (J. Chem. Eng. Data **8** [1963] 113/6). – [3] L. C. W. Baker, G. Foster, W. Tan, F. Scholnick, T. P. McCutcheon (J. Am. Chem. Soc. **77** [1955] 2136/42). – [4] E. Matijević, M. Kerker (J. Am. Chem. Soc. **81** [1959] 5560/6). – [5] J. R. Keller, E. Matijević, M. Kerker (J. Phys. Chem. **65** [1961] 56/8).

[6] S. P. Rozman (Zh. Neorgan. Khim. **16** [1971] 567/8; Russ. J. Inorg. Chem. **16** [1971] 303/4).

10.3.3 $Li_3[Cr(OH)_6Mo_6O_{18}]\cdot10H_2O$ ($=3Li_2O\cdot Cr_2O_3\cdot12MoO_3\cdot26H_2O$)

The salt can be prepared by adding Li_2CO_3 to an aqueous solution of the 6-molybdochromic acid in the stoichiometric molar ratio and slowly evaporating the solution. After recrystallization the decahydrate is obtained.

The d values and the IR bands between 3600 and 700 cm^{-1} are listed; the IR spectrum between 1000 and 700 cm^{-1} is plotted. The spectrum is similar to the other alkali salts except for the Na salt.

The TG curve has two steps corresponding to the loss of $10H_2O$ in the range of 45 to 200°C with a dehydration enthalpy of $\Delta H=80$ kcal/mol and to the loss of $3H_2O$ at 210°C with $\Delta H=59$ kcal/mol. The last three water molecules must be considered as constitutional water as in the other alkali salts, because after their loss the salt is decomposed irreversibly to the tetramolybdate. The DTA curve shows the endothermic peaks of the two dehydration steps, a strong exothermic peak at 365°C corresponding to the tetramolybdate crystallization, a weak exothermic reaction at 485°C with unknown products, and the endothermic melting peak at 565°C. The DTA and TG curves up to 600 and 400°C are given in the paper.

A. La Ginestra, R. Cerri, F. Giannetta, P. Fiorucci (J. Therm. Anal. **2** [1970] 107/17).

10.3.4 $Na_3[Cr(OH)_6Mo_6O_{18}]\cdot nH_2O$ ($=3Na_2O\cdot Cr_2O_3\cdot12MoO_3\cdot(2n+6)H_2O$), $n=13, 16$

The compound can be obtained by the same method as the Li compound (see above) or (according to [3]) from a boiled solution of $Na_2MoO_4\cdot2H_2O$ and $Cr(NO_3)_3\cdot9H_2O$ at pH 4.5. Recrystallization at 0°C gives the salt with 15 to $16H_2O$, and between room temperature and 0°C the 13-hydrate. Under normal temperature and humidity conditions both hydrates lose water. The thermogravimetric curve of the 13-hydrate has two steps corresponding to the loss of 11 and $5H_2O$; the center of the intermediate plateau lies at 160°C. The latter step of $5H_2O$ is found in the dehydration curves of all hydrates of the sodium salt [1].

For the 13-hydrate, an X-ray structure determination was made by Perloff, space group $P\bar{1}$-C_i^1 (No. 2), $R=0.041$, however, the data are not published [2].

References:

[1] A. La Ginestra, R. Cerri, F. Giannetta, P. Fiorucci (J. Therm. Anal. **2** [1970] 107/17). – [2] H. T. Evans, Jr. (Perspect. Struct. Chem. **4** [1971] 1/59, 11). – [3] A. Perloff (Inorg. Chem. **9** [1970] 2228/39, 2228).

10.3.5 $Na_3[Cr(OH)_6Mo_6O_{18}]\cdot8H_2O$ ($=3Na_2O\cdot Cr_2O_3\cdot12MoO_3\cdot22H_2O$)

To obtain the octahydrate a mixture of the aqueous solution of $Na_2MoO_4\cdot2H_2O$, adjusted to pH 4.5 with HNO_3, and of $Cr(NO_3)_3\cdot9H_2O$ is boiled, filtered while hot, and set aside until the purple sodium salt has precipitated [1, 2]. Another method begins with a solution of the free 6-molybdochromic acid and adds the stoichiometric amount of sodium carbonate; the mixture is evaporated slowly at room temperature until the salt is precipitated. Recrystallization at room temperature gives the octahydrate [3].

Slowly grown crystals show a tabular habit with a tendency to elongate in the [001] direction. The dominant form is {010}, the secondary forms are {100} and {001}. The crystals have perfect (010) cleavage [2].

The crystals are triclinic, space group $P\bar{1}$-C_i^1 (No. 2). The lattice parameters are a = 10.9080(4), b = 10.9807(4), c = 6.4679(2) Å, α = 107.594(2)°, β = 84.438(2)°, γ = 112.465(3)° at 25°C; Z = 1 [2]. The d values are listed in [3]. The crystal structure has been elucidated by three-dimensional X-ray investigation. Atomic parameters are listed in the paper (R = 0.033) [2]. The anion has the Anderson-Evans [5, 6] configuration, see **Fig. 85** (from [4]). The disk-like $[Cr(OH)_6Mo_6O_{18}]^{3-}$ anions are stacked along the c axis. These stacks are tilted parallel to the (2, 8, 11) plane, and are held together through bonds between anionic oxygen and octahedrally coordinated Na atoms and through H bonds of the water of crystallization. The MoO_6 octahedra are considerably distorted displacing the Mo atoms towards the exterior of the complex [2]. Average distances are (from [4]): Cr-O_a = 1.976 (6×), Mo-O_a = 2.295 (12×), Mo-O_b = 1.938 (12×), Mo-O_c = 1.705 Å (12 ×). The range of nearest neighbor Mo-Mo distances is 3.309 to 3.351 Å and the Cr-Mo distances vary from 3.303 to 3.349 Å. The range of Na-O distances is 2.307 to 2.536 Å [2].

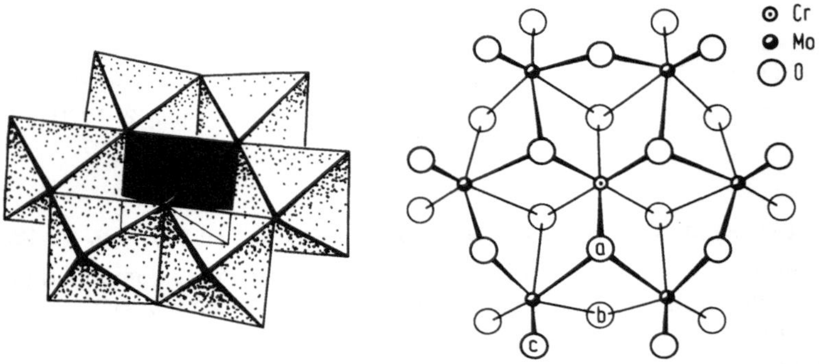

Fig. 85. View of the $[Cr(OH)_6Mo_6O_{18}]^{3-}$ anion, coordination polyhedra (left) and model parameters (right), from [4].

The hydrogen atoms could not be located directly, but considerations about charge balance and H bonding arguments make it reasonably certain that the H atoms of the anion are bonded to the oxygen atoms of the CrO_6 octahedron. The postulated H bonds in the structure are inferred from short oxygen-oxygen distances [2].

The measured density is 2.950, the calculated density 2.995 g/cm³ [2]. The IR bands between 3600 and 700 cm⁻¹ are listed and the spectrum is plotted between 1000 and 700 cm⁻¹. It differs from those of the other alkali compounds [3].

Drying at room temperature yields the dihydrate [2]. TG and DTA analyses show that the compound contains a total of 11 water molecules which are lost in two steps: first $6H_2O$ up to 110 to 200°C with a dehydration enthalpy ΔH = 54 kcal/mol and then $5H_2O$ up to 250 to 300°C with ΔH = 78.7 kcal/mol. The last dehydration reaction is observed in all hydrates of the sodium salt and consists of the loss of water of hydration and of constitutional water, which cannot be separated thermoanalytically [2, 3]. Above 350°C the thermal behavior is very similar to that of the other alkali salts of the 6-molybdochromic acid: a strong exothermic reaction at 370°C with ΔH = 13 kcal/mol, a weak exothermic peak at 470°C with ΔH = 13 kcal/mol, and the melting peak at 515°C with ΔH = 24 kcal/mol (see p. 351); for plots see the paper [3].

References:

[1] G. A. Tsigdinos (Diss. Boston Univ. 1961; Diss. Abstr. **22** [1961] 732). – [2] A. Perloff (Inorg. Chem. **9** [1970] 2228/39). – [3] A. La Ginestra, R. Cerri, F. Giannetta, P. Fiorucci (J.

Therm. Anal. **2** [1970] 107/17). – [4] H. T. Evans, Jr. (Perspect. Struct. Chem. **4** [1971] 1/59, 25). – [5] J. S. Anderson (Nature **140** [1937] 850).

[6] H. T. Evans, Jr. (J. Am. Chem. Soc. **70** [1948] 1291/2).

10.3.6 $Na_3[Cr(OH)_6Mo_6O_{18}]\cdot nH_2O$ (=$3Na_2O\cdot Cr_2O_3\cdot 12MoO_3\cdot (2n+6)\ H_2O$), n=1, 2, 5

The octahydrate, left in a container which is not completely closed, forms the pentahydrate, which then dehydrates in four steps according to the TG and DTA curves. The first three steps correspond to a total of three molecules of water, the last one with a loss of $5H_2O$ (water of crystallization and constitutional water) is identical with the final step of the other sodium hydrates; the DTA range above 350°C is identical with the other alkali salts (see p. 351) [1].

The thermoanalytical curves of all higher hydrates of the sodium salt show the formation of a stable dihydrate, which decomposes irreversibly at 250 to 300°C losing $5H_2O$ (water of crystallization and constitutional water) [1, 2].

A monohydrate is obtained by a method which was developed to synthesize very pure sodium molybdochromate, with yields of about 80%. After the formation of the compound in a hot solution of Na_2MoO_4 and $Cr(NO_3)_3$ at pH 4.4, acetone is added at room temperature to the reaction mixture, separating the product as a viscous liquid. The simple salts like $NaNO_3$ remain in the aqueous phase and are decanted. The acetone extraction is repeated and the oil dried in vacuum at 105°C. The elemental analysis agrees with the formula of the monohydrate. Cryoscopic measurements in fused $Na_2SO_4\cdot 10H_2O$ give a molal freezing point depression K_0 = 3.45 K/mol at infinite dilution corresponding to 1.02 anions [3].

References:

[1] A. La Ginestra, R. Cerri, F. Giannetta, P. Fiorucci (J. Therm. Anal. **2** [1970] 107/17). – [2] A. Perloff (Inorg. Chem. **9** [1970] 2228/39, 2229). – [3] O. W. Rollins (J. Inorg. Nucl. Chem. **33** [1971] 75/80).

10.3.7 $K_3[Cr(OH)_6Mo_6O_{18}]\cdot 7H_2O$ (=$3K_2O\cdot Cr_2O_3\cdot 12MoO_3\cdot 20H_2O$)

This salt is synthesized from the free molybdochromic acid and K_2CO_3 in the same manner as the other alkali salts of the molybdochromic acid (see p. 351) and may be recrystallized from water [1, 3, 4].

The X-ray powder diagram is similar to those of the heptahydrates of NH_4 and Rb. The d values are listed in the paper [4]. The XPS data (related to the C 1s = 285.0 eV) are: Cr $2p_{3/2}$ = 577.9, Mo $3d_{5/2}$ = 233.0, O 1s = 531.1, and K $2p_{3/2}$ = 293.0 eV. The Mo value is identical for all isostructural compounds of the type $[X(OH)_6Mo_6O_{18}]^{n-}$ (X = Al, Ga, Fe, Co, Ni, etc.), the O value is independent from the water content, and the Cr value is slightly higher than in the oxide Cr_2O_3 (576.8 eV), corresponding more to the value of $Cr(OH)_3$ [2].

The IR spectra of the isostructural $[X(OH)_6Mo_6O_{18}]^{n-}$ heteropoly compounds are very similar. The IR data for $K_3[Cr(OH)_6Mo_6O_{18}]\cdot 7H_2O$ are: 1615 ($\delta(H_2O)$), 1060, 1015 (δ(OH)), 943, 925 ($\nu(MoO_2)$), 890, 655, 575, 545, 490, 410 cm^{-1} (ν(MoOMo) and ν(MoOCr)) [2]. Tabulated IR bands from 3600 to 820 cm^{-1} and a plot between 1000 and 700 cm^{-1} are given in [4].

The TG data show that dehydration takes place in four steps. At 40°C the loss of $4H_2O$ begins, at 100°C of 1.5 H_2O, at 120°C of other 1.5 H_2O, and at 200°C of 3 H_2O. The corresponding dehydration enthalpies are ΔH = 27, 20, 30, and 34.5 kcal/mol, respectively. The last three water molecules are constitutional as in the hydrates of the analogous Rb and Cs salts; their loss is irreversible as experiments of rehydration and with D_2O have shown. In the DTA curve four endothermic peaks corresponding to the four dehydration steps are observed; above 350°C the curve is qualitatively identical with the other alkali 6-molybdochromates. There are two exothermic peaks at 340 and 420°C (leaving the Cr species totally insoluble in water) and the melting peak at 540 to 590°C (see p. 351) [1, 4].

References:

[1] A. La Ginestra, R. Cerri (Gazz. Chim. Ital. **95** [1965] 26/32). – [2] L. P. Kazanskii, S. Holguin Quinones, B. N. Ivanov-Emin, L. A. Filatenko (Koord. Khim. **4** [1978] 1676/83; Soviet J. Coord. Chem. **4** [1978] 1279/86). – [3] L. C. W. Baker, G. Foster, W. Tan, F. Scholnick, T. P. McCutcheon (J. Am. Chem. Soc. **77** [1955] 2136/42). – [4] A. La Ginestra, R. Cerri, F. Giannetta, P. Fiorucci (J. Therm. Anal. **2** [1970] 107/17).

10.3.8 $(NH_4)_3[Cr(OH)_6Mo_6O_{18}]\cdot nH_2O$ ($=3(NH_4)_2O\cdot Cr_2O_3\cdot 12MoO_3\cdot(2n+6)H_2O$), n = 4, 5, 7

The pink **heptahydrate** can be prepared from the free acid by adding NH_4Cl or $(NH_4)_2CO_3$ and recrystallizing [1, 2].

The crystals are small, flat rhombs. Single crystal investigations show them to be triclinic; lattice parameters a = b = 12.1, c = 10.7 Å, $\alpha = 91°7'$, $\beta = 112°20'$, $\gamma = 72°9'$; Z = 2 ($(NH_4)_3[Cr(OH)_6Mo_6O_{18}]\cdot 7H_2O$), space group $P\bar{1}$-C_i^1 (No. 2) (from [6]) [3]. The d values are listed. The IR bands between 3600 and 800 cm^{-1} and a plot between 1000 and 700 cm^{-1} are given. The X-ray powder diagram and the IR spectrum show the NH_4 compound to be isomorphous with the K and Rb compound [2]. At 20°C the ^1H NMR spectrum of the solid salt shows, besides the N-H protons of the ammonium ion, a signal at $\Delta H = 3.5 \pm 0.3$ G indicating the existence of OH groups in the $[Cr(OH)_6Mo_6O_{18}]^{3-}$ anion [4]. The measured density is 2.866, the calculated density 2.602 g/cm^3 [3, 6]. From magnetic measurements at 31°C the specific susceptibility $\chi = 5.17 \times 10^{-6}$ cm^3/g and the molar susceptibility $\chi_m = 12710 \times 10^{-6}$ cm^3/mol are obtained. The magnetic moment (4.0 μ_B) corresponds to the octahedral coordination of the tervalent Cr ion with d^2sp^3 bonds in the anion [5]. The thin crystal plates have a refractive index of 1.72 [7].

The thermogravimetric curve shows six steps, the first four ones correspond to the loss of water, beginning with $3H_2O$ at 40°C ($\Delta H = 23$ kcal/mol), 2.5 H_2O at 85°C (39.2 kcal/mol), 1.5 H_2O at 100°C (16 kcal/mol), and 3 H_2O at 140°C (24 kcal/mol). These last three water molecules are constitutional and comparable to the last dehydration step in all alkali salts of the molybdochromic acid except the sodium salt. The two other steps at 200 to 270°C and 300 to 350°C are caused by synchronous deammoniation and dehydration reactions ($\Delta H = 28$ and 21 kcal/mol, respectively). Beyond these six steps the DTA curve has a strong exothermic peak at 480°C (formation of $Cr_2(MoO_4)_3$ and crystalline MoO_3); for plots see the paper [2].

A **pentahydrate** is described, obtained by the method of Hall (see "Molybdän", 1935, p. 387), which ordinarily results in the formation of the heptahydrate. Cryoscopic measurements of this salt in fused $Na_2SO_4\cdot 10H_2O$ give a molal freezing point depression of $K_0 = 13.6$ K/mol at infinite dilution corresponding to 4.02 ions per formula [8].

The thermoanalysis shows that the heptahydrate loses $3H_2O$ at a slightly elevated temperature (40°C) [2]. Therefore the **tetrahydrate** may be formed at temperatures somewhat higher than room temperature [1].

Aqueous Solutions. The osmotic coefficients φ of the ammonium salt have been measured at 37°C in the molal range from 0.002 to 0.02. From these coefficients the activity coefficients $\gamma_\pm$

are derived and tabulated together with φ in steps of 0.01 to 0.02. They agree with the calculated values of the Debye-Hückel theory for 1:3 electrolytes with a maximum deviation of 3%. The values for $\gamma_\pm$ decrease considerably with increasing concentration (0.787 at 0.003 m, 0.586 at 0.02 m), which is typical for polyvalent salts. An interionic distance of 2.9 Å has been calculated [9].

In a saturated solution of $(NH_4)_3[Cr(OH)_6Mo_6O_{18}]$ the concentration of the cation ("K") has been varied and the concentration of the heteropoly anion ("A") determined at constant ionic strength. A plot of the logarithm of the two concentrations is linear with a slope of −3 according to the relation $\log A = \log L - 3 \log K$ (L = solubility product of $(NH_4)_3[Cr(OH)_6Mo_6O_{18}]$), or $L = K^3 \cdot A$. This is another proof of the tribasicity of the heteropoly anion [10]. The same result is obtained from coagulation curves of a silver bromide sol in statu nascendi in the presence of small amounts of $(NH_4)_3[Cr(OH)_6Mo_6O_{18}]$. These curves are obtained by measuring the turbidity of the system. The coagulation concentration of an electrolyte like AgBr depends on the charge of the counter ions in the solution; thus, comparison of systems with known and unknown species reveals the charge of the counter ion [11].

References:

[1] L. C. W. Baker, B. Loev, T. P. McCutcheon (J. Am. Chem. Soc. **72** [1950] 2374/7). – [2] A. La Ginestra, R. Cerri, F. Giannetta, P. Fiorucci (J. Therm. Anal. **2** [1970] 107/17). – [3] C. W. Wolfe, M. L. Block, L. C. W. Baker (J. Am. Chem. Soc. **77** [1955] 2200). – [4] T. Wada (Compt. Rend. B **263** [1966] 51/4). – [5] P. Rây, A. Bhaduri, B. Sarma (J. Indian Chem. Soc. **25** [1948] 51/6).

[6] J. D. H. Donnay, H. M. Ondik (Crystal Data, Vol. 2: Inorganic Compounds, Washington, D.C., 1973, p. A-44). – [7] H. T. Hall, H. Eyring (J. Am. Chem. Soc. **72** [1950] 782/90, 784). – [8] O. W. Rollins (J. Inorg. Nucl. Chem. **33** [1971] 75/88). – [9] E. Meyer, Jr., R. Huckfeldt (J. Phys. Chem. **74** [1970] 164/7). – [10] P. Souchay (Bull. Soc. Chim. France **1951** 932/8).

[11] E. Matijević, M. Kerker (J. Am. Chem. Soc. **81** [1959] 5560/6).

10.3.9 $(CH_3NH_3)_3[Cr(OH)_6Mo_6O_{18}] \cdot 8H_2O$, $(t\text{-}C_4H_9NH_3)_3[Cr(OH)_6Mo_6O_{18}]$, and $[C(NH_2)_3]_3[Cr(OH)_6Mo_6O_{18}] \cdot 4H_2O$

The pink monomethyl ammonium salt is prepared by adding an aqueous solution of CH_3NH_3Cl to a solution of the free molybdochromic acid. Well-crystallized preparations are obtained by recrystallization [1].

For the t-butyl ammonium salt an acetic acid/methanolic solution of $(t\text{-}C_4H_9NH_3)_2MoO_4$ is mixed with a methanolic solution of $Cr(NO_3)_3 \cdot 3H_2O$; the precipitate is washed with ethanol and dried in vacuum over P_2O_5. The compound can be recrystallized from methanol but with poor yield. The IR spectrum from 1010 to 310 cm^{-1} and the Raman spectrum from 956 to 109 cm^{-1} are tabulated. The compound is soluble in water, methanol, and dimethylsulfoxide [2].

The guanidinium salt is obtained from an aqueous solution of guanidinium paramolybdate and $Cr(NO_3)_3 \cdot 9H_2O$ heated for some hours. At room temperature deep pink, well-defined crystals precipitate, which may be recrystallized from water. X-ray studies on a single crystal show that the compound is triclinic; lattice parameters a = 8.84, b = 11.64, c = 8.01 Å (each ±0.3%), $\alpha = 91°9'$, $\beta = 74°25'$, $\gamma = 96°5'$ (each ±5′); Z = 1. The pycnometric density is 2.661 g/cm^3 [3].

References:

[1] L. C. W. Baker, B. Loev, T. P. McCutcheon (J. Am. Chem. Soc. **72** [1950] 2374/7). – [2] J. Fuchs, I. Brüdgam (Z. Naturforsch. **32b** [1977] 403/7). – [3] O. W. Rollins (J. Inorg. Nucl. Chem. **33** [1971] 75/80).

10.3.10 $Rb_3[Cr(OH)_6Mo_6O_{18}]\cdot 7H_2O$ (=$3Rb_2O\cdot Cr_2O_3\cdot 12MoO_3\cdot 20H_2O$)

This salt can be synthesized like the other alkali salts of the 6-molybdochromic acid from the free acid and Rb_2CO_3 (see p. 351). The X-ray powder and IR spectra show that it is isomorphous with the analogous K and NH_4 salts; lists of d values and IR bands are given. Also the TG and DTA data of the K and Rb salts are nearly identical. The four dehydration steps corresponding to the loss of 4, 1.5, 1.5, and $3H_2O$ begin at 35, 70, 110, and 170°C with the enthalpies $\Delta H = 22$, 21, 12, and 27 kcal/mol, respectively. The last three H_2O are constitutional. For plots see the paper.

A. La Ginestra, R. Cerri, F. Giannetta, P. Fiorucci (J. Therm. Anal. **2** [1970] 107/17).

10.3.11 $Cs_3[Cr(OH)_6Mo_6O_{18}]\cdot 5H_2O$ (=$3Cs_2O\cdot Cr_2O_3\cdot 12MoO_3\cdot 16H_2O$)

This salt is synthesized like the other alkali salts of the 6-molybdochromic acid from the free acid and Cs_2CO_3 (see p. 351). The d values and the IR bands (3550 to 815 cm^{-1}) are listed; the spectrum between 1000 and 700 cm^{-1} is plotted. The TG curve of dehydration has three steps corresponding to the loss of 3.5, 1.5, and $3H_2O$ beginning at 30, 105, and 160°C, respectively, with the enthalpies $\Delta H = 35$, 29, and 22 kcal/mol. As in the other alkali salts (except sodium) the last three H_2O are constitutional. The DTA curve shows the three endothermic dehydration peaks. Above 350°C the curve is equal to the other alkali salts; for plots see the paper.

A. La Ginestra, R. Cerri, F. Giannetta, P. Fiorucci (J. Therm. Anal. **2** [1970] 107/17).

10.3.12 $K_5[CrMo_{12}O_{40}]\cdot 36H_2O$ (=$5K_2O\cdot Cr_2O_3\cdot 24MoO_3\cdot 72H_2O$)

This compound is obtained as orange needles from an aqueous solution of $K_2Cr_2O_7$ at 80°C after gradually adding together solutions of molybdic acid and 20% H_2O_2, and leaving the mixture for 3 d in vacuum. It is recrystallized from water.

The dimensions of the unit cell from single crystal X-ray diffraction measurements are $a = 11.62$, $b = 26.12$, $c = 8.62$ Å, $\alpha = \beta = \gamma = 90°$; $Z = 2$. The pycnometric density is 3.38 g/cm³. The EPR spectrum indicates that Cr is in the oxidation state +3 with a d^3 configuration and a strongly distorted O_h symmetry. The experimentally found value of the factor g_{eff} is 1.9919 (77 K), and the zero field splitting coefficient $D = 0.16$ cm^{-1}.

In vacuum the compound loses $12H_2O$ at 80°C and $7H_2O$ above 290°C. At 450 ± 5°C it decomposes completely.

H. C. Mishra, S. K. Roy, A. N. Ojha (J. Indian Chem. Soc. **55** [1978] 307/8).

10.3.13 $(NH_4)_6[Cr^{VI}Mo_6O_{24}]\cdot 10H_2O$ (=$3(NH_4)_2O\cdot CrO_3\cdot 6MoO_3\cdot 10H_2O$)

This is the only 6-molybdochromate(VI) compound that has been described. The orange compouna was obtained from an acetic acid mixture of $(NH_4)_2MoO_4$ and $K_2Cr_2O_7$ at pH 2.8. The solution was evaporated to crystallization on a water bath and left overnight.

H. C. Mishra, S. K. Roy, A. N. Ojha (J. Indian Chem. Soc. **57** [1980] 929).

10.3.14 $Ge_xMo_yCr_z$ Type Heteropoly Compounds

Well-defined ternary heteropoly compounds containing Ge, Mo, and Cr have not been isolated, but there are some indications that they are formed in aqueous mixtures of the three components.

In contrast to Al^{3+}, Mn^{2+}, Co^{2+}, Ni^{2+}, and Cu^{2+}, the addition of a Cr^{3+} salt to a solution of molybdogermanic acid up to a molar ratio Mo : Cr = 6 at pH 2 and room temperature reduces the optical density indicating the competitive formation of the more stable binary 6-molybdochromate complex [1]. But if these mixtures are heated at 80 to 90°C for 7 h, the fall of the optical density in the region $\lambda < 440$ nm is accompanied by an increase at $\lambda > 440$ nm, indicating the formation of a ternary heteropoly compound (molar ratios investigated are Ge : Mo:Cr = 1:10 :1, 1:10 : 2, 1:10 : 3; $[Ge^{IV}] = 10^{-3}$ g-atom/L) [2].

Also extraction experiments with a 1 : 3 mixture of chloroform and butanol show the formation of ternary hetero compounds. Solutions containing Ge^{IV} and Mo^{VI} in the molar ratio Ge : Mo = 1:12 and variable portions of Cr^{III} at pH 2.2 (1N H_2SO_4) have been extracted after heating for 7 h. Quantitative analysis of the organic phase shows that only Cr-deficient compounds are extracted. From the initial ratios Ge : Mo : Cr = 1:12 :1, 1:12 : 2, and 1:12 : 3, one obtains in the organic phase 1:10 : 0.4, 1:8 : 0.3, and 1:8 : 0.3, respectively. Since neither the simple Cr salt nor the molybdochromate is extracted by this organic phase, the results indicate the formation of ternary Ge-Mo-Cr heteropoly compounds [3].

References:

[1] Z. F. Shakhova, E. N. Dorokhova (Zh. Neorgan. Khim. **10** [1965] 2060/4; Russ. J. Inorg. Chem. **10** [1965] 1121/4). – [2] L. I. Lebedeva, F. A. Muradova (Zh. Neorgan. Khim. **16** [1971] 1056/8; Russ. J. Inorg. Chem. **16** [1971] 561/3). – [3] L. I. Lebedeva, F. A. Muradova, G. N. Prokof'eva (Zh. Neorgan. Khim. **16** [1971] 2025/6; Russ. J. Inorg. Chem. **16** [1971] 1081/2).

Table of Conversion Factors

Following the notation in Landolt-Börnstein [7], values which have been fixed by convention are indicated by a bold-face last digit. The conversion factor between calorie and Joule that is given here is based on the thermochemical calorie, $cal_{th\,ch}$, and is defined as 4.184**0** J/cal. However, for the conversion of the "Internationale Tafelkalorie", cal_{IT}, into Joule, the factor 4.1868 J/cal is to be used [1, p. 147]. For the conversion factor for the British thermal unit, the Steam Table Btu, BTU_{ST}, is used [1, p. 95].

Force	N	dyn	kp
1 N (Newton)	1	10^5	0.1019716
1 dyn	10^{-5}	1	1.019716×10^{-6}
1 kp	**9.80665**	**9.80665** $\times 10^5$	1

Pressure	Pa	bar	kp/m²	at	atm	Torr	lb/in²
1 Pa (Pascal) = 1 N/m²	1	10^{-5}	1.019716×10^{-1}	1.019716×10^{-5}	0.986923×10^{-5}	0.750062×10^{-2}	145.0378×10^{-6}
1 bar = 10^6 dyn/cm²	10^5	1	10.19716×10^3	1.019716	0.986923	750.062	14.50378
1 kp/m² = 1 mm H_2O	**9.80665**	**0.980665** $\times 10^{-4}$	1	10^{-4}	0.967841×10^{-4}	0.735559×10^{-1}	1.422335×10^{-3}
1 at = 1 kp/cm²	**0.980665** $\times 10^5$	**0.980665**	10^4	1	0.967841	735.559	14.22335
1 atm = **760** Torr	1.0132**5** $\times 10^5$	1.0132**5**	1.033227×10^4	1.033227	1	760	14.69595
1 Torr = 1 mm Hg	133.3224	1.333224×10^{-3}	13.59510	1.359510×10^{-3}	1.315789×10^{-3}	1	19.33678×10^{-3}
1 lb/in² = 1 psi	6.89476×10^3	68.9476×10^{-3}	703.069	70.3069×10^{-3}	68.0460×10^{-3}	51.7149	1

Work, Energy, Heat	J	kWh	kcal	Btu	MeV
1 J (Joule) = 1 Ws = 1 Nm = 10^7 erg	1	2.778×10^{-7}	2.39006×10^{-4}	9.4781×10^{-4}	6.242×10^{12}
1 kWh	**3.6** $\times 10^6$	1	860.4	3412.14	2.247×10^{19}
1 kcal	4184.**0**	1.1622×10^{-3}	1	3.96566	2.6117×10^{16}
1 Btu (British thermal unit)	1055.06	2.93071×10^{-4}	0.25164	1	6.5858×10^{15}
1 MeV	1.602×10^{-13}	4.450×10^{-20}	3.8289×10^{-17}	1.51840×10^{-16}	1

1 eV ≙ 23.0578 kcal/mol = 96.473 kJ/mol

Power	kW	PS	kp m/s	kcal/s
1 kW = 10^{10} erg/s	1	1.35962	101.972	0.239006
1 PS	0.73550	1	**75**	0.17579
1 kp m/s	9.80665 $\times 10^{-3}$	0.01333	1	2.34384×10^{-3}
1 kcal/s	4.18**40**	5.6886	426.650	1

References:

[1] A. Sacklowski, Die neuen SI-Einheiten, Goldmann, München 1979. (Conversion tables in an appendix.)
[2] International Union of Pure and Applied Chemistry, Manual of Symbols and Terminology for Physicochemical Quantities and Units, Pergamon, London 1979; Pure Appl. Chem. **51** [1979] 1/41.
[3] The International System of Units (SI), National Bureau of Standards Spec. Publ. 330 [1972].
[4] H. Ebert, Physikalisches Taschenbuch, 5th Ed., Vieweg, Wiesbaden 1976.
[5] Kraftwerk Union Information, Technical and Economic Data on Power Engineering, Mülheim/Ruhr 1978.
[6] E. Padelt, H. Laporte, Einheiten und Größenarten der Naturwissenschaften, 3rd Ed., VEB Fachbuchverlag, Leipzig 1976.
[7] Landolt-Börnstein, 6th Ed., Vol. II, Pt. 1, 1971, pp. 1/14.
[8] ISO Standards Handbook 2, Units of Measurement, 2nd Ed., Geneva 1982.

Key to the Gmelin System of Elements and Compounds

System Number	Symbol	Element
1		Noble Gases
2	H	Hydrogen
3	O	Oxygen
4	N	Nitrogen
5	F	Fluorine
6	**Cl**	**Chlorine**
7	Br	Bromine
8	I	Iodine
	At	Astatine
9	S	Sulfur
10	Se	Selenium
11	Te	Tellurium
12	Po	Polonium
13	B	Boron
14	C	Carbon
15	Si	Silicon
16	P	Phosphorus
17	As	Arsenic
18	Sb	Antimony
19	Bi	Bismuth
20	Li	Lithium
21	Na	Sodium
22	K	Potassium
23	NH_4	Ammonium
24	Rb	Rubidium
25	Cs	Caesium
	Fr	Francium
26	Be	Beryllium
27	Mg	Magnesium
28	Ca	Calcium
29	Sr	Strontium
30	Ba	Barium
31	Ra	Radium
32	**Zn**	**Zinc**
33	Cd	Cadmium
34	Hg	Mercury
35	Al	Aluminium
36	Ga	Gallium

HCl → 6; $ZnCl_2$ → 32

System Number	Symbol	Element
37	In	Indium
38	Tl	Thallium
39	Sc, Y La–Lu	Rare Earth Elements
40	Ac	Actinium
41	Ti	Titanium
42	Zr	Zirconium
43	Hf	Hafnium
44	Th	Thorium
45	Ge	Germanium
46	Sn	Tin
47	Pb	Lead
48	V	Vanadium
49	Nb	Niobium
50	Ta	Tantalum
51	Pa	Protactinium
52	**Cr**	**Chromium**
53	Mo	Molybdenum
54	W	Tungsten
55	U	Uranium
56	Mn	Manganese
57	Ni	Nickel
58	Co	Cobalt
59	Fe	Iron
60	Cu	Copper
61	Ag	Silver
62	Au	Gold
63	Ru	Ruthenium
64	Rh	Rhodium
65	Pd	Palladium
66	Os	Osmium
67	Ir	Iridium
68	Pt	Platinum
69	Tc	Technetium[1]
70	Re	Rhenium
71	Np, Pu...	Transuranium Elements

$CrCl_2$ → 52; $ZnCrO_4$ → 52

Material presented under each Gmelin System Number includes all information concerning the element(s) listed for that number plus the compounds with elements of lower System Number.

For example, zinc (System Number 32) as well as all zinc compounds with elements numbered from 1 to 31 are classified under number 32.

[1] A Gmelin volume titled "Masurium" was published with this System Number in 1941.

A Periodic Table of the Elements with the Gmelin System Numbers is given on the Inside Front Cover